The
Herpesviruses
Volume 3

THE VIRUSES

Series Editors
HEINZ FRAENKEL-CONRAT, *University of California*
Berkeley, California

ROBERT R. WAGNER, *University of Virginia School of Medicine*
Charlottesville, Virginia

THE HERPESVIRUSES
Volumes 1–3 • Edited by Bernard Roizman
Volume 4 • Edited by Bernard Roizman and Carlos Lopez

THE REOVIRIDAE
Edited by Wolfgang K. Joklik

THE PARVOVIRUSES
Edited by Kenneth I. Berns

THE ADENOVIRUSES
Edited by Harold S. Ginsberg

THE VIRUSES: Catalogue, Characterization, and Classification
Heinz Fraenkel-Conrat

The Herpesviruses

Volume 3

Edited by
BERNARD ROIZMAN
University of Chicago
Chicago, Illinois

PLENUM PRESS • NEW YORK AND LONDON

Library of Congress Cataloging in Publication Data

Main entry under title:

The Herpesviruses.

(The Viruses)
Includes bibliographies and indexes.
1. Herpesvirus, diseases—Collected works. 2. Herpesviruses—Collected works. I. Roizman, Bernard, 1929–   . [DNLM: 1. Herpesviridae. QW 165.5.H3 H5637]
RC147.H6H57   1982   616.9′25   82-15034
ISBN 0-306-41778-2 (v. 3)

© 1985 Plenum Press, New York
A Division of Plenum Publishing Corporation
233 Spring Street, New York, N.Y. 10013

Printed in the United States of America

Contributors

Tamar Ben-Porat, Department of Microbiology, Vanderbilt University School of Medicine, Nashville, Tennessee 37232

Gabriella Campadelli-Fiume, Institute of Microbiology and Virology, University of Bologna, 40126 Bologna, Italy

Terence J. Hill, Department of Microbiology, University of Bristol Medical School, Bristol BS8 1TD, England

Albert S. Kaplan, Department of Microbiology, Vanderbilt University School of Medicine, Nashville, Tennessee 37232

Lenore Pereira, Viral and Rickettsial Disease Laboratory, California Department of Health Service, Berkeley, California 94704

William E. Rawls, Department of Pathology, McMaster University, Hamilton, Ontario, Canada L8N 3Z5

Franca Serafini-Cessi, Institute of General Pathology, University of Bologna, 40126 Bologna, Italy

Patricia G. Spear, Department of Molecular Genetics and Cell Biology, The University of Chicago, Chicago, Illinois 60637

Mary J. Tevethia, Department of Microbiology and Cancer Research Center, The Pennsylvania State University College of Medicine, Hershey, Pennsylvania 17033

Edward K. Wagner, Department of Molecular Biology and Biochemistry, University of California, Irvine, California 92717

Richard J. Whitley, Department of Pediatrics and Microbiology, University of Alabama School of Medicine, Birmingham, Alabama 45294

Preface

A great truth is a truth whose opposite is also a great truth.

Thomas Mann (Essay on Freud, 1937)

This volume centers on pseudorabies (PRV), herpes simplex viruses 1 and 2 (HSV-1 and HSV-2), and human cytomegalovirus (CMV) and fulfills three objectives.

The chapters on the epidemiology and latency of HSV, and on the glycoproteins specified by HSV and CMV, set the stage for the discussions of the immunobiology and pathogenesis of human herpesvirus infections in Volume 4. The epidemiology of HSV is the basis of our understanding of the spread and survival of this virus in the human populations. Central to the epidemiology of HSV and its pathogenesis in humans is the ability of the virus to remain in a latent state for the life of its host. The viral membrane glycoproteins are among the most interesting virion proteins, primarily because of their critical role in the initiation of infection. Since they are the surface membrane proteins of the virion and appear on the surface of productively infected cells, they are also the obvious if not the exclusive targets of the immune response.

The chapters on the transforming potential of HSV and CMV, and on the role of HSV in human cancer, deal with challenging problems requiring rather different experimental tools. The question of whether herpesviruses cause cancer can be answered only by careful, painstaking epidemiological studies whereas the analytical techniques required for understanding the mechanisms involved in oncogenesis fall largely within the domain of molecular biology. The hypothesis that there is a causal association between HSV and cervical cancer commemorated its 16th birthday in 1984. The ability of HSV and CMV to transform cells morphologically has also been known for many years. Probably few fields of herpesvirus research have generated as much effort as the attempts to elucidate the role of these viruses in human cancer and the involvement of viral genetic information in morphologic transformation. The chapters provide a critical evaluation of these complex and critical fields of research.

The chapters dealing with the transcription of HSV and with the molecular biology of PRV are the forerunners of Volume 5 which will focus on the molecular biology of herpesviruses. It is on PRV, however, that I would like to focus my remarks.

The studies on PRV by Albert Kaplan and Tamar Ben-Porat were initiated three decades ago. Their contributions to the biology of PRV provided important comparative data on PRV and more significant, a conceptual framework for the studies on the molecular biology of herpesviruses in general. Today, these seminal studies illustrate both the unity and diversity of the biology of herpesviruses. I am grateful for their contribution; without it this series would lack one of the most illustrious contributions to the current knowledge of the molecular biology of herpesviruses.

I would like to be the first to acknowledge my indebtedness to the contributors of this volume for conveying not just the facts, but also the excitement of their fields.

Bernard Roizman

Chicago, Illinois

Contents

Chapter 2

Individual HSV Transcripts: Characterization of Specific Genes

Edward K. Wagner

Chapter 3

Molecular Biology of Pseudorabies Virus

Tamar Ben-Porat and Albert S. Kaplan

Chapter 4

Herpes Simplex Virus Latency

Terence J. Hill

Chapter 5

Herpes Simplex Viruses and Their Roles in Human Cancer

William E. Rawls

Chapter 6

Transforming Potential of Herpes Simplex Viruses and Human Cytomegalovirus

Mary J. Tevethia

Chapter 9

Glycoproteins Specified by Human Cytomegalovirus

Leonore Pereira

CHAPTER 1

Epidemiology of Herpes Simplex Viruses

RICHARD J. WHITLEY

I. INTRODUCTION

Epidemiology treatises are notoriously dull as they recite statistics on the incidence and prevalence of infection as well as the implications of these infections on the salubrity of the community. In these commentaries, the data presented and their evaluation, which is often subject to multiple interpretations, invariably reflect the bias of the author. In studies of herpes simplex viruses (HSV), these biases are particularly apparent. The field has been influenced by many investigators who have performed excellent research aimed toward the understanding of the physiology of these viruses and their interactions with the host. Every attempt will be made to distinguish sound and precise clinical observations from those that warrant further clarification. Controversial issues range from the very simplistic, such as the incidence of symptomatic versus asymptomatic primary infections, to the more complex and unique propensities of these viruses, namely their apparent ability to cause recurrent disease and incrimination as oncogenic agents.

This chapter will review the epidemiology of HSV infections of man and will provide a background of historical developments, including the clinical and social significance of these infections, in order to assess the scientific advances of our understanding of these infections. A distinction between primary, nonprimary initial, recurrent, and exogenous (re)infections from both a clinical and a laboratory standpoint becomes essential to the epidemiology of these viruses. Clinical, virologic, and

RICHARD J. WHITLEY • Department of Pediatrics and Microbiology, University of Alabama School of Medicine, Birmingham, Alabama 45294.

serologic assessments that directly or indirectly incriminate HSV as the causes of disease will be discussed. Major controversies requiring pursuit will be explored relative to the application of methods developed in the laboratories of our colleagues in molecular virology. Through knowledge gained from application of these tools in future studies, it will be possible to understand more rationally the epidemiology of these viruses and, consequently, to develop regimens for prevention and/or specific therapeutic intervention in human disease, the ultimate goal of biomedical investigators.

II. HISTORICAL BACKGROUND

The impact of HSV infections on man has been well documented historically. Record of human HSV infections, particularly of spreading cutaneous lesions thought to be of herpetic etiology, dates to ancient Greek times (Nahmias and Dowdle, 1968) and particularly to the writing of Hippocrates (Wildy, 1973). The association between fever and mouth lesions was attributed to Herodotus (Mettler, 1947). Many of these original observations were predicated on Gallen's premise that the lesions themselves were an attempt by the body to rid itself of evil humors and, perhaps, resulted in the name of herpes excretins. However, these descriptions probably bear little resemblance to those of 20th century infections (Beswick, 1962). As noted by Wildy (1973), Shakespeare was no doubt cognizant of recurrent labial lesions as was recounted in *Romeo and Juliet* where Queen Mab, the midwife of the fairies, stated:

O'er ladies lips, who straight on kisses dream which oft the angry Mab with blisters plagues, because their breaths with sweetmeats tainted are.

It was not until the 18th century that Astruc, physician to the King of France, drew the appropriate correlation between herpetic lesions and genital infection (Hutfield, 1966). By the early 19th century, the vesicular nature of lesions associated with herpetic infections was well characterized; however, it was not until 1893 that Vidal specifically recognized human transmission of HSV infections from one individual to another (Wildy, 1973).

Observations at the beginning of the 20th century brought an end to the early imprecise descriptive era of HSV infections. First, histopathologic studies described the multinucleated giant cells associated with all herpesvirus infections (Unna, 1896). Second, the unequivocal infectious nature of HSV was recognized by Lowenstein (1919) who demonstrated that virus from the lesions of human herpes keratitis and the vesicles of herpes labialis produced lesions on the rabbit cornea. Vesicle fluid from patients with herpes zoster failed to produce similar dendritic lesions. In fact, these observations were actually attributed to earlier investigations by Gruter (1920) who performed virtually identical experiments around 1910 but did not report them until later.

Investigations reported between 1920 and the late 1960s focused on the biologic properties of these viruses and the natural history of human disease. During this period the host range of HSV was expanded to include a variety of laboratory animals, chick embryos, and, ultimately, cell culture systems. Expanded animal studies demonstrated that transmission of human virus to the rabbit resulted not only in corneal disease, but could also produce encephalitis (Doerr, 1920).

Important studies performed by one group (Andrews and Carmichael, 1930) revealed the presence of neutralizing antibodies to HSV in serum of previously infected adult patients. Subsequently, some of these patients developed recurrent lesions albeit less severe. Only individuals with neutralizing antibodies developed these recurrent lesions, a paradoxical finding given the classical lessons of infectious diseases whereby antibodies are usually associated with protection from disease. By the late 1930s it was well recognized that infants with severe stomatitis shed a virus thought to be HSV (Dodd *et al.*, 1938), subsequently developed neutralizing antibodies during the convalescent period (Burnet and Williams, 1939), and later could have apparent recurrent lesions.

The medical literature of the 1940s and 1950s was replete with articles describing such specific disease entities as primary HSV infections on mucous membranes as with gingivostomatitis, involving the skin such as eczema herpeticum (Seidenberg, 1941), primary keratoconjunctivitis (Gallardo, 1943), and herpes simplex encephalitis (Smith *et al.*, 1941). Furthermore, the clinical spectrum of HSV infection was expanded to include Kaposi's varicella-like eruption and neonatal disease.

Over the past 15 years major laboratory findings have built a foundation for the recent clinical evaluations of HSV infections. Certainly, one cornerstone of this foundation was the independent observation by two investigators that antigenic differences existed between HSV strains. Although suggested by Lipschitz (1921) on clinical grounds over 60 years ago and by others from laboratory observations (Schneweis and Bradiz, 1961; Plummer, 1964), it was not until 1968 (Nahmias and Dowdle, 1968) that well-defined antigenic and biologic differences were demonstrated between HSV, type 1 (HSV-1) and type 2 (HSV-2). These latter two investigators demonstrated that HSV-1 was more frequently associated with nongenital infection while HSV-2 was associated with genital disease, an observation that has set the stage for many clinical, serologic, immunologic, and epidemiologic studies.

III. DESCRIPTIVE EPIDEMIOLOGY

HSV are distributed worldwide and have been reported in both developed and underdeveloped countries, including remote Brazilian tribes (Black, 1975). Animal vectors for human HSV have not been described and, therefore, man remains the sole reservoir for transmission of these viruses to other humans. Virus is transmitted from infected to susceptible individuals during close personal contact. There is no seasonal variation in the incidence of infection. Because infection is rarely fatal and because these viruses become latent, it is estimated that over one-third of the world's population has recurrent infections and, therefore, the capability of transmitting HSV during episodes of productive infection.

HSV infections have historically been considered to be of biologic interest, but of little clinical significance. The associated diseases these agents cause, with attendant clinical manifestations, range from the usual case of mild illness, even undiscernable in many patients, to sporadic, severe, and life-threatening disease in a very few children and adults.

Two apparently independent phenomena are drawing increasing attention to the need for control of these infections. On the one hand, enhanced scientific awareness has developed because of a proliferation of knowledge regarding the molecular structure and function of these viruses, particularly as regards the physiology of host–virus interactions and recent advances in the clinical applications of antiviral chemotherapy. On the other hand, the lay public has become increasingly aware of the social significance of genital herpetic infections, neonatal herpes, and cervical carcinoma and, now, openly discusses its prevalence. As a consequence, renewed interest has been generated in the epidemiology of infections caused by HSV.

A. Epidemiology of HSV-1

Although HSV-1 and HSV-2 are usually transmitted by different routes and involve different areas of the body (HSV-1 above the waist and HSV-2 below the waist), there is a great deal of overlap between the expression of these viruses. The mouth and lips are clearly the most common sites of HSV-1 infections; however, any organ can become infected with HSV-1. Children, particularly those less than 5 years of age, are most often affected; yet, primary infections can also occur in older individuals. Great variability exists in clinical symptomatology of primary HSV-1 infections, ranging from being totally asymptomatic to combinations of fever, sore throat, ulcerative and vesicular lesions, gingivostomatitis, edema, localized lymphadenopathy, anorexia, and malaise. The incubation period ranges from 2 to 12 days with a mean of about 4 days. The duration of disease is generally 10 to 16 days. A clinical distinction should be drawn between intraoral lesions and lip lesions indicative of presumed primary and recurrent infection, respectively (Douglas and Couch, 1970).

Precise epidemiologic studies of HSV-1 infections, as well as those caused by HSV-2, are difficult to perform on clinical grounds alone because of the frequency of asymptomatic acquisition. Moreover, antigenic cross-reactivity between the two virus strains only further confuses the

issue, as will be discussed below. Nevertheless, tools applied to epidemiologic investigations include clinical evidence of infection, serologic
surveys to determine prevalence of antibodies connotated with infection,
and serial attempts at virus isolation to determine the frequency of occurrence and associated clinical findings.

There have been few studies using both serologic and virologic tools
for the determination of the incidence of primary HSV-1 infections. In
the absence of such detailed investigations, seroepidemiologic assessments provide an acceptable and the most commonly employed approach
to determine the occurrence of infection according to antibody prevalence. Such serologic surveys have included a variety of methods, although none is currently practical for quantitating type-specific antibodies. These assays include neutralization, complement fixation, passive
hemagglutination, immune adherence, indirect immunofluorescence,
and, more recently, the enzyme-linked immunosorbent assay (ELISA),
among others. Kinetic microneutralization determinations have been utilized in an attempt to calculate potency of HSV-1 and HSV-2 antigens
and, thereby, determine specific antibodies (Plummer, 1973). However,
the presence of common HSV-1 and HSV-2 antibodies has created difficulty in distinguishing specific antibody responses, especially for recurrent infections (Schneweis and Nahmias, 1971). Nevertheless, a primary
antibody response can be easily distinguished from a recurrent infection
or reinfection with the opposite major serotype because of the absence
of preexisting antibodies. Fortunately, recent elucidations of the glycoprotein composition of these viruses and the application of that knowledge to the development of type-specific antibody assays should provide
more precise tools for future surveys (Pereira et al., 1976; Eberle and
Courtney, 1981; Arvin et al., 1983). Furthermore, utilization of immune
blotting technology may also be of assistance (Bernstein et al., 1983).

Geographic location, socioeconomic status, and age influence to a
considerable extent the frequency of HSV infection, regardless of the
mode of assessment. These associations were brought to light by several
investigators (Dodd et al., 1938; Scott et al., 1941) and these data were
recently summarized by Rawls and Campione-Piccardo (1981). Utilizing
antibody prevalence, in developing countries, seroconversion occurs early
in life with some differences between study populations. In Brazilian
Indians, HSV antibodies were present in over 95% of children by the age
of 15 (Black et al., 1974). Similarly, serologic studies performed in New
Orleans demonstrated acquisition of antibodies in over 90% of children
by the age of 15 (Buddingh et al., 1953). In developing countries such as
Uruguay or in lower socioeconomic populations in the central United
States, the appearance of antibodies occurred at similar but lower frequencies (Rawls and Campione-Piccardo, 1981). These lower rates of acquisition were particularly evident in the poor black communities of
Atlanta and Houston (Rawls et al., 1969; Nahmias et al., 1970). By 5 years
of age, approximately one-third of patients had seroconverted; this fre-

quency increased to 70–80% by early adolescence. Predictably, presumed middle-class individuals of industrialized societies acquired antibodies even later in life. Seroconversion occurred over the first 5 years of life in 20% of children, followed by no significant increases until the second and third decades of life at which time the prevalence of antibodies increased to 40 and 60%, respectively (Sawanabori, 1973; Wentworth and Alexander, 1971). Further support for the relationship of socioeconomic status to acquisition of infection is offered by a variety of other studies (Smith *et al.*, 1967; McDonald *et al.*, 1974; Glezen *et al.*, 1975).

The influence of age is underscored by two other studies, in particular (Juretic, 1966; De Giordano *et al.*, 1970). A thorough clinical–serologic study of primary herpetic oral infection was performed in Yugoslavia. Of the 18,730 children attending outpatient clinics, evidence of oral herpetic infection was found in approximately 13% over a 10-year period. Children from 1 to 2 years of age were most commonly afflicted, accounting for over half of all cases. Children less than 6 months of age were uninfected in this study. With increasing age, the frequency of viral isolation decreased. No differences in sex or seasonal variation were detected. Indirectly, it is implied that most children had asymptomatic infection.

From these data, primary oral infection in older individuals appears to occur less commonly. However, in one thorough study of university students, seroconversion of susceptibles occurred at an annual frequency of approximately 10% (Glezen *et al.*, 1975), a figure three times higher than that estimated by Rawls and Campione-Piccardo (1981). In the college student population the major clinical manifestation of disease was an upper respiratory tract infection, particularly pharyngitis, as noted by others as well (Evans and Dick, 1964).

These studies demonstrate the significantly lower prevalence of antibodies in the relatively middle and upper socioeconomic classes. Primary infection occurs much earlier in life, generally very early, in children of underdeveloped countries as well as those of lower socioeconomic classes, whereas in developed countries and more affluent classes, primary infection may be delayed until adolescence or, perhaps, even adulthood. Frequency of direct person-to-person contact, indicative of crowding encountered with lower socioeconomic status, appears to be the major mediator of infection.

Following the implication of HSV as the etiologic agent responsible for gingivostomatitis (Dodd *et al.*, 1938) and its confirmation by Burnet and Williams (1939), virologic screening became a useful tool for epidemiologic surveys. Dodd *et al.* (1938) determined that primary infection led to the shedding of virus in mouth and stool and from the former site for as long as 23 days but on the average 7–10 days. Neutralizing antibodies began to appear between days 4 and 7 after clinical onset of disease, and peaked in approximately 3 weeks. In these studies, virus could be isolated from the saliva of asymptomatic children (Buddingh *et al.*, 1953). Viral shedding was documented in approximately 20% (15 of 72) of the children in the age range of 7 months to 2 years. Interestingly, in spite of no overall race or sex differences for the entire study, for the young age group, black children had higher rates of both neutralizing antibodies and virus isolation than did their Caucasian counterparts. Occurrence of virus shedding in children less than 6 months of age was uncommon, as noted in more recent studies (Juretic, 1966). In older children, 3 to 14 years of age, presumed asymptomatic shedding was documented in 18% (18 of 99). Virus retrieval decreased with advancing age in the Buddingh studies such that over 15 years of age, the frequency of excretion was 2.7% (5 of 185 individuals tested). These frequencies of shedding are very similar to more contemporary cross-sectional surveys, ranging from 2 to 5% (Stern *et al.*, 1959; Lindgren *et al.*, 1968; Cesario *et al.*, 1969; Kloene *et al.*, 1970).

A major question that continues to arise is the nature of infection itself; specifically, is infection synonymous with symptomatic disease? The epidemiologic consequences are obvious. With asymptomatic infection, either primary or recurrent, a reservoir exists for transmission of virus to susceptible individuals. The ratio of symptomatic to asymptomatic cases can only be determined when serologic and virologic tools are applied prospectively to targeted patient populations. Unfortunately, only few such studies exist. Cesario *et al.* (1969) reported a study of 70 seronegative children who had evidence of a primary infection. Only 6 of the 70 children, or less than 10% had clinical symptoms associated with illness. In an Australian study, Anderson and Hamilton (1949) found that 67.4% of seronegative children (29 of 43) developed HSV antibodies over a period of 1 year; 69% of the cases had symptomatic infections. Although the data from these two studies indicate that primary infection with HSV-1 can be associated with no symptoms of illness, the frequency of its occurrence remains for more precise determination.

Early epidemiologic studies have failed to precisely define the intraoral sites of virus replication with primary infection or, for that matter, with recurrent infection either. Perhaps such definition would clarify severe manifestations of disease such as herpes simplex encephalitis, often thought to progress from viral replication at the site of the olfactory bulb or correlate with frequency of recurrent lesions. Moreover, since the reporting of these original studies, it has been suggested that the epidemiology of infection is changing (as reflected by acquisition of primary infection later in life) because of improved standards of living. It is clear that many of these studies will be repeated to consider these among other possibilities.

B. Epidemiology of Genital HSV Infections

Although most genital HSV infections are caused by HSV-2, a recognizable and variable proportion is attributable to HSV-1 (Naib *et al.*,

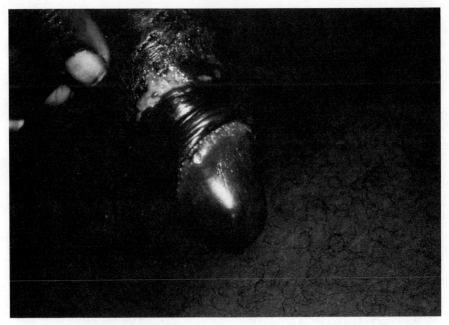

FIGURE 1. Ulcerative lesions of primary herpes simplex genitalis.

1973; Smith *et al.*, 1976; Wolontis and Jeansson, 1977; Kalinyak *et al.*, 1977; Corey *et al.*, 1983a). This distinction in virus type is not insignif- icant as genital HSV-1 infections are usually less severe clinically and are less prone to recur (Corey *et al.*, 1983a). Sexual transmission is the primary route of the spread of HSV-2 (Josey *et al.*, 1966, 1968, 1972; Parker and Banatvala, 1967; Deardourff *et al.*, 1974). The major line of evidence supporting sexual transmission of HSV includes the demonstrated high risk of infection following sexual contact with patients having confirmed disease. In the male, genital HSV infections are most often associated with vesicular lesions superimposed upon an erythematous base, usually appearing on the glans penis or the penile shaft, as shown in Fig. 1. In the female, lesions may involve the vulva, perineum, buttocks, cervix, and/or vagina. In the female, the lesions usually are excruciatingly sen- sitive to the touch, as suggested by Fig. 2, and associated with discharge. Primary infections independent of sex can be associated with fever, dy- suria, localized inguinal adenopathy, and malaise. As recently reviewed, the severity of primary infection and its association with complications are statistically higher in women than men for unknown reasons (Corey *et al.*, 1983a). Systemic complaints are common in both sexes, approach- ing 70% of all cases. The mean duration of disease in patients with pri- mary genital HSV infections is 19 days. The most common complications include aseptic meningitis (8%) and extragenital lesions (20%). Nonpri- mary but initial genital infection (preexisting antibody) is less severe as

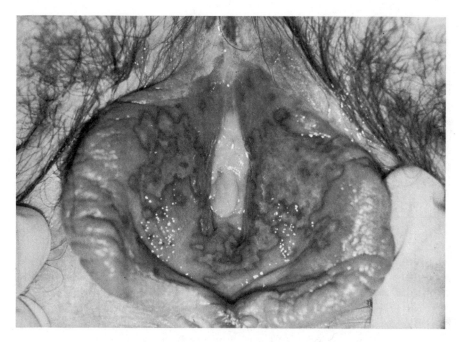

FIGURE 2. Bilateral labial lesions of primary herpes simplex genitalis.

healing is of shorter duration. Several studies estimate that the risk of susceptible females contracting HSV from infected males is 80% (Nahmias *et al.*, 1969; Rawls *et al.*, 1971, 1972; Poste *et al.*, 1972). Following primary genital herpetic infection, complications have occurred such as sacral radioculomyelitis, which can lead to urinary retention, neuralgias, and meningoencephalitis (Terni *et al.*, 1971; Skoldenberg *et al.*, 1975; Hinthorn *et al.*, 1976; Hevron, 1977; Caplan *et al.*, 1977). Primary perianal and anal HSV-2 infections (Jacobs, 1976), as well as associated proctitis (Goodell *et al.*, 1983), are becoming increasingly more common in male homosexuals. As with HSV-1 infections, many primary HSV-2 illnesses are subclinical, involving the mouth (Hale *et al.*, 1953) or the uterine cervix (Yen *et al.*, 1965; Josey *et al.*, 1966, 1968).

Sexually transmitted disease clinics provide the basis for prevalence data on genital HSV infections, particularly those in the United States, England and Sweden (Beilby *et al.*, 1968; Josey *et al.*, 1972; Nahmias *et al.*, 1973). Current estimates of the number of new cases of genital HSV infections number approximately 200,000–300,000 individuals per year (Nahmias *et al.*, 1973), which is probably a conservative estimate at best. These data were derived from the calculation that for every 5–10 cases of gonorrhea, 1 case of genital HSV infection exists (Nahmias *et al.*, 1973; Jeansson and Molin, 1974). A recent study by Corey *et al.* (1981) suggests a ratio of 1 case of genital herpes infection to 2.2 cases of gonorrhea. Such

estimates would strikingly increase the frequency of primary disease. One
difficulty in determining actual prevalence stems from the lack of a national reporting system. Current estimates of individuals with genital
herpetic infection in the United States alone range from as low as 2–4
million to as high as 10–20 million Americans (Corey et al., 1981).

Cytopathologic screening by Papanicolaou staining of clinical specimens adds another epidemiologic tool for the study of these problems
(Josey et al., 1968, 1972; Vesterinen et al., 1977). If these methods are
employed along with virus isolation, the incidence of genital HSV infections ranges from 0.09 to 0.24% in normal women (Kleger et al., 1968;
Naib et al., 1969; Wolinska and Melamed, 1970; Jeansson and Molin,
1970). In contrast, rates of infection in individuals routinely attending
sexually transmitted disease clinics vary between 0.02 and 3.3% to even
7.0% depending upon the population studied (Duenas et al., 1972; Naib
et al., 1973).

Women appear to have the highest rates of infection, particularly
prostitutes and others with multiple sex partners. An interesting observation has been made relating the incidence of HSV shedding to the age
of prostitutes. The most active prostitutes (age range 20–29 years) were
most likely to excrete HSV (12%), while those in older age groups had a
decreased frequency (6%). Of note, one study found that the incidence
of genital HSV infections in both indigent women and those of middle
and upper socioeconomic classes appeared significantly lower than that
found among women attending sexually transmitted disease clinics, 0.3
and 0.2% respectively (Ng et al., 1970). As with HSV-1 infections of the
mouth, HSV-2 can be excreted in the absence of symptoms at the time
of primary or recurrent infection (Rattray et al., 1978; Ekwo et al., 1979);
this occurs more frequently with HSV-2 and, again, a silent reservoir for
transmission is created.

In spite of the difficulty encountered in determining type-specific
antibodies, the appearance of HSV-2 antibodies reflects the time of exposure or, more simply, the acquisition of infection. The appearance of
antibodies can be positively correlated with the onset of sexual activity
(Rawls et al., 1969; Nahmias et al., 1970b; Adam et al., 1979), although
crowded living conditions may indirectly contribute to antibody prevalence (Becker, 1966; Naib et al., 1973). If HSV-2 antibodies are sought in
healthy women, there is a wide discrepancy in prevalence, ranging from
10% in Americans to 77% in Ugandans (Rawls et al., 1972). Because the
infection occurs primarily in adolescents and young adults, antibody prevalence rates increase most rapidly in those aged 20–29 years, followed by
those under 20 years of age (Bolognese et al., 1976). Predictably, those
individuals with multiple sexual partners have the highest frequency of
antibodies (Rawls et al., 1976).

Conversely, antibodies to HSV-2 are virtually nonexistent in nuns
(Rawls et al., 1969; Nahmias et al., 1970b; Duenas et al., 1972). As an
interesting and provocative side-issue, a previous history of antibodies to

either HSV-1 or HSV-2 may have an ameliorative effect on the expression of clinical disease (Nahmias *et al.*, 1970c; Rawls *et al.*, 1971; Kaufman *et al.*, 1973; Corey *et al.*, 1981). This suggestion is of importance as it relates to vaccine development to at least ameliorate if not prevent primary infection.

C. Treatment of Primary Infections

Although the focus of this review is the epidemiology of HSV infections, therapy cannot be ignored. No conclusive data are available for therapy of primary herpes simplex gingivostomatitis. Topical therapy of primary genital HSV infections has been shown to have some clinical usefulness. In a study of 77 patients with the first episode of genital herpetic infection, topical acyclovir applied six times daily reduced the duration of viral shedding, alleviated local symptoms, and accelerated time to crusting by mean times of approximately 3, 2, and 4 days, respectively. Topical therapy, however, did not significantly shorten the duration of total disease or decrease the appearance of new lesions. These data did not appear to correlate to sex or preexisting antibodies to HSV. Furthermore, therapy did not alter the frequency or severity of recurrences (Corey *et al.*, 1982, 1983c).

With the apparent difficulties of skin penetration of nucleoside analogs applied topically, two alternate systemic routes of therapy have been studied; these are oral and intravenous administration of acyclovir. Preliminary findings from these studies are encouraging for skin healing. A study on the effects of intravenous administration of acyclovir at 15 mg/kg per day for 5 days to 31 hospitalized patients with initial disease demonstrated a reduction in viral shedding by a mean of 11 days ($p = 0.001$), in symptoms by a mean of 5 days and accelerated time to total healing (Corey *et al.*, 1983b). Furthermore, complications were eliminated in those receiving the drug. In placebo recipients, complications included extragenital disseminated lesions (2 of 16 patients) and urinary retention (2 of 16). These data clearly indicate enhanced healing when compared to data derived from the topical trial; however, cost-effectiveness of i.v. therapy, which requires 5 days of hospitalization, can only be justified if recurrences are prevented or for therapy of severe complications. In the follow-up of these 31 patients, 60% of acyclovir recipients versus 88% of placebo recipients had a recurrence within 10 months of infection. These data can be transposed to a recurrence frequency per month of 0.26 and 0.50 for acyclovir and placebo recipients, respectively.

A larger number of patients are being assessed to determine the effect of therapy on recurrences. In this trial there was no evidence of significant clinical or laboratory toxicity nor the appearance of resistant virus in either randomization group. A virtually identical study has been performed in Europe. The data are essentially identical for clearance of virus from lesions, total healing time, and resolution of all symptoms; however, no reference is made to the frequency of recurrence.

A similar study design—double-blind, placebo-controlled—has been completed for oral therapy of initial genital HSV infections. This study involved 48 patients receiving 200 mg acyclovir or placebo five times daily for 10 days (Bryson et al., 1983). The data from this trial are similar to those resulting from i.v. administration of drug; thereby, a much less costly modality of therapy is provided. As with the i.v. studies, no effect was found on the frequency of recurrence. Here also, however, the physician must be concerned with the possibility of drug resistance following repeated exposures to drug.

Although the results of several trials substantiate the antiviral effects of acyclovir for genital HSV infections, the effect focuses on acceleration of acute disease. This capability in and of itself represents a major advance in therapy of viral disease. Yet, its clinical usefulness may be short-lived as recurrences do not appear to be prevented and resistant virus could pose additional problems.

IV. RECURRENT INFECTION VERSUS REINFECTION WITH HSV

A. Latency

The observation that lesions caused by HSV can reappear at a site of prior occurrence introduces a fascinating biologic phenomenon, namely, the apparent ability of the infection to recur following a period of presumed latency. The mechanism by which this phenomenon occurs is not totally understood; several controversial explanations have been offered. Ultimately, one must question whether a defined episode is the result of reactivation of latent virus, subclinical chronic production of virus, or reinfection with an exogenous strain of HSV either of the same or opposite serotype. Several reviews of this subject have addressed these issues in detail (Roizman, 1968, 1971; Terni, 1971; Nahmias and Roizman, 1973; Stevens et al., 1975; Pagano, 1975; Baringer, 1976). The fundamental data upon which our understanding of latency and, therefore, reactivation are founded warrant brief review for epidemiologic purposes.

The demonstration of latency is relatively new. Animal model studies provide one dimension in our understanding of this phenomenon. Several investigators have described the ability of HSV to remain latent for prolonged periods of time within nervous tissue of animal model systems (Stevens and Cook, 1971, 1974; Nesburn et al., 1972; Baringer and Swoveland, 1973; Stevens et al., 1975; Hill, 1981). Although these studies are not totally conclusive, data have shown that virus can only be retrieved from the ganglion rather than the peripheral nerve itself. These results have been obtained from animals infected at peripheral sites, generally the foot pad or ear.

From human and animal model studies, latency follows primary infection, as is best understood (Bastian *et al.*, 1972; Baringer and Swoveland, 1973; Stevens, 1975), by transmission of virus or its genome or a part thereof, via sensory nerve pathways to ganglion sites (Hill, 1981). Once within the ganglion, viral DNA can be detected intracellularly but, with existing tools, no other form of infectious virus or products of replication can be demonstrated (Stevens, 1975). Latent virus has been retrieved from the trigeminal, sacral, and vagal ganglia (Bastian *et al.*, 1972; Baringer and Swoveland, 1973; Baringer, 1974; Warren *et al.*, 1977). Reactivation of latent virus appears dependent upon an intact anterior nerve route and peripheral nerve pathways. The resulting recurrent infection frequently involves the same site, but only a very limited portion of the dermatome. No data exist to support the existence of virus peripherally in a latent state (Rustigian *et al.*, 1966).

In man the frequency of isolation of HSV from ganglia has been described by Baringer (Baringer and Swoveland, 1973; Baringer, 1974), among others. When trigeminal and sacral ganglia were explanted with human embryonic lung cells, cytopathic effect could be detected within the newly growing cells. Evidence of virus replication, however, may not appear until cells have been subcultured and passed. In one study, virus isolation was attempted from the trigeminal ganglia of 90 cadavers and was documented in 44 individual situations. In 26 of these cases, viruses were isolated bilaterally from both ganglia. Similarly, when 68 sacral ganglia were removed and cocultivated, only 9 were positive for HSV, and all but one were HSV-2. Baringer extended these observations to successfully retrieve virus from the thoracic ganglia of one patient whose trigeminal ganglia yielded HSV-1 bilaterally (Baringer and Swoveland, 1973).

These findings, regarding the retrieval of virus from the trigeminal ganglia, help explain the observation of vesicles that recur at the same site. Nevertheless, one cannot ignore the question: can exogenous reinfection with the same or an opposite strain of virus account for "apparent" recurrent HSV infections? Clinical evidence as early as 1968 (Nahmias and Dowdle, 1968) has suggested that this is a real possibility. However, late in 1979, speculation regarding exogenous reinfections was based upon the prior observation that it was possible to establish genital reinfection with HSV-2 in mice and *Cebus* monkeys who had recovered from prior genital inoculation at the same site (London *et al.*, 1971). Unequivocal demonstration of exogenous reinfection was possible only with the development of restriction endonuclease technology. It is worthwhile to briefly review its thesis and application to clinical–epidemiologic studies.

Hayward *et al.* (1975) found that fragments resulting from the cleavage of HSV DNA by restriction endonuclease enzymes become fingerprints for the virus. Analyses of numerous HSV-1 and HSV-2 isolates from a variety of clinical situations and geographic areas demonstrated that epidemiologically unrelated strains yielded distinct HSV DNA fragment patterns. In contrast, fragments of HSV DNA derived from the same individual obtained years apart, mothers and their newborns, or sexual partners, or following short and long passages *in vitro* had identical fragments after restriction endonuclease cleavage.

This technique has been applied to specimens collected from a variety of patients with a genital HSV infection to assess potential strain differences or similarities with various endonucleases. Analyses of patterns of the HSV DNA fragments revealed that isolates from the same patients or their respective sexual partners could be either the same or different (Buchman *et al.*, 1979). Stated differently, a given patient might well have nonidentical isolates obtained from lesions at adjacent sites. This finding implies that an individual can be infected with multiple HSV-2 strains at different sites following prior infection with a genetically different strain. No doubt, similar findings will become apparent for orolabial HSV-1 infections when properly studied. The utilization of endonuclease restriction enzymes has unequivocally demonstrated that a target organ, in this case the genitalia, is not immune to reinfection. The frequency of this occurrence in large-scale studies has not been established. From a public health viewpoint this is an important issue because if such events do in fact occur, regardless of incidence, techniques of vaccination will have to be varied from those currently available to render immunity to target organs. Obviously, an exogenous source of virus does not explain the majority of instances of apparent reactivation.

Reinfection with the same strain of HSV can occur as by autoinoculation at a distant site. Thus, HSV could be mechanically transmitted from one site to either an adjacent site or a distal one. These instances have been reported in cases from mouth to genital transmission (Nahmias *et al.*, 1970c) or intentional inoculation of vesicle fluid to "bolster immunity" (Teissier *et al.*, 1926; Lazer, 1956; Goldman, 1961; Blank and Haines, 1973).

At the present no data support the contention that recurrent infections are the result of chronic viral production at the site of recurrence. Skin biopsies performed at the site of recurrent lesions but between episodes of disease for purposes of virus isolation have not resulted either in the retrieval of virus or in the demonstration of its presence by standard laboratory procedures (Roizman, 1968, 1971). However, during the prodrome of recurrent herpes labialis, HSV can be retrieved by needle biopsy or cell scrapings from apparently normal tissue (Spruance *et al.*, 1977).

B. Recurrent HSV-1 Infections

The largest reservoir of HSV infections in the community is that associated with recurrent herpes labialis and genitalis. The onset of recurrent orolabial lesions is heralded by a prodrome of pain, burning, tingling, or itching, which generally lasts for less than 6 hr followed within 24–48 hr by vesicles (Young et al., 1976; Spruance et al., 1977). Vesicles appear most commonly at the vermillion border of the lip and persist in most patients only 48 hr at the longest. The total area of involvement usually is localized, being less than 100 mm², and lesions progress to the ulcerative and crusting stage within 48 hr. Pain is most severe at the outset and resolves quickly over 96–120 hr. Similarly, the loss of virus from lesions decreases with progressive healing (Spruance et al., 1977; Bader et al., 1978). Healing is rapid, generally being complete in 8–10 days. The frequency of recurrences varies among individuals (Spruance et al., 1977). The factors responsible for recurrences are both highly variable and poorly defined.

Studies performed to assess the frequency and severity of recurrent infection are limited. Rawls and Campione-Piccardo (1981) have reviewed several studies performed to determine the nature of recurrent infection; however, the studies encompass admittedly limited socioeconomic and geographic groups. Baseline frequencies for recurrence can be gleaned from a study performed in Philadelphia. A positive history of recurrent herpes labialis was noted in 38% of 1800 students attending the University of Pennsylvania graduate education program (Ship et al., 1960, 1961). New lesions occurred at a frequency of 1 per month in 5% of the students and at intervals of 2–11 months in 34% of the students. Recurrence of 1 per year or less often was found in 61%. The frequency of recurrence is relatively constant, approximately 33% in a series of other studies (Ship et al., 1967, 1977; Friedman et al., 1977). Exceptions to this 33% mean frequency of recurrence are studies by Young et al. (1976) in Ann Arbor, Michigan, and Embil et al. (1975), which record lower frequencies of recurrence, approximately 16%, but under vastly different socioeconomic conditions, namely, of upper versus lower social groups, respectively. In comparative studies done within one community in this country, Ship et al. (1977) found that recurrences occur more frequently among the more socially privileged. These findings are somewhat perplexing because recurrence should reflect the population previously infected. Clearly, the distinction between socioeconomic status, age of acquisition, and prevalence of infection versus the frequency of recurrence warrants further investigative pursuit.

Factors associated with the recurrence of HSV-1 lesions are highly stereotyped and include stress, fatigue, menses, and exposure to sunlight (Segal et al., 1974; Ship et al., 1977). In addition, fever has been associated with recurrent lesions. Greenberg et al. (1969) demonstrated that recurrent lesions of herpes labialis were three times more frequent in febrile patients than in nonfebrile controls. The mechanisms by which these factors lead to production of infectious virus are not known.

Several interesting studies have provided insight into stimuli associated with recurrences. In two studies (Carlton and Kilbourne, 1952;

Ellison *et al.*, 1959), it was possible to demonstrate an incidence of herpetic lesions in the mouth or palate of at least 90% in patients undergoing nerve root section for trigeminal neuralgia. Lesions appeared in the immediate postoperative period. These findings have been further verified by Pazin *et al.* (1978, 1979) in assessments of virus excretion following microvascular surgery to alleviate tic douloureux. Clearly, in these situations it is unreasonable to think that exogenous viral infection could account for the recurrent disease. Reactivation by axonal injury has been further investigated by Walz *et al.* (1974), who reactivated a latent ganglionic infection following section of the peripheral nerve with the resultant appearance of virus within the ganglia 3–5 days after the surgical manipulation. Similarly, if an attempt is made to excise lesions, it will fail as vesicles will recur adjacent to the site of excision (Kibrick and Gooding, 1965).

As with primary infections recurrent disease may occur in the absence of clinical symptoms. Asymptomatic excretion of HSV in normal children following recurrence is approximately 1% (Hellgren, 1962; Haynes *et al.*, 1968) and varies from approximately 1 to 5% in the normal adult (Lindgren *et al.*, 1968; Sheridan and Herrmann, 1971; Glezen *et al.*, 1975). Interestingly, it has been noted that approximately 1% of pregnant women and nursery personnel excrete HSV at any point in time. This source of virus may pose risks for transmission to the newborn (Buchman *et al.*, 1978). Asymptomatic excretion of virus is not limited to the healthy adult as demonstrated by the studies of Pass *et al.* (1979) whereby excretion of HSV in renal transplant recipients, without signs or symptoms of disease, occurs in nearly one-third of infected patients. Furthermore, studies of closed populations have documented a high frequency of recurrent HSV excretion in the absence of symptoms (Cesario *et al.*, 1969).

C. Therapy of Recurrent HSV-1 Infections

Treatment of recurrent herpes labialis infections has not been rewarding. A major focus of antiviral research in the United States has been attempts to develop therapeutically useful and simple modalities of therapy for herpes labialis and genital HSV infections. As reviewed by Overall (1980), numerous compounds have been tested for these diseases with little evidence of success with one recent exception. No doubt these failures are in large part related to the inability of these nucleoside derivatives to penetrate the skin and reach the site of viral replication.

In acyclovir treatment of active recurrent herpes labialis, no benefit could be demonstrated for time to total healing, crusting, or loss of pain; however, there was a trend to accelerated loss of virus from lesions (Spruance and Crumpacker, 1982). A follow-up study to determine the effect of therapy instituted during the prodrome phase of illness similarly failed to show evidence of a drug effect (Spruance *et al.*, 1983).

D. Recurrent HSV-2 Infections

As with orolabial lesions, recurrent genital infection is the largest reservoir of HSV-2. Factors responsible for reactivation of genital HSV infections cannot be unequivocally gleaned from retrospective studies. Recurrent HSV-2 infection can be either symptomatic or asymptomatic as with HSV-1; however, recurrence is usually associated with a shorter duration of viral shedding and fewer lesions (Corey et al., 1981). Constitutional symptoms are uncommon with recurrent disease. The frequency of recurrences varies somewhat between males and females with calculations of 2.7 and 1.9 per 100 patient days, respectively (Corey et al., 1981). Overall, several studies have implicated a frequency of recurrence as high as 60% (Adam et al., 1979; Chang et al., 1974). These rates must be verified by additional studies and in a larger series of patients as they appear exceedingly high, a legitimate possibility based upon selection of a nonrepresentative subpopulation. The type of genital infection, HSV-1 versus HSV-2, appears predictive of the frequency of recurrence (Corey et al., 1981), such that HSV-1 appears to recur less frequently. Further epidemiologic investigations of genital HSV infections must address such issues as the age of acquisition, frequency of intercourse, number of sexual partners, and prior antibody status as well as the influence of socioeconomic conditions. In addition, the severity and duration of initial disease correlated with recurrences as well as the frequency and significance of exogenous infection will have to be clarified.

E. Therapy of Recurrent HSV-2 Infections

A study of 111 patients was similarly conducted for the therapeutic effectiveness of acyclovir on the management of recurrent genital herpes infections (Corey et al., 1982). The data suggest an effect on shedding of virus from lesions and, perhaps, the time to crusting of lesions in men but not women. Overall, topical acyclovir had no effect on the symptoms of disease or healing time in either sex, with the exception of crusting in women. Similarly, no effect on recurrences was apparent. All the therapeutic trials with acyclovir from one center were recently reviewed (Corey et al., 1983b). An extension of this study to determine therapeutic effectiveness when drug application is introduced during the prodrome has been completed and similarly appears to show no benefit (Luby et al., 1984).

These trials encapsulate the difficulties encountered in topical studies of antiviral agents. Specifically, issues regarding penetration of nucleoside derivatives to the site of viral replication and the clinical value of therapy become foremost questions. In that recurrences are not prevented, improved therapeutic modalities must be developed. Furthermore, the potential for development of drug resistance is apparent in a disease such as genital HSV infection with a known propensity for re-currence and, therefore, the probable resistance increases with repeated exposure to medications following each episode (Crumpacker *et al.*, 1982). Such resistant viruses have been documented, following repeated systemic therapy with acyclovir. The frequency of development of re-sistance and the potential pathogenicity of these isolates will require further extensive investigation. Nevertheless, one must weigh topical therapy of HSV infections against the need to reserve such potentially useful drugs for truly life-threatening disease lest problems of drug re-sistance encountered in the management of bacterial disease occur also with viral infections.

Only recently have field trials been initiated with phosphonoformate, a compound developed in Sweden, for topical therapy of both herpes la-bialis and genital HSV infections. These trials have employed cross-over study designs and can only be considered preliminary because of the lim-ited numbers of patients studied and, consequently, an inability to ap-propriately stratify analyses for demographic characteristics as well as for treatment responses. Regardless, these initial studies demonstrate a trend toward accelerated healing, particularly loss of virus from lesions and time to total crusting. A larger, collaborative trial has recently been initiated to assess clinical usefulness in Scandinavian countries. With phosphonoformate, as with acyclovir, resistant strains of virus have been encountered and, clearly, this possibility must lead to careful definition of frequency of occurrence. Should this occur even as frequently as 1–2% following a course of therapy, serious consideration must be given to the possible resultant harm that could result from drug application to a usually benign illness.

V. DISEASES OF EPIDEMIOLOGIC IMPORTANCE

A. Herpes Keratoconjunctivitis

Infections of the eye are usually caused by HSV-1 beyond the newborn age (Ostler, 1976; Binder, 1977). Approximately 300,000 cases are diag-nosed yearly and ocular HSV infections are second only to trauma as the cause of corneal blindness in this country. Primary herpetic keratocon-junctivitis is associated with either unilateral or bilateral conjunctivitis which can be follicular in nature followed soon thereafter by preauricular adenopathy. HSV infection of the eye is also associated with photophobia, tearing, eyelid edema, and chemosis with the pathognomonic findings of branching dendritic lesions, as shown in Fig. 3, or less commonly but with advanced disease a geographic ulcer of the cornea, as shown in Fig. 4. Healing of the cornea can take as long as 1 month even with appropriate antiviral therapy.

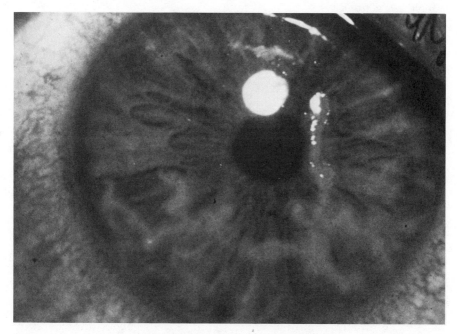

FIGURE 3. Dendritic herpes keratitis.

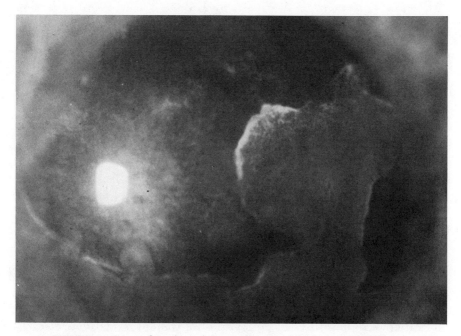

FIGURE 4. Geographic ulcer of the cornea caused by herpes simplex virus.

Recurrent herpes infections of the eye are common. Most frequently, they are unilateral in involvement, but a small percentage of cases involve both eyes. Characteristically, either dendritic ulcerations or stromal involvement appears. Visual acuity is decreased in the presence of the ulcers; and with progressive stromal involvement, opacification of the cornea may occur. Repeated individual attacks may last for weeks or even months following appropriate antiviral therapy. The route and pathogenesis of infection remain unknown. Clearly, therapeutic attempts warrant further refinement. At present, however, three compounds have been licensed for therapy of these diseases: idoxuridine, trifluorothymidine, and vidarabine ophthalmic preparations. Progressive disease can result in visual loss and even rupture of the globe. A major goal for clinical and molecular research is the development of more efficacious agents for the management or even prevention of this disease.

B. Skin Infection

Skin infections caused by HSV generally manifest as eczema herpeticum in patients with underlying atopic dermatitis (Pugh et al., 1955; Wheeler and Abele, 1966; Terezhalmy et al., 1979). The lesions can either be localized resembling herpes zoster with a dermatomal distribution or disseminated. The latter occurs commonly in Kaposi's varicellalike eruption (Ruchman et al., 1947). HSV infections of the digits, Whitlow ulcers, have been reported, particularly among dentists (Rosato et al., 1970). Whitlow lesions are commonly caused by HSV-1; however, an increasing incidence of HSV-2 Whitlows has been recognized in the community.

The prevalence of skin infections in Skaraborg, Sweden, among approximately 7500 individuals over 7 years of age was found to be about 1% (Hellgren, 1962). In another Swedish study performed in dermatology clinics, 2% of men and 1.5% of women attending clinics in Gothenburg over a 6-year period had evidence of herpetic skin infections (Eilard and Hellgren, 1965). In addition to individuals with atopic disease, patients with skin abrasions or burns appear particularly susceptible to HSV-1 and HSV-2 infections and some may develop disseminated infection (Foley et al., 1970). Disseminated HSV infections have also been reported among wrestlers, being referred to as herpes gladitorium (Selling and Kibrick, 1960; Wheeler and Cabraniss, 1965). Other skin disorders associated with extensive cutaneous lesions include: Darier's disease and Sezary's syndrome (Hitselberger and Burns, 1961; Hazen and Eppes, 1977). As would be predicted, localized recurrence followed by a second episode of dissemination has been observed (Orenstein et al., 1974).

C. Infections of the Immunocompromised Host

Patients compromised by either immune therapy, underlying disease, or malnutrition are at increased risk for HSV infections. Renal transplant

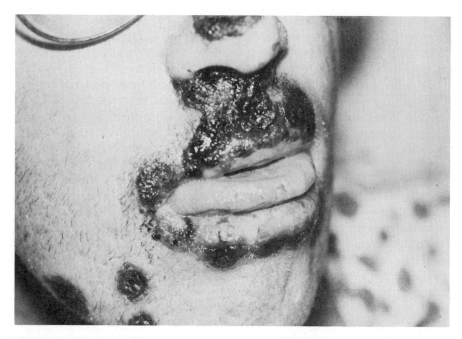

FIGURE 5. Cutaneous dissemination of herpes simplex virus in a renal transplant recipient.

and cardiac recipients are at particular risk for increased severity of HSV infections (Logan *et al.*, 1971; Muller *et al.*, 1972; Pass *et al.*, 1978). An example of cutaneous dissemination following shaving, in a renal transplant recipient, is shown in Fig. 5. Often these infections are asymptomatic as in the normal host. In these latter two patient populations, the presence and quantity of antibodies before treatment predict the individual at greatest risk for recurrence (Pass *et al.*, 1979). These patients may develop progressive disease involving the respiratory tract, esophagus, or even the gastrointestinal tract (Korsager *et al.*, 1975; Montgomerie *et al.*, 1969). The severe nature of progressive disease in these patients appears to be directly related to the degree of immunosuppressive therapy employed (Rand *et al.*, 1977). Reactivation of latent HSV infections in these patients can occur at multiple sites and healing in these patients with severe progressive disease occurs over an average of 6 weeks (Whitley *et al.*, 1984).

D. Infections of the Newborn

Ongoing evaluations of neonatal HSV infections by Nahmias *et al.* (1983) estimate that the incidence of disease is approximately 1 in 5000 to 10,000 deliveries. Overall, approximately 120 to 150 cases are thought to occur each year. Recently, in one country, an increasing incidence has been documented (Sullivan-Boyai *et al.*, 1983). The primary route of transmission of virus to the newborn appears to be contact of the child with infected genital secretions during delivery; however, *in utero* acquisition of infection by hematogenous routes has been reported. The major epidemiologic problem worthy of attention regarding neonatal HSV infections is identification of the woman with genital herpes at risk of transmitting the infection to her offspring. Such identification may lead to early intervention by cesarean section and, hopefully, prevention of newborn disease. A recent study performed to evaluate antiviral chemotherapy of this disease (Whitley *et al.*, 1980b) also verifies a previous observations by Nahmias *et al.* (1970a) and Tejani *et al.* (1979) that genital herpetic infection in the pregnant woman is more frequently asymptomatic than symptomatic. In women with suspected genital herpesvirus infections, every attempt must be made to define the presence or absence of virus excretion at the time of delivery. If viral diagnostic facilities are available, viral cultures should be obtained from the genital tract of suspected viral excretors. Lacking viral isolation tests, it is possible to identify the presence of HSV infection by Papanicolaou smear of the involved area, as noted previously. In children born with suspected HSV infections, it is imperative to obtain a history from the baby's parents or their sexual partners regarding evidence of genital vesicles, genital pain, dysuria, or penile lesions.

Another problem of potential epidemiologic importance is nosocomial transmission of HSV within nurseries (Francis *et al.*, 1975). The only truly precise means for assessment of transmission would be the application of restriction endonucleases for fingerprinting of viral isolates, as has already been employed in a limited study (Linnemann *et al.*, 1978). More recently, direct evidence for nosocomial transmission has been documented in several babies receiving care in a single high-risk nursery (Hammerberg *et al.*, 1983). Verification of this observation will have an obvious impact on care of newborns with HSV infections.

The risk of neonatal infection following delivery through an infected birth canal appears dependent in part, and logically so, upon the duration of ruptured membranes (Nahmias *et al.*, 1983). If membranes are ruptured for longer than 4 hr, the risk of neonatal infection was found to be 50% in one small study. The investigators, recognizing the limited number of patients available for the study, suggest this risk may be excessive. Currently, it is recommended that women should undergo cesarean section if they are excreting HSV at the onset of labor and the membranes have been ruptured for less than 4 hr. Verification of incidence of HSV excretion at the time of delivery and the risks to the newborn can only be assessed prospectively; such studies are in progress. Apparently, infants born to mothers with primary HSV infections of the cervix and vagina late in gestation are at greater risk for developing severe infection than those born to women with recurrent genital excretion. No doubt, this is related to the quantity of virus present in the genital tract during primary

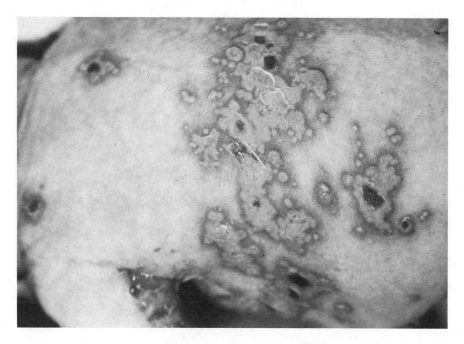

FIGURE 6. Skin vesicles of neonatal herpes simplex infection.

infection and its site of replication. Maternal viremia with either HSV-1 or HSV-2 early in gestation may lead to fetal wastage or birth defects (Nahmias *et al.*, 1971; Florman *et al.*, 1973; Komorous *et al.*, 1977).

Clinical disease of the newborn with HSV results in a broad spectrum of involvement ranging from skin vesicles to disseminated organ involvement. Importantly, disease can occur in the absence of skin vesicles. Skin vesicles, as shown in Fig. 6, are the hallmark of infection and are present in 60–86% of all babies with infection. The vesicles themselves may precede the appearance of more severe forms of infection (e.g., pneumonia, hepatitis, disseminated intravascular coagulopathy). The eye is a site commonly involved. Clinical manifestations of disease reflect the sites of involvement. Skin vesicles, keratoconjunctivitis, seizures, poor feeding, lethargy, bleeding diatheses are common findings and are usually present in association with hepatosplenomegaly. When the CNS is involved, neurologic signs predominate (Nahmias *et al.*, 1970a, 1983; Hanshaw, 1973; Florman *et al.*, 1973; Komorous *et al.*, 1977). If death ensues in babies with CNS disease, cortical destruction of brain tissue is invariable (Fig. 7). There does not appear to be a distinct difference in clinical manifestations of disease with either HSV-1 or HSV-2.

A true disseminated infection involves target organs of the body and, in particular, the liver, adrenals, and lungs with or without CNS involvement and with or without skin involvement. This form of disease, in its

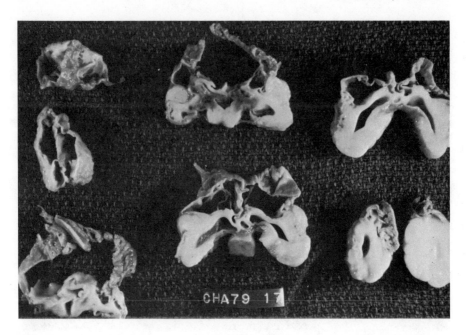

FIGURE 7. Hydroencephaly caused by neonatal herpes simplex encephalitis. Reprinted with permission from *Pediatrics in Review* 2(9):261, 1981.

untreated state, is associated with a mortality of over 80% (Whitley *et al.*, 1980a; Nahmias *et al.*, 1983). Mortality with skin, mouth, or eye disease alone is low. However, frequently (>60% of babies), disease begins locally and progresses to other organs. Regardless of clinical presentation, neurologic impairment is common. It has been estimated that only approximately 18% of infected newborns at the most are free of neurologic sequelae.

Vidarabine therapy of neonatal HSV infection leads to statistically significant decreased mortality and improved morbidity (Whitley *et al.*, 1980a, 1983). Mortality in babies with disseminated or CNS infection is decreased to 38% overall as shown in Fig. 8, and the number of babies who develop normally increases at least threefold.

Ongoing epidemiologic evaluations must lead to appropriate methods of prevention and to improved therapy. Even with current awareness of HSV infections of the newborn, diagnosis is frequently missed because of the absence of vesicles and the lack of apparent symptoms or signs of target organ involvement.

E. Infection of the Nervous System

Herpes simplex encephalitis is one of the most devastating of all HSV infections (Fig. 9). It is considered the most common cause of sporadic,

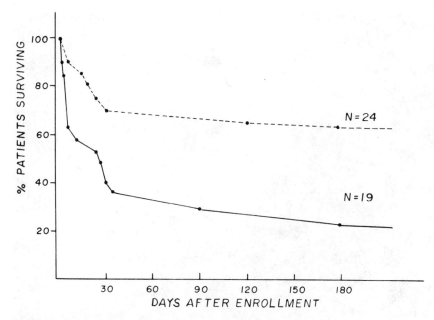

FIGURE 8. Survival following disseminated or CNS neonatal HSV infection treated with vidarabine (.....) or a placebo (____) p = 0.028. Vidaribine versus placebo; p = 0.014. Reprinted with permission from *Pediatrics* 1981.

fatal encephalitis in this country (Olson *et al.*, 1967). The Centers for Disease Control estimate the incidence of herpes simplex encephalitis to be approximately 40–50 cases per year; however, disease occurrence is undoubtedly higher based upon data collected by the National Institute of Allergy and Infectious Diseases Antiviral Study Group (Whitley *et al.*, 1981). Importantly, the actual incidence remains for definition as no national reporting system exists for any HSV infection, even those that are life-threatening as herpes simplex encephalitis. The manifestations of this infection in the older child and adult include primarily a focal encephalitis associated with fever, altered consciousness, bizarre behavior, disordered mentation, and localized neurologic findings. These clinical findings generally are associated with evidence of localized temporal lobe disease by neurodiagnostic procedures (Whitley *et al.*, 1977, 1981).

A major debate at the present time focuses on the pathogenesis of the infection, particularly as it relates to the source of virus responsible for brain disease. Recently, HSV isolates obtained from both the brain and the lip or mouth of patients with virologically confirmed herpes simplex encephalitis were examined by endonuclease restriction enzymes for identity (Whitley *et al.*, 1982). These studies demonstrated that only approximately 65% of patients excreting virus from two sites had identical isolates. Thus, virus excreted from the mouth of patients with herpes simplex encephalitis may be identical to that of the brain or may be

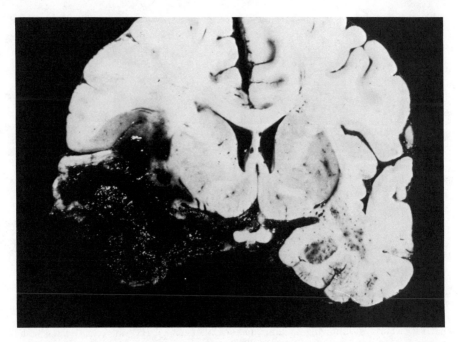

FIGURE 9. Necrotizing lesion of herpes simplex encephalitis. Reprinted with permission from *Pediatrics in Review* 2(9):**260, 1981.**

entirely different. The significance and frequency of the latter finding will be pursued.

The issue of the nature of the brain infection, i.e., primary or recurrent, has been considered time and again among both infectious diseases and neurology specialists. The Collaborative Antiviral Study Group observed that approximately half of the patients with herpes simplex encephalitis, confirmed by brain biopsy and virus isolation, appear to have primary infection and the remaining patients appear to have recurrent infection (Whitley *et al.*, 1981). The basis for primary and recurrent infections hinges upon the presence of neutralizing antibodies in the serum (screened at 1 : 10 dilutions) obtained at the onset of the CNS disease (Nahmias and Whitley, 1982). The relative role of exogenous versus endogenous HSV in causing herpes simplex encephalitis in individuals with preexisting antibodies also warrants delineation, particularly in light of the aforementioned endonuclease restriction enzyme patterns of the paired brain and orolabial isolates.

Through the Collaborative Antiviral Study Group, it has been possible to demonstrate the therapeutic value of vidarabine for biopsy-proven herpes simplex encephalitis (Whitley *et al.*, 1977, 1981). Therapy decreased the mortality from 70% to 38%. Outcome is influenced by age and level of consciousness at the outset, as shown in Fig. 10. Furthermore,

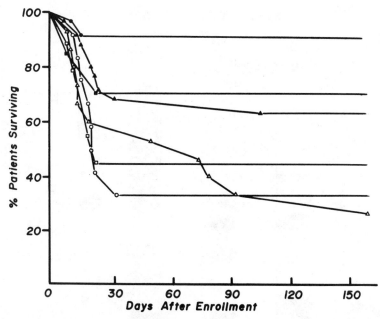

FIGURE 10. Influence of age and level of consciousness on survival with vidarabine therapy of herpes simplex encephalitis (●——●) lethargic, <30, N = 23; (■——■) semicomatose, <30, N = 7; (▲——▲) lethargic, ≥30, N = 26; (□——□) comatose, <30, N = 10; (○——○) semicomatose, ≥30, N = 13; (△——△) comatose, ≥30, N = 14. Reprinted with permission from *N. Engl. J. Med.* 1981.

these same latter two variables influence morbidity (Fig. 11). Thus, although therapy is clearly effective, obvious improvement in both diagnosis and therapy is mandatory.

One of many areas for further investigation is the elucidation of the spectrum of HSV infections of the brain. Current technology allows for unequivocal diagnosis only by brain biopsy. Therefore, noninvasive diagnostic procedures must be developed. Such procedures include HSV-specified thymidine kinase, DNA polymerase, and DNase (Cheng *et al.*, 1979) or detection of HSV-excreted glycoprotein (Chen *et al.*, 1978) in the cerebrospinal fluid. The true spectrum of HSV infections of the CNS will be defined only with the availability of such tools. The development of all these assays is under way and we hope that technologic breakthroughs will allow for their application to epidemiologic studies within the next several years.

F. Other Neurologic Syndromes

In addition to herpes simplex encephalitis, these viruses can involve virtually all anatomic areas of the nervous system. Craig and Nahmias

28   RICHARD J. WHITLEY

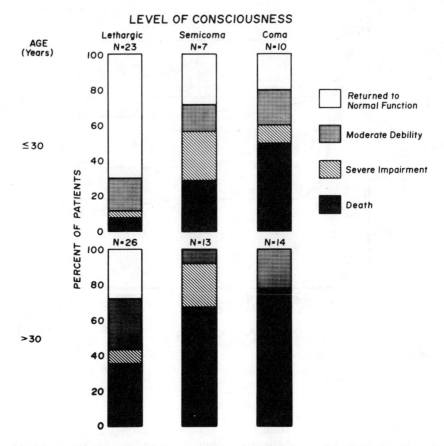

FIGURE 11. Influence of age and level of consciousness at the time of initiation of vidarabine therapy on outcome in herpes simplex encephalitis. Reprinted with permission from *N. Engl. J. Med.* 1981.

(1973) have reviewed the association of HSV with meningitis, myelitis, radiculitis, etc. The relationship between HSV infections of the brain and chronic degenerative disease, psychiatric disorders, or Bell's palsy requires further definition (Constantine *et al.*, 1968; Cleobury *et al.*, 1971; McCormick, 1972).

G. Unusual Outbreaks of Infection

Clustered outbreaks of human HSV infections have been reported in the literature (Scott, 1957; Juretic, 1966); however, there is no indication from either clinical or molecular epidemiologic studies that these viruses cause epidemic diseases. Most of the studies reported involve families where several individuals within the family suffered from HSV infection at approximately the same time. No index case could be identified, although frequently recurrent labial lesions of one family member were incriminated. Outbreaks within hospitals have been identified; however, here also, no clear epidemic nature of the disease could be determined. The frequent occurrence of HSV infections among renal transplant recipients within the 4-week period immediately after surgery has been reported (Pass *et al.*, 1979). Pugh *et al.* (1955) described an outbreak of eczema herpeticum in a group of hospitalized patients within an 8-day period. In this latter case, hospital conditions, in particular the lack of attention to nursing details (e.g., handwashing), were incriminated as being responsible for virus transmission. Outbreaks of herpetic stomatitis have been reported within orphanages (Hale *et al.*, 1953; Juretic, 1966) where the attack rate for clinically apparent primary infections was approximately 75% of the susceptible patients.

Endonuclease restriction enzyme technology has been employed to study a limited number of clustered outbreaks of HSV infections. Genetic analysis (Buchman *et al.*, 1978) of the isolates from a Louisville Pediatric Intensive Care Unit permitted tracing of the spread of virus. These studies demonstrated transmission of an identical strain of HSV-1 within the unit, a strain distinct from one simultaneously isolated but not transmitted. Similarly, presumed nosocomial transmission of virus from one baby to another (Linnemann *et al.*, 1978) has been shown. Although this study showed that an identical virus infected the infants, it was impossible to define the route or vector of transmission. Presumably, the father of one of the two infants had recurrent herpes labialis that was the source of infection for his child. Nevertheless, the second child had to become infected by vector spread, most likely a nurse or medical person in the Intensive Care Unit. These investigations have been further pursued by two reports of nongenital acquisition of HSV by the newborn (Yeager *et al.*, 1983; Douglas *et al.*, 1983).

When an outbreak of herpes simplex encephalitis occurred in 14 patients in Boston in the summer of 1977, restriction endonuclease assessment failed to show genetic identity among any of the strains isolated from 11 patients, all of whom were proven by brain biopsy to have herpes simplex encephalitis (Hammer *et al.*, 1980). This particular outbreak was unrelated to a single strain of virus epidemically passed through the community. Restriction endonuclease analyses are powerful tools for comparison of genetic differences among viral strains; however, the source of virus cannot be identified by such a procedure. The source and routes of virus spread remain for identification by the epidemiologists.

VI. HSV INFECTIONS AND CERVICAL CARCINOMA

A. General Association

As early as a century ago, it was noted that cervical carcinoma occurred rarely in nuns, and at a much higher incidence in married versus unmarried women (Rigani-Stern, 1842). Nahmias and colleagues associated HSV infection of the genital tract with cervical carcinoma (Naib et al., 1969). These original observations have been reviewed on several occasions (Nahmias et al., 1970c; Nahmias and Roizman, 1973; Roizman and Frenkel, 1973; Rawls et al., 1977; Nahmias and Sawanabori, 1978), and will not be detailed here. The fundamental epidemiologic observation upon which this premise is based is that women with genital HSV infection had an increased incidence of cervical carcinoma (Adam et al., 1972; Kao et al., 1974). As the cervix is the most common site of infection in the female genital tract, even more so than the external genitalia (Yen et al., 1965; Josey et al., 1966, 1968), a natural association was drawn between HSV-2 infection and the appearance of malignancy of the cervix. As with many other observations, data demonstrating the causative role of HSV-2 for cervical carcinoma remain unavailable; moreover, several inconsistencies exist in the hypothesis. Supportive evidence for the association between HSV-2 and cervical carcinoma was reviewed and the issues brought into sharp focus (Roizman and Frenkel, 1973). Proof that HSV-2 is the causative agent of cervical carcinoma must not simply rest on the correlation of genital infection with an increased incidence of malignancy. This is particularly the case with an agent that establishes latency within sensory root ganglia of nervous tissue with the propensity to recur and when it represents but one of multiple microbial organisms of the genital tract.

From an epidemiologic standpoint, two avenues support the incrimination of HSV-2: (1) the descriptive epidemiology and (2) the molecular support for the putative role of HSV-2 and cervical carcinoma.

B. Epidemiology of Cervical Carcinoma

Following the original observation of Naib et al. (1969), the hypothesis relating HSV-2 with cervical carcinoma was verified in a larger series of patients (Naib et al., 1969, 1973). Most of the conclusions were predicated on the evidence of active viral infection at the time of diagnosis in patients with cervical carcinoma, presumably the result of reactivated infection (Nahmias et al., 1973).

Subsequent follow-up seroepidemiologic studies have been the cornerstone for the hypothesis that HSV-2 is the cause of cervical carcinoma (Catalano and Johnson, 1971; Centifanto et al., 1971; Smith et al., 1972b, 1977; Nahmias et al., 1973; Janda et al., 1973; Rawls et al., 1973; Adam et al., 1974; Adelusi et al., 1975; Ito et al., 1976; Figueroa and Zambrana, 1976; Choi et al., 1977; Pasca et al., 1975). Significant discrepancies exist in reported series regarding the quantity and frequency of neutralizing antibodies present in the serum of women with cervical carcinoma. Studies have demonstrated an incidence of neutralizing antibodies varying widely from as low as 2% (Rawls et al., 1970) to as high as 100% (Royston and Aurelian, 1970a; Christenson and Esmark, 1976, 1977). Interestingly, one study demonstrated no difference between the frequency or level of neutralizing antibodies in patients with cervical carcinoma and controls (Rawls *et al.*, 1973). All of these studies employed a neutralization procedure to detect antibodies; however, because of the cross-reactivity of HSV-1 and HSV-2 strains, type-specific antibody assays were not, nor could be, performed. Thus, because HSV-1 antibodies can neutralize small quantities of HSV-2 antigens and thereby inhibit the appearance of HSV-2-specific antibodies, a correlation may well have been missed. Only recently have seroepidemiologic studies utilizing determinations reported to be type-specific and also highly sensitive been performed (Matson *et al.*, 1981). These studies further support the observations of the Emory investigators.

The fallacies in utilizing patterns of neutralizing antibodies as a means of establishing cause and effect in women with cervical carcinoma have been addressed in detail (Roizman and Frenkel, 1973). Several demographic problems prevent precise data interpretation obtained from the clinical milieu. Moreover, suggestions in the literature imply that sexual activity in general, rather than acquisition of genital HSV-2 specifically, may be the most important determinant in acquisition of cervical carcinoma. Another difficulty in assessing epidemiologic studies of cervical carcinoma is the role of other venereal infections in the expression of malignancy (Freedman *et al.*, 1974; Kessler, 1976). When attempts are made to balance all population variables, it still appears that HSV-2 is associated with the development of cervical carcinoma (Kessler *et al.*, 1974).

Most of the seroepidemiologic studies performed to date have utilized neutralization antibodies as the marker of correlation between infection and malignancy as reviewed by Melnick and Adam (1978). Indirect evidence suggests an alternative marker, namely, that the tumor cells themselves may express an antigen with characteristics distinct from those resulting in neutralizing antibodies. This particular tumor antigen or antigens might represent the expression of a specific portion of the viral genome coding for polypeptides identified after release from infected cells. A precedent for such findings with other virus-transformed cells exists whereby the antigens excreted are nonstructural viral proteins.

Four antigens have been described by three groups of investigators. First, in the early 1970s Royston and Aurelian (1970b) and Aurelian (1976) attempted to detect antigens in exfoliated cells with an immunofluorescence assay from women with preinvasive and invasive cervical carcinoma. However, the work remains unverified. The most extensive studies performed were those by Tarro and Sabin (1970), in which they thought a labile virion antigen could be detected. Initial work by these investigators demonstrated the presence of antibodies to these antigens in patients with nine different kinds of tumors including cervical carcinoma.

More extensive studies, however, by the same investigators failed to confirm the original observations (Sabin and Tarro, 1973).

Hollinshead *et al.* (1973) and Notter *et al.* (1973) found a complement-fixing antigen in sera of patients with head and neck cancer as well as uterine cancer. Aurelian *et al.* (1974) attempted an alternative approach to the seroepidemiology of cervical carcinoma by employing a virion polypeptide named AG4 (Aurelian and Strnad, 1976) that was considered equivalent to VP4 (Pereira *et al.*, 1976). Antibodies to this antigen were present in approximately 85% of cervical cancer cases in Baltimore (Aurelian *et al.*, 1977). However, in Japanese patients with squamous cell carcinoma, only 47% of cases had antibodies to this antigen (Kawana *et al.*, 1976). Careful attempts to verify the origin of AG4 as identical to VP4 have not been convincing and, in fact, these polypeptides are more likely structural glycosylated polypeptides as reported by Kaplan *et al.* (1975). Thus, these studies can be considered as failing to show evidence of a unique tumor antigen present in the serum of patients with cervical carcinoma.

C. Molecular Epidemiology of HSV-2 Antigens in Cervical Carcinoma

Additional attempts to incriminate HSV with cervical carcinoma have been predicated upon the demonstration of viral nucleic acids or specific antigens within the cancer cells. The finding of such determinants is based upon the concept that HSV-2 infections are not invariably lethal to the cell, but that infection can be abortive, resulting in cells being transformed with persistence of gene products of this abortive infection resident within the cell. In the first series of experiments performed by Frenkel *et al.* (1972) and later reports by Roizman and Frenkel (1973), it was possible to demonstrate the presence of both viral DNA and RNA, yet only a small fragment of the viral gene. These procedures were performed by purification of nucleic acids extracted from the tumor and followed by renaturation kinetics with viral DNA labeled by repair synthesis. A portion of the viral DNA appeared to be covalently linked with intact DNA. Roizman and Frenkel (1973) specifically addressed the problems associated with this assay in their review of HSV infection and cervical carcinoma. These problems included the inability of the labeled probes to reassociate properly, and the finding of a fragment rather than the entire genome of HSV-2, creating a problem of sensitivity.

If HSV-2 mRNA is sought utilizing different techniques, 63% (5 of 8) of specimens were positive by *in situ* hybridization (Jones *et al.*, 1978). Expanded studies in search of HSV-2-specific glycoproteins in tumor cells (Cabral *et al.*, 1981) or limited regions of the HSV-2 (McDougall *et al.*, 1981) are in progress and provide encouraging data. Preliminary data from both laboratories further incriminate HSV-2 as a factor in the development of cervical carcinoma.

Current information does not allow for conclusive statements as to the relationship between HSV of the genital tract and cervical carcinoma. These studies are proceeding as rapidly as technology will permit. The association between HSV-2 and cervical carcinoma has further been confused by the recent implications of papillomaviruses as causative for this malignancy (Durst et al., 1983). Further technological developments will be essential to resolve this problem.

VII. CONCLUSION

A wealth of information has been developed in the last decade on HSV and the infections they cause. Current developments in molecular virology have brought to the forefront technology that can be applied to epidemiologic studies of human HSV infections. Such tools include among others, the utilization of restriction endonuclease enzymes and monoclonal antibodies (Pereira and Baringer, 1981), definition of type-specific polypeptides, and efforts to define the nature and mechanism of latency. The development of improved methods, the application of new techniques, and the resultant clinical epidemiologic determinations are crucial to understanding more clearly the natural history of these diseases and generating adequate measures of prevention. As such, the obvious solution to the ambiguous relationship of genital HSV-2 infection and cervical carcinoma is the eradication of HSV infections, thereby erasing a presumed factor or cofactor responsible for this malignancy.

Other diseases that justify prevention and/or therapeutic intervention have been described in this chapter and range from the uncommon, but often fatal diseases, e.g., herpes simplex encephalitis and neonatal herpes simplex, to the very common, usually benign, but all too frequently psychologically debilitating diseases such as herpes labialis and genital herpes. Methods of prevention by vaccine remain the dream of all biomedical investigators, yet a safe subunit vaccine remains to be developed. On the other hand, therapeutic intervention is being successfully explored and is applied when feasible but must be improved. Further improvements in therapy will occur in association with the development of rapid and specific diagnostic methods, particularly as in the case of herpes simplex encephalitis and asymptomatic genital HSV infections.

A few potentially useful antiviral compounds have appeared and are being tested in controlled trials. These compounds include adenine arabinoside (9β-D-arabinofuranosyladenine, vidarabine, vira-A). Our understanding of HSV replication will lead to the targeted development of specific inhibitors of viral replication, one of these being acyclovir (Elion et al., 1977; Shaeffer et al., 1978). It is hoped that the creative application of laboratory probes and the ever-increasing knowledge of the structure, function, and replication *in vitro* of the study of human herpesvirus diseases will help resolve many controversial and empirical problems.

ACKNOWLEDGMENTS. The author wishes to gratefully acknowledge the critical reviews of this chapter by C. A. Alford, Jr., A. J. Nahmias, S. Stagno, and P. Griffiths as well as grant support for research discussed from the National Institute of Allergy and Infectious Diseases (N01-AI-12667).

REFERENCES

Adam, E., Kaufman, R. H., Melnick, J. L., Levy, A. H., and Rawls, W. E., 1972, Seroepidemiologic studies of herpesvirus type-2 and carcinoma of the cervix, *Am. J. Epidemiol.* **96**:427.

Adam, E., Rawls, W. E., and Melnick, J. L., 1974, The association of herpesvirus type-2 infection and cervical cancer, *Prev. Med.* **3**:122.

Adam, E., Kaufman, R. H., Mirkovic, R. R., and Melnick, J. L., 1979, Persistence of virus shedding in asymptomatic women after recovery from herpes genitalis, *Obstet. Gynecol.* **54**:171.

Adelusi, B., Osunkoya, B. O., and Fabiyi, A., 1975, Antibodies to herpes virus type-2 in carcinoma of the cervix uteri in Ibavan, Nigeria, *Am. J. Obstet. Gynecol.* **123**:758.

Anderson, S. G., and Hamilton, J., 1959, The epidemiology of primary herpex simplex infection, *Med. J. Aust.* **1**:308.

Andrews, C. H., and Carmichael, E. A., 1930, A note on the presence of antibodies to herpesvirus in post-encephalitic and other human sera, *Lancet* **1**:857.

Arvin, A. M., Koropchak, C. M., Yeager, A. S., and Pereira, L., 1983, The detection of type specific antibody to herpes simplex virus type 1 by radioimmunoassay using HSV-1 glycoprotein C purified with monoclonal antibody, *Infect. Immun.* **40**:184.

Aurelian, L., 1976, Sexually transmitted cancers? The case for genital herpes, *J. Am. Vener. Dis. Assoc.* **2**:10.

Aurelian, L., and Strnad, B. C., 1976, Herpesvirus type-2 related antigens and their relevance to humoral and cell-mediated immunity in patients with cervical cancer, *Cancer Res.* **36**:810.

Aurelian, L., Strandberg, J. E., and Marcus, R. L., 1974, Neutralization, immunofluorescence and complement fixation tests in identification of antibody to a herpesvirus type-2 induced, tumor-specific antigen in sera from squamous cervical carcinoma, *Prog. Exp. Tumor Res.* **19**:165.

Aurelian, L., Strnad, B. C., and Smith, M. F., 1977, Immunodiagnostic potential of a virus-coded, tumor-associated antigen (Ag4) in cervical cancer, *Cancer* **39**:1834.

Bader, C., Crumpacker, C. S., Schnipper, L. E., Ransil, B., Clark, J. E., Arndt, K., and Freedberg, I. M., 1978, The natural history of recurrent facial-oral infection with herpes simplex virus, *J. Infect. Dis.* **138**:897.

Baringer, J. R., 1974, Recovery of herpes simplex virus from human sacral ganglions, *N. Engl. J. Med.* **291**:828.

Baringer, J. R., 1976, The biology of herpes simplex virus infection in humans, *Surv. Ophthalmol.* **21**(2):171.

Baringer, J. R., and Swoveland, P., 1973, Recovery of herpes simplex virus from human trigeminal ganglions, *N. Engl. J. Med.* **288**:648.

Bastian, F. O., Rabson, A. S., and Yee, C. L., 1972, Herpesvirus hominis: Isolation from human trigeminal ganglion, *Science* **178**:306.

Becker, W. W., 1966, The epidemiology of herpesvirus infection in three racial communities in Cape Town, *S. Afr. Med. J.* **40:**109.

Beilby, J. O. W., Cameron, C. H., Catterall, R. O., and Davidson, D., 1968, Herpesvirus hominis infection of the cervix associated with gonorrhea, *Lancet* **1:**1065.

Bernstein, D., Garratty, E., Lovett, M., and Bryson, Y., 1983, Comparison of Western blot analysis (WBA), microneutralization (MN) for determination of HSV-1 and HSV-2 antibody, 23rd Annual Interscience Conference on Antimicrobial Agents and Chemotherapy, Abstract No. 747.

Beswick, T. S. L., 1962, The origin and the use of DAO herpes, *Med. Hist.* **6:**214.

Binder, P. A., 1977, Herpes simplex keratitis, *Surv. Ophthalmol.* **21:**313.

Black, F. L., 1975, Infectious diseases in primitive societies, *Science* **187:**515.

Black, F. L., Hierholzer, W. J., Pinheiro, F., Evans, A. S., Woodall, J. P., Opton, E. M., Emmons, J. E., West, B. S., Edsall, G., Downs, W. G., and Wallace, G. D., 1974, Evidence for persistence of infectious agents in isolated human populations, *Am. J. Epidemiol.* **100:**230.

Blank, H., and Haines, H. G., 1973, Experimental human reinfection with herpes simplex virus, *J. Invest. Dermatol.* **61:**233.

Bolognese, R. J., Corson, S. L., Fuccillo, D. A., Traub, R., Moder, F., and Sever, J. L., 1976, Herpesvirus hominis type-2 infections in asymptomatic pregnant women, *Obstet. Gynecol.* **48:**507.

Bryson, Y. J., Dillon, M., Lovett, M., Acura, G., Taylor, S., Cherry, J. D., Johnson, B. L., Wiesmeier, E., Growdon, W., Creagh-Kirk, T., and Keeney, R., 1983, Treatment of first episodes of genital herpes simplex virus infections with oral acyclovir. A randomized double-blind controlled trial in normal subjects, *N. Engl. J. Med.* **308:**916.

Buchman, T. G., Roizman, B., Adams, G., and Stover, B. H., 1978, Restriction endonuclease fingerprinting of herpes simplex DNA: A novel epidemiological tool applied to a nosocomial outbreak, *J. Infect. Dis.* **138:**488.

Buchman, T. G., Roizman, B., and Nahmias, A. J., 1979, Demonstration of exogenous reinfection with herpes simplex virus type-2 by restriction endonuclease fingerprinting of viral DNA, *J. Infect. Dis.* **140:**295.

Buddingh, G. J., Schrum, D. I., Lanier, J. C., and Guidy, D. J., 1953, Studies of the natural history of herpes simplex infections, *Pediatrics* **11:**595.

Burnet, F. M., and Williams, S. W., 1939, Herpex simplex: New point of view, *Med. J. Aust.* **1:**637.

Cabral, G. A., Lumpkin, C. L., Fry, D. G., Mercer, L. J., Gopelrud, D. R., and Marciano-Cabral, M., 1981, Identification of herpesvirus antigens in human cervical and vulvar carcinoma cells, *International Workshop on Herpesviruses* p. 218.

Caplan, L. R., Kleeman, F. J., and Berg, S., 1977, Urinary retention probably secondary to herpes genitalis, *N. Engl. J. Med.* **297:**920.

Carlton, C. A., and Kilbourne, E. D., 1952, Activation of latent herpes simplex by trigeminal sensory-root section, *N. Engl. J. Med.* **246:**172.

Catalano, L. W., Jr., and Johnson, L. D., 1971, Herpesvirus antibody and carcinoma *in situ* of the cervix, *J. Am. Med. Assoc.* **217:**447.

Centifanto, Y. M., Hildebrandt, R. J., Held, B., and Kaufman, H. E., 1971, Relationship of herpes simplex genital infection and carcinoma of the cervix: Population studies, *Am. J. Obstet. Gynecol.* **110:**690.

Cesario, T. C., Poland, J. D., Wulff, H., Chin, T. D., and Wenner, H. A., 1969, Six years experiences with herpes simplex virus in a children's home, *Am. J. Epidemiol.* **90:**416.

Chang, T. W., Fiumara, N. J., and Weinstein, L., 1974, Genital herpes: Some clinical and laboratory observations, *J. Am. Med. Assoc.* **229:**554.

Chen, A. B., Ben-Porat, T., Whitley, R. J., and Kaplin, A., 1978, Purification and characterization of proteins excreted by cells infected with herpes simplex virus and their use in diagnosis, *J. Virol.* **91:**234.

Cheng, Y. C., Hoffmann, P. J., Ostrander, M., Grill, S., Caradonna, S., Tsou, J., Chen, J. Y., Gallagher, M. R., and Flanagan, T. D., 1979, Properties of herpesvirus-specific thymidine kinase, DNA polymerase and DNAse and their implication in the development of specific antiherpes agents, *Adv. Ophthalmol.* **38**:173.

Choi, N. W., Skettigara P. T., Abu-Zeid, H. A. H., and Nelson, N. A., 1977, Herpesvirus infection and cervical anaplasia: A seroepidemiological study, *Int. J. Cancer* **19**:167.

Christenson, B., and Esmark, A., 1976, Long-term follow-up studies on herpes simplex antibodies in the course of cervical cancer. II. Antibodies to surface antigen of herpes simplex virus infected cells, *Int. J. Cancer* **17**:318.

Christenson, B., and Esmark, A., 1977, Long-term follow-up studies on herpes simplex antibodies in the course of cervical cancer: Patterns of neutralizing antibodies, *Am. J. Epidemiol.* **105**:296.

Cleobury, J. F., Skinner, G. R. B., and Thoules, M. D., 1971, Association between psychopathic disorder and serum antibody to herpes simplex virus (type 1), *Br. Med. J.* **1**:438.

Constantine, V. S., Francis, R. D., and Montes, L. F., 1968, Association of herpes simplex with neuralgia, *J. Am. Med. Assoc.* **205**:131.

Corey, L., Holmes, K., Benedetti, J., and Critchlow, C., 1981, Clinical course of genital herpes: Implications for therapeutic trials, in: *The Human Herpesviruses: An Interdisciplinary Perspective* (A. Nahmias, W. R. Dowdle, and R. Schinazi, eds), pp. 496–502, Elsevier/North-Holland, Amsterdam.

Corey, L., Nahmias, A. J., Guinan, M., Benedetti, J., Critchlow, C., and Holmes, K., 1982, A trial of topical acyclovir in genital herpes simplex virus infections, *N. Engl. J. Med.* **306**:1313.

Corey, L., Adams, H., Brown, Z., and Holmes, K., 1983a, Genital herpes simplex virus infections: clinical manifestations, course and complications. *Annals. Intern. Med.* **98**:958.

Corey, L., Fife, K. H., Benedetti, J. K., Winter, C. A., Fahnlander, A., Connor, J. D., Hintz, M. A., and Holmes, K. K., 1983b, Intravenous acyclovir for the treatment of primary genital herpes, *Ann. Intern. Med.* **98**:914–921.

Corey, L., Benedetti, J., Critchlow, C., Mertz, G., Douglas, J., Fife, K., Fahnlander, A., Remington, M. L., Winter, C., and Dragavon, J., 1983c, Treatment of primary first episode genital herpes simplex virus infections with acyclovir: Results of topical, intravenous, and oral therapy, *J. Antimicrob. Chemother.* **12**:79.

Crain, C. P., and Nahmias, A. J., 1973, Different patterns of neurologic involvement with herpes simplex virus types 1 and 2: Isolation of herpes simplex virus 2 from the buffy coat of two adults with meningitis, *J. Infect. Dis.* **127**:365.

Crumpacker, C. S., Schnipper, L. E., Marlowe, S. K., Kowalsky, P. N., Hershey, B. J., and Levin, M. J., 1982, Resistance to antiviral drugs of herpes simplex virus isolated from a patient treated with acyclovir, *N. Engl. J. Med.* **305**:343.

Deardourff, S. L., Deture, F. A., Drylie, D. M., Centifanto, Y., and Kaufman, H., 1974, Association between herpes virus hominis type-2 and the male genitourinary tract, *J. Virol.* **112**:126.

De Giordano, H. M., Banza, C. A., Russi, J. C., Campione-Piccardo, J., Somano, R. E., and Tosi, H. C., 1970, Prevalence of herpes simplex virus infection, *Arch. Pediatr. Urug.* **41**:107.

Dodd, K., Johnston, L. M., and Buddingh, G. J., 1938, Herpetic stomatitis, *J. Pediatr.* **12**:95.

Doerr, R., 1920, Sitzungsberichte der Gesellschaft der Schweizerischen Augenartzte Diskussion, *Klin. Monatsbl. Augenheilkd.* **65**:104.

Douglas, J., Schmidt, O., and Corey, L., 1983, Acquisition of neonatal HSV-1 infection from a paternal source contact, *J. Pediatr.* **103**:908–910.

Douglas, R. G., Jr., and Couch, R. B., 1970, A prospective study of chronic herpes simplex virus infection and recurrent herpes labialis in humans, *J. Immunol.* **104**:289.

Duenas, A., Adam, E., Melnick, J. L., and Rawls, W. E., 1972, Herpes virus type-2 in a prostitute population, *Am. J. Epidemiol.* **95**:483.

Durst, M., Gissman, L., Ikenberg, H., and Hausen, H., 1983, A papillomavirus DNA from a cervical carcinoma and its prevalence in cancer biopsy samples from different geographic regions, *Proc. Natl. Acad. Sci. USA* **80**:3812–3815.

Eberle, R., and Courtney, R., 1981, Assay of type-specific and type-common antibodies to herpes simplex virus types 1 and 2 in human sera, *Infect. Immun.* **31**:1062.

Eilard, U., and Hellgren, L., 1965, Herpes simplex: A statistical and clinical investigation based on 669 patients, *Dermatology* **130**:101.

Ekwo, E., Wong, Y. W., and Myers, M., 1979, Asymptomatic cervicovaginal shedding of herpes simplex virus, *Am. J. Obstet. Gynecol.* **134**:102.

Elion, G. B., Furman, P. A., Fyfe, J. A., de Miranda, P., Beauchamp, P. L., and Schaeffer, H., 1977, Selectivity of action of an antiherpetic agent, 9-(2-hydroxyethoxymethyl) guanine, *Proc. Natl. Acad. Sci. USA* **74**:12.

Ellison, S. A., Carlton, C. A., and Rose, H. M., 1959, Studies of recurrent herpes simplex infections following section of the trigeminal nerve, *J. Infect. Dis.* **105**:161.

Embil, J. A., Stephens, G., and Manuel, F. R., 1975, Prevalence of recurrent herpes labialis and aphthous ulcers among young adults in six continents, *Can. Med. Assoc. J.* **113**:627.

Evans, A. S., and Dick, E. C., 1964, Acute pharyngitis and tonsillitis in University of Wisconsin students, *J. Am. Med. Assoc.* **190**:699.

Figueroa, M., and Zambrana, A., 1976, El Herpes genital en Honduras y su relacion con el carcinom del cerix uterino, *Rev. Latinoum. Microbiol.* **18**:111.

Florman, A., Gershon, A. A., Blackett, P. R., and Nahmias, A. J., 1973, Intrauterine infection with herpes simplex virus: Resultant of congenital malformation, *J. Am. Med. Assoc.* **225**:129.

Foley, F. D., Greenwald, K. A., Nash, G., and Pruitt, B. A., 1970, Herpesvirus infection in burned patients, *N. Engl. J. Med.* **282**:652.

Francis, D. P., Herrmann, K. L., MacMahon, J. R., Chivigny, K. H., and Sanderlin, K. C., 1975, Nosocomial and maternally acquired herpesvirus hominis infections, a report of four fatal cases in neonates, *Am. J. Dis. Child.* **129**:889.

Freedman, R. S., Joosting, A. C., Ryan, J. T., and Koni, S., 1974, A study of associated factors, including genital herpes, in black women with cervical carcinoma in Johannesburg, S. *Afr. Med. J.* **48**:1747.

Frenkel, N., Roizman, B., Cassai, E., and Nahmias, A., 1972, A DNA fragment of herpes simplex 2 and its transcription in human cervical cancer tissue, *Proc. Natl. Acad. Sci. USA* **69**:3784.

Friedman E., Katcher, A. H., and Brightman, V. J., 1977, Incidence of recurrent herpes labialis and upper respiratory infection: A prospective study of the influence of biologic, social, and psychologic predictors, *Oral Surg. Oral. Med. Oral Pathol.* **43**:873.

Gallardo, E., 1943, Primary herpes simplex keratitis: Clinical and experimental study, *Arch. Ophthalmol.* **30**:217.

Glezen, W. P., Fernald, G. W., and Lohr, J. A., 1975, Acute respiratory disease of university students with special references to the etiologic role of herpesvirus hominis, *Am. J. Edpiemiol.* **101**:111.

Goldman, L., 1961, Reactions by autoinoculation for recurrent herpes simplex, *Arch. Dermatol.* **84**:1025.

Goodell, S. E., Quinn, T. C., Mkrtichian, E., Schuffler, M. D., Holmes, K. K., and Corey, L., 1983, Herpes simplex virus proctitis in homosexual men, *N. Engl. J. Med.* **308**:868–871.

Greenberg, M. S., Brightman, V. J., and Ship, I. I., 1969, Clinical and laboratory differentiation of recurrent intraoral herpes simplex virus infections following Fever, *J. Dent. Res.* **48**:385.

Gruter, W., 1920, Experimentelle und Klinische untersuchungen uber den sogenannten herpes cornea, *Ber. Dtsch. Ophthalmol. Ges.* **42**:162.

Hale, B. D., Reindtorff, R. C., Walker, L. C., and Roberts, A. N., 1953, Epidemic herpetic stomatitis in an orphanage nursery, *J. Am. Med. Assoc.* **183**:1068.

Hammer, S. M., Buchman, T. G., D'Angelo, L. J., Karchmer, A. W., Roizman, B., and Hirsch, M. S., 1980, Temporal cluster of herpes simplex encephalitis investigation by restriction endonuclease cleavage of viral DNA, *J. Infect. Dis.* **141**:479.

Hammerberg, O., Wahs, J., Chernesky, M., Luchsinger, I., and Rawls, W., 1983, An outbreak of herpes simplex virus type 1 in an intensive care nursery, *Pediatr. Infect. Dis.* **2:**290.

Hanshaw, J. W., 1973, Herpesvirus hominis infections in the fetus and newborn, *Am. J. Dis. Child.* **126:**546.

Haynes, R. E., Azimi, P., and Cramblett, H., 1968, Fatal herpes virus hominis infections in children, *J. Am. Med. Assoc.* **206:**312.

Hayward, G. S., Frenkel, N., and Roizman, B., 1976, Anatomy of herpes simplex virus DNA: strain difference and heterogenicity in the locations of restriction endonuclease cleavage sites, *Proc. Natl. Acad. Sci. USA* **72:**1768.

Hazen, P. G., and Eppes, R. B., 1977, Eczema herpeticum caused by herpesvirus type-2, a case in a patient with Darier's disease, *Arch. Dermatol.* **113:**1085.

Hellgren, L., 1962, The prevalence of some skin diseases and joint disease in total populations in different areas of Sweden, *Proc. Dermatol. Soc.* **155:**162.

Hevron, J. E., Jr., 1977, Herpes simplex virus type-2 meningitis, *Obstet. Gynecol* **49:**622.

Hill, T. J., 1981, Mechanisms involved in recurrent herpes simplex, in: *The Human Herpesviruses: An Interdisciplinary Perspective* (A. Nahmias, W. R. Dowdle, and R. Schinazi, eds.), pp. 241–244, Elsevier/North-Holland, Amsterdam.

Hinthorn, D. R., Baker, L. H., and Romig, D. A., 1976, Recurrent conjugal neuralgia caused by herpesvirus hominis type-2, *J. Am. Med. Assoc.* **236:**587.

Hitselberger, J. F., and Burns, R. E., 1961, Darier's disease: Report of a case complicated by Kaposi's varicellaform eruption, *Arch. Dermatol.* **83:**425.

Hollinshead, A. C., Lee, O., Chretien, P. B., Tarpley, J. L., Rawls, W. E., and Adam, E., 1973, Antibodies to herpesvirus nonvirion antigens in squamous carcinoma, *Science* **182:**713.

Hutfield, D. C., 1966, History of herpes genitalis, *Br. J. Vener. Dis.* **42:**263.

Ito, H., Tsutsui, F., Kurihara, S., Akabovahi, T., Tobe, T., and Chiaki, N., 1976, Serum antibodies to herpesvirus early antigens in patients with cervical carcinoma determined by anticomplement immunofluorescence technique, *Int. J. Cancer* **18:**557.

Jacobs, E., 1976, Anal infections caused by herpes simplex virus, *Dis. Colon Rectum* **19:**151.

Janda, Z., Kanka, J., Vonka, V., and Svoboda, B., 1973, A study of herpes simplex type-2 antibody status in groups of patients with cervical neoplasia in Czechoslovakia, *Int. J. Cancer* **12:**626.

Jeansson, S., and Molin, L., 1970, Genital herpes virus hominis infection: A venereal disease?, *Lancet* **1:**1064.

Jeansson, S., and Molin, L., 1974, On the occurrence of genital herpes simplex virus infection: Clinical and virologic findings and relation to gonorrhea, *Arch. Dermatol.* **54:**479.

Jones, K. W., Fenoglio, C. M., Shevchuk-Chaban, M., Mailtand, N. J., and McDougall, J. K., 1978, Detection of herpes simplex virus type-2 mRNA in human cervical biopsies by *in situ* cytological hybridization, *IRAC Sci. Publ.* **24**(2):917.

Josey, W. E., Nahmias, A. J., and Naib, Z. M., 1966, Genital herpes infection in the female, *Am. J. Obstet. Gynecol.* **96:**493.

Josey, W. E., Nahmias, A. J., and Naib, Z., 1968, Genital infection with type-2 herpesvirus hominis, *Am. J. Obstet. Gynecol.* **101:**718.

Josey, W. E., Nahmias, A. J., and Naib, Z. M., 1972, The epidemiology of type-2 (genital) herpes simplex virus infection, *Obstet. Gynecol. Surv. Suppl.* **27:**295.

Juretic, M., 1966, Natural history of herpetic infection, *Helv. Pediatr. Acta* **21:**356.

Kalinyak, J. E., Fleagle, G., and Docherty, J. J., 1977, Incidence and distribution of herpes simplex virus types 1 and 2 from genital lesions in college women, *J. Med. Virol.* **1:**173.

Kao, C. L., Chen, Y.H., Tang, M. C., Wen, H. K., and Hsia, S., 1974, Association of herpes simplex type 2 virus with cervical cancer of the uterus, *J. Formosan Med. Assoc.* **73:**122.

Kaplan, A. S., Erickson, J. S., and Ben-Porat, T., 1975, Synthesis of proteins in cells infected with herpesvirus. X. Proteins excreted by cells infected with herpes simplex virus types 1 and 2, *Virology* **64:**132.

Kaufman, R. H., Gardner, H. L., Rawls, W. E., Dixon, R. E., and Young, R. L., 1973, Clinical features of herpes genitalis, *Cancer Res.* **33:**1446.

Kawana, T., Cornish, J. D., Smith, M. F., and Aurelian, L., 1976, Frequency of antibody to a virus-induced tumor-associated antigen (Ag-4) in Japanese sera from patients with cervical cancer and controls, *Cancer Res.* **36**:1910.

Kessler, I. I., 1976, Human cervical cancer as a venereal disease, *Cancer Res.* **36**:783.

Kessler, I. I., Kulcar, Z., Rawls, W. E., Smerdel, S., Strnad, M., and Lilienfeld, A. M., 1974, Cervical cancer in Yugoslavia. II. Epidemiologic factors of possible etiologic significance, *J. Natl. Cancer Inst.* **52**:369.

Kibrick, S., and Gooding, G. W., 1965, Pathogenesis of infection with herpes simplex virus with special reference to nervous tissue, NINDB Monograph Vol. II, National Institute of Neurological Diseases and Blindness, pp. 143–154.

Kleger, B., Prier, J. E., Rosato, D. J., and McGinnis, A. E., 1968, Herpes simplex infection of the female genital tract. I. Incidence of infection, *Am. J. Obstet. Gynecol.* **102**:745.

Kloene, W., Bang, F. B., Chakraborty, S. M., Cooper, M. R. Kulemann, H., Shah, K. V., and Ota, M., 1970, A two year respiratory virus survey in four villages in Weor Bengal, India, *Am. J. Epidemiol.* **92**:307.

Komorous, J. M., Wheeler, C. E., Briggaman, R. A., and Caro, I., 1977, Intrauterine herpes simplex infections, *Arch. Dermatol.* **113**:919.

Korsager, B., Spencer, E. S., and Mordhorst, C. H., 1975, Herpesvirus hominis infections in renal transplant recipients, *Scand. J. Infect. Dis.* **7**:11.

Lazar, M. P., 1956, Vaccination for recurrent herpes simplex infection: Initiation of a new disease site following the use of unmodified material containing the live virus, *Arch. Dermatol.* **73**:70.

Lindgren, K. M., Douglas, R. G., Jr., and Couch, R. B., 1968, Significance of herpesvirus hominis in respiratory secretions of man, *N. Engl. J. Med.* **276**:517.

Linnemann, C. C., Jr., Buchman, T. G., Light, I. J., Ballard, J. L., and Roizman, B., 1978, Transmission of herpes simplex virus type-1 in a nursery for the newborn: Identification of viral species isolated by DNA fingerprinting, *Lancet* **1**:964.

Lipschitz, B., 1921, Untersuchungen uber die atiologie der krankheiten der herpesgruppe (herpes zoster, genitalis and febrilis), *Arch. Dermatol. Res.* **136**:428.

Logan, W. S., Tindall, J. P., and Elson, M. L., 1971, Chronic cutaneous herpes simplex, *Arch. Dermatol.* **103**:606.

London, W. T., Catalano, L. W., Jr., Nahmias, A. J., Fuccillo, D. A., and Sever, J. L., 1971, Genital herpes virus type-2 infection of monkeys, *Obstet. Gynecol.* **37**:501.

Lowenstein, 1919, Aetiologische untersuchungen uber den fieberhaften, herpes, *Muench. Med. Wochenschr.* **66**:769.

Luby, J. P., Gnann, J. W., Jr., Alexander, J., Hatcher, V. A., Friedman-Kien, A. E., Keyserling, H., Nahmias, A. J., Mills, J., Schachter, J., Douglas, J. N., Corey, L., and Sacks, S. L., 1984, A study of patient initiated topical acyclovir versus placebo in the therapy of recurrent genital herpes, *J. Infect. Dis.* (in press).

McCormick, D. P., 1972, Herpes simplex virus as a cause of Bell's palsy, *Lancet* **1**:937.

McDonald, A. D., Williams, M. C., and West, R., 1974, Neutralizing antibodies to herpes virus types 1 and 2 in Montreal women, *Am. J. Epidemiol.* **100**:124.

McDougall, J. K., Crum, C. P., Levine, R. V., Richart, R. M., and Fenoglio, C. M., 1981, Expression of limited regions of the HSV-2 genome in cervical carcinoma, *International Workshop on Herpesviruses* p. 219.

Matson, D. O., Adam, E., Melnick, J. L., and Dressman, G. R., 1981, Prevalence of antibodies to herpes simplex virus measured with a type specific radioimmunoassay in cervical neoplasia—Case control studies, in: *The Human Herpesviruses: An Interdisciplinary Perspective* (A. Nahmias, W. Dowdle, and R. Schinazi, eds.), p. 628, Elsevier/North-Holland, Amsterdam.

Melnick, J. L., and Adam, E., 1978, Epidemiological approaches to determining whether herpes virus is the etiological agent of cervical cancer, *Prog. Exp. Tumor Res.* **21**:49.

Mettler, C., 1947, in: *History of Medicine*, p. 356, McGraw–Hill (Blakiston), New York.

Montgomerie, J. Z., Croxson, M. C., Becroft, D. M. O., Doak, P. B., and North, J. D. K., 1969, Herpes simplex virus infection after renal transplantation, *Lancet* **2**:867.

Muller, S. A., Herrmann, F. C., and Winkelman, R. K., 1972, Herpes simplex infections in hematologic malignancies, *Amer. J. Med.* **52:**102.

Nahmias, A. J., and Dowdle, W. R., 1968, Antigenic and biologic differences in herpesvirus hominis, *Prog. Med. Virol.* **10:**110.

Nahmias, A. J., and Roizman, B., 1973, Infection with herpes simplex viruses 1 and 2, *N. Engl. J Med.* **289:**667, 719, 781.

Nahmias, A. J., and Sawanabori, S., 1978, The genital herpes–cervical cancer hypothesis—10 years later, *Prog. Exp. Tumor Res.* **21:**177.

Nahmias, A. J., Dowdle, W. R., Naib, Z. M., Josey, W. E., McClone, L., and Domestick, G., 1969, Genital infection with type-2 herpesvirus hominis, a commonly occurring venereal disease, *Br. J. Vener. Dis.* **45:**294.

Nahmias, A. J., Josey, W. E., Naib, Z. M., Luce, C., and Duffey, C., 1970a, Antibodies to herpesvirus hominis type 1 and 2 in humans. I. Patients with genital herpetic infections, *Am. J. Epidemiol.* **91:**539.

Nahmias, A. J., Josey, W. E., Naib, Z. M., Luce, C. F., and Fuest, B., 1970b, Antibodies to herpesvirus hominis types 1 and 2 in humans. II. Women with cervical cancer, *Am. J. Epidemiol.* **91:**547.

Nahmias, A. J., Alford, C., and Korones, S., 1970c, Infection of the newborn with herpes virus hominis, *Adv. Pediatr.* **17:**185.

Nahmias, A. J., Josey, W. E., Naib, Z. M., Freeman, M. G., Fernandez, R. J., and Wheeler, J. H., 1971, Perinatal risk associated with maternal genital herpes simplex virus infection, *Am. J. Obstet. Gynecol.* **110:**825.

Nahmias, A. J., Von Reyn, C. F., Josey, W. E., Naib, Z. M., and Hutton, R. D., 1973, Genital herpes simplex virus infection and gonorrhea: Association and analogies, *Br. J. Vener. Dis.* **49:**306.

Nahmias, A. J., Whitley, R. J., Vistine, A. N., Takei, Y., Alford, C. A., Jr., and the NIAID Collaborative Antiviral Study Group, 1982, Herpes simplex encephalitis: laboratory evaluations and their diagnostic significance, *J. Infect. Dis.* **145:**829.

Nahmias, A. J., Keyserling, H. L., and Kerrick, G. M., 1983, Herpes simplex, in: *Infectious Diseases of the Fetus and Newborn Infant,* (J. Remington and J. Klein, eds.), pp. 636–676, Saunders, Philadelphia.

Naib, Z. M., Nahmias, A. J., Josey, W. E., and Kramer, J. H., 1969, Genital herpes infection: Association with cervical dysplasia and carcinoma, *Cancer* **23:**940.

Naib, Z. M., Nahmias, A. J., Josey, W. E., and Zaki, S. A., 1973, Relation of cytohistopathology of genital herpesvirus infection to cervical anaplasia, *Cancer* **33:**1452.

Nesburn, A. B., Cook, M. L., and Stevens, J. G., 1972, Latent herpes simplex virus isolation from rabbit trigeminal ganglia between episodes of recurrent ocular infection, *Arch. Ophthalmol.* **88:**412.

Ng, A. B. P., Reagin, J. W., and Yen, S. S., 1970, Herpes genitalis—Clinical and cytopathologic experience with 256 patients, *Obstet. Gynecol.* **36:**645.

Notter, M. F. D., Docherty, J. J., Mortel, R., and Hollinshead, A. C., 1978, Detection of herpes simplex virus tumor-associated antigen in uterine cervical cancer tissue: Five case studies, *Gynecol. Oncol.* **6:**574.

Olson, L. C., Buescher, E. L., and Artenstein, M. S., 1967, Herpesvirus infections of the human central nervous system, *N. Engl. J. Med.* **277:**1271.

Orenstein, J. M., Castadot, M. J., and Wilens, S. T., 1974, Fatal herpes hepatitis association with Pemphigus vulgaris and steroids in an adult, *Hum. Pathol.* **5:**489.

Ostler, H. B., 1976, Herpes simplex: The primary infection, *Surv. Ophthalmol.* **21:**91.

Overall, J. C., Jr., 1980, Antiviral chemotherapy of oral and genital herpes simplex virus infections, in *The Human Herpesviruses: An Interdisciplinary Perspective* (A. J. Nahmias, W. R. Dowdle, and R. E. Schinazi, eds.), pp. 447–465, Elsevier/North-Holland, Amsterdam.

Pacsa, A. S., Kummerlander, L., Pejtsik, B., and Poli, K., 1975, Herpes virus antibodies and antigens in patients with cervical anaplasia and in controls, *N. Natl. Cancer Inst.* **55:**775.

Pagano, J. S., 1975, Diseases and mechanisms of persistent DNA virus infection: Latency and cellular transformation, *J. Infect. Dis.* **132:**209.

Parker, J. D., and Banatvala, J. E., 1967, Herpes genitalis: Clinical and virologic studies, *Br. J. Vener. Dis.* **43:**212.

Pass, R. F., Long, W. K. Whitley, R. J., Soong, S. J., Diethelm, A. G., Reynolds, D. W., and Alford, C. A., 1978, Productive infection with cytomegalovirus and herpes simplex in renal transplant recipients, role of source of kidney, *J. Infect. Dis.* **137:**556.

Pass, R. F., Whitley, R. J., Whelchel, J. D., Diethelm, A. G., Reynolds, D. W., and Alford, C. A., 1979, Identification of patients with increased risk of infection with herpes simplex virus after renal transplantation, *J. Infect. Dis.* **140:**487.

Pazin, G. J., Ho, M., and Jannetta, P. J., 1978, Herpes simplex reactivation after trigeminal nerve root decompression, *J. Infect. Dis.* **138:**405.

Pazin, G. J., Armstrong, J. A., Lam, M. T., Tarr, G. C., Jannetta, P. J., and Ho, M., 1979, Prevention of reactivation of herpes simplex virus infection by human leukocyte interferon after operation on the trigeminal route, *N. Engl. J. Med.* **301:**225.

Pereira, L., and Baringer, J. R., 1981, Use of monoclonal antibody to identify the herpes virus glycoprotein antigens, in: *The Human Herpesviruses: An Interdisciplinary Perspective* (A. J. Nahmias, W. R. Dowdle, and R. Schinazi, eds.), p. 642, Elsevier/North-Holland, Amsterdam.

Pereira, L., Cassai, E., Jones, R. W., Roizman, B., Terni, M., and Nahmias, A. J., 1976, Variability in the structural polypeptides of herpes simplex virus-1 strains: Potential application in molecular epidemiology, *Infect. Immun.* **13:**211.

Plummer, G., 1964, Serological comparison of the herpesviruses, *Br. J. Exp. Pathol.* **45:**135.

Plummer, G., 1973, A review of the identification and titration of antibodies to herpes simplex viruses types 1 and 2 in human sera, *Cancer Res.* **33:**1469.

Poste, G., Hawkins, D. F., and Thomlinson, J., 1972, Herpesvirus hominis infection of the female genital tract. *Obstet. Gynecol.* **40:**871.

Pugh, R. C. B., Dudgeon, J. A., and Bodia, M., 1955, Kaposi's varicella-form eruption (eczema herpeticum) with typical and atypical visceral necrosis, *J. Pathol. Bacteriol.* **69:**67.

Rand, K. H., Rasmussen, L. E., and Pollard, R. B., 1977, Cellular immunity and herpesvirus infections in cardiac-transplant patients, *N. Engl. J. Med.* **296:**1372.

Rattray, M. C., Corey, L., Reeves, W. C., Vontver, L. A., and Holmes, K. K., 1978, Recurrent genital herpes among women: Symptomatic versus asymptomatic viral shedding, *Br. J. Vener. Dis.* **54:**262.

Rawls, W. E., and Campione-Piccardo, J., 1981, Epidemiology of herpes simplex virus type 1 and 2, in: *The Human Herpesviruses: An Interdisciplinary Perspective* (A. Nahmias, W. Dowdle, and R. Schinazi, eds.), pp. 137–152. Elsevier/North-Holland, Amsterdam.

Rawls, W. E., and Gardner, H. L., 1972, Herpes genitalis: Venereal aspects, *Clin. Obstet. Gynecol.* **15:**913.

Rawls, W. E., Tompkins, W. A., and Melnick, J. L., 1969, The association of herpesvirus type 2 and carcinoma of the uterine cervix, *Am. J. Epidemiol.* **89:**547.

Rawls, W. E., Iwamoto, K., Adam, E., Melnick, J. L., and Green, G. H., 1970, Herpesvirus type 2 antibodies and carcinoma of the cervix, *Lancet* **2:**1142.

Rawls, W. E., Gardner, H. L., Flanders, R. W., Lowry, S. P., Kaufman, R. H., and Melnick, J. L., 1971, Genital herpes in two social groups, *Am. J. Obstet. Gynecol.* **110:**682.

Rawls, W. E., Adam, E., and Melnick, J. L., 1972, Geographical variation in the association of antibodies to herpesvirus type 2 and carcinoma of the cervix, in: *Oncogenesis and Herpesviruses* (P. M. Biggs, G. de Thé, and L. N. Paynes, eds.), pp. 424–427, Scientific Publication II, International Agency for Research on Cancer, Lyon, France.

Rawls, W. E., Adam, E., and Melnick, J. L., 1973, An analysis of seroepidemiological studies of herpesvirus type 2 and carcinoma of the cervix, *Cancer Res.* **33:**1477.

Rawls, W. E., Garfield, C. H., Seth, P., and Adam, E., 1976, Serological and epidemiological considerations of the role of herpes simplex virus type 2 in cervical cancer, *Cancer Res.* **36:**829.

Rawls, W. E., Bacchetti, S., and Graham, F. L., 1977, Relation of simplex viruses to human malignancies, *Curr. Top. Microbiol. Immunol.* **7**:71.

Rigani-Stern, D., 1842, Fatti statistici relativi alle malattie cancrose che servirono da base alle poche cose detta dal dott, G. Servre, *Prog. Pathol. Ther. Ser.* **2**:507.

Roizman, B., 1968, An inquiry into the mechanisms of recurrent herpes infections in man, in: *Perspectives in Virology IV* (M. Pollard, ed.), pp. 283–304, Harper & Row, New York.

Roizman, B., 1971, Herpesviruses, man, and cancer. Persistency of the virus of love in: *Microbes and Life*, pp. 187–214, Columbia University Press, New York.

Roizman, B., and Frenkel, N., 1973, Does genital herpes cause cancer, A midway assessment, in: *Sexually Transmitted Disease* (R. D. Catterall and C. S. Nicol, eds.), p. 151, Academic Press, New York.

Rosato, F. E., Rosato, E. F., and Plotkin, S. A., 1970, Herpetic paronychia—An occupational hazard of medical personnel, *N. Engl. J. Med.* **283**:804.

Royston, I., and Aurelian, L., 1970a, The association of genital herpesvirus with cervical atypia and carcinoma in situ, *Am. J. Epidemiol.* **91**:531.

Royston, I., and Aurelian, L., 1970b, The association of genital herpesvirus antigens in exfoliated cells from human cervical cancer, *Proc. Natl. Acad. Sci. USA* **67**:204.

Ruchman, I., Welsh, A. L., and Dodd, K., 1947, Kaposi-varicelliform eruption: Isolation of virus of herpes simplex from cutaneous lesions of three adults and one infant, *Arch. Dermatol. Syphilol.* **56**:846.

Rustigian, R., Smulow, J. B., and Tye, R., 1966, Studies of latent infection of skin and oral mucosa in individuals with recurrent herpes simplex, *J. Invest. Dermatol.* **47**:218.

Sabin, A. B., and Tarro, G., 1973, Herpes simplex and herpes genitalis viruses in etiology of some human cancers, *Proc. Natl. Acad. Sci. USA* **70**:3225.

Sawanabori, S., 1973, Acquisition of herpes simplex virus infection in Japan, *Acta Paediatr. Jpn. Overseas Ed.* **15**:16.

Schaeffer, H. J., Beauchamp, P. L., de Miranda, P., Elion, G. B., and Collins, P., 1978, 9-(2-hydroxyethoxymethyl) guanine activity against viruses in the herpes group, *Nature (London)* **272**:583.

Schneweis, K. E., and Nahmias, A. J., 1961, On the atability of three strains of herpes simplex virus at low temperatures, *Z. Hyg. Infektionskr.* **183**:556.

Schneweis, K. E., and Nahmias, A. J., 1971, Antigens of herpes simplex virus types 1 and 2—Immunodiffusion and inhibition passive hemagglutination studies, *Z. Immunitaetsforsch. Exp. Klin. Immunol.* **141**:471.

Scott, T. F., 1957, Epidemiology of herpectic infections, *Am. J. Ophthalmol.* **43**:134.

Scott, T. F., Steigman, A. J., and Convey, J. H., 1941, Acute infectious gingivostomatitis: Etiology, epidemiology, and clinical pictures of a common disorder caused by the virus of herpes simplex, *J. Am. Med. Assoc.* **117**:999.

Segal, A. L., Katcher, A. H., Brightman, V. J., and Miller, M. F., 1974, Recurrent herpes labialis, recurrent aphthous ulcers and the menstrual cycles, *J. Dent. Res.* **53**:797.

Seidenberg, S., 1941, Zuraetiologie der pustulosis vacciniformis acuta, *Schweiz. Z. Pathol. Bakteriol.* **4**:398.

Selling, B., and Kibrick, S., 1960, An outbreak of herpes simplex among wrestlers (herpes gladiatorum), *N. Engl. J. Med.* **270**:979.

Sheridan, P. J., and Herrmann, E. C., Jr., 1971, Intraoral lesions of adults associated with herpes simplex virus, *Oral Surg. Oral Med. Oral Pathol.* **32**:390.

Ship, I. I., Morris, A. L., Durocher, R. T., and Burket, L. W., 1960, Recurrent aphthous ulcerations and recurrent herpes labialis in a professional school student population. I. Experience, *Oral Surg. Oral Med. Oral Pathol* **13**:1191.

Ship, I. I., Morris, A. L., Durocher, R. T., and Burket, L. W., 1961, Recurrent aphthous ulcerations and recurrent herpes labialis in a professional school student population. IV. Twelve month study of natural disease patterns, *Oral Surg. Oral Med. Oral Pathol.* **14**:39.

Ship, I. I., Brightman, V. J., and Laster, L. L., 1967, The patient with recurrent aphthous ulcers and the patient with recurrent herpes labialis: A study of two population samples, *J. Am. Dent. Assoc.* **75**:645.

Ship, I. I., Miller, M. F., and Ram, C., 1977, A retrospective study of recurrent herpes labialis (RHL) in a professional population, 1958–1971, *Oral Surg. Oral Med. Oral Pathol.* **44**:723.

Skoldenberg, B., Jeansson, S., and Wolontis, S., 1975, Herpes simplex virus type 2 and acute aseptic meningitis: Atypical features of cases with isolation of herpes simplex virus from cerebrospinal fluids, *Scand. J. Infect. Dis.* **7**:227.

Smith, I. W., Peutherer, J. F., and MacCallum, F. O., 1967, The incidence of herpesvirus hominis antibody in the population, *J. Hyg.* **65**:395.

Smith, I. W., Adam, E., Melnick, J. L., and Rawls, W. E., 1972a, Use of the ^{51}Cr release test to demonstrate patterns of antibody response in humans to herpesvirus types 1 and 2, *J. Immunol.* **109**:554.

Smith, I. W., Lowry, S. P., Melnick, J. L., and Rawls, W. E., 1972b, Antibodies to surface antigens of herpesvirus type 1 and 2 infected cells among women with cervical cancer and control women, *Infect. Immun.* **5**:305.

Smith, I. W., Peutherer, J. F., and Robertson, D. H., 1977, Virological studies in genital herpes [letter], *Lancet* **2**:1089.

Smith, M. G., Lennette, E. H., and Reames, U. R., 1941, Isolation of the virus of herpes simplex and the demonstration of intranuclear inclusions in a case of acute encephalitis, *Am. J. Pathol.* **17**:55.

Spruance, S. T., Overall, J. C., Jr., and Kern, E. R., 1977, The natural history of recurrent herpes simplex labialis—Implications for antiviral therapy, *N. Engl. J. Med.* **297**:69.

Spruance, S. T., and Crumpacker, C. S., 1982, Topical 5% acyclovir in polyethylene glycol for herpes simplex labialis: Antiviral effect without clinical benefit, *Am. J. Med.* **73**:315.

Spruance, S. T., Crumpacker, C. S., Schnipper, L. E., Kern, E. R., Marlow, S., Modlin, J., Arndt, K. A., and Overall, J. C., Jr., 1982, Topical 10% acyclovir in polyethylene glycol for herpes simplex labialis: results of treatment begun in prodrome and erythema stages, 22nd Annual Interscience Conference on Antimicrobial Agents and Chemotherapy, Miami.

Stern, H., Elek, S. D., Miller, D. M., and Anderson, H. F., 1959, Herpetic Whitlow, a form of cross-infection in hospitals, *Lancet* **2**:871.

Stevens, J. C., 1975, Latent herpex simplex virus and the nervous system, *Curr. Top. Immunol.* **70**:31.

Stevens, J. G., and Cook, M. L., 1971, Latent herpes simplex virus in spinal ganglia, *Science* **173**:843.

Stevens, J. G., and Cook, M. L., 1974, Latent herpes simplex virus in sensory ganglia, *Perspect. Virol.* **8**:171.

Stevens, J. G., Cook, M. L., and Jordan, M. C., 1975, Reactivation of latent herpes simplex virus after pneumococcal pneumonia in mice. *Infect. Immun.* **11**:635.

Sullivan-Boyai, J., Hull, H., Wilson, C., and Corey, L., 1983, Neonatal herpes simplex virus infection in King County, Washington, *J. Am. Med. Assoc.* **250**:3059.

Tarro, G., and Sabin, A. B., 1970, Virus specific labile, nonvirion antigen in herpesvirus infected cells, *Proc. Natl. Acad. Sci. USA* **65**:735.

Teissier, P., Gastinel, P., and Reilly, J., 1926, L'herpes experimental humain: L'inoculabilite dulvirus herpetique, *J. Physiol. Pathol. Gen.* **24**:271.

Tejani, N., Klein, S. W., and Kaplan, M., 1979, Subclinical herpes simplex genitalis infections in the perinatal period, *Am. J. Obstet. Gynecol.* **135**:547.

Terezhalmy, G. T., Tyler, M. T., and Ross, G. R., 1979, Eczema herpeticum: Atopic dermatitis complicated by primary herpetic gingivostomatitis, *Oral Surg. Oral Med. Oral Pathol.* **48**:513.

Terni, M., 1971, Infection with the virus of herpes simplex—The recrudescence of the disease and the problem of latency, *G. Mal. Infett. Parasit.* **23**:433.

Terni, M., Carcialanza, D., Cassai, E., and Kieff, E., 1971, Aseptic meningitis in association
    with herpex progenitalis, N. Engl. J. Med. 285:503.
Unna, P. G., 1896, The Histopathology of the Diseases of the Skin (translated by N. Walker),
    W. F. Clay, ed., Macmillan Co., New York.
Vesterinen, E., Purola, E., and Saksela, E., 1977, Clinical and virological findings in patients
    with cytologically diagnosed gynecologic herpes simplex infections, Acta Cytol. 21:299.
Walz, M. A., Price, R. W., and Notkins, A. L., 1974, Latent ganglionic infection with herpes
    simplex virus types 1 and 2: Viral reactivation in vivo after neurectomy, Science
    184:1185.
Warren, K. G., Gilden, D. H., and Brown, S. M., 1977, Isolation of herpes simplex virus from
    human trigeminal ganglia, including ganglia from one patient with multiple sclerosis,
    Lancet 2:637.
Wentworth, B. B., and Alexander, E. R., 1971, Seroepidemiology of infections due to members
    of the herpesvirus group, Am. J. Epidemiol. 94:496.
Wheeler, C. E., Jr., and Abele, D. C., 1966, Eczema herpeticum, primary and recurrent, Arch.
    Dermatol. 93:162.
Wheeler, C. E., Jr., and Cabraniss, W. H., Jr., 1965, Epidemic cutaneous herpes simplex in
    Wrestlers (herpes gladiatorum), J. Am. Med. Assoc. 194:993.
Whitley, R. J., Soong, S. J., Dolin, R., Galasso, G. J., Chien, L. T., Alford, C. A., Jr., and the
    NIAID Collaborative Antiviral Study Group, 1977, Adenine arabinoside therapy of bi-
    opsy-proved herpes simplex encephalitis, J. Engl. J. Med. 297:289.
Whitley, R. J., Nahmias, A. J., Soong, S. J., Galasso, G. J., Fleming, C. L., Alford, C. A., Jr.,
    and the NIAID Collaborative Antiviral Study Group, 1980a, Vidarabine therapy of ne-
    onatal herpes simplex virus infection, Pediatrics 66:495.
Whitley, R. J., Nahmias, A. J., Visintine, A. M., Fleming, C. L., Alford, C. A., Jr., and the
    NIAID Collaborative Antiviral Study Group, 1980b, The natural history of herpes sim-
    plex virus infection of mother and newborn, Pediatrics 66:489.
Whitley, R. J., Soong, S. J., Hirsch, M. S., Karchmer, A. W., Dolin, R., Galasso, G., Dunnick,
    J. K., Alford, C. A., and the NIAID Collaborative Antiviral Study Group, 1981, Herpes
    simplex encephalitis: Vidarabine therapy and diagnostic problems, N. Engl. J. Med.
    304:313.
Whitley, R. J., Lakeman, A. D., Nahmias, A. J., and Roizman, B., 1982, DNA restriction
    enzyme analysis of herpes simplex virus isolates obtained from patients with enceph-
    alitis, N. Engl. J. Med. 307:1060.
Whitley, R. J., Yeager, A., Kartus, P., Bryson, Y., Connor, J. D., Alford, C. A., Nahmias, A.
    J., and Soong, S. J., 1983, Neonatal herpes simplex virus infection: Follow-up evaluation
    of vidarabine therapy, Pediatrics 72:778.
Whitley, R. J., Spruance, S., Hayden, F. G., Overall, J., Alford, C. A., Jr., Gwaltney, J. M.,
    Jr., Soong, S. J., and the NIAID Collaborative Antiviral Study Group, 1984, Vidarabine
    therapy for mucocutaneous herpes simplex virus infections in the immunocomprom-
    ised host, J. Infect. Dis. 72:125.
Wildy, P., 1973, Herpes: History and classification, in: The Herpesviruses (A. S. Kaplan, ed.),
    p. 1, Academic Press, New York.
Wolinska, W. A., and Melamed, M. R., 1970, Herpes genitalis in women attending planned
    parenthood of New York City, Acta Cytol. 14:239.
Wolontis, S., and Jeansson, S., 1977, Correlation of herpes simplex virus types 1 and 2 with
    clinical features of infection, J. Infect. Dis. 135:28.
Yeager, A. S., Ashley, R. L., and Corey, L., 1983, Transmission of herpes simplex virus from
    father to neonate, J. Pediatr. 103:905.
Yen, S. S. C., Reagan, J. W., and Rosenthal, M. S., 1965, Herpes simplex infection in the
    female genital tract, Obstet. Gynecol. 25:479.
Young, S. K., Rowe, N. H., and Buchanan, R. A., 1976, A clinical study for the control of
    facial mucocutaneous herpes virus infections. I. Characterization of natural history in
    a professional school population, Oral Surg. Oral Med. Oral Pathol. 41:498.

CHAPTER 2

Individual HSV Transcripts
Characterization of Specific Genes

EDWARD K. WAGNER

I. INTRODUCTION

An increasingly complete picture of the phenomenology of animal virus gene expression during productive infection is at hand. In the case of herpesviruses [particularly herpes simplex virus type 1 (HSV-1)], this picture is dependent on the revolution in molecular biology resulting from restriction enzyme analysis of viral DNA and, more recently, from the use of recombinant DNA technology for the construction and analysis of fine probes of HSV-1 DNA transcription. These powerful techniques, along with techniques available from the parallel revolution in immunology and assays of biological activities, suggest that a detailed mechanistic description of the intricacies of HSV replication is technically feasible.

Workers in the U.S. and Great Britain have been able to use fine restriction enzyme and polypeptide analysis of HSV-1 and HSV-2 intertypic recombinants to locate many HSV marker proteins on the viral genome (Marsden et al., 1978; Morse et al., 1978; Ruyechan et al., 1979; reviewed by Spear and Roizman, 1980; reviewed less extensively by Halliburton, 1980). Marker rescue has an increasing role in defining specific HSV genes (Stow et al., 1978; Parris et al., 1980). Hybridization of viral RNA present at specific stages of infection to separated restriction fragments of HSV DNA has allowed localization of specific viral transcripts on the viral genome (reviewed by E. Wagner et al., 1981; Wagner, 1983). The in vitro translation of resolvable viral mRNA species has allowed

EDWARD K. WAGNER • Department of Molecular Biology and Biochemistry, University of California, Irvine, California 92717.

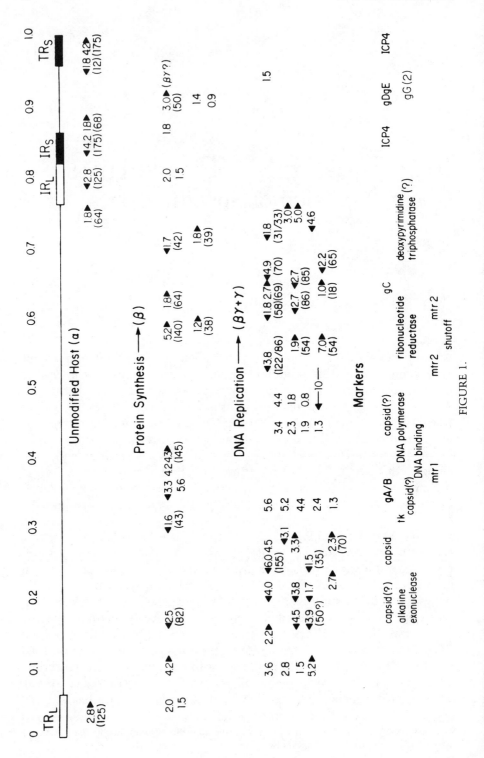

FIGURE 1.

some correlation of these mRNAs with viral gene products seen *in vivo* (Preston, 1977; Cremer *et al.*, 1977, 1978; Anderson *et al.*, 1980b; Frink *et al.*, 1983). As discussed below, immunological identification of *in vitro* translation products and the ability to recover measurable enzymatic activity by translation of purified mRNAs in amphibian oocytes have allowed in some cases complete identification of specific mRNA species. This, along with the precise mRNA localization techniques, can lead to definition of the actual coding frames for individual HSV proteins. Such data, when combined with the high-resolution genetic and immunological studies designed to examine functional domains of specific proteins, will provide the ultimate resolution of genetic maps required for full characterization of the biology of HSV replication. A moderate resolution map of HSV-1 mRNAs present at various stages of infection, along with some of the biological markers correlated with these regions of the genome, is shown in Fig. 1. Specific contributions of many workers are acknowledged in the legend to this figure.

By application of these powerful techniques to studies on the replication of HSV, one can now ask mechanistically meaningful questions concerning HSV gene expression. There are already sufficient data available to allow a general description of HSV genes. Further, experiments by McKnight (McKnight, 1982; McKnight and Kingsbury, 1982) and others have allowed the functional definition of HSV promoters. Mackem and Roizman (1982a) have shown that some HSV genes, at least, have control elements functionally separable from the promoter region. These control elements have a role in the modulation of viral gene expression.

←

FIGURE 1. Location of HSV-1 mRNA species abundant at the three stages of replication. This figure summarizes studies in this laboratory (reviewed in Wagner, 1983; E. Wagner *et al.*, 1981) and described briefly here. Also included are results of experiments currently in progress. Individual mRNA species are localized to the nearest restriction fragment or junction of two fragments found to have significant homology with them. The size of RNA, including the approximately 200-base poly(A) tail, is indicated above the location in kilobases. The direction of transcription, where known, is indicated by arrows pointing toward the 3' end in the P arrangement of HSV-1. The size of polypeptides encoded by isolated mRNA species (where determined) is shown in thousands of daltons below the mRNA species in question. Locations of other markers were determined from data described in the following references: Betz *et al.* (1983), D. Bzika and D. Person (unpublished results), Camacho and Spear (1978), Chartrand *et al.* (1979, 1980), Clements *et al.* (1977), Conley *et al.* (1981) R. Costa, K. Draper, L. Banks, K. Powell, R. Eisenberg, G. Cohen, and E. Wagner (unpublished results), R. Costa, K. Draper, and E. Wagner (unpublished results), Cremer *et al.* (1977), Crumpacker *et al.* (1980), DeLuca *et al.* (1982), Dixon and Schaffer (1980), Dixon *et al.* (1983), Docherty *et al.* (1981), Fenwick *et al.* (1979), Frink *et al.* (1983), Galloway *et al.* (1982b) L. Holland, R. Sandri-Goldin, M. Levine, and J. Glorioso (unpublished results), Huszar and Bacchetti (1983), Jariwalla *et al.* (1980), Jones *et al.* (1977), Lee *et al.* (1982a,b), Lemaster and Roizman (1980), McKnight (1980), Maitland and McDougall (1977), Marsden *et al.* (1978), Morse *et al.* (1978), Para *et al.* (1982, 1983b), Powell *et al.* (1981), Preston (1979b), Preston and Cordingley (1982), Preston *et al.* (1983), Rafield and Knipe (1984), Reyes *et al.* (1979), Ruyechan *et al.* (1979), Sharp *et al.* (1983), Spang *et al.* (1983), M. J. Wagner *et al.* (1981), Watson *et al.* (1979, 1981a, 1982), Weller *et al.* (1983), Wigler *et al.* (1977), Wohlrab *et al.* (1982), Zweig *et al.* (1983).

All these contributions make it evident that a molecular description of those factors involved in regulation of HSV gene expression is attainable.

II. HSV mRNA

A. General Properties

Most studies have been carried out using one strain or another of HSV-1; however, the very close similarity between HSV-1 and HSV-2 suggests that generalizations will cover both. But of course, specific points may differ in important details. HSV-1 mRNA shares general properties with host cell mRNA (i.e., it is synthesized in the nucleus, is polyadenylated on the 3' end and capped on the 5' end, and is internally methylated) (Wagner and Roizman, 1968a; Bachenheimer and Roizman, 1972; Silverstein et al., 1973, 1976; Bartkoski and Roizman, 1976; Moss et al., 1977; Stringer et al., 1977).

It is interesting to note that although HSV-1 mRNA is in many ways like its host cell mRNA, it differs in degree of splicing. Although some HSV-1 mRNAs are spliced during their biogenesis (Watson et al., 1981a; Frink et al., 1981a, 1983), many others do not appear to be. Splices have been detected in mRNAs of both Epstein–Barr virus (EBV) and human cytomegalovirus (CMV) (Heller et al., 1982; Stinski et al., 1983; van Santen et al., 1983). Whether the relatively low frequency of splices in HSV mRNAs is a reflection of its very rapid replication in host cells or is due to other aspects of its biology is unclear at this time.

B. Nuclear Forms of HSV mRNA

The relatively low frequency of HSV splicing and the fact that transcriptional control sequences map juxtapositioned to structural genes (see below) both lead to the prediction that HSV nuclear mRNA precursors will generally be close to the size of the mature product. However, it has long been observed that HSV RNA species with normal sizes greater than 10 kilobases (kb) can be isolated from the nucleus (Wagner and Roizman, 1969b). In the case of very large nuclear RNA detected under conditions where only α (immediate–early, see below) mRNA was being synthesized, Anderson et al. (1980a) found this RNA hybridized to the same restriction fragments as the α mRNAs seen in the cytoplasm. They concluded that much of this apparently large RNA was an artifact due to trapping of HSV RNA in larger nuclear species. This conclusion was based on the fact that mRNA species greater than 10 kb in length would by necessity hybridize to neighboring or other restriction fragments than those seen.

R. Frink and E. Wagner (unpublished results) carried out some experiments to examine the properties of high-molecular-weight HSV nu-

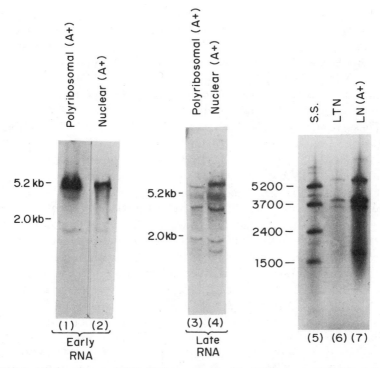

FIGURE 2. Nuclear forms of HSV RNA encoded by HindIII fragment K (0.527–0.592 map units) are of the same general size as seen on polyribosomes. General methods for RNA transfer (Northern blots) and S1 nuclease mapping are reviewed by Wagner (1983).

clear RNA in general. In unpublished experiments, they used guanidine–urea extraction of nuclear RNA (cf. Berk *et al.*, 1979) to examine the size distribution of nuclear RNA via Northern blot and S1 nuclease analysis. The mRNAs encoded by HindIII fragment K (0.527–0.592 map units) were used as a standard. Both early and late after infection, RNA transfer (Northern blots) showed that the poly(A+) nuclear RNA contained mRNA species identical to polyribosomal poly(A) mRNA (see Fig. 2, tracks 1–4). The picture with total nuclear RNA was complicated by the fact that late (but not early) after infection, there was considerable heterodispersely migrating RNA of large size (not shown). S1 analysis of hybrids between ^{32}P-labeled HindIII fragment K DNA and poly(A) and total nuclear RNA, however, showed only the four major species identified with polyribosomal (A+) mRNA (6.0, 4.3, 3.7, and 1.7 kb; Fig. 2, tracks 6 and 7). Similar conclusions were obtained with mRNA from BamHI fragment I (0.60–0.64 map units). Thus, it was concluded that no specific RNA species other than those seen on polyribosomes were present in readily detectable amounts in the nucleus.

In a second set of experiments, R. Frink and E. Wagner (unpublished results) examined whether any members of the overlapping mRNAs of

HindIII fragment K were kinetic precursors to others. Very short ^{32}P pulses were carried out and nuclear poly(A) mRNA was isolated using preparative hybridizations. This material was then subjected to size fractionation. At the shortest pulses (ca. 5 min), the proportions of 7-, 5.2-, and 1.9-kb mRNAs isolable using a specific HSV-1 DNA fragment were the same as for longer pulses. It was concluded that no readily detectable precursor–product relationship existed between these mRNAs.

These results suggest that, in general, there is no large amount of processing of nuclear RNA forms of HSV mRNAs prior to their appearance as mRNA. One possible origin of some of the very-high-molecular-weight nuclear RNA observed (Wagner and Roizman, 1969a,b; Jacquemont and Roizman, 1975; Jacquemont et al., 1980; R. Frink and E. Wagner, unpublished results) is inefficient termination of transcripts originating at many sites upstream of the transcription unit in question.

The close correspondence in size between nuclear precursor and mature HSV mRNA does not apply to every HSV mRNA species. At least one transcription unit with a sizeable intron has been detected (see below). This can be expected to have a nuclear precursor considerably larger than the mature mRNA. A. Deatly, L. Feldman, and T. Ben-Porat (unpublished results) found a lack of appreciable amounts of large pseudorabies virus nuclear precursors to specific viral mRNA. This suggests that the situation with many herpesviruses is similar. However, Heller et al. (1982) have clearly identified specific large species of EBV RNA that are confined to the nucleus and appear to be precursors to mature EBV mRNA. This example reemphasizes the point that specific questions concerning the biogenesis of particular HSV or other herpesvirus mRNAs really require a precise description of the specific transcription unit. With such, one can readily posit the potential range of sizes of nuclear precursors.

III. RNA EXPRESSED IN THE INFECTED CELL

A. Temporal Regulation of HSV Gene Expression

The viral proteins seen in HSV-1-infected cells were differentiated into three general groups: α, β, and γ, based on both their kinetics of and requirements for synthesis (Honess and Roizman, 1974, 1975). This "cascade" pattern of HSV protein expression is generally mirrored in the other herpesviruses. This cascade is more complex than a simple threefold one and Honess and Roizman differentiated subclasses of the major groupings according to whether specific proteins could be readily distinguished at early times but became more abundant following viral DNA replication, etc. As discussed in Spear and Roizman (1980), such proteins are considered "intermediate" or βγ.

The temporal expression of HSV-1 mRNA generally mirrors the cascade of protein synthesis. For this reason, viral mRNAs are classified here as α, β, or γ, depending upon their kinetics of expression. This classification is not entirely satisfactory as it ignores the fact that mRNAs for certain proteins may be expressed well before the mature protein is recognized. However, mRNAs expressed abundantly only following DNA replication (late) can be subclassified according to whether they are detectable prior to DNA synthesis (βγ) or only after it (γ). Other schemes of temporal expression can be used, but the current classification has the value that it generally can lead to an inference concerning possible function of proteins encoded by specific kinetic classes of mRNA. It is well to remember, however, that in the present review, the kinetic classifications apply only to the mRNA expressed.

Although HSV-1 is the most extensively studied of the herpesviruses, many of the general features of RNA expression during the HSV-1 replication cycle have been described in other herpesviruses. Rakusanova *et al.* (1971) and Ben-Porat *et al.* (1974) demonstrated requirements for protein synthesis for full early RNA expression and for transition from early to late phases of expression in pseudorabies virus. Huang *et al.* (1971) and Cohen *et al.* (1975, 1977) demonstrated early and late phases of RNA expression of equine abortion (herpes) virus. Recent work on CMV (DeMarchi *et al.*, 1980; Wathem *et al.*, 1981) showed that the limited, immediate–early transcription patterns seen for HSV-1 were also seen with this virus.

Each transcription state is characterized by an increasing complexity of viral mRNA expressed (reviewed by Spear and Roizman, 1980; Wagner, 1983). In the first stage, α, or immediate–early mRNA species are seen. These can be expressed abundantly without *de novo* protein synthesis (i.e., in an unmodified host cell nucleus) (Jean *et al.*, 1974; Kozak and Roizman, 1974). Abundant members of this class of mRNA are quite limited, map in regions of the HSV-1 genome at or near the long and short repeat regions, and encode only a limited number of polypeptides *in vivo* and *in vitro* (Honess and Roizman, 1975; Marsden *et al.*, 1976, 1978; Clements *et al.*, 1977, 1979; Jones *et al.*, 1977; Watson and Clements, 1978; Holland *et al.*, 1979; Preston, 1979a; Watson *et al.*, 1979; Anderson *et al.*, 1980a; Mackem and Roizman, 1980; Beck and Millette, 1981; Preston, 1981).

As discussed below (Section V.B), α proteins appear to have regulatory roles in later HSV gene expression. In the case of pseudorabies (herpes) virus, an α gene has been shown to be able to replace the immediate–early gene function (E1a) in adenovirus infection (Feldman *et al.*, 1982; Imperiale *et al.*, 1983). S. Bachenheimer (unpublished results) noted similar finding using the HSV ICP4 function. It would thus appear that such regulation may be at a common point in very different nuclear replicating DNA viruses. If this is the case, the generality is not complete, as adenovirus E1a function will not readily substitute for HSV-1 α-gene expression (Batterson and Roizman, 1983).

Two α mRNAs encoding different polypeptides are chimeric in that their 5' ends map in the short repeat region, but their 3' ends map in different ends of the short unique region (Clements *et al.*, 1979; Watson *et al.*, 1979; Anderson *et al.*, 1980a). The coding sequence for their identical 5' ends contains an intron, which is about 150 bases in length beginning approximately 260 bases from the 5' ends of the mRNAs (Watson *et al.*, 1981a,b).

Following expression of one or several HSV-1 α proteins, a more complex population of viral mRNA becomes abundant prior to viral DNA replication (Wagner, 1972; Wagner *et al.*, 1972; Murray *et al.*, 1974; Swanstrom and Wagner, 1974; Swanstrom *et al.*, 1975). Those mRNAs expressed after the α mRNAs, but prior to viral DNA replication, comprise the β or early class. These β viral mRNAs map throughout the HSV-1 genome in noncontiguous regions. However, only a limited number of readily resolvable species are found (Stringer *et al.*, 1978; Holland *et al.*, 1979; Jones and Roizman, 1979). As described in a following section, a number of β mRNAs were rigorously mapped.

Many β mRNAs encode proteins involved with priming the cell for viral DNA replication. Early work from a number of laboratories suggested that without viral DNA replication, the β viral mRNA and protein populations appear to persist (Wagner, 1972; Swanstrom and Wagner, 1974; Powell *et al.*, 1975; Swanstrom *et al.*, 1975; Ward and Stevens, 1975; Stringer *et al.*, 1977). However, Honess and Roizman (1974) found reduced but significant late protein synthesis occurring in the absence of readily detectable viral DNA replication. They concluded that viral DNA replication was not strictly coupled to expression of late genes, although abundance of late genes was.

Such disagreement reflected the difficulty in fully inhibiting HSV DNA replication in all cells using reasonable criteria and drug levels, and it reflected differences in criteria used to define mRNA and protein populations. An illustration of the former problem can be found in the fact that Wagner *et al.* (1972) and Swanstrom and Wagner (1974) used hydroxyurea and mitomycin C to inhibit viral DNA synthesis in early studies with HeLa cells, yet Cohen *et al.* (1977) found the drug ineffective in studies on equine herpesvirus. With HeLa cells, R. Swanstrom and E. Wagner (unpublished results) found cytosine arabinoside (Ara-C) to allow significant viral DNA replication under conditions where total DNA replication in cells was reduced by over 99%.

Because of these discrepancies, Holland *et al.* (1979, 1980) and Jones and Roizman (1979) carefully reexamined the role of viral DNA replication in late HSV-1 mRNA transcription using a number of criteria, including quantitative hybridization, specific mRNA species identifiable by Southern blot hybridization of size-fractionated RNA, or the size of polypeptides resolved by *in vitro* translation of purified viral mRNA. Both groups concluded that normal late mRNA expression requires viral DNA replication in addition to β polypeptides. This conclusion is in general agreement with data presented by Conley *et al.* (1981) and Pedersen *et al.* (1981).

As discussed above, late HSV-1 mRNA appears concomitantly with viral DNA replication (Frenkel and Roizman, 1972; Wagner, 1972; Wagner *et al.*, 1972). Two subclasses of late mRNA are readily distinguishable. There is a group that is detectable in the absence of viral DNA replication, the "leaky-late" or βγ mRNAs; and a group that cannot be seen at all in the cytoplasm in the absence of viral DNA replication, the "true-late" or γ mRNAs. Both groups of mRNA species encode a large number of polypeptides (Holland *et al.*, 1980), many of which presumably are structural proteins of the virion. As described in the next section, workers in this laboratory precisely mapped a number of βγ and γ mRNAs. Several late mRNAs are spliced (R. Costa, K. Draper, and E. Wagner, unpublished results) and one family of γ mRNAs contains spliced and unspliced members (Frink *et al.*, 1981a, 1983).

As the HSV-2 genome is generally homologous to HSV-1 (Kieff *et al.*, 1972; Kudler *et al.*, 1983; K. Draper, R. Frink, M. Swain, D. Galloway, and E. Wagner, unpublished results), one would expect the temporal regulation to be the same and generally this is the case. Details, however, may well differ. In one area, which has been carefully compared between HSV-1 and HSV-2 (ca. 0.59–0.62 map units), there is evidence that the time of expression of some homologous HSV genes may differ between the types (Anderson *et al.*, 1981; Galloway *et al.*, 1982a,b; Jenkins *et al.*, 1982). Further, Pereira *et al.* (1977) and Fenwick and Clark (1982) have shown that amino acid analogs and cycloheximide reversal result in different patterns of post α genes being expressed in HSV-1- and HSV-2-infected cells.

B. Effect of HSV Infection on Host Cell RNA

Studies on the inhibition of host cell macromolecular synthesis following HSV infection go back at least 20 years, well into the "dark ages" prior to the availability of more modern biological techniques. Host cell polysomes are dispersed, and stable RNA synthesis is rapidly inhibited after infection (Sydiskis and Roizman, 1967; Wagner and Roizman, 1969a; Sasaki *et al.*, 1974). Stringer *et al.* (1977) showed that viral mRNA on polyribosomes becomes a major class early after infection so that, by late times, as much as 90% of newly synthesized polyribosomal mRNA in infected cells is viral. This disaggregation of host polysomes appears to be due to the presence of a virion component (Nishioka and Silverstein, 1978). Generally, HSV-2 strains shut off host cell function much more rapidly than do HSV-1 strains.

Studies on specific cellular mRNAs showed a rapid shutoff of synthesis and apparent degradation of cellular mRNA (Pizer and Beard, 1976; Nishioka and Silverstein, 1977; Stenberg and Pizer, 1982). Recently, Fenwick and Clark (1982) and Hill *et al.* (1983) reported that with HSV-2 the virion component itself can rapidly inhibit host protein synthesis. This virion component also functions in HSV-1 infection, but Read and Frenkel (1983) described mutants of HSV-1 that were defective in this virion-associated function. They also found that the function was not strictly required for virus replication in cell culture, and that a second more general shutoff of host function also occurred following HSV infection.

Although specific mechanisms for host shutoff are not fully known, it is clear that a definable viral function is involved (Fenwick *et al.*, 1979). Read and Frenkel (1983) suggested that the virion-associated shutoff function is involved at a posttranscriptional level. Recently, Bartkoski (1982) showed a change in the protein associated with polysomes following HSV infection. The role of such proteins in the shutoff of host cell protein synthesis is unknown, but could play a role.

C. Gross Localization of HSV Transcripts

Even before the general availability of recombinant DNA techniques for use with animal virus genomes, it was determined that HSV-1 mRNA transcripts map discretely along the genome. The discrete localization of the α HSV-1 mRNAs has been discussed above. Location of β and γ mRNAs was more difficult and with lower resolution, but basic patterns were discernible.

Clements *et al.* (1977) could not distinguish discrete locations for β mRNAs using RNA hybridized to Southern blots of large restriction fragments; however, Stringer *et al.* (1978) were able to use electron microscopy to locate discrete areas of the HSV-1 genome forming abundant R-loops with early RNA. Later, Holland *et al.* (1979) located 13 abundant β mRNAs within specific restriction fragments on the HSV-1 genome. These data conclusively demonstrated that HSV-1 β genes are not highly clustered.

Anderson *et al.* (1979) used hybrid selection and Southern blot analysis to precisely map a large number of HSV-1 mRNAs abundant following DNA replication. Further, work with a specific portion of the HSV-1 genome (*Hind*III fragment K, 0.527–0.592 map units) demonstrated that β and γ mRNAs can share partial colinearity (Anderson *et al.*, 1980a,b, 1981). The location of β mRNAs quite near βγ and γ mRNAs was also found in several other regions of the genome (Frink *et al.*, 1981a; Hall *et al.*, 1982; Sharp *et al.*, 1983). The use of specific restriction fragments to isolate mRNA also allowed tentative assignment of specific *in vitro* translation products to specific isolated mRNA species.

A moderate-resolution "transcription map" of the HSV-1 genome derived from these studies has been presented (E. Wagner *et al.*, 1981; Wagner, 1983), and an updated version of this map is shown in Fig. 1. The potential correlation of some mRNAs with specific biological functions is clear even at this resolution.

The picture with HSV-2 is, of course, not as fully developed. The DNA sequences between the strains are generally homologous. Oakes *et al.* (1976) showed that there was fairly general RNA homology between HSV-1 and HSV-2, as would be expected from intertypic recombination data (see below). As mentioned above, several groups have shown that the mRNA transcription patterns for HSV-1 and HSV-2 between 0.51 and 0.62 map unit are generally homologous and that the polypeptides encoded are similar in size (Anderson *et al.*, 1981; Docherty *et al.*, 1981; Draper *et al.*, 1982; Galloway *et al.*, 1982b; Jenkins *et al.*, 1982. Reyes *et al.* (1982) and M. Swain and D. Galloway (unpublished results) have shown the base sequence for the thymidine kinase (TK) gene of HSV-1 and HSV-2 to be similar and to map similarly.

K. Draper, R. Frink, M. Swain, D. Galloway, and E. Wagner (unpublished results) have carried out some comparative Southern blot and nucleotide sequence analyses between 0.59 and 0.70 map units. In this region, the sequence conservation is greatest in the protein-coding sequences, and there is considerable divergence outside them. In several instances, however, there are abrupt divergences within coding regions also. The fact that not all protein-coding regions of the genome are highly homologous is in keeping with the original studies of Kieff *et al.* (1972). More recently, Kudler *et al.* (1983) found regions of high homology interspersed with nonhomologous regions, especially in the short unique region using electron microscopic heteroduplex mapping. The region between 0.61 and 0.71 map units in a mosaic of regions of high homology, low homology, and no detectable homology. The size of the probes used by K. Draper *et al.* (unpublished results) indicated that regions of no appreciable homology can be as long as 1 kb or greater. As stated, the nonhomologous regions can interrupt specific protein-coding frames. A prime example is the case of the translation frame from HSV-1 glycoprotein C (gC), where about half appears highly homologous to a comapping region of HSV-2, and the other half has much lower homology. It is clear, then, that generalizations assuming virtual identity of gene functions between the two subtypes should be treated cautiously.

IV. DETAILED ANALYSIS OF SPECIFIC HSV TRANSCRIPTS

A. Isolation, Localization, and Translation of Specific HSV-1 mRNA Species

The data reviewed in the preceding section indicate that the general patterns of HSV-1 mRNA expression are clear. It also is clear that detailed information concerning HSV-1 gene function and mechanisms of control of gene expression must come from the analysis of specific viral genes and the mRNAs and proteins they encode. The use of hybrid selection, size fractionation, and *in vitro* translation to isolate and characterize viral mRNA was pioneered by workers using somewhat more tractable viruses, such as adenovirus (Anderson *et al.*, 1974; Lewis *et al.*, 1975). In theory, recombinant DNA technology is not a requirement for such studies as discussed above. However, the purity and large quantities of recombinant DNA available make its use a practical necessity. Enquist *et al.* (1979) cloned HSV-1 and made their γ clones immediately available to all workers. It is appropriate to note that workers in a number of other laboratories have reported their cloning of all or essentially all of the HSV genome (Galloway and Swain, 1980; Post *et al.*, 1980; Goldin *et al.*, 1981; Matz *et al.*, 1983). M. Levine, B. Roizman, and others have been generous with HSV-1 clones, and D. Galloway and P. Spear, among others, have been quite generous with their HSV-2 clones. At present, a number of different clones of HSV-1 and HSV-2 restriction fragments from various strains are readily available.

Procedures for specific viral mRNA isolation and characterization used in this laboratory were previously outlined (E. Wagner *et al.*, 1981; Wagner, 1983). Variations on these methods are used in many laboratories, but for simplicity's sake, this laboratory's general procedures are given as a guide. Hybrid selection of HSV-1 mRNA species is conveniently carried out using DNA fragments bound to diazotized cellulose powder (Noyes and Stark, 1975; Anderson *et al.*, 1979). Hybridization near the T_m of the DNA in high (80%) formamide concentration (Casey and Davidson, 1977) minimizes background. Following preparative hybridization of fragment-specific mRNA, it can be translated directly or size-fractionated using methylmercury-containing agarose gels for electrophoresis (Bailey and Davidson, 1976). These gels have the advantage that specific viral mRNA species can be eluted and then translated (Anderson *et al.*, 1980a; Costa *et al.*, 1981; Frink *et al.*, 1981a; Hall *et al.*, 1982). Reselection of mRNA after elution from gels by oligo-dT chromatography is valuable. Similar procedures using DNA bound to nitrocellulose or diazotized paper have been developed (Conley *et al.*, 1981; Preston and McGeoch, 1981).

RNA or Northern blots are useful for the localization of HSV-1 mRNA species using radioactive probes made to cloned DNA fragments. The method is so convenient that fairly detailed localization of transcripts of many refractory herpesviruses, such as EBV, has been reported (Hummel and Kieff, 1982; van Santen *et al.*, 1983). Cloned DNA fragments ranging from 1- to 3-kb pairs are the most suitable sizes of probes for "walking" down a region of the viral genome. Use of limited areas of overlap in these probes tends to minimize error.

Northern blots using early HSV mRNA allows assignment to the temporal class of individual species. Some of these mRNAs overlap and some translate into the same polypeptides, and thus appear to be "redundant." Mechanisms for generation of overlapping mRNAs are discussed below. Overlapping mRNAs often are not of the same temporal class. Examples of this abound in the HSV-1 genome. In the case of HSV-2, Northern blot surveys of transcript location imply a similar situation, but here, more detailed analyses will need to be carried out to confirm this conclusion.

Translation of specific HSV-1 mRNA is efficiently carried out using either reticulocyte lysates or by microinjecting amphibian oocytes. The latter method is laborious, but well suited for assays of biological activity. Further, it can be used as a coupled transcription–translation system if the proper DNA fragment is injected. Translation of total mRNA from a region is feasible and can be used to generally locate viral genes. In general, however, such techniques are most readily interpretable when biological activity can be measured, such as for thymidine kinase or the alkaline exonuclease (Preston, 1977; Preston and Cordingley, 1982), or where a specific immunoprecipitation can be carried out (Lee et al., 1982b; Costa et al., 1983, 1984; Frink et al., 1983).

Translation of size-fractionated, region-specific mRNAs affords both the most unambiguous assignment of polypeptide products to mRNA and a good check on possible mRNA redundancy. Denaturing methylmercury agarose gel electrophoresis gives excellent resolution of even small mRNAs. However, even translation of size-fractionated mRNA does not always yield completely unambiguous results, as witnessed by the consistent finding from this laboratory that the 3.8-kb mRNA mapping in the left half of HindIII fragment K encodes both a 122,000-dalton and an 86,000-dalton polypeptide (Anderson et al., 1981). Preston and McGeoch (1981) found that purified mRNA from BamHI fragment P, which encodes TK, also encodes a 39,000-dalton polypeptide, and sequence analysis indicates both polypeptides are from the same mRNA. Marsden et al. (1983) showed that these polypeptides can arise in vivo via alternative use of initiation codons. It is really not clear whether there is a biological function to such degeneracy. It will be seen below, however, that even more complex patterns can be detected.

Despite these ambiguities, the ability to isolate, locate, and translate HSV mRNAs generally leads to readily interpretable map locations for viral transcripts and the polypeptides they encode. This type of information was used to construct the moderate-resolution map of Fig. 1 and allows the construction of high-resolution maps of individual HSV mRNAs described below.

B. High-Resolution Mapping of HSV-1 mRNA

The use of S1 nuclease and exonuclease VII to digest hybrids between viral RNA and defined fragments of DNA was developed by Berk and Sharp (1977) to provide a reliable and (fairly) rapid means of precisely locating viral transcripts and locating splices in them. Provided the proper standards and separation techniques are used, resolution can be taken to the nucleotide level (see below). A good basic outline of the method has been described by Flint and Broker (1981). Although the size of the HSV genome is so large as to necessitate at best "deliberate" progress, the general lack of splicing makes things somewhat easier. The highest level resolution of S1 digests of HSV-1 mRNA hybrid species was used to locate the TK gene on its DNA sequence (McKnight, 1980; M. J. Wagner et al., 1981), to precisely locate splices in the two 1.8-kb (immediate–early) genes (Watson et al., 1981b), to analyze the precise sequence of several α-gene promoters (Mackem and Roizman, 1982a,b,c), and to locate the 5' ends of model β and γ mRNAs for comparative purposes (Frink et al., 1981b; Hall et al., 1982; Costa et al., 1983; and see below). Such resolution is sufficient to allow one to see detailed overlap and to begin sequence studies on mRNAs of interest. Such a map is shown in Fig. 3. For completeness, data from several other laboratories are included. These data, with complete transcription maps of the short region of the genome (D. McGeoch, A. Dolan, S. Donald, and F. Rixon, unpublished data), provide a reasonably complete transcription map for HSV-1. It should be noted that the size of the coding regions for specific mRNAs is 200 bases shorter than the size of the mRNAs themselves (Fig. 1) due to the lack of poly(A) tails. Procedures used in this laboratory are published (cf. Frink et al., 1981a; Hall et al., 1982), and several examples follow.

This laboratory used 5' and 3' end-labeled HSV-1 DNA probes for hybridization to define direction of transcription of transcripts and these also allow a very precise characterization of overlapping mRNA sequences. Although many overlapping mRNA "families" were seen, most showed no evidence of splices. There were, however, exceptions. These included the two 1.8-kb α mRNAs for ICP22 and ICP47 characterized by Watson et al. (1981a,b) and subsequently by Rixon and Clements (1982). The transcription unit encoding gC (0.63–0.65 map units) has several low-abundance spliced members (Frink et al., 1981a, 1983). Several other mRNAs have 5' ends that indicate short noncontinuities near their 5' ends. These are indicated by brackets in Fig. 3. To date, however, only one mRNA, a 2.8-kb one mapping between 0.185 and 0.225 map units, has a large intron. Here, R. Costa, K. Draper, and E. Wagner (unpublished data) have tentatively determined it to be on the order of 4 kb. Other exceptions will occur, but at this time, no region of the HSV genome appears to encode transcripts as extensively spliced as has been seen in some other viral systems.

Most HSV-1 mRNAs mapped do not show significant complementary overlaps with others, but there are exceptions. Frink et al. (1981a) reported one minor 2.7-kb mRNA mapping in HindIII fragment L (0.592–0.647 map units) that had a complementary overlap with a major β mRNA, as well as with the mRNA encoding gC. One group of mRNAs in EcoRI fragment I (0.633–0.721 map units) was carefully analyzed. The

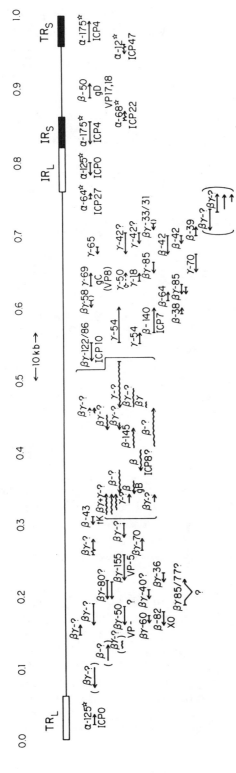

FIGURE 3. High-resolution map of highly resolved HSV-1 mRNAs. The times of appearance of mRNAs are shown at the top along with the size of polypeptide products they translate *in vitro* where known. The direction of transcription is indicated. In addition to data from this laboratory, data are interpreted from other laboratories for TK (McKnight, 1980), for gD (Watson *et al.*, 1982, 1983), for α mRNAs (Mackem and Roizman, 1982a,b,c), and for the 3.1-kb mRNA mapping to the left of TK (Sharp *et al.*, 1983). The region between 0.3 and 0.4 map units was mapped by L. Holland, R. Sandri-Goldin, A. Goldin, J. Glorioso, and M. Levine (unpublished results). Transcripts indicated by brackets are mapped only by high-resolution Northern blot analysis.

5' ends of this group of mRNAs showed complementary overlap, and it was found that potential protein-coding frames did not overlap (Hall *et al.*, 1982). Other individual cases will remain as intriguing enigmas until fully characterized by sequence analysis, etc.

V. SPECIFIC HSV GENES

A. Correlation of HSV Biological Markers with Specific mRNAs

One use of the data precisely locating HSV transcripts on the viral genome is to identify them with biological functions. Some information for such correlation can be inferred from intertypic recombination data, while other information can be obtained from studies on biological markers. Such markers include the location of drug resistance, assays of enzymatic activity, immunological methods, etc. Table I lists the present state of knowledge concerning the correlation of biological markers with specific HSV transcripts. Credit for individual assignments are found in the text of following sections. There are several nomenclatures available for listing HSV proteins (cf. Marsden *et al.*, 1978; Morse *et al.*, 1978). None is entirely satisfactory. In order to maintain consistency, the terminology used here is that described by Spear and Roizman (1980). In Table I, the identity of proteins is referred to as infected cell protein number (ICP-) where known, or virion-associated protein number (VP-), or if applicable, both are used.

B. α Genes

It is very clear that the 4.2-kb α mRNA mapping wholly in the short repeat region encodes an important regulatory protein of nominal (unphosphorylated) molecular weight of 160,000 (ICP4). This phosphorylated protein concentrates in the nucleus of infected cells (Pereira *et al.*, 1977; Preston, 1979b). A partial nucleotide sequence of the DNA encoding this protein has been published (Murchie and McGeoch, 1982). Identification of the mRNA encoding this protein comes from direct *in vitro* translation of size-fractionated 4.2-kb α mRNA from cells held at the immediate–early stage of replication by cycloheximide block (Clements *et al.*, 1979; Watson *et al.*, 1979) and from *in vitro* translation of such hybrid-selected mRNA using DNA from the short repeat region (Anderson *et al.*, 1980b). Several groups (Courtney *et al.*, 1976; Knipe *et al.*, 1978; Marsden *et al.*, 1978; Preston *et al.*, 1978; Dixon and Schaffer, 1980; Preston, 1981) described mutants shown by them, or subsequently by others, to map in the region encoding the 4.2-kb mRNA. It is evident that at the nonpermissive temperature (39°C), the immediate–early transcription pattern persists in spite of available *de novo* protein synthesis (Knipe *et al.*, 1978;

Watson and Clements, 1978; Holland *et al.*, 1979; Preston, 1979a; Dixon and Schaffer, 1980). Further, Watson and Clements (1980) found that the immediate–early transcription pattern can be reestablished upon temperature upshift. Such a result suggests that ICP4 is a protein required continuously throughout infection. It has recently been shown that the DNA-binding capacity of the cytoplasmic form of ICP4 requires combination with a cellular protein (Freeman and Powell, 1982). This suggests the possibility that ICP4's regulatory function requires the activity of cellular proteins, but the insolubility of nuclear ICP4 will make this difficult to confirm directly.

Other α proteins appear to be candidates for regulatory roles. ICP0, -22, and -27 are phosphorylated and are transported to the nucleus (Pereira *et al.*, 1977). The 12,000-dalton ICP47, on the other hand, is not phosphorylated and is cytoplasmic (Preston, 1979a; Marsden *et al.*, 1982). Mutants mapping in these genes have not been described and recent work of Post and Roizman (1981) indicated that the intact 68,000-dalton gene product (ICP22) is not required for lytic infection in cultured cells. More recently, however, I. Halliburton, L. Post, and B. Roizman (unpublished results) reported that truncated forms of this protein retaining an essential function could not be excluded.

Finally, it should be noted that Davison *et al.* (1981) used some nice techniques to place only one copy of ICP0 in the long repeat and showed that such a virus is viable. Similarly, adroit genetic techniques were used by Poffenberger *et al.* (1983) to demonstrate that ICP4 need be present only in the terminal repeat for HSV viability. Thus, only one functional copy of these genes is necessary for infection.

C. HSV DNA Replication Machinery

The fact that HSV encodes so much of its own machinery for DNA replication, which must be early functions, means that a number of enzymological and DNA-associating functions involved in DNA replication are potentially identifiable. Several specific DNA replication functions are carefully mapped, and the transcripts encoding several of these are at least partially characterized.

HSV infection leads to the induction of alkaline exonuclease activity, which has been shown to be virally coded (Morrison and Keir, 1968; Hoffman and Cheng, 1977; Franke *et al.*, 1978; Moss *et al.*, 1979; Strobel-Fidler and Franke, 1980). Recently, Preston and Cordingley (1982) used enzymatic assay and hybrid-arrested translation to locate the coding region for the alkaline exonuclease (ICP18) encoded by HSV-1 in the region 0.15–0.2 map units. Costa *et al.* (1983) used a monoclonal antibody to the HSV-2 enzyme characterized by Banks *et al.* (1983) to identify the HSV-1 transcript encoding this enzyme as a 2.5-kb unspliced β mRNA mapping between 0.16 and 0.175 map units.

TABLE I. Location of Specific Protein Markers on the HSV Genome

Peptide No.	Function	Nominal size (daltons)	Location	Size (kb) of encoding transcripts[a]	Temporal class of mRNA
ICP4	Regulation	160,000 (unphosphorylated)	0.82–0.86 0.96–1	4.2	α
ICP0	?	110,000	0.02–0.04 0.78–0.80	2.8	α
ICP22	?	68,000	0.86–0.87	1.8	α
ICP27	?	64,000	0.74–0.75	1.8	α
ICP47	?	12,000	0.95–0.96	1.8	α
ICP18	Alkaline exonuclease	82,000	0.16–0.175	2.5	β
ICP36	Thymidine kinase	43,000	0.30–0.31	1.5	β
ICP8 (ICP11–12 in HSV-2)	DNA binding	127,000	0.37–0.39	4.2 (?)	β
?	DNA polymerase	145,000	0.41–0.43	4.5	β
ICP6 (ICP10 in HSV-2)	Ribonucleotide reductase	148,000 and/or 38,000	0.57–0.60 0.59–0.60	5.2 and/or 1.5	β
?	Deoxypyrimidine triphosphatase	?	?	?	β (?)
gA/B (type 1) (VP7–8.5)	Glycoprotein	125,000/120,000 (glycosylated) ~100,000 in vitro translation value	0.35–0.38	3.3	β

gC (type 1) (VP8)	Glycoprotein	128,000 (glycosylated) (60,000 amino acid residues)	0.63–0.65	2.7	γ
gC (type 2) (formerly HSV-2 gF) (VP-?)	Glycoprotein	80,000 (glycosylated)	0.63–0.65	2.7	?
gG (type 2) (formerly HSV-2 gC) (VP-?)	Glycoprotein	130,000	0.65–0.69	?	?
gD (type 1) (VP17–18)	Glycoprotein	60,000 (glycosylated) (46,000 amino acid residues)	0.91–0.924	3.0	β or βγ
gE (type 1) (VP12.3–12.6)	Glycoprotein	87,000	0.93–0.95	?	β or βγ
VP5	Capsid	155,000	0.25–0.29	6.0	βγ
VP-?	Capsid (vertex)	~50,000	0.175–0.19 (?)	3.9 (?)	βγ
VP19C	Capsid (interior)	~55,000	0.58–0.6	?	?
VP24	Capsid (?)	26,000	0.46–0.52	?	βγ or γ (?)
VP22a (ICP37)	Capsid(?)	37,000	0.32–0.35	?	βγ or γ (?)

[a] Includes about 200-base poly(A) tail.

The TK gene is a favorite subject for study due to its tremendous utility as a selectable marker (Kit and Dubbs, 1963; Munyon *et al.*, 1971; Bacchetti and Graham, 1977; Maitland and McDougall, 1977; Wigler *et al.*, 1977; Graham *et al.*, 1980; Pellicer *et al.*, 1980). Its mRNA was identified by biological assay of size-fractionated infected cell mRNA (Cremer *et al.*, 1977, 1978; Preston, 1977), assay of hybrid-selected mRNA translation products (Preston and McGeoch, 1981), and assay of coupled transcription–translation products synthesized in amphibian oocytes microinjected with cloned HSV DNA restriction fragments (McKnight and Gavis, 1980; Cordingley and Preston, 1981). This enzyme is encoded by a 1.2-kb coding sequence mapping between 0.30 and 0.31 map units, and as discussed below, the expression of the gene in transfected cells allowed some careful analysis of HSV promoters.

Reyes *et al.* (1982) compared the control sequences and precise location of the 5′ ends of the HSV-1 and HSV-2 mRNAs encoding the enzyme and demonstrated exact colinearity. Further, Swain and Galloway (1983) sequenced the HSV-2 TK gene and demonstrated its general similarity to the type 1 enzyme. There are, however, a significant number of base changes between strains. The regulatory sequences appeared the most conserved, with the least conservation in the 3′ end of the genes.

DNA-binding proteins have been identified in HSV-1- and HSV-2-infected cells (Bayliss *et al.*, 1975; Purifoy and Powell, 1976). A major DNA-binding protein (ICP8 in HSV-1 and ICP11–12 in HSV-2) was shown to be involved in viral DNA replication (Powell *et al.*, 1981) and with the expression of late HSV genes, either via this function or via another (Conley *et al.*, 1981). The gene for this function was mapped in the region 0.37–0.39 map units by marker rescue (Conley *et al.*, 1981; Dixon *et al.*, 1983; Spang *et al.*, 1983; Weller *et al.*, 1983). The β mRNAs of sufficient size to encode such a protein were located in this region (Holland *et al.*, 1979). L. Holland, R. Sandri-Goldin, M. Levine, and J. Glorioso (unpublished results) located a 4.2-kb β mRNA mapping between 0.38 and 0.41 map units as a probable candidate. Rafield and Knipe (1984) translated a 127,000-dalton protein *in vitro* using mRNA purified with DNA mapping at 0.386–0.417 and 0.361–0.386 map units. This protein can be immunoprecipitated with a monoclonal antibody specific for HSV-1 ICP8. Taken together, these data strongly suggest the 4.2-kb β mRNA encodes HSV-1 ICP8.

In HSV-1, the HSV DNA polymerase marker (Keir *et al.*, 1966; Hay *et al.*, 1971; Powell and Purifoy, 1977; Purifoy *et al.*, 1977) was mapped by virtue of the association of phosphonoacetic acid sensitivity with it to the region around 0.40–0.42 map units (Chartrand *et al.*, 1979, 1980; Crumpacker *et al.*, 1980). HSV DNA polymerase is a single polypeptide of about 145,000 daltons (Powell and Purifoy, 1977) with at least two functional domains (Coen and Schaffer, 1980). The general map location of the enzymatic function led E. Wagner *et al.* (1981) to suggest that a 4.5-kb β mRNA isolable using *Eco*RI fragment M (0.40–0.45 map units) might encode the HSV-1 DNA polymerase as it could encode a polypeptide of the expected size.

Ribonucleotide reductase induced by HSV infection was originally described by Cohen and co-workers (Cohen, 1972; Cohen et al., 1974; Ponce de Leon et al., 1977) and partially purified by Huszar and Bacchetti (1981). Dutia (1983) mapped the ribonucleotide reductase of HSV to the region 0.57–0.60 map units, which encodes two overlapping early mRNAs in both HSV-1 and HSV-2 (Anderson et al., 1981; Frink et al., 1981a; Draper et al., 1982; Jenkins et al., 1982). A major early protein mapping in this area is ICP6 in HSV-1 and ICP10 in HSV-2. Recently, Huszar and Bacchetti (1983) used a monoclonal antibody to the HSV-2 ribonucleotide reductase and showed that it precipitates both a 38,000-dalton and a 144,000-dalton HSV-2 protein from HSV-2-infected cells. The size of this protein suggests that it is HSV-2 ICP10. This also suggests that one or both of the β mRNAs mapped in this region encode(s) the enzyme.

These results confirm the findings of Galloway et al. (1982b) showing that the monoclonal antibody made against the 38,000-dalton polypeptide (discussed above) cross-reacts with a 144,000-dalton polypeptide in HSV-2. Nucleotide sequence data from HSV-1 (Clements and McLauchlan, 1981; Draper et al., 1982) showed that there is not a shared translation reading frame between the two mRNAs. Recently, M. Swain and D. Galloway (unpublished data) showed that this is also the case with HSV-2. These data can most readily be reconciled by postulating a shared epitope between two interactive proteins involved in enzymatic activity. The high proline content of many HSV proteins, however, or other nonspecific effects, could complicate the picture. Comparative nucleotide sequence analysis indicated that the 38,000-dalton polypeptides from HSV-1 and HSV-2 should share much of their amino acid sequences, although there are important differences. Showalter et al. (1981) isolated monoclonal antibodies to large HSV proteins (including ICP6); S. Bacchetti and co-workers (unpublished results) used two of these to precipitate both the 144,000-dalton and the 38,000-dalton HSV-1 reductases. Further, one of these antibodies (48S) cross-reacted with the enzymes from both viral serotypes. Detailed analysis of the protein subunits of this complex enzyme may lead to further insight concerning any interaction between these two polypeptides.

HSV infection induces high levels of deoxypyrimidine triphosphatase. The intracellular localization of the HSV-1- and HSV-2-induced enzymes differ and Wohlrab et al. (1982) reported the use of intertypic recombinants to locate the function controlling this localization to around 0.67–0.68 map units. Although these workers suggested that the gene controlling location differs from position of the structural gene, the simplest interpretation is the contrary. Interestingly, the area in question is near a region that does not contain a large amount of homology between HSV-1 and HSV-2 (R. Frink and E. Wagner, unpublished results). This region (0.69–0.71 map units) does contain two prominent β transcripts in HSV-1 (Hall *et al.*, 1982). Identification of the actual transcript encoding the enzymes will require further work.

D. HSV Glycoproteins

Infection with HSV induces a number of glycoproteins that are present at the surface of the infected cell and in the enveloped virions. HSV glycoproteins are subject to a separate chapter in this review and it suffices here to state that at least four specific glycoprotein genes are identified as encoded by HSV-1 and five by HSV-2. At least four of the glycoproteins (gA/B, gC, gD, and gE) share some immunological cross-reactivity between serotypes (Eisenberg *et al.*, 1980, 1982; Eberle and Courtney, 1982; Para *et al.*, 1982; Zweig *et al.*, 1983; Zezulak and Spear, 1983).

The gene for the gA/B complex (Eberle and Courtney, 1980) was mapped in the region of 0.30–0.40 map unit by analysis of intertypic recombinants (Marsden *et al.*, 1978; Ruyechan *et al.*, 1979). It thus maps in a region known to be involved with cell fusion (Ruyechan *et al.*, 1979). Studies by groups including M. Levine and S. Person precisely located the gA/B gene to 0.35–0.37 map units using high-resolution marker rescue. These data are described in DeLuca *et al.* (1982) and Holland *et al.* (1983). Pereira *et al.* (1982b) used a series of intertypic recombinants to locate the gene near here, but slightly to the left (0.37–0.38 map units). As marker rescue and intertypic recombinant maps do not always fully correspond, this amount of uncertainty is understandable.

Rafield and Knipe (1984) used DNA encompassing the region 0.343–0.361 map unit to isolate mRNA that *in vitro* translates into polypeptides of about 100,000 daltons precipitable with polyclonal antisera to gA/B. Holland *et al.* (1979) located a 3.3-kb mRNA in the general region encoding the gA/gB polypeptide. More recently, L. Holland, R. Sandri-Goldin, M. Levine, and J. Glorioso (unpublished results) localized the transcript to the specific region of 0.35–0.37 map units. B. Bzik and S. Person (unpublished results) obtained similar results. Interestingly, this mRNA is a β transcript, although it is not clear that full expression of the gA/B function is early. It is expected that further data will confirm the identification of this as the transcript for gA/B.

HSV-1 gC appears to be dispensable in tissue culture (Keller *et al.*, 1970; Heine *et al.*, 1974). The position of HSV-1 gC was roughly mapped by studies with intertypic recombinants between 0.53 and 0.64 map units. Lee *et al.* (1982b) used *in vitro* translation of hybrid-selected mRNA followed by immunoprecipitation to locate the coding region to between 0.62 and 0.64 map units. Frink *et al.* (1983) demonstrated that a 2.7-kb transcript mapping between 0.63 and 0.65 map units encodes gC, and nucleotide sequence analysis demonstrated that the actual coding frame maps between 0.63 and 0.64. The amino acid sequence of gC derived from the nucleotide sequence data demonstrates the expected membrane insertion and signal sequences, as well as several potential glycosylation sites in the NH_2-terminal region. The very high proline content of the polypeptide has been suggested as the reason that the predicted residue molecular weight for gC (ca. 60,000) is less than its nominal gel value (69,000). In addition to the major gC protein, Frink et al. (1983) were able to show that several minor spliced mRNAs processed from the primary transcript appeared to encode truncated forms of the polypeptide that were weakly reactive with a polyclonal antibody to HSV-2 envelope glycoprotein.

In HSV-2, a protein designated as gC was located to the right of HSV-1 gC (Ruyechan et al., 1979). This protein appears unrelated to HSV-1 gC. A fifth glycoprotein of HSV-2 (gF) was described by Balachandran et al. (1981). Zweig et al. (1983) and Zezulak and Spear (1983) recently isolated monoclonal antibodies cross-reactive between this glycoprotein and type 1 gC. These results suggest why some monoclonal antibodies made to HSV-1 gC can cross-react with antigens on HSV-2-infected cells, as reported by Pereira et al. (1982a). Para et al. (1983) and Zezulak and Spear (1983, 1984) mapped this HSV-2 glycoprotein to a region that is colinear with HSV-1 gC. This tends to confirm partial homology between HSV-1 gC and HSV-2 gF. Recently, K. Draper, R. Frink, M. Swain, D. Galloway, and E. Wagner (unpublished results) found that there is an HSV-2 mRNA species roughly colinear with the HSV-1 gC mRNA, and of about the same size.

Because of the finding that HSV-2 has an added glycoprotein and that HSV-2 gF appears related to HSV-1 gC, Para et al. (1983) and Zezulak and Spear (1983, 1984) suggested that the HSV-2 gC and gF proteins be renamed to make the nomenclature self-consistent. Members of the 1983 International Herpesvirus Workshop tentatively agreed that the former HSV-2 gC be renamed gG, and that HSV-2 gF be renamed gC to indicate correspondence to the HSV-1 glycoproteins. To date, no HSV-1 counterpart of the HSV-2 gG has been reported, although several groups have mapped the glycoprotein in the short region of the HSV genome (Marsden et al., 1978; B. Roizman, B. Norrild, C. Chan, and L. Pereira, unpublished results).

gD maps in the short unique region of the HSV genome (Marsden et al., 1978; Ruyechan et al., 1979) and was located to the region 0.91–0.924 map units (Ruyechan et al., 1979; Lee et al., 1982a). The location of the βγ gE was mapped into the short unique region (Para et al., 1982; Hope et al., 1982), and subsequently, between 0.924 and 0.95 map units (Lee et al., 1982a). Watson et al. (1982, 1983) have presented a precise location for the mRNA encoding HSV-1 gD and a nucleotide sequence for this gene. The predicted amino acid sequence shows, as with gC, membrane insertion and anchor sequences, as well as potential glycosylation sites. Again, a high proline content suggests that the predicted residue molecular weight is at variance with the value derived from denaturing gel electrophoresis. The mRNA for gD has been reported to be a β one as its synthesis is unaffected by the presence of Ara-C to inhibit HSV DNA replication (Watson et al., 1983). However, as mentioned above, Ara-C is not the drug of choice for inhibiting HSV-1 DNA synthesis. Watson et al. (1983) confirmed the temporal classification of Cohen et al. (1980a) for gD. However, Gibson and Spear (1983) found very reduced levels of synthesis of gD mRNA in the absence of DNA replication. This would suggest that the mRNA for gD is of the βγ class. It is worth mentioning that Holland et al. (1980) found an mRNA of a size and location consistent with its encoding gD and displaying unusual kinetics of synthesis in the presence of DNA synthesis inhibitors. Finally, Ikura et al. (1983) have suggested multiple forms of the gD mRNA related by splicing. Although this observation requires more detailed analysis for confirmation, it is clear that transcription in the region encoding gD is not simple. With this gene, the general problem of fully categorizing HSV mRNAs as to a specific temporal class is well illustrated. The most convincing data will be from a comparison of the kinetic synthesis of gD mRNA with other known β and βγ ones. Until such is reported, uncertainty will remain.

The mRNA encoding gE has not been characterized as yet. The kinetics of appearance of the protein (Balachandran et al., 1981) may correlate with the kinetics of synthesis of specific mRNAs located in the short unique region (Holland et al., 1979).

E. Structural Proteins

There are a large number of structural proteins for the HSV virion besides the glycoproteins. These generally are described in the review by Spear and Roizman (1980). The nonenvelope virion proteins can be operationally divided into nucleocapsid-associated and tegument-associated (Gibson and Roizman, 1972). The latter are presumably found in the region between the envelope and the nucleocapsid. Cohen et al. (1980b) prepared polyclonal antibodies against seven isolated nucleocapsid proteins and showed that some of these could be located at specific sites on the nucleocapsid by immune electronmicroscopy (Vernon et al., 1981). R. Costa, G. Bernstein, and E. Wagner (unpublished results) found that at least five of these antibodies will precipitate appropriately sized in vitro translation products from total infected cell poly(A) mRNA. These antibodies have been useful in confirming the identity of the mRNA for the major 155,000-dalton capsid protein (ICP5, VP5) (Costa et al., 1984), and to tentatively identify a transcript encoding a capsid protein of approximately 50,000-daltons (Costa et al., 1983).

The major HSV capsid protein (ICP5, VP5) was located to the region between 0.2 and 0.3 map units by studies with intertypic recombinants (Marsden et al., 1978; Morse et al., 1978). Costa et al. (1981) located a major 6-kb mRNA mapping between 0.25 and 0.29 map units that yielded a 155,000-dalton polypeptide when translated *in vitro*. This, and the fact that hybrid-arrested translation indicated that this was the only significant region of the genome encoding such a polypeptide, led to the suggestion that this 6-kb mRNA encodes VP5. Recently, Costa *et al.* (1984) showed that antibody made against VP5 precipitated the 155,000-dalton translation product of this 6-kb mRNA. Further, tryptic peptides corresponded between translation product and authentic VP5. Thus, the identification appears confirmed.

Frink *et al.* (1981b) located the 5′ end of the 6-kb mRNA encoding VP5 as being 65 bases to the right of the *Bam*HI site at 0.276. However, Costa *et al.* (1984) showed this to be the 5′ end of a minor colinear species and the 5′ end of the most abundant 6-kb mRNA actually is 75 bases to the left of this site, or about 10 bases to the left of the *Bam*HI site. The presence of overlapping mRNAs with 5′ ends very near each other is very common with HSV-1 and possible roles of such mRNAs are discussed in a later section.

Costa *et al.* (1983) tentatively concluded that the mRNA encoding a capsid protein of about 50,000 daltons was one of two colinear mRNAs of 4.5- and 3.9-kb in size mapping between 0.16 and 0.19 on the HSV-1 genome. Sequence analysis of the 5′ ends of these two mRNAs suggested that the smaller encoded the 50,000-dalton protein. This identification was based on weak immunoprecipitation of the *in vitro* translation product of this mRNA with an antibody made against a 50,000-dalton capsid protein (Cohen *et al.*, 1980b), which Vernon *et al.* (1981) found to be located on the vertices of the capsid icosohedra. Lemaster and Roizman (1980) reported the mapping of a structural protein of about 50,000 daltons in the region encoding the 3.9-kb transcript in question, and Costa *et al.* (1983) concluded that the capsid protein might be VP19C. This conclusion was apparently in error, however, as D. Braun, W. Batterson, and B. Roizman (unpublished results) have located VP19C elsewhere and have reported it to be a DNA-binding protein located in the interior of the capsid. Taken together, these results suggest that there are several capsid proteins of nearly the same size, but with different functions. It should be emphasized, however, that the small amount of the *in vitro* translation products precipitable by the polyclonal antibody does not allow one to completely rule out the mapped 50,000-dalton polypeptide being another HSV-1 protein of about the same size as VP19C, yet not associated with the virion.

The precise location and size of mRNAs encoding other capsid and tegument proteins also have yet to be rigorously established. Lemaster and Roizman (1980) mapped several to locations near the right and left repeat sequences. Further, Knipe *et al.* (1981) located a ts mutation (tsB7) mapping between 0.46 and 0.52 map unit. This has the phenotype of a lesion in a capsid protein. The phenotype was confirmed by Batterson *et al.* (1983). Examination of the size of polypeptides encoded by this region

(ICP2, -6, -10, -32, -43, and -44) suggests that the 26,000-dalton capsid proteins identified by Cohen *et al.* (1980b) (VP24?) could map in this region. Data has been presented demonstrating that a 37,000-dalton capsid protein (the p40 of Zweig *et al.*, 1980) maps in the region around TK (ca. 0.32 map unit) (Preston *et al.*, 1983). This protein, which appears to be VP22, was shown to be a capsid protein by Gibson and Roizman (1972). It corresponds to ICP35 located in this region by Conley *et al.* (1981) and more recently confirmed by Braun *et al.* (1983).

F. Genes Involved with Morphological Transformation

Tremendous excitement comes from the finding that HSV-1 and HSV-2 genes can cause morphological transformation (mtr) (Duff and Rapp, 1971, 1973; Rapp and Li, 1974). Such transformation is both a convenient and a fashionable biological marker, especially in view of early and more recently confirmed reports that HSV DNA and RNA could be seen in cervical cancer tissue (Frenkel *et al.*, 1972; McDougall *et al.*, 1980). Such data correlate well with epidemiological surveys linking HSV infection with cervical carcinomas (Naib *et al.*, 1966; Nahmias *et al.*, 1970). Interestingly, it is one of the few functions that does not appear to map in the same region in HSV-1 and HSV-2. The HSV-1 *mtr* (*mtr1*) region maps between 0.30 and 0.45 map units (Camacho and Spear, 1978; Reyes *et al.*, 1979). No demonstration of a stably maintained cell line derived from this region is at hand. Further, attempts to use cloned HSV-1 DNA fragments for transformation have not been successful. It would appear that the mtr1 gene product is not sufficient for required maintenance of the transformed phenotype.

A major *mtr2* region lies in HSV-2 *Bgl*II fragment N (Reyes *et al.*, 1979; Galloway and McDougall, 1981). This region appears to contain mRNAs and proteins homologous to the well-characterized transcripts from the corresponding region of HSV-1, although differences in time of expression and abundance may exist. Monoclonal antibodies directed toward the only early protein encoded by this region (the 38,000-dalton one) did not detect this protein in transformed cells (Galloway *et al.*, 1982b) or in cervical carcinoma tissue (McDougall *et al.*, 1982). However, a protein of 38,000 daltons was seen in cells transformed with UV-inactivated HSV-2 (Suh *et al.*, 1980; Suh, 1982), and it has been suggested that a 35,000- to 38,000-dalton polypeptide is involved in some way with transformation (Docherty *et al.*, 1981) and cervical carcinomas (Gilman *et al.*, 1980). The evidence relating the 38,000-dalton protein with transformation is not convincing. The region of HSV-2 encoding this protein is able to transform, but the smallest portions of viral DNA with transforming activity on 3T3 cells map to the right of this protein. D. Galloway, S. Weinheimer, M. Swain, and J. McDougall (unpublished results) have identified a 300-nucleotide-long segment of DNA in the right-hand portion of *Bgl*II fragment N as the transforming fragment. This DNA is capable of forming a stem–loop structure, which could act as an insertional element in host cell DNA.

HSV-2 DNA encompassed in *Bgl*II fragment C (just to the left of the mtr2 discussed above) also has been reported to transform cells (Jariwalla *et al.*, 1980). Some cells transformed by inactivated HSV-2 appear to express HSV-2 ICP10 (Strnad and Aurelian, 1976), as do cervical carcinoma cells. This protein corresponds to HSV-1 ICP6, which is encoded by the major 5.2-kb β mRNA described by Anderson *et al.* (1981), and a major portion of this protein is encoded by *Bgl*II fragment C. However, the mRNA coding the HSV-2 polypeptide has yet to be rigorously characterized.

Huszar and Bacchetti (1983) suggested that the mapping of HSV-2 ribonucleotide reductase through this junction region may be significant to HSV-2 transformation. However, as the proteins of this area are similar between the strains, the transformation effect may be only incidentally involved with the enzymatic activity of the protein. Recent data on retroviral genes, of course, indicated that even very small changes can be significant in cell immortalization. But the left portion of the *Bgl*II fragment C region seems to be important in cell immortalization, while ICP10 would appear to map to the right. However, the right-hand portion of this DNA fragment has been reported by Jariwalla *et al.* (1980) to be important in tumorigenicity.

At this time, then, although the regions of the HSV-2 DNA inducing *mtr* are readily definable, it is clear that identification of an HSV "transforming" protein is not at hand. Indeed, a recent review by Galloway and McDougall (1983) provided strong argument against a single "transforming" protein being the cause of HSV mtr. These workers favor a "hit-and-run" mechanism, and their data discussed above surely suggest this.

HSV-1 and HSV-2 infections activate endogenous mouse retrovirus synthesis (Duff and Rapp, 1975; Hampar *et al.*, 1976; Reed and Rapp, 1976). This activation is presumably a result of the action of specific HSV gene products (at least specific regions of the HSV-1 genome mediate this activation). To date, the best locations in HSV-1 for this activation function are 0.30–0.32, 0.46–0.49, and 0.90–0.95 map unit (Boyd *et al.*, 1980). Only one of these regions corresponds to the *mtr1* region, so there may be little relation between this activation and *mtr1* itself.

G. HSV Gene Expression during Latency

Genes controlling the latent phase of HSV infection are extremely important biological markers. The whole topic of HSV latency is covered in another chapter, but as attempts have been made to correlate specific genes with this phase of infection, some points are briefly covered here. It is well established that latent HSV is harbored in neurons in the peripheral nerve ganglia (Stevens and Cook, 1973; Stevens, 1975, 1980). Latency requires the virus to be able to multiply in mice. Further, it is clear that herpes genes are required to induce latency (Lofgren *et al.*, 1977). The form of viral DNA in latently infected cells has not been established with certainty. Recently, however, Rock and Fraser (1983) reported that in CNS neurons of mice, viral DNA is seen in structures that lack restriction fragments from the ends of the viral DNA. Such could be circles, concatamers, or both. The relatively large copy number of viral genomes seen, however, is consistent with some viral DNA replication taking place. Generalization of these data is further complicated by the fact that latent HSV from the CNS does not reactivate.

Viral RNA and gene products have been detected in latently infected neurons. Recently, Galloway *et al.* (1979, 1982a) reported that RNA mapping in the left 30% of the long unique region of HSV-1 could be detected in sections of human autopsy neurons that were assumed to contain latent virus. Interpretation is complicated by the possibility of reactivation, especially in light of the fact that the most abundant mRNA in HSV-infected cells maps in this region (Costa *et al.*, 1981).

There is some evidence that the α-protein ICP4 can be detected in latently infected trigeminal ganglia of experimental animals (Green *et al.*, 1981). The fact that ICP4 interacts with host cell proteins (Freeman and Powell, 1982) and has been reported to induce cellular stress protein (Notarianni and Preston, 1982) is suggestive evidence that expression of ICP4 could be involved in maintaining latency by induction of a specific cellular defense response to the presence of the HSV gene product. Again, spontaneous reactivation could lead to artifacts and Galloway *et al.* (1979, 1982a) did not consistently detect mRNA for ICP4 in latently infected neurons.

HSV gene products can be detected in persistently infected cultured neural cells (Leung *et al.*, 1980). Cultured cell models for herpesvirus latency (Wigdahl *et al.* 1982; Youssoufian *et al.*, 1982) may have an impact in assigning the latency function to a specific gene product. Wigdahl *et al.* (1983) found the major form of viral DNA in such cells to be linear, but they did not exclude the presence of other forms.

VI. A TYPICAL HSV-1 TRANSCRIPTION UNIT

A. Control Regions

The lack of splices in most HSV-1 mRNAs and the fact that a very small segment of viral DNA (*Bam*HI fragment Q) can transduce expressible TK activity (Maitland and McDougall, 1977; Wigler *et al.*, 1977) demonstrated that this enzyme can be expressed and function in uninfected cells. This basal expression of the viral TK is greatly amplified following superinfection with HSV (Leiden *et al.*, 1976). Further, selective removal of sequences immediately upstream of the cap site for the TK gene lead to loss of basal expression and virus-induced increased expression (McKnight and Gavis, 1980; McKnight and Kingsbury, 1982; Minson et al., 1982; Smiley et al., 1983). Thus, a promoter region involving both regulatory sequences and RNA polymerase recognition sequences can be identified directly upstream of the HSV TK promoter. As discussed below, this is a common feature of HSV structural genes.

There are functional similarities between the promoter regions of HSV transcription units and cellular ones. HSV genes are transcribed with cellular RNA polymerase II (Alwine et al., 1974; Ben-Zeev and Becker, 1977; Costanzo et al., 1977). Both α and β gene promoters function in transfected cells. McKnight and co-workers' studies on the expression of TK (McKnight and Gavis, 1980; McKnight et al., 1981; McKnight, 1982; McKnight and Kingsbury, 1982) are the most exhaustive on the expression of β promoters in uninfected cells. Several other β gene products have been at least transiently detected in transformed or transfected cells. These include gA/B (Camacho and Spear, 1978; Lewis et al., 1982), gD (Reed et al., 1976), the ICP10 of HSV-2 (Flannery et al., 1977; Jariwalla et al., 1980; Lewis et al., 1982), the DNA-binding protein of HSV-2 (ICP11–12) (Dreesman et al., 1980), TK (MacNab and Timbury, 1976; Rapp and Westmoreland, 1976), and the 38,000 to 40,000-dalton β polypeptide (Suh et al., 1980; Docherty et al., 1981; Galloway et al., 1982b).

The uninfected cell can also express α gene products. Mackem and Roizman (1982a,b,c) and Post et al. (1981, 1982) placed several α promoters upstream of the TK gene and demonstrated that TK activity was present in transfected cells. Post et al. (1982) also showed that an α promoter can mediate the expression of chicken ovalbumin in transfected cells. Finally, antigens suggested to be α polypeptides have been identified in transfected cells (Middleton et al., 1982).

Frink et al. (1981b) and Draper et al. (1982) showed that a "Manley" (Manley et al., 1980) transcription system from uninfected HeLa cells can accurately recognize and initiate transcription at several β genes. In addition, Read and Summers (1982) demonstrated accurate initiation of the β TK gene. These data suggest that the HeLa cell polymerase system generally recognized β promoters in vitro.

There are structural similarities between early and late HSV promoters (see below), but it is not clear whether transfected or transformed cells can readily express late genes (βγ or γ). There is little evidence in the literature suggesting that late genes are readily detectable in such cells; however, Sandri-Goldin et al. (1983) have reported that several γ genes mapping in EcoRI fragment F of HSV-1 are transcribed in cell lines carrying integrated copies of these genes. This observation suggests that promoters for some, but not all, late genes are recognized by cellular RNA polymerase II.

Variability in recognition of late promoters by uninfected cells is also seen in vitro. Frink et al. (1981b) found that the uninfected Manley cell

system did not initiate detectable transcripts from two model late promoters. One of these is the promoter for VP5; and J. Smiley and D. Dennis
(unpublished results) showed that stably integrated genes for HSV-1 TK
to which the VP5 promoter was fused do not express the gene from this
promoter until superinfection by HSV. In contrast, Read and Summers
(1982) reported initiation of transcription from a late promoter near TK,
so the ability of the Manley system to recognize late promoters may be
a function of the specific promoter in question. A recent report by Wright
et al. (1983) showing that a putative late HSV promoter was expressed
in murine erythroleukemia cells undergoing chemically induced differentiation suggests that the cell's state is also important. The comparative
sequence analysis of early and late promoters does not suggest an obvious
basis for discrimination and individual cases may be of value in studies
on factors involved in control of late gene expression.

K. Draper and E. Wagner (unpublished results) developed a Manley-
type system from HSV-infected HeLa cells and found it to be considerably
more active than the uninfected system. It also recognizes transcription
initiation sites not detectably utilized by the uninfected system, and it
appears to recognize at least one late promoter tested. These data suggest
that although there may be exceptions, late promoters are generally not
in complete functional equivalence to early ones in vivo and in vitro.

HSV genes are controlled during their expression by viral products.
Mackem and Roizman (1982c) and Post et al. (1981) showed that HSV α
promoters are up-regulated by a viral product. This product does not appear to be ICP4, for the up-regulation occurred at nonpermissive temperatures using a ts mutant of ICP4 (ts 502Δ305). These workers suggest
that a viral structural protein is involved in the amplification of α genes.
This was confirmed by Batterson and Roizman (1983) using a ts mutant
(tsB7), which is a lesion in a viral capsid protein and which does not induce
α genes at the nonpermissive temperature.

Although transcription of α genes can be increased above a basal level
by other viral products, the abundance of α transcripts is low during normal infection. The rate of synthesis of α mRNAs does not increase during
the productive infection cycle (Anderson et al., 1980b). Control of the
rate of synthesis of mRNA encoding ICP4 and other α genes requires
functional ICP4, for ts mutants of ICP4 do not shut off α transcription
(Holland et al., 1979; Preston, 1979b; Watson and Clements, 1978). Thus,
ICP4 is an autoregulatory protein analogous to T-antigen of papovaviruses.

Whatever the control factors are for up- and down-regulation, it is
clear that some feature of the α promoters must be responsive to them
(Herz and Roizman, 1983). Mackem and Roizman (1982c) have presented
comparative sequence analysis of the region spanning ca. -300 to -400
and $+100$ to $+150$ bases at the 5' ends of three α mRNAs (ICP0, -4,
-27). The DNA encoding the 5' ends and control regions of genes for ICP4,
-22, and -48 have been sequenced by Murchie and McGeoch (1982). All

control regions show a "TATA" box sequence about 28–30 bases upstream of the 5' end of the mRNAs (Benoist and Chambon, 1981; Benoist *et al.*, 1980), and ICP0, -22, and -48 have sequences related to the nominal "CAT" box 60–80 bases upstream (Mathis and Chambon, 1981). Mackem and Roizman (1982c) pointed out "AT-rich" regions and "GC-rich" regions capable of self-pairing in the putative control sequences. Presumably, nucleotide sequence substitution experiments will lead to a finer definition of the sequences involved in α regulation. It should be noted that Cordingley *et al.* (1983) reported that sequences responsive to superinfection can be identified in the region between -174 and -331 bases upstream of the 5' end of the 4.2-kb α mRNA encoding ICP4, while bases between -38 and -108 are required for basal levels of expression in transfected cells.

In the normal course of infection, β genes require functional ICP4 for expression (Kit *et al.*, 1978; Watson and Clements, 1978, 1980; Preston, 1979b; Leung *et al.*, 1980). In cells biochemically transformed with TK, the enzymatic activity is up-regulated by superinfecting virus (Leiden *et al.*, 1976). It is clear in this case that functional ICP4 is required for this up-regulation. The mechanism of ICP4 control is as yet unclear, but it could involve interaction with a regulatory portion of β gene promoters. Smiley *et al.* (1983) found regions between -200 to -80 and -70 to -12 bases upstream of the TK transcription start that are involved in the positive regulation event. Data by Minson *et al.* (1982) demonstrated important regulatory sequences upstream of the promoter of HSV-2 TK. Further work in both systems will lead to better definition of regulatory/control sequences.

A comparative sequence analysis of the regions to about 200 bases upstream of the 5' transcription start of five β mRNAs has been carried out by K. Draper, R. Costa, R. Frink, and E. Wagner (unpublished results). Methods for such were described in a previous review (Wagner, 1983). These data were combined with published data for HSV-1 and HSV-2 TK (McKnight, 1980; E. Wagner *et al.*, 1981; Reyes *et al.*, 1982; Swain and Galloway, 1983), and are summarized in Fig. 4. All regions have identifiable "TATA" box sequences, usually 28–30 bases upstream of the mRNA 5' end. Most have recognizable variants of "CAT" at about -60 and again around -90 bases, and many are distinguishable by an "AC-rich" string *somewhere* between -90 and -120 bases. It is notable that McKnight and co-workers find this region of the TK genome to be vital to TK expression in transformed cells. A more detailed discussion of comparative sequence analysis from this laboratory is presented in the next section.

The factors controlling βγ and γ gene expression are of great interest. Again, functional ICP4 is required for late gene transcription (Watson and Clements, 1980). Conley *et al.* (1981) reported that a mutation in the major DNA-binding protein (ICP8) reduces the expression of late genes. They noted that the involvement of the ICP8 protein in DNA replication

EARLY (BETA) HSV PROMOTERS

-120   -90   -60   -30   -1

Alkaline exonuclease HSV-1 — Costa et al., (1983) — Transcribed from left to right — Cap at 0.174--2.3kb
CCAACACCCACGGCCTGGCGTATGACGTCC CAGAGGGCATCCGGCGGCCACCTCCGCAATC CCAAGAATTCGGCGCGCGTTTACGGATCGGT GTATAAATTACCAGCACACACACAAGGCGA

Thymidine kinase HSV-1 — McKnight (1980); M. Wagner et al., (1981) — Transcribed right to left — Cap at 0.315--1.3kb
CGCATATTAAGGTGACGCGTGTGGCCTCGA

Thymidine kinase HSV-2 — Reyes et al. (1982) — Cap at 0.315--1.3kb
CTATGATGACAAACCCCGCCCAGCGTCT TGTCATTGGCGAATTCGAAACACGCAGATGC AGTCGGGGGCGGCGGCCGGTCCCAGGTCCACTT CGCATATTAAGGTGACGCGGTGGCCTCGA
CCAGGGATGACGCACACCTCCCAACGTTT TGTCATTGGCGAATTCGAAACACGCAGATGC AGTCTGGGCGGGCGGCCCGAGGTCCACTT

Ribonucleotide reductase HSV-1 — Frink et al. (1981) — Transcribed from left to right — Cap at 0.565--5kb
CATGGAAGAACACACCCCCGACTCAGG ACATCGGCGTGTCCTTTGGGTTCACTGA AACTGGTCCGCGCCCACCCCTGCGCGATG TGGATAAAAAGCCAGCGCGGGTGGTTTGGG

Ribonucleotide reductase HSV-1 — Draper et al. (1982) — Transcribed from left to right — Cap at 0.590--1.3kb
TCGACCATAGCCAATCATGACCCTGTATG TCACGGAGAGAGGCGGACGGGACCCTCCCAG CCTCCACCCTGGTCGCCTTCTGGTCCACG CATATAAGGCGGACTAAAAACAGGGATGT

Ribonucleotide reductase HSV-2 — Swain and Galloway (unpublished) — Cap at 0.590--1.3kb
TTGATCACAGCCAATCCATGACTCTGTATG TCACAGAGAAGGCGGACGGACGCTCCCG CCTCCACCCTGGTCCGCCTTCTCGTCCACG CATATAAGCGGCCTGAAGACGGGGATGT

mRNA encoding 42k protein HSV-1 — Hall et al. (1982) — Transcribed from right to left — Cap at 0.698--3.6(1.3)kb
CCCCTTCTTCGCGGAAACCGAGACCGTTTG GGGGCGTGTCGTTTCTTGGCCCCTGGGGA TTGGTTAGACCCATGGGTTGCCATATATG CACTTCCTATAAGACTCTCCCCACCGCCCC

mRNA encoding 39k protein HSV-1 — Hall et al. (1982) — Transcribed from left to right — Cap at 0.699--1.5kb
CACCCCTCACCCCACACAGGGCGGGTTCAG GCGTGCCCGGCAGCAGTAGCCTCTGGCAG ATCTGACAGACGTGCGATAATACACACG CCCATCGAGGCCATGCTACATAAAAGGGC

gD HSV-1 — Watson et al. (1982) — Transcribed from left to right — Cap at 0.90--2.5kb
GTGTGACACTATCGTCCATACCGACCACAC CGACGAACCCCTAAGGGGGAGGGGCCATTT TACGAGGAGGAGGTATAACAAAGTCTGT CTTTAAAAAGCAGGGGTAGGGAGTTGTTC

FIGURE 4. Comparative analysis of nucleotide sequences upstream of the 5′ ends of several early (β) mRNAs.

and the requirement for DNA replication in late gene expression mean that the effect of the mutation could be either a specific effect of ICP8 or late expression due to a loss of DNA replication.

As there are at least two classes of late genes (βγ and γ), comparative sequence analysis of late promoters to identify class-specific sequences may be complicated. In spite of this caveat, K. Draper *et al.* (unpublished results) compiled enough sequence data upstream from the 5' ends of late mRNAs to make some general points (Fig. 5). First, "TATA" box sequences are usually present about 30 bases 5' of the mRNA start. Second, a rather "AT-rich" (for HSV DNA, at least) "CAT" box consensus is usually found around −60 bases. Further, the "AC-rich" string at −90 to −120 bases seen with β mRNAs is generally missing and a rather "AT-rich" region between ca. −80 and −100 bases is often seen. Three of the best characterized late promoters were more extensively analyzed, as is discussed in the next section.

There are some notable exceptions to the general position of "TATA" boxes in late promoters. One particular late mRNA, the 3.9-kb one, which appears to encode the 50,000-dalton capsid polypeptide, has its 5' end contiguous with an obvious "TATA" box (Costa *et al.*, 1983). Some mRNAs appear to have multiple 5' ends. A 4.6-kb βγ mRNA mapping between 0.64 and 0.67 map units encoding a 70,000-dalton polypeptide was characterized by Hall *et al.* (1982). This mRNA's 5' end was precisely located on a sequence ladder to be 237 bases to the right of a *Sal*I site at 0.674 (into *Sal*I fragment J'). This position, which is fully confirmatory of that found by Hall *et al.* (1982), lies at least 180 bases to the 3' side of the nearest recognizable "TATA" box (K. Draper and E. Wagner, unpublished results). The proximal region upstream of the 5' end does contain several long "T" strings. Such a position could reflect either an atypical promoter, a short but not readily detectable splice at the 5' end of this mRNA, or an artifact of the nuclease digestion due to the high AT content of the DNA in this region.

In view of the large number of HSV promoters, it is certainly not surprising that exceptions to general rules will be found as they have in other systems. Full characterization of both βγ and γ promoters will be facilitated by the development of some types of assayable marker under late control. Such could be a late promoter linked with an assayable enzyme or protein, as suggested by Post *et al.* (1981) and used by J. Smiley and D. Dennis (unpublished results; see above). Construction of such markers is technically feasible.

B. HSV Control Regions as Eukaryotic Promoters

The ease of manipulation of the HSV-1 TK gene and assay of the enzyme have made it a fashionable model for a eukaryotic promoter. In several papers, McKnight and colleagues (McKnight, 1980, 1982; Mc-

LATE HSV PROMOTERS

	-120	-90	-60	-30	-1
mRNA colinear with alk. exo.	CCACCTTCCGTCCCCTGACCCCCACCCC	Transcribed from right to left AGACGACGTCAGCTGTGGACCGAGCTCCC	Cap at 0.15--1.9kb ATTCGCCCGTTAACCCCCACGTGATCAGC	Costa et al., (1983) ACGGCACCGACACCGCAGACGAAAAGCCCC	
mRNA encoding ca. 50k protein	AGCACCCAAGATCCTCGGTTTGTTGGCGCC	Transcribed from right to left TTTATGGCTGCAAAGGCGGCCCACTTGGAA	Cap at 0.186--3.9kb TTGGAGGCGGCGGCTAAAGTCCCGCGCGCGC	Costa, Draper, Wagner (in prep) TTAGAGATGATGGCACAGCGCGCGCACCTGT	
mRNA colinear with above	CCGTACCAACTCGGGGGCCAGGAATTCCAG	Transcribed from right to left CCTGGCCGTGGTGGTCGCCCCGAAATCACG	Cap at 0.188--4.1kb CCCCTTTAGTTGCGCGGCCACCAGGCTATT	Costa, Draper, Wagner (in prep) AAACAGCAGCCGCCGCCACGGCCGAGAA	
Spliced mRNA	ATTTTAACACAGGTCGCGGCTGTGTCCATC	Transcribed from left to right ATCTCTAAGCGCGCGGGACTTTAGCCGC	Cap at 0.187--2.5kb GCCTCCAATTCCAAGTGGGCCGCCTTTGCA	Costa, Draper, Wagner (in prep) GCCATAAAGGCGCCAACAAACCAGAGGATCT	
mRNA encoding 40k protein HSV-1	CCCCCGGCGACCTCCCCGCCCGCTTTTG	Transcribed from right to left CCGACTCCGACGCATTACGTTTTTGACTA	Cap at 0.204--1.4kb CTACAGCACAGCGGAGACACGCTCGCGGCT	Costa, Draper, Wagner (in prep) TAACAATCGTCCAATCGCCGTGGCGATGGA	
mRNA encoding 80k protein HSV-1	CCCCGTCATCGTGATGAAAACGGGCTTGT	Transcribed from right to left GAGGATATAACAAGAACAGGCCGTGGCGTT	Cap at 0.220--3.8kb TGTGTGCGTCACCACCCTCGGCATGTGATC	Costa, Draper, Wagner (in prep) ATCGCATATATAGGTCACCACGTTGAGAAG	
mRNA colinear with above	AAAAAGCTCGTACTGCCTTCCCGTTGTTG	Transcribed from right to left GTGGACGACGAAGATAATCTTGCTGTTG	Cap at 0.219--3.6kb GCCTTGTTGAGAAAGCCCATAATCGTCTGG	Costa, Darper, Wagner (in prep) ACCGCATCCGGCGAATAAAGTTGGCCTCG	
mRNA encoding VP-5 HSV-1	GCTTTGCCGCCTCTGCCAATTTCTTCCTGC	Transcribed from right to left ACGCTTTTGGACCAGGGCCATCTTGAATGC	Cap at 0.265--6kb ACCCGTCGGGTTCTAACGGGGGTGGGCGG	Costa et al., (1984) GGGGGTATATAAGGCCTGGGATCCCACGT	
mRNA encoding gC HSV-1	TTGATATATTTTCAATAAAAGGCATTAGT	Transcribed from left to right CCCAAGACCCGCGGTGTGTGATGATTTCG	Cap at 0.630--2.5kb CCATAACACCCAAACCCGATGGGGCCCG	Frink et al (1983) GGTATAAATTCCGGAAGGGGACACGGGCTA	
mRNA encoding 17.8k protein	CGCATCGGCGGTAACGCGAGACCCCCCGTT	Colinear with 3' end of above ACCTTTTAATATCTATATAGTTTGGTCCC	Cap at 0.640--0.73kb CCCTTCTATCCGCCACCGCTGGGCGGCTA	Frink et al (1983) TAAAGCGCCACCCTCTTCCCTCCCTCAGGTC	
mRNA encoding 65k protein HSV-1	GCGGGAGAACTGCGTTTTTTTGCGCGGC	Transcribed from right to left CCCGTCGCTCCCGTCCAT6TCCATCGCGAG	Cap at 0.675--4.4kb ACCGCGCTGTGGGGTCTTTTTCTTTTTTT	Draper and Wagner (unpublished) CACCGCGTGTGGGGTCTTTTTCTTTTTTT	
mRNA encoding 65k protein HSV-1	GGTTGGACGCCGCCCTCGGTTCGCCTTCA	Transcribed from right to left CGTGACAGGGACAATGTGGGGGAAGTCAC	Cap at 0.688--1.4kb GAGGTACGGGGCGGGGCCCGTGCGGGGTTGCTT	Draper and Wagner (unpublished) AAATGCGGGGTGGCGACCACGGGTGTCAT	
mRNA colinear with early RNA	CGTGAACAATCGCGCAAAGTCAGCCGGCA	Transcribed from left to right TAGCCATTCGCAGGTCCAGAGAGACGCGGCC	Cap at 0.702--4.0(1.7kb) CGACGGCCCATCCGGAGTCCCGCTGACCT	Hall et al. (1982) TCGGCATAAAGCCACCGCGCGCCGCCTGTTGA	

FIGURE 5. Comparative analysis of nucleotide sequences upstream of the mapped 5' ends of several late (βγ and γ) HSV mRNAs.

Knight and Gavis, 1980; McKnight et al., 1981; McKnight and Kingsbury, 1982) defined three regions of this promoter important in its recognition by unmodified cellular enzymes. There are the "TATA" box sequences (ca. -16 to -32 bases), a region between -61 and -47 bases, and finally, a region between -105 and -80 bases. Sweet et al. (1982) showed that regions of the HSV TK promoter are hypersensitive to nuclease digestion during transcription, as are other putative cellular promoters.

Obviously, one must treat generalizations with caution, yet there are surely some features of HSV promoters that must be recognized by polymerase II. Common sequence features between many classes of promoters have been interpreted as strongly suggestive evidence of common function. Examples for HSV promoters would be "TATA" and "CAT" box sequences (such as they are) that have been noted in the preceding section. Such do not occur in all HSV promoters, however. Other common features may also occur, yet they are not strikingly obvious in casual perusals of given promoters.

Symmetry elements in eukaryotic promoters appear to have an important function in determining promoter "strength." The availability of computer programs for comparative analysis of promoters has been of value in searching for symmetry in identified HSV promoters. E. Wagner (unpublished work) used a matrix comparison program developed by J. Coffin of the Tufts University Medical School Cancer Research Center as a modification of sequencing programs originally developed by Larson and Messing (1982) for the Apple II computer. Matrix analysis allows a base-by-base comparison between any two sequences or their complements to look for regions of repeat or inverse repeats. The criteria of match can be altered at will so that long stretches of perfect match, or of partial match (i.e., 15 of 20 bases, etc.), can be scored.

The DNA sequences 200 bases upstream of the 5' ends of some well-characterized HSV mRNAs were analyzed. These were all the β mRNAs of Fig. 4 and the mRNAs for the HSV-1 155,000-dalton capsid protein (VP5), gC, and the 18,000-dalton protein 3' to the gC translational frame (Fig. 5). A search was conducted for the presence of (1) direct repeats, (2) common sequences, and (3) inverse repeats. This analysis was illuminating as much for the lack of specificity of certain features as for their presence.

No direct repeats longer than 6 or 8 bases are seen. No longer repeats are seen if the mismatch frequency is increased to 15% (i.e., 13 matches of 15 bases, etc.). Such data indicate that the "enhancer" sequences of long perfect repeats seen in the SV40 early promoter or the promoters in the long terminal repeat (LTR) of retroviruses are not present.

Short imperfect repeats (10 of 15 bases in a stretch) are frequently found in HSV promoters. The "AC-rich" region seen in the β promoters about 100 bases upstream of the mRNA 5' end is often, but not always, part of such repeat sequences. Examples of such imperfect repeats in the promoters for the mRNAs encoding HSV-1 alkaline exonuclease and

HSV-1 TK are shown in Fig. 6. It should be noted that in the case of TK, the prototypical β promoter, this symmetry is in the region 50–70 bases upstream of the mRNA cap site, not 100–110 bases.

Although suggestive, this type of symmetry is not confined to promoters. It occurs throughout other regions of the HSV-1 genome and is especially striking in the NH_2-terminal portion of the gC translational reading frame. Therefore, although imperfect direct repeats have been implicated in the expression of β-globin (Dierks et al., 1983) and their generation can increase TK expression (McKnight, 1982), they are not restricted to HSV promoters. Further, the extent of symmetry does not readily correlate with mRNA abundance as there is a very extensive region in the γ HSV-1 gC promoter 110–150 bases upstream of the mRNA cap site, as well as 80–110 bases upstream of the cap site for the β HSV-1 mRNA encoding the 38,000-dalton polypeptide, but a less extensive one in the promoter for the extremely abundant 6-kb mRNA encoding the major capsid protein. It should be further noted that the promoter for the low-abundance 1.2-kb (β) HSV-1 mRNA, encoding the 38,000-dalton protein correlated with ribonucleotide reductase (see above), does not show any striking symmetry at this matching criterion.

The question of whether there are common sequences besides "TATA" and "CAT" boxes in all eukaryotic promoters is an open one. Farrell et al. (1983) suggested that the sequence "GGGGTGTGGCC," its reverse, or its complement is common to eukaryotic promoters in general. E. Wagner (unpublished work) analyzed the promoter sequences described above for the presence of this sequence or its variants using a criterion of 8 bases out of 11 fit. As shown in Table II, the sequence or its variants do occur in most of the viral promoters analyzed. The most frequently occurring variant was the inverse complement. However, this sequence family is not confined to promoter regions. The sequence itself occurs three times in the region stretching from 500 to 200 bases upstream of the cap site of the HSV-1 gC mRNA and four times in the first 500 bases downstream of the cap site for this mRNA. At the other extreme, no variant of the sequence occurs in the promoter region for the mRNA encoding HSV-1 alkaline exonuclease, but its inverse does occur in the structural gene for the protein about 400 bases downstream of the cap site.

Using a match criterion of 8 bases in 10, analysis revealed that the 15-nucleotide sequence, "CAAACCCCGCCCAGC," which occurs 110 bases upstream of the cap site for the HSV-1 TK mRNA, is observed at random locations in the 200-base promoter sequences under study. Again, such occurrences may be significant, but they are not confined to promoter regions. A match at the selected criterion occurs four times in a 100-base stretch of the translational reading frame of the HSV-1 gC mRNA. It should be noted that given the high G + C content of HSV-1 DNA and the high G + C content of the putative common sequences, the observed match frequency is not altogether unexpected.

TABLE II. Number of Occurrences (8 of 11 Minimum fit) of a Putative Common Eukaryotic Promoter Oligonucleotide Element in the 200-Base Promoter Region Upstream of Specific HSV Transcripts

mRNA	GGGGTGTGGCC	CCGGTGTGGGG	GGCCACACCCC	CCCCACACCGG
Alkaline exonuclease (β) (type 1)	0	0	0	0
Major capsid protein (βγ) (type 1)	1	2	0	1
Thymidine kinase (β)				
Type 1	2	0	0	0
Type 2	2	0	1	0
ICP6 (140,000 daltons) (β) (type 1)	0	0	3	1
38,000-dalton early protein (β)				
Type 1	0	1	2	2
Type 2	0	1	3	2
gC (γ)	3	4	1	1
18,000-dalton protein (γ) (type 1)	0	1	1	1
39,000-dalton early protein (β) (type 1)	0	0	4	4

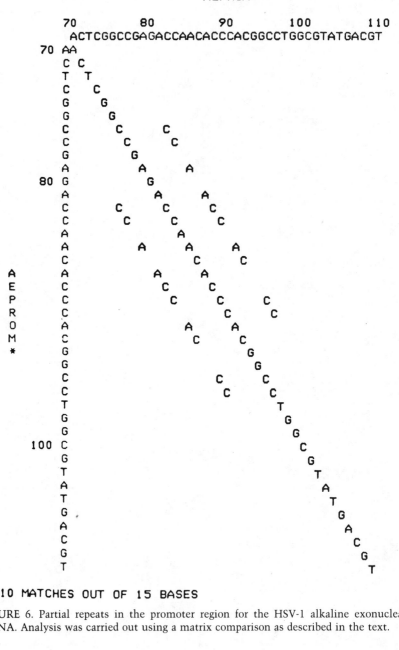

10 MATCHES OUT OF 15 BASES

FIGURE 6. Partial repeats in the promoter region for the HSV-1 alkaline exonuclease mRNA. Analysis was carried out using a matrix comparison as described in the text.

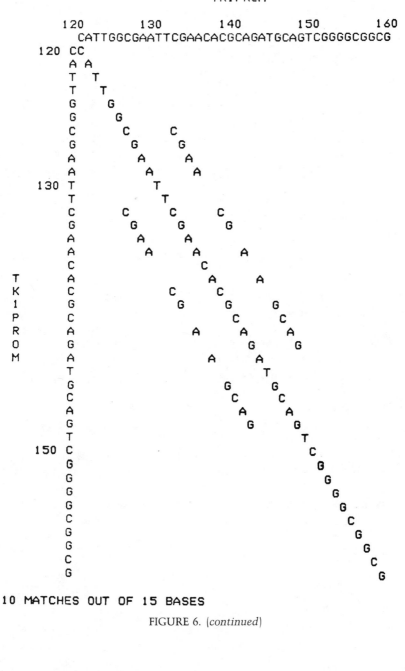

FIGURE 6. (continued)

If the above variations of direct repeats are frequent enough to suggest functional correlation with HSV-1 promoters, a search for lengthy inverse repeats suggests that these are not common features. Setting a match criterion of 9 bases in 10, an inverse repeat sequence occurs in all the sequences of HSV-1 DNA available for analysis at about once per 150–200 nucleotides. This frequency is no higher in the promoter sequences described than it is in any other region of the DNA. This suggests that regulatory interpretations for pallindromes, such as observed by Reyes *et al.* (1982) for the HSV TK genes, should be viewed with some scepticism.

As Mackem and Roizman (1982c) pointed out, many possible self-pairing structures with both high- and low-match criteria are present in HSV DNA upstream of viral α mRNAs. Even with the use of promoter modification analysis, it is not readily apparent that unequivocal assignment of function to any or all of these will be possible. Interpretation of comparative structural data from promoters would be aided by the characterization of highly sequence-specific HSV DNA-binding proteins. It is hoped that such will be forthcoming.

C. Properties of "Typical" HSV Structural Genes

Sufficient sequence data are available to form a picture of a typical HSV gene. The nontranslated leader sequences range from about 80 bases with gD to better than 200 bases for several partial sequences examined (K. Draper, R. Frink, G. Devi, and E. Wagner, unpublished results). A good average is around 150 bases based on data of TK, gC, and the 38,000- to 40,000-dalton protein mRNA encoded around 0.58–0.60 map units in HSV-1.

HSV translation initiation signals are often, but not always, of the canonical form "Pu-base-base-A-U-G-G" identified by Kozak (1981). It is interesting that in the case of HSV-1 TK, which has the sequence "CGUAUGG" as its first translation initiation signal, there is good evidence that TK translation can begin at any one of the first three "AUG" codons in phase at the 5' end of the mRNA (Marsden *et al.*, 1983). The significance of this finding is unclear, but it does demonstrate that assigning a single initiation codon may be an oversimplification.

The high G + C content of HSV DNA is reflected in the high predicted proline content of several HSV proteins. The 38,000-dalton HSV-1 polypeptide encoded by the 1.2-kb β mRNA mapping at 0.58–0.60 map units has a proline content of 5.6%. The content is 5.3% for the corresponding HSV-2 protein. HSV-1 and HSV-2 TK have a proline content of 7.7%. The high proline content may reflect possible function as both HSV-1 gC and gD have very high proline contents (13 and 10.9%, respectively), which suggests that they have a very open configuration. The 18,000-dalton polypeptide encoded by the 1-kb mRNA mapping under the 3' end of the 2.7-kb gC mRNA has a proline content of 7.8%, which

TABLE III. Apparent Codon Use-Frequency Predicted for Several HSV Proteins[a]

Thymidine kinase

	HSV-1				HSV-2			
	T	C	A	G	T	C	A	G
First	4.3	10.6	6.6	11.8	3.8	10.9	5.7	13.0
Second	8.8	10.3	7.8	6.5	8.8	10.7	7.1	6.7
Third	4.5	13.9	3.0	11.9	4.1	14.4	1.8	13.1

HSV-1 gC "family"

	gC				18,000-dalton peptide			
	T	C	A	G	T	C	A	G
First	5.2	9.5	7.6	11.1	5.9	8.7	5.2	13.5
Second	7.4	12.4	6.9	6.7	8.9	9.8	5	9.6
Third	2.7	15.6	2.5	12.5	4.3	15.2	2.6	11.1

38,000-dalton ribonucleotide reductase

	HSV-1				HSV-2			
	T	C	A	G	T	C	A	G
First	6.1	9.1	7.4	10.8	6.1	9.0	6.9	11.3
Second	9.9	8.4	9.7	5.3	10.1	8.6	9.4	5.3
Third	3.4	17.7	2	10.2	2.3	19.2	1	10.8

Glycoprotein D

	T	C	A	G
First	5.1	10.3	7.4	10.5
Second	8.8	9.8	8.8	6
Third	2.8	16.3	2.5	11.7

[a] Based on sequence data referred to in the text.

indicates that the extremely high proline content for gC is confined to that gene and does not reflect an extremely high gC content in the DNA mapping in this region.

Table III is a compilation of an average codon use-frequency for several HSV proteins. This indicates the expected high usage of "GC-containing" triplets. Beyond this, however, it is notable that not all "GC-rich" codons are used at the same frequency. These data may be useful in analysis of potential translation reading frames in HSV DNA sequences.

The length of sequence following the translation terminator codon (trailer) and the actual end of the mRNA can be rather short. It is about

80 bases for HSV-1 TK, and 50 bases for the 1.2-kb mRNA at 0.58–0.60 map units and for the 1-kb β mRNA underlying the 3' end of the mRNA for gC. In many cases, however, the trailer sequence can be extremely long. Such length arises if an mRNA initiates in the interior of another mRNA beyond the translation terminator of the latter. An example is the 5.2-kb β mRNA that overlies the 1.2-kb one. Sequence analysis shows chain terminators in all three reading frames upstream of the 5' end of the 1.2-kb mRNA. Therefore, the trailer for the 5.2-kb mRNA is over 1000 bases. Similar findings have been reported by Clements and Mc-Lauchlan (1981). Of course, if the overlapping mRNA is quite long, very long trailers can occur. The 7-kb γ mRNA whose 5' end is upstream of the 5.2-kb mRNA, would appear to have a trailer of 5-kb in length.

D. Transcription Termination

The sequence "AATAAAA" appears to be a nominal transcription terminator signal for eukaryotic mRNAs (Proudfoot and Brownlee, 1976). This sequence, or a near variant, ends HSV-1 TK mRNA and is found near the 3' end of all the HSV-1 mRNAs examined by workers in this laboratory. In areas of the HSV genome where there is little or no transcriptional read through, the signal is followed by a region of 100–300 bases where there are several areas of very high "AT-rich" sequences. B. Roizman (unpublished results) found that these regions contain short inverted repeat sequences, which are important to efficient polyadenylation. Both these features would tend to "open" the DNA helix, especially in the high G + C environment of the HSV genome. Other signal sequences may also be involved (J. Whitton, F. Rixon, and J. Clements, unpublished results). Often, but not always, efficient polyadenylation regions serve for transcripts from both strands.

VII. GENERATION OF HSV TRANSCRIPT "FAMILIES"

A. Multiple Promoters

The simple model for a typical HSV-1 gene described above suggests that the virus is aptly named in terms of mRNA packaging, if not in biology, genome arrangement, or other parameters. In spite of this simple picture, it is clear from the high-resolution maps of Fig. 3 that overlapping HSV-1 mRNA families encoding the same or different polypeptides are common. Results from this laboratory indicate that a number of mechanisms for the generation of these "families" are operating in HSV transcription (Fig. 7).

One method of generating overlapping mRNAs is by having a second promoter nearby upstream or downstream of the principal one. Such a

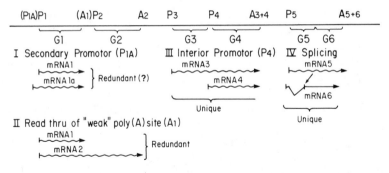

FIGURE 7. Mechanisms for generation of partially overlapping HSV mRNA "families." Details are described in Section VII. (I) Cases are seen where the 5′ ends of overlapping mRNAs map very close to each other (ca. 200–300 bases). The larger mRNAs could encode a unique, small polypeptide or could be redundant with the smaller one. (II) In several instances, mRNAs have coterminal 5′ ends, but the larger continues past the poly(A) site of the shorter and terminates at the next poly(A) site (Anderson *et al.*, 1981; Hall *et al.*, 1982). Such mRNAs should be redundant as the 5′ reading frame would be expected to be translated in both. (III) In many cases, a partially overlapping mRNA will be generated by transcription from a promoter within the transcribed portion of another mRNA. Such mRNAs can encode unique, distinct proteins (Draper *et al.*, 1982; Frink *et al.*, 1983) or proteins utilizing partially overlapping shared reading frames (Costa *et al.*, 1983). (IV) Splicing can generate mRNA families transcribed from the same promoter but varying in the extent of usage of a translational reading frame, or possibly using an alternate reading frame of different phase than the first (Frink *et al.*, 1982).

promoter could have the same or different temporal regulation. If the promoter affected the length of the leader sequence, the two mRNAs would encode the same polypeptide products and would be redundant. Nucleotide sequence data, when analyzed for potential reading frames, indicate that such redundancy can occur. For example, the two approximately 3.8-kb mRNAs mapping between 0.22 and 0.24 map units differ in length by about 150 bases, yet no translation reading frame occurs within 200 bases of the 5′ end of the longer of the two mRNAs. Any biological benefit of such redundancy to HSV is obscure. It should be noted that further investigation of such cases is necessary. Nucleotide sequencing of cDNA transcripts spanning the 5′ end of such mRNAs should be done to ensure that one is not related to the other via a process of differential splicing. Further, it must be established that two such promoters are actually independently functioning and not interactive in some way.

There is no *a priori* reason why such multiple promoters need generate redundant mRNAs, and situations may be found where an upstream promoter leads to an mRNA molecule with additional encoded amino acids or an altered reading frame. On the basis of *in vitro* translation, none of the multiple mRNAs above falls into such a category, but *in vitro* translation can lead to artifacts. Nucleotide sequence analysis has shown that several of these colinear mRNAs with juxtaposed 5' ends could potentially encode different proteins. The best example is found with the colinear γ and βγ mRNAs mapping between 0.67 and 0.7 map units on the HSV-1 genome (Hall *et al.*, 1982). Here, the analysis of reading frames suggests that the larger mRNA could encode a unique 100-amino-acid protein. Another potential example of this is seen where the minor 6-kb mRNA 65 bases to the right of the *Bam*HI site at 0.266 map unit could lead to a unique translation frame encoding about 80 amino acids, while the major start near the *Bam*HI site does not lead to an open reading frame for at least 150 bases. Although nucleotide sequence analysis is suggestive evidence, only the demonstration that two specific proteins are in fact encoded by two nearly colinear mRNAs will validate a biological function for close juxtaposition of HSV promoters.

B. Inefficient Termination

Another mechanism for the generation of multiple mRNAs is found when an mRNA species is inefficiently terminated at a given polyadenylation site. Then, transcription proceeds downstream to the next polyadenylation site. Here, two resolvable mRNA species are found that are colinear at their 5' ends, encode the same polypeptide and translation termination signals, but the largest mRNA contains nontranslated sequences beyond the nominal polyadenylation sites. An example is seen with 1.9- and 7-kb γ mRNA species mapping between 0.55 and 0.60 map units (Anderson *et al.*, 1981). A second example is seen at the 5' end of the colinear mRNAs mapping near the *Bgl*II site at 0.7 map units (Hall *et al.*, 1982). Here, the inefficient polyadenylation signal, along with two close promoters, leads to the rather bizarre situation of having four nearly colinear mRNAs.

The reason for an inefficient polyadenylation near the 3' end of certain mRNAs may lie in the sequence of DNA around the termination region. As discussed in Section VI.D, several sequence features appear to be required for efficient polyadenylation. Also, as the polyadenylation signal appears to be "AATAAAA" or its complement, a "TATA" box resembling this might lead to polyadenylation. Such a speculation requires comparative sequence data for confirmation, but the fact that the "TATA" box for the 5.2-kb mRNA mapping between 0.56 and 0.6 map units is "AAATAAAAA" and maps near the 3' end of the inefficiently terminated 1.9-kb γ mRNA between 0.55 and 0.56 map units is at least suggestive evidence.

C. Interior Promoters

There are many cases where the promoter for an HSV mRNA lies in the interior of another mRNA, but there is no evidence for transcription termination of the latter. One well-characterized example is the protein sharing the COOH-terminal two-thirds of alkaline exonuclease (Costa *et al.*, 1983). This latter peptide could be detected by *in vitro* translation, but the smaller peptide has not been unequivocally identified in infected cells.

D. Splicing

In the case of the mRNA family expressed from the region of the genome encoding HSV-1 gC, Frink *et al.* (1983) clearly demonstrated that splicing can generate overlapping viral mRNAs. The low abundance of such spliced mRNAs has made the assignment of a specific biological role to them very difficult. At present, the best evidence that they do have a function is that there are specific open translation frames, which could be utilized by one or another of the mRNAs. The problem of assigning a function to spliced variants in mRNA families is not confined to the HSV system, as a perusal of the papovavirus and adenovirus literature readily demonstrates. It should be noted, however, that in the case of HSV-1, there are several other regions of the genome that encode a high complexity of mRNAs that could well indicate variable splicing patterns. Further assessment of the importance of this mechanism in HSV mRNA biogenesis awaits detailed investigation of these regions.

E. Do the Proteins Encoded by Overlapping HSV mRNA Families Have Related Functions?

The most obvious case of clustering of related HSV gene functions is the situation of the α genes where all are located in or near repeat sequences. This clustering may reflect features of the mode of expression of these genes or coordinated function. There is no obvious clustering of β, βγ, or γ genes. In the case of related function, such as capsid proteins, or potentially related functions, such as glycoproteins, the genes can be widely dispersed. In the case of the DNA polymerase complex, although the DNA-binding protein (ICP8) and DNA polymerase functions map near each other, the alkaline exonuclease that partially purifies with this complex (Banks *et al.*, 1983) maps 30 kb away (Preston and Cordingley, 1982; Costa *et al.*, 1983). Again, such data suggest that related functions need not be encoded on closely linked genes.

In spite of the demonstrated lack of clustering of many generally related functions, proteins encoded by partially overlapping mRNA families share an antigenic determinant in at least one instance. These are the two overlapping β mRNAs mapping around 0.56–0.59 map units. As outlined above, these encode proteins with at least one shared epitope (Huszar and Bacchetti, 1983; D. Galloway and L. Goldstein, unpublished results). These mRNAs do not, however, appear to share translational reading frames, and they appear to be under individual promoter control. There are no data available suggesting how the two proteins are related, and the large amount of proline present in most HSV proteins (see Section VI-C) could result in a "trivial" explanation for the immunological cross-reactivity.

In one case, it can be seen that polypeptides encoded by overlapping mRNAs do *not* share readily detectable immunological cross-reactivity. The 730-base γ mRNA underlying the 3' end of the 2.7-kb mRNA for HSV-1 gC at 0.640–0.647 encodes an 18,000-dalton polypeptide. Polyclonal antibodies against envelope proteins that efficiently react with gC do not interact with this 18,000-dalton polypeptide (Frink *et al.*, 1983). Further, no antigenic cross-reactivity was seen by Pereira *et al.* (1982a) using monoclonal antibodies. This does not, however, preclude the proteins having some cooperative function in the infected cell.

If a functional relationship between proteins encoded by overlapping mRNA families is demonstrated to be a common feature of HSV gene packaging, then important clues may be available concerning the function of many viral proteins. It will be of interest to determine how mRNAs under separate promoter control could be maintained in physical proximity.

ACKNOWLEDGMENTS. The work originating from this laboratory described here was supported by grants from the National Cancer Institute (CA-11861), the American Cancer Society (MV-159), and the Cancer Research Coordinating Committee of the University of California.

I thank S. Bacchetti, T. Ben-Porat, B. Dutia, N. Fraser, L. Holland, D. Knipe, M. Levine, S. Person, L. I. Pizer, C. M. Preston, F. Rapp, B. Roizman, J. Smiley, M. Zweig, and their colleagues, associates, and co-workers for sharing unpublished and prepublished data with me. Special thanks should go to P. Spear of the University of Chicago, D. Galloway of the Fred Hutchinson Cancer Center, and H. Marsden of the MRC Virology Unit in Glasgow for their critical evaluation of portions of this review. I am also grateful to the series editor for his criticism which strengthened many portions of this review. Similar thanks go to my colleagues who are currently collaborating with me (E. Barquist, G. Bernstein, R. Costa, K. Draper, R. Frink, L. Hall, and M. Rice here at the University of California, Irvine; G. Cohen and R. Eisenberg at the University of Pennsylvania; W. Steinhart at Bowdoin College; and L. Banks and K. Powell at the University of Leeds). I also thank J. Coffin of Tufts University Medical School for his generosity in providing this laboratory with sequence analysis programs for the Apple II computer. Finally, I thank J. W. for manuscript preparation assistance.

REFERENCES

Alwine, J. C., Steinhart, W. L., and Hill, C. W., 1974, Transcription of herpes simplex type 1 DNA in nuclei isolated from infected HEp-2 and KB cells, *Virology* **60:**302.

Anderson, C. W., Lewis, J. B., Atkins, J. F., and Gesteland, R. F., 1974, Cell free synthesis of adenovirus 2 proteins programmed by fractionated messenger RNA: A comparison of polypeptide products and messenger RNA lengths, *Proc. Natl. Acad. Sci. USA* **71:**2756.

Anderson, K. P., Stringer, J., Holland, L., and Wagner, E., 1979, Isolation and localization of herpes simplex virus type 1 mRNA, *J. Virol.* **30:**805.

Anderson, K. P., Costa, R., Holland, L., and Wagner, E., 1980a, Characterization and translation of HSV-1 mRNA abundant in the absence of de novo protein synthesis, *J. Virol.* **34:**9.

Anderson, K. P., Holland, L., Gaylord, B., and Wagner, E., 1980b, Isolation and translation of mRNA encoded by a specific region of the herpes simplex virus type 1 genome, *J. Virol.* **33:**749.

Anderson, K. P., Frink, R., Devi, G., Gaylord, B., Costa, R., and Wagner, E., 1981, Detailed characterization of the mRNA mapping in the HindIII fragment K region of the HSV-1 genome, *J. Virol.* **37:**1011.

Bacchetti, S., and Graham, F., 1977, Transfer of the gene for thymidine kinase to thymidine kinase-deficient human cells by purified herpes simplex viral DNA, *Proc. Natl. Acad. Sci. USA* **74:**1590.

Bachenheimer, S., and Roizman, B., 1972, Ribonucleic acid synthesis in cells infected with herpes simplex virus. VI. Polyadenylic acid sequences in viral messenger ribonucleic acid, *J. Virol.* **10:**875.

Bailey, J. M., and Davidson, N., 1976, Methylmercury as a reversible denaturing agent for agarose gel electrophoresis, *Anal. Biochem.* **70:**75.

Balachandran, N., Harnish, D., Killington, R., Bacchetti, S., and Rawls, W., 1981, Monoclonal antibodies to two glycoproteins of herpes simplex virus type 2, *J. Virol.* **39:**438.

Banks, L., Purifoy, D. J. M., Hurst, P.-F., Killington, R. A., and Powell, K., 1983, Herpes simplex virus nonstructural proteins. 4. Purification of the virus induced exonuclease and characterization of the enzyme using monoclonal antibodies, *J. Gen. Virol.* **64:**2249.

Bartkoski, M. J., Jr., 1982, Polysome-associated proteins in herpes simplex virus-infected cells, *J. Virol.* **43:**357.

Bartkoski, M., Jr., and Roizman, B., 1976, RNA synthesis in cells infected with herpes simplex virus. XIII. Differences in the methylation patterns of viral RNA during the reproductive cycle, *J. Virol.* **20:**583.

Batterson, W., and Roizman, B., 1983, Characterization of the herpes simplex virion associated factor responsible for induction of α genes, *J. Virol.* **46:**371.

Batterson, W., Furlong, D., and Roizman, B., 1983, Molecular genetics of herpes simplex virus. VIII. Further characterization of a temperature-sensitive mutant defective in release of viral DNA and in other stages of the viral reproductive cycle, *J. Virol.* **45:**397.

Bayliss, G. J., Marsden, H. S., and Hay, J., 1975, Herpes simplex virus proteins: DNA-binding proteins in infected cells and in the virus structure, *Virology* **68:**124.

Beck, T., and Millette, R., 1981, In vitro transcription of HSV-1 DNA by RNA polymerase II from HEp-2 cells, in: *International Workshop on Herpesviruses* (A. S. Kaplan, M. La Placa, F. Rapp, and B. Roizman, eds.), p. 62, Esculapio, Bologna, Italy.

Benoist, C., and Chambon, P., 1981, In vivo sequence requirements of the SV40 early promoter region, *Nature* **290:**304.

Benoist, C., O'Hare, K., Breathnach, R., and Chambon, P., 1980, The ovalbumin gene-sequence of putative control regions, *Nucleic Acids Res.* **8:**127.

Ben-Porat, T., Jean, J.-H., and Kaplan, A. S., 1974, Early functions of the genome of herpesvirus. IV. Fate and translation of immediate-early viral RNA, *Virology* **59:**524.

Ben-Zeev, A., and Becker, Y., 1977, Requirement of host cell RNA polymerase II in the replication of herpes simplex virus in alpha-amanitin-sensitive and -resistant cell lines, *Virology* **76:**246.

Berk, A. J., and Sharp, P. A., 1977, Sizing and mapping of early adenovirus mRNAs by gel electrophoresis of S1 endonuclease-digested hybrids, *Cell* **12:**721.

Berk, A. J., Lee, F., Harrison, T., Williams, J., and Sharp, P. A., 1979, Pre-early adenovirus 5 gene product regulates synthesis of early viral messenger RNAs, *Cell* **17:**935.

Betz, J. L., Hill, T. M., Pizer, L. I., Peake, M. L., and Sadler, J. R., 1983, Transcription from the BamHI J fragment of HSV-1 (KOS), *J. Virol.* **47:**238.

Boyd, A. L., Enquist, L., Vande Woude, G. F., and Hampar, B., 1980, Activation of mouse retrovirus by herpes simplex virus type 1 cloned DNA fragments, *Virology* **103:**228.

Braun, D. K., Pereira, L., Norrild, B., and Roizman, B., 1983, Application of denatured, electrophoretically separated, and immobilized lysates of herpes simplex virus-infected cells for detection of monoclonal antibodies and for studies of the properties of viral proteins, *J. Virol.* **46:**103.

Camacho, A., and Spear, P. G., 1978, Transformation of hamster embryo fibroblasts by a specific fragment of the herpes simplex virus genome, *Cell* **15:**993.

Casey, J., and Davidson, N., 1977, Rate of formation and thermal stabilities of RNA:DNA and DNA:DNA duplexes at high concentrations of formamide, *Nucleic Acids Res.* **4:**1539.

Chartrand, P., Stow, N. D., Timbury, M. C., and Wilkie, N. M., 1979, Physical mapping of paa^r mutations of herpes simplex virus type 1 and type 2 by intertypic marker rescue, *J. Virol.* **31:**265.

Chartrand, P., Crumpacker, C. S., Schaffer, P. A., and Wilkie, N. M., 1980, Physical and genetic analysis of the herpes simplex virus DNA polymerase locus, *Virology* **103:**311.

Clements, J. B., and McLauchlan, J., 1981, Control regions involved in the expression of two 3' co-terminal early mRNAs, in: *International Workshop on Herpesviruses* (A. S. Kaplan, M. La Placa, F. Rapp, and B. Roizman, eds.), p. 57, Esculapio, Bologna, Italy.

Clements, J. B., Watson, R. J., and Wilkie, N. M., 1977, Temporal regulation of herpes simplex virus type 1 transcription: Location of transcripts on the viral genome, *Cell* **12:**275.

Clements, J. B., McLauchlan, J., and McGeoch, D. J., 1979, Orientation of herpes simplex virus type 1 immediate early mRNA's, *Nucleic Acids Res.* **7:**77.

Coen, D. M., and Schaffer, P. A., 1980, Two distinct loci confer resistance to acycloguanosine in herpes simplex virus type 1, *Proc. Natl. Acad. Sci. USA* **77:**2265.

Cohen, G., 1972, Ribonucleotide reductase activity in synchronized KB cells infected with herpes simplex virus, *J. Virol.* **9:**408.

Cohen, G., Factor, M., and Ponce de Leon, M., 1974, Inhibition of HSV-2 replication by thymidine, *J. Virol.* **14:**20.

Cohen, G., Long, D., and Eisenberg, R., 1980a, Synthesis and processing of glycoproteins gD and gC of herpes simplex virus type 1, *J. Virol.* **36:**429.

Cohen, G., Ponce de Leon, M., Diggelmann, H., Lawrence, W. C., Vernon, S. K., and Eisenberg, R. J., 1980b, Structural analysis of the capsid polypeptides of herpes simplex virus types 1 and 2, *J. Virol.* **34:**521.

Cohen, J. C., Randall, C. C., and O'Callaghan, D. J., 1975, Transcription of equine herpesvirus type 1: Evidence for classes of transcripts differing in abundance, *Virology* **68:**561.

Cohen, J. C., Perdue, M. L., Randall, C. C., and O'Callaghan, D. J., 1977, Herpesvirus transcription: Altered regulation induced by FUdR, *Virology* **76:**621.

Conley, A. J., Knipe, D. M., Jones, P. C., and Roizman, B., 1981, Molecular genetics of herpes simplex virus. VII. Characterization of a temperature-sensitive mutant produced by in vitro mutagenesis and defective in DNA synthesis and accumulation of gamma polypeptides, *J. Virol.* **37:**191.

Cordingley, M. G., and Preston, C. M., 1981, Transcription and translation of the herpes simplex virus type 1 thymidine kinase gene after microinjection into *Xenopus laevis* oocytes, *J. Gen. Virol.* **54:**409.

Cordingley, M. G., Campbell, M. E., and Preston, C. M., 1983, Functional analysis of a herpes simplex virus type 1 promoter: Identification of far upstream regulatory sequences, *Nucleic Acids Res.* **11:**2347.

Costa, R. H., Devi, B. G., Anderson, K. P., Gaylord, B. H., and Wagner, E. K., 1981, Characterization of a major late herpes simplex virus type 1 mRNA, *J. Virol.* **38**:483.

Costa, R., Draper, K., Banks, L., Powell, K., Cohen, G., Eisenberg, R., and Wagner, E., 1983, High resolution characterization of HSV-1 transcripts encoding alkaline exonuclease and a 50,000 dalton protein tentatively identified as a capsid protein, *J. Virol.* **48**:591.

Costa, R., Cohen, G., Eisenberg, R., Long, D., and Wagner, E., 1984, A direct demonstration that the abundant 6 kb HSV-1 mRNA mapping between 0.23–0.27 encodes the major capsid protein VP-5, *J. Virol.* **49**:285.

Costanzo, F., Campadelli-Fiume, G., Foa-Tomasi, L., and Cassai, E., 1977, Evidence that herpes simplex virus DNA is transcribed by cellular RNA polymerase B, *J. Virol.* **21**:996.

Courtney, R. J., Schaffer, P. A., and Powell, K. L., 1976, Synthesis of virus-specific polypeptides by temperature-sensitive mutants of herpes simplex virus type 1, *Virology* **75**:306.

Cremer, K. J., Summers, W. C., and Gesteland, R. F., 1977, Cell-free synthesis of herpes simplex virus proteins, *J. Virol.* **22**:750.

Cremer, K., Bodemer, M., and Summers, W., 1978, Characterization of the mRNA for herpes simplex virus thymidine kinase by cell-free synthesis of active enzyme, *Nucleic Acids Res.* **5**:2333.

Crumpacker, C. S., Chartrand, P., Subak-Sharpe, J. H., and Wilkie, N. M., 1980, Resistance of herpes simplex virus to acycloguanosine—Genetic and physical analysis, *Virology* **105**:171.

Davison, A. J., Marsden, H. S., and Wilkie, N. M., 1981, One functional copy of the long terminal repeat gene specifying the immediate-early polypeptide 1E 110 suffices for a productive infection of human foetal lung cells by herpes simplex virus, *J. Gen. Virol.* **55**:179.

DeLuca, N., Bzik, D. J., Bond, V. C., Person, S., and Snipes, W., 1982, Nucleotide sequences of herpes simplex virus type 1 (HSV-1) affecting virus entry, cell fusion, and production of glycoprotein gB (VP7), *Virology* **122**:411.

DeMarchi, J. M., Schmidt, C. A., and Kaplan, A. S., 1980, Patterns of transcription of human cytomegalovirus in permissively infected cells, *J. Virol.* **35**:277.

Dierks, P., van Ooyen, A., Cochran, M. D., Dobkin, C., Reiser, J., and Weissman, C., 1983, Three regions upstream from the cap site are required for efficient and accurate transcription of the rabbit β-globin gene in mouse 3T6 cells, *Cell* **32**:695.

Dixon, R. A. F., and Schaffer, P. A., 1980, Fine-structure mapping and functional analysis of temperature-sensitive mutants in the gene encoding the herpes simplex virus type 1 immediate early protein VP175, *J. Virol.* **36**:189.

Dixon, R. A. F., Sabourin, D. J., and Schaffer, P. A., 1983, Genetic analysis of temperature-sensitive mutants which define the genes for the major herpes simplex virus type 2 DNA-binding protein and a new late function, *J. Virol.* **45**:343.

Docherty, J. J., Subak-Sharpe, J. H., and Preston, C. M., 1981, Identification of a virus-specific polypeptide associated with a transforming fragment (BglII-N) of herpes simplex virus type 2 DNA, *J. Virol.* **40**:126.

Draper, K. G., Frink, R. J., and Wagner, E. K., 1982, Detailed characterization of an apparently unspliced β herpes simplex virus type 1 gene mapping in the interior of another, *J. Virol.* **43**:1123.

Dreesman, G. R., Burek, J., Adam, E., Kaufman, R. H., Melnick, J. L., Powell, K. L., and Purifoy, D. J. M., 1980, Expression of herpes virus-induced antigens in human cervical cancer, *Nature* **283**:591.

Duff, R., and Rapp, F., 1971, Oncogenic transformation of hamster cells after exposure to herpes simplex virus type 2, *Nature New Biol.* **233**:48.

Duff, R., and Rapp, F., 1973, Oncogenic transformation of hamster embryo cells after exposure to inactivated herpes simplex virus type 1, *J. Virol.* **12**:209.

Duff, R., and Rapp, F., 1975, Quantitative assay for transformation of 3T3 cells by herpes simplex virus type 2, *J. Virol.* **15**:490.

Dutia, B. M., 1983, Ribonucleotide reductase induced by herpes simplex virus has a virus-specified constituent, *J. Gen. Virol.* **64:**513.

Eberle, R., and Courtney, R. J., 1980, gA and gB glycoproteins of herpes simplex virus type 1: Two forms of a single polypeptide, *J. Virol.* **36:**665.

Eberle, R., and Courtney, R. J., 1982, Multimeric forms of herpes simplex virus type 2 glycoproteins, *J. Virol.* **41:**348.

Eisenberg, R. J., Ponce de Leon, M., and Cohen, G. H., 1980, Comparative structural analysis of glycoprotein gD of herpes simplex virus types 1 and 2, *J. Virol.* **35:**428.

Eisenberg, R. J., Ponce de Leon, M., Pereira, L., Long, D., and Cohen, G. H., 1982, Purification of glycoprotein gD of herpes simplex virus types 1 and 2 by use of monoclonal antibody, *J. Virol.* **41:**1099.

Enquist, L., Madden, M., Schiop-Stansly, P., and Vande Woude, G., 1979, Cloning of herpes simplex type 1 DNA fragments in a bacteriophage lambda vector, *Science* **203:**541.

Farrell, P. J., Deininger, P. L., Bankier, A., and Barrell, B., 1983, Homologous upstream sequences near Epstein–Barr virus promoters, *Proc. Natl. Acad. Sci. USA* **80:**1565.

Feldman, L. T., Imperiale, M. J., and Nevins, J. R., 1982, Activation of early adenovirus transcription by the herpesvirus immediate early gene: Evidence for a common cellular control factor, *Proc. Natl. Acad. Sci. USA* **79:**4952.

Fenwick, M. L., and Clark, J., 1982, Early and delayed shut-off of host protein synthesis in cells infected with herpes simplex virus, *J. Gen. Virol.* **61:**121.

Fenwick, M., Morse, L. S., and Roizman, B., 1979, Anatomy of herpes simplex virus DNA. XI. Apparent clustering of functions affecting rapid inhibition of host DNA and protein synthesis, *J. Virol.* **29:**825.

Flannery, V. L., Courtney, R. J., and Schaffer, P. A., 1977, Expression of an early, nonstructural antigen of herpes simplex virus in cells transformed in vitro by herpes simplex virus, *J. Virol.* **21:**284.

Flint, S., and Broker, T., 1981, Lytic infection by adenoviruses, in: *Molecular Biology of Tumor Viruses,* Part 2, *DNA Tumor Viruses* (J. Tooze, ed.), pp. 443–546, Cold Spring Harbor Laboratory, New York.

Franke, B., Moss, H., Timbury, M., and Hay, J., 1978, Alkaline DNAse activity in cells infected with a temperature-sensitive mutant of HSV type 2, *J. Virol.* **26:**209.

Freeman, M. J., and Powell, K. L., 1982, DNA-binding properties of a herpes simplex virus immediate early protein, *J. Virol.* **44:**1084.

Frenkel, N., and Roizman, B., 1972, Ribonucleic acid synthesis in cells infected with herpes simplex virus. VI. Control of transcription and of RNA abundance, *Proc. Natl. Acad. Sci. USA* **69:**2654.

Frenkel, N., Roizman, B., Cassai, E., and Nahmias, A., 1972, A herpes simplex 2 DNA fragment and its transcription in human cervical cancer tissue, *Proc. Natl. Acad. Sci. USA* **69:**3784.

Frink, R. J., Anderson, K. P., and Wagner, E. K., 1981a, Herpes simplex virus type 1 HindIII fragment L encodes spliced and complementary mRNA species, *J. Virol.* **39:**559.

Frink, R. J., Draper, K. G., and Wagner, E. K., 1981b, Uninfected cell polymerase efficiently transcribes early but not late herpes simplex virus type 1 mRNA, *Proc. Natl. Acad. Sci. USA* **78:**6139.

Frink, R. J., Eisenberg, R., Cohen, G., and Wagner, E. K., 1983, Detailed analysis of the portion of the herpes simplex virus type 1 genome encoding glycoprotein C, *J. Virol.* **45:**634.

Galloway, D., and McDougall, J., 1981, Transformation of rodent cells by a cloned DNA fragment of herpes simplex virus type 2, *J. Virol.* **38:**749.

Galloway, D., and McDougall, J., 1983, The oncogenic potential of herpes simplex viruses: Evidence for a "hit and run" mechanism, *Nature* **302:**21.

Galloway, D. A., and Swain, M., 1980, Cloning of herpes simplex virus type 2 DNA fragments in a plasmid vector, *Gene* **11:**253.

Galloway, D. A., Fenoglio, C., Shevchuk, M., and McDougall, J. K., 1979, Detection of herpes simplex RNA in human sensory ganglia, *Virology* **95:**265.

Galloway, D. A., Fenoglio, C. M., and McDougall, J. K., 1982a, Limited transcription of the herpes simplex virus genome when latent in human sensory ganglia, *J. Virol.* **41**:686.

Galloway, D. A., Goldstein, L. C., and Lewis, J. B., 1982b, Identification of proteins encoded by a fragment of herpes simplex virus type 2 DNA that has transforming activity, *J. Virol.* **42**:530.

Gibson, M. G., and Spear, P. G., 1983, Insertion mutants of herpes simplex virus have a duplication of the glycoprotein D gene and express two different forms of glycoprotein D, *J. Virol.* **48**:396.

Gibson, W., and Roizman, B., 1972, Proteins specified by herpes simplex virus. VIII. Characterization and composition of multiple capsid forms of subtypes 1 and 2, *J. Virol.* **10**:1044.

Gilman, S. C., Docherty, J. J., Clarke, A., and Rawls, W. E., 1980, Reaction patterns of herpes simplex virus type 1 and type 2 proteins with sera of patients with uterine cervical carcinoma and matched controls, *Cancer Res.* **40**:4640.

Goldin, A. L., Sandri-Goldin, R. M., Levine, M., and Glorioso, J. C., 1981, Cloning of herpes simplex virus type 1 sequences representing the whole genome, *J. Virol.* **38**:50.

Graham, F., Bacchetti, S., McKinnon, R., Stanners, C., Cordell, B., and Goodman, H., 1980, Transformation of mammalian cells with DNA using the calcium technique, in: *Introduction of Macromolecules into Viable Mammalian Cells*, (R. Beserga, ed.) pp. 3–25, Liss, New York.

Green, M. T., Courtney, R. J., and Dunkel, E. C., 1981, Detection of an immediate early herpes simplex virus type 1 polypeptide in trigeminal ganglia from latently infected animals, *Infect. Immun.* **34**:987.

Hall, L. M., Draper, K. G., Frink, R. J., Costa, R. H., and Wagner, E. K., 1982, Herpes simplex virus mRNA species mapping in EcoRI fragment I, *J. Virol.* **43**:594.

Halliburton, I., 1980, Intertypic recombinants of herpes simplex virus, *J. Gen. Virol.* **48**:1.

Hampar, B., 1983, Herpes simplex virus type 2 glycoprotein gF and type 1 glycoprotein gC have related antigenic determinants, *J. Virol.* **47**:185.

Hampar, B., Aaronson, S. A., Derge, J. G., Chakrabarty, M., Showalter, S. D., and Dunn, C. Y., 1976, Activation of an endogenous mouse type C virus by ultraviolet irradiated herpes simplex virus types 1 and 2, *Proc. Natl. Acad. Sci. USA* **73**:646.

Hay, J., Moss, H., and Halliburton, I., 1971, Induction of deoxyribonucleic acid polymerase and deoxyribonuclease activities in cells infected with herpes simplex virus type 2, *Biochem. J.* **124**:64.

Heine, J. W., Honess, R. W., Cassai, E., and Roizman, B., 1974, Proteins specified by herpes simplex virus. XII. The virion polypeptides of type-1 strains, *J. Virol.* **14**:640.

Heller, M., van Santen, V., and Kieff, E., 1982, Simple repeat sequence in Epstein–Barr virus DNA is transcribed in latent and productive infections, *J. Virol.* **44**:311.

Herz, C., and Roizman, B., 1983, The α promoter regulator-ovalbumin chimeric gene resident in human cells is regulated like the authentic α 4 gene after infection with herpes simplex virus 1 mutants in α 4 gene, *Cell* **33**:145.

Hill, T. M., Sinden, R. R., and Sadler, J. R., 1983, Herpes simplex virus types 1 and 2 induce shutoff of host protein synthesis by different mechanisms in Friend erythroleukemia cells, *J. Virol.* **45**:241.

Hoffman, P., and Cheng, Y., 1977, The deoxyribonuclease induced after infection of KB cells by herpes simplex virus type 1 or type 2, *J. Biol. Chem.* **253**:3557.

Holland, L. E., Anderson, K. P., Stringer, J. A., and Wagner, E. K., 1979, Isolation and localization of HSV-1 mRNA abundant prior to viral DNA synthesis, *J. Virol.* **31**:447.

Holland, L. E., Anderson, K. P., Shipman, C., Jr., and Wagner, E. K., 1980, Viral DNA synthesis is required for the efficient expression of specific herpes simplex virus type 1 mRNA species, *Virology* **101**:10.

Holland, T., Sandri-Goldin, R., Holland, L. E., Marlin, S. D., Levine, M., and Glorioso, J., 1983, Physical mapping of the mutation in an antigenic variant of HSV type 1 using an immunoreactive plaque assay, *J. Virol.* **46**:649.

Honess, R. W., and Roizman, B., 1974, Regulation of herpesvirus macromolecular synthesis. I. Cascade regulation of the synthesis of three groups of viral proteins, *J. Virol.* **14**:8.

Honess, R. W., and Roizman, B., 1975, Regulation of herpesvirus macromolecular synthesis: Sequential transition of polypeptide synthesis requires functional viral polypeptides, *Proc. Natl. Acad. Sci. USA* **72**:1276.

Hope, R. G., Palfreyman, J., Suh, M., and Marsden, H. S., 1982, Sulphated glycoproteins induced by herpes simplex virus, *J. Gen. Virol.* **58**:399.

Huang, H., Szabocsik, J., Randall, C., and Gentry, G., 1971, Equine abortion (herpes) virus specific RNA, *Virology* **45**:381.

Hummel, M., and Kieff, E., 1982, Epstein–Barr virus RNA. VIII. Viral RNA in permissively infected B95-8 cells, *J. Virol.* **43**:262.

Huszar, D., and Bacchetti, S., 1981, Partial purification and characterization of the ribonucleotide reductase induced by herpes simplex virus infection of mammalian cells, *J. Virol.* **37**:580.

Huszar, D., and Bacchetti, S., 1983, Is ribonucleotide reductase the transforming function of herpes simplex virus 2?, *Nature* **302**:76.

Ikura, K., Betz, J. L., Sadler, J. R., and Pizer, L. I., 1983, RNAs transcribed from a 3.6-kilobase SmaI fragment of the short unique region of the herpes simplex virus type 1 genome, *J. Virol.* **48**:460.

Imperiale, M. J., Feldman, L. T., and Nevins, J. R., 1983, Activation of gene expression by adenovirus and herpesvirus regulatory genes acting in *trans* and by a *cis*-acting adenovirus enhancer element, *Cell* **35**:127.

Jacquemont, B., and Roizman, B., 1975, Ribonucleic acid synthesis in cells infected with herpes simplex virus: Characterization of viral high molecular weight nuclear RNA, *J. Gen. Virol.* **29**:155.

Jacquemont, B., Garcia, A., and Huppert, J., 1980, Nuclear processing of viral high-molecular-weight RNA in cells infected with herpes simplex virus type 1, *J. Virol.* **35**:382.

Jariwalla, R. J., Aurelian, L., and Ts'o, P. O. P., 1980, Tumorigenic transformation induced by a specific fragment of DNA from herpes simplex virus type 2, *Proc. Natl. Acad. Sci. USA* **77**:2279.

Jean, J.-H., Ben-Porat, T., and Kaplan, A. S., 1974, Early functions of the genome of herpesvirus. III. Inhibition of the transcription of the viral genome in cells treated with cycloheximide early during the infective process, *Virology* **59**:516.

Jenkins, F. J., Howett, M. K., Spector, D. J., and Rapp, F., 1982, Detection by RNA blot hybridization of RNA sequences homologous to the BglII-N fragment of herpes simplex virus type 2 DNA, *J. Virol.* **44**:1092.

Jones, P. C., and Roizman, B., 1979, Regulation of herpesvirus macromolecular synthesis. VIII. The transcription program consists of three phases during which both extent of transcription and accumulation of RNA in the cytoplasm are regulated, *J. Virol.* **31**:299.

Jones, P. C., Hayward, G. S., and Roizman, B., 1977, Anatomy of herpes simplex virus DNA. VII. α RNA is homologous to noncontiguous sites in both the L and S components of viral DNA, *J. Virol.* **21**:268.

Keir, H., Hay, J., Morrison, J., and Subak-Sharpe, H., 1966, Altered properties of deoxyribonuclease from cultured cells infected with herpes simplex virus, *Nature* **210**:369.

Keller, J. M., Spear, P. G., and Roizman, B., 1970, The proteins specified by herpes simplex virus. III. Viruses differing in their effects on the social behavior of infected cells specify different membrane glycoproteins, *Proc. Natl. Acad. Sci. USA* **65**:865.

Kieff, E., Hoyer, B., Bachenheimer, S., and Roizman, B., 1972, Genetic relatedness of type 1 and type 2 herpes simplex viruses, *J. Virol.* **9**:738.

Kit, S., and Dubbs, D., 1963, Acquisition of thymidine kinase activity by herpes simplex mouse fibroblast cells, *Biochem. Biophys. Res. Commun.* **11**:55.

Kit, S., Dubbs, D. R., and Schaffer, P. A., 1978, Thymidine kinase activity of biochemically transformed mouse cells after superinfection by thymidine kinase-negative, temperature-sensitive, herpes simplex virus mutants, *Virology* **85**:456.

Knipe, D. M., Ruyechan, W. T., Roizman, B., and Halliburton, I., 1978, Molecular genetics of herpes simplex virus: Demonstration of regions of obligatory and nonobligatory identity within diploid regions of the genome by sequence replacement and insertion, *Proc. Natl. Acad. Sci. USA* **75**:3896.

Knipe, D. M., Batterson, W., Nosal, C., Roizman, B., and Buchan, A., 1981, Molecular genetics of herpes simplex virus. VI. Characterization of a temperature-sensitive mutant defective in the expression of all early viral gene products, *J. Virol.* **38**:539.

Kozak, M., 1981, Possible role of flanking nucleotides in recognition of the AUG initiator codon by eukaryotic ribosomes, *Nucleic Acids Res.* **9**:5233.

Kozak, M., and Roizman, B., 1974, Regulation of herpesvirus macromolecular synthesis: Nuclear retention of nontranslated viral RNA sequences, *Proc. Natl. Acad. Sci. USA* **71**:4322.

Kudler, L., Jones, T. R., Russell, R. J., and Hyman, R. W., 1983, Heteroduplex analysis of cloned fragments of HSV DNAs, *Virology* **124**:86.

Larson, R., and Messing, J., 1982, Apple II software for MI3 shotgun DNA sequencing, *Nucleic Acids Res.* **10**:39.

Lee, G. T.-Y., Para, M. F., and Spear, P. G., 1982a, Location of the structural genes for glycoproteins gD and gE and for other polypeptides in the S component of herpes simplex virus type 1 DNA, *J. Virol.* **43**:41.

Lee, G. T.-Y., Pogue-Geile, K. L., Pereira, L., and Spear, P. G., 1982b, Expression of herpes simplex virus glycoprotein C from a DNA fragment inserted into the thymidine kinase gene of this virus, *Proc. Natl. Acad. Sci. USA* **79**:6612.

Leiden, J. M., Buttyan, R., and Spear, P. G., 1976, Herpes simplex virus gene expression in transformed cells. I. Regulation of the viral thymidine kinase gene in transformed L cells by products of superinfecting virus, *J. Virol.* **20**:413.

Lemaster, S., and Roizman, B., 1980, Herpes simplex virus phosphoproteins. II. Characterization of the virion protein kinase and of the polypeptides phosphorylated in the virion, *J. Virol.* **35**:798.

Leung, W.-C., Dimock, K., Smiley, J. R., and Bacchetti, S., 1980, Herpes simplex virus thymidine kinase transcripts are absent from both nucleus and cytoplasm during infection in the presence of cycloheximide, *J. Virol.* **36**:361.

Lewis, J. B., Atkins, J. F., Anderson, C. W., Baum, P. R., and Gesteland, R. F., 1975, Mapping of late adenovirus genes by cell-free translation of RNA selected by hybridization to specific DNA fragments, *Proc. Natl. Acad. Sci. USA* **72**:1344.

Lewis, J. G., Kucera, L. S., Eberle, R., and Courtney, R. J., 1982, Detection of herpes simplex virus type 2 glycoproteins expressed in virus-transformed rat cells, *J. Virol.* **42**:275.

Lofgren, K. W., Stevens, J. G., Marsden, H. S., and Subak-Sharpe, J. H., 1977, Temperature sensitive mutants of herpes simplex virus differ in the capacity to establish latent infections in mice, *Virology* **76**:440.

McDougall, J. K., Crum, C. P., Fenoglio, C. M., Goldstein, L. C., and Galloway, D. A., 1982, Herpesvirus-specific RNA and protein in carcinoma of the uterine cervix, *Proc. Natl. Acad. Sci. USA* **79**:3853.

Mackem, S., and Roizman, B., 1980, Regulation of herpesvirus macromolecular synthesis: Transcription-initiation sites and domains of α genes, *Proc. Natl. Acad. Sci. USA* **77**:7122.

Mackem, S., and Roizman, B., 1982a, Differentiation between α promoter and regulator region of herpes simplex virus 1: The functional domains and sequence of a movable α regulator, *Proc. Natl. Acad. Sci. USA* **79**:4917.

Mackem, S., and Roizman, B., 1982b, Regulation of α genes of herpes simplex virus: The α 27 gene promoter-thymidine kinase chimera is positively regulated in converted L cells, *J. Virol.* **43**:1015.

Mackem, S., and Roizman, B., 1982c, Structural features of the herpes simplex virus α gene 4, 0, and 27 promoter-regulatory sequences which confer α regulation on chimeric thymidine kinase genes, *J. Virol.* **44**:939.

Macnab, J. C. M., and Timbury, M. C., 1976, Complementation of ts mutants by a herpes simplex virus ts-transformed cell line, *Nature (London)* **261**:233.

McKnight, S. L., 1980, The nucleotide sequence and transcript map of the herpes simplex virus thymidine kinase gene, *Nucleic Acids Res.* **8**:5949.

McKnight, S. L., 1982, Functional relationships between transcriptional control signals of the thymidine kinase gene of herpes simplex virus, *Cell* **31**:355.

McKnight, S. L., and Gavis, E. R., 1980, Expression of the herpes thymidine kinase gene in *Xenopus laevis* oocytes: An assay for the study of deletion mutants constructed in vitro, *Nucleic Acids Res.* **8**:5931.

McKnight, S. L., and Kingsbury, R., 1982, Transcriptional control signals of a eukaryotic protein-coding gene, *Science* **217**:316.

McKnight, S. L., Gavis, E. R., and Kingsbury, R., 1981, Analysis of transcriptional regulatory signals of the HSV thymidine kinase gene: Identification of an upstream control region, *Cell* **25**:385.

Maitland, N. J., and McDougall, J. K., 1977, Biochemical transformation of mouse cells by fragments of herpes simplex virus DNA, *Cell* **11**:233.

Manley, J., Fire, A., Cano, A., Sharp, P., and Gefter, M., 1980, DNA-dependent transcription of adenovirus genes in a soluble whole cell extract, *Proc. Natl. Acad. Sci. USA* **77**:3855.

Marsden, H., Crombie, I., and Subak-Sharpe, J., 1976, Control of protein synthesis in herpesvirus-infected cells: analysis of the polypeptides induced by wild-type and 16 temperature-sensitive mutants of HSV strain 17, *J. Gen. Virol.* **31**:347.

Marsden, H. S., Stow, N. D., Preston, V. G., Timbury, M. C., and Wilkie, N. M., 1978, Physical mapping of herpes simplex virus-induced polypeptides, *J. Virol.* **28**:624.

Marsden, H. S., Lang, J., Davison, A. J., Hope, R. G., and MacDonald, D. M., 1982, Genomic location and lack of phosphorylation of the HSV immediate-early polypeptide IE 12, *J. Gen. Virol.* **62**:17.

Marsden, H., Haarr, L., and Preston, C., 1983, Processing of HSV proteins and evidence that translation of thymidine kinase mRNA is initiated at three separate AUG codons, *J. Virol.* **46**:434.

Mathis, D. J., and Chambon, P., 1981, The SV40 early region TATA box is required for accurate in vitro initiation of transcription, *Nature* **290**:310.

Matz, B., Subak-Sharpe, J. H., and Preston, V. G., 1983, Physical mapping of temperature-sensitive mutations of herpes simplex virus type 1 using cloned restriction endonuclease fragments, *J. Gen. Virol.* **64**:2261.

McDougall, J. K., Galloway, D. A., Purifoy, D. J. M., Powell, K. L., Richart, R. M., and Fenoglio, C. M., 1980, Herpes simplex virus expression in latently infected ganglion cells and in cervical neoplasia, in: *Viruses in Naturally Occurring Cancers*, Cold Spring Harbor Conferences on Cell Proliferation, Cold Spring Harbor Laboratory, New York.

Middleton, M. H., Reyes, G. R., Ciufo, D. M., Buchan, A., Macnab, J. C. M., and Hayward, G. S., 1982, Expression of cloned herpesvirus genes. I. Detection of nuclear antigens from herpes simplex virus type 2 inverted repeat regions in transfected mouse cells, *J. Virol.* **43**:1091.

Minson, A., Bell, S. E., and Bastow, K., 1982, Correlation of the virus sequence content and biological properties of cells carrying the herpes simplex virus type 2 thymidine kinase gene, *J. Gen. Virol.* **58**:127.

Morrison, J., and Keir, H., 1968, A new DNA exonuclease in cells infected with herpesvirus: Partial purification and properties of the enzyme, *J. Gen. Virol.* **3**:337.

Morse, L. S., Pereira, L., Roizman, B., and Schaffer, P. A., 1978, Anatomy of herpes simplex virus (HSV) DNA. X. Mapping of viral genes by analysis of polypeptides and functions specified by HSV-1 × HSV-2 recombinants, *J. Virol.* **26**:389.

Moss, B., Gershowitz, A., Stringer, J. R., Holland, L. E., and Wagner, E. K., 1977, 5'-terminal and internal methylated nucleosides in herpes simplex virus type 1 mRNA, *J. Virol.* **23**:234.

Moss, H., Chartrand, P., Timbury, M., and Hay, J., 1979, Mutant of HSV type 2 with temperature-sensitive lesions affecting virion thermo-stability and DNAse activity: Identification of the lethal mutation and physical mapping of the nuc⁻ lesion, *J. Virol.* **32:**140.

Munyon, W., Kraiselburd, E., Davis, D., and Mann, J., 1971, Transfer of thymidine kinase to thymidine kinaseless L cells by infection with ultraviolet-irradiated herpes simplex virus, *J. Virol.* **7:**813.

Murchie, M. J., and McGeoch, D. J., 1982, DNA sequence analysis of an immediate-early gene region of the herpes simplex virus type 1 genome (map coordinates 0.950 to 0.978), *J. Gen. Virol.* **62:**1.

Murray, B. K., Benyesh-Melnick, M. N., and Biswal, N., 1974, Early and late viral-specific polyribosomal RNA in herpes virus-1 and -2-infected rabbit kidney cells, *Biochim. Biophys. Acta* **361:**209.

Nahmias, A. J., Josey, W. E., Naib, Z. M., Luce, C. F., and Guest, B. A., 1970, Antibodies to herpesvirus hominis types 1 and 2 in humans. II. Women with cervical cancer, *Am. J. Epidemiol.* **91:**547.

Naib, Z. M., Nahmias, A. J., and Josey, W. E., 1966, Cytology and histopathology of cervical herpes simplex infection, *Cancer* **19:**1026.

Nishioka, Y., and Silverstein, S., 1977, Degradation of cellular mRNA during HSV infection, *Proc. Natl. Acad. Sci. USA* **74:**2370.

Nishioka, Y., and Silverstein, S., 1978, Requirement of protein synthesis for the degradation of host mRNA in Friend erythroleukemia cells infected with herpes simplex virus type 1, *J. Virol.* **27:**619.

Notarianni, E. L., and Preston, C. M., 1982, Activation of cellular stress protein genes by herpes simplex virus temperature-sensitive mutants which overproduce immediate early polypeptides, *Virology* **123:**113.

Noyes, B. E., and Stark, G. R., 1975, Nucleic acid hybridization using DNA covalently coupled to cellulose, *Cell* **5:**301.

Oakes, J. E., Hyman, R. W., and Rapp, F., 1976, Genome location of polyadenylated transcripts of herpes simplex virus type 1 and type 2 DNA, *Virology* **75:**145.

Para, M. F., Goldstein, L., and Spear, P. G., 1982, Similarities and differences in the Fc-binding glycoprotein (gE) of herpes simplex virus types 1 and 2 and tentative mapping of the viral gene for this glycoprotein, *J. Virol.* **41:**137.

Para, M. F., Zezulak, K. M., Conley, A. J., Weinberger, M., Snitzer, K., and Spear, P. G., 1983, Use of monoclonal antibodies against two 75,000-molecular-weight glycoproteins specified by herpes simplex virus type 2 in glycoprotein identification and gene mapping, *J. Virol.* **45:**1223.

Parris, D. S., Dixon, R. A. F., and Schaffer, P. A., 1980, Physical mapping of herpes simplex virus type 1 ts mutants by marker rescue: Correlation of the physical and genetic maps, *Virology* **100:**275.

Pedersen, M., Talley-Brown, S., and Millette, R. L., 1981, Gene expression of herpes simplex virus. III. Effect of arabinosyladenine on viral polypeptide synthesis, *J. Virol.* **38:**712.

Pellicer, A., Robins, D., Wold, B., Sweet, R., Jackson, J., Lowy, I., Roberts, J. M., Sim, G. K., Silverstein, S., and Axel, R., 1980, Altering genotype and phenotype by DNA-mediating gene transfer, *Science* **209:**1414.

Pereira, L., Wolff, M., Fenwick, M., and Roizman, B., 1977, Regulation of herpes macromolecular synthesis. V. Properties of α polypeptides made in HSV-1 and HSV-2 infected cells, *Virology* **77:**733.

Pereira, L., Dondero, D., Gallo, D., Devlin, V., and Woodie, J., 1982a, A serological analysis of herpes simplex virus types 1 and 2 with monoclonal antibodies, *Infect. Immun.* **35:**363.

Pereira, L., Dondero, D., and Roizman, B., 1982b, Herpes simplex virus glycoprotein gA/B: Evidence that the infected Vero cell products comap and arise by proteolysis, *J. Virol.* **44:**88.

Pizer, L. I., and Beard, P., 1976, The effect of herpes virus infection on mRNA in polyoma virus-transformed cells, *Virology* **75:**477.

Poffenberger, K. L., Tabares, E., and Roizman, B., 1983, Characterization of a viable, non-inverting herpes simplex virus 1 genome derived by insertion and deletion of sequences at the junction of components L and S, *Proc. Natl. Acad. Sci. USA* **80**:2690.

Ponce de Leon, M., Eisenberg, R., and Cohen, G., 1977, Ribonucleotide reductase from herpes simplex virus (type 1 and 2) infected and uninfected KB cells: Properties of the partially purified enzymes, *J. Gen. Virol.* **36**:163.

Post, L. E., and Roizman, B., 1981, A generalized technique for deletion of specific genes in large genomes: α gene 22 of herpes simplex virus 1 is not essential for growth, *Cell* **25**:227.

Post, L. E., Conley, A. J., Mocarski, E. S., and Roizman, B., 1980, Cloning of reiterated and nonreiterated herpes simplex virus 1 sequences as *Bam*HI fragments, *Proc. Natl. Acad. Sci. USA* **77**:4201.

Post, L. E., Mackem, S., and Roizman, B., 1981, Regulation of α genes of herpes simplex virus: Expression of chimeric genes produced by fusion of thymidine kinase with α gene promoters, *Cell* **24**:555.

Post, L. E., Norrild, B., Simpson, T., and Roizman, B., 1982, Chicken ovalbumin gene fused to a herpes simplex α promoter and linked to a thymidine kinase gene is regulated like a viral gene, *Mol. Cell. Biol.* **2**:233.

Powell, K. L., and Purifoy, D. J. M., 1977, Nonstructural proteins of herpes simplex virus. I. Purification of the induced DNA polymerase, *J. Virol.* **24**:618.

Powell, K. L., Purifoy, D. J. M., and Courtney, R. J., 1975, The synthesis of herpes simplex virus proteins in the absence of virus DNA synthesis, *Biochem. Biophys. Res. Commun.* **66**:262.

Powell, K. L., Littler, E., and Purifoy, D. J. M., 1981, Nonstructural proteins of herpes simplex virus. II. Major virus-specific DNA-binding protein, *J. Virol.* **39**:894.

Preston, C. M., 1977, Cell-free synthesis of herpes simplex virus-coded pyrimidine deoxy-ribonucleoside kinase enzyme, *J. Virol.* **23**:455.

Preston, C. M., 1979a, Control of herpes simplex virus type 1 mRNA synthesis in cells infected with wild-type virus or the temperature-sensitive mutant tsK, *J. Virol.* **29**:275.

Preston, C. M., 1979b, Abnormal properties of an immediate early polypeptide in cells infected with the herpes simplex virus type 1 mutant, *J. Virol.* **32**:357.

Preston, C. M., and Cordingley, M. G., 1982, mRNA- and DNA-directed synthesis of herpes simplex virus-coded exonuclease in *Xenopus laevis* oocytes, *J. Virol.* **43**:386.

Preston, C. M., and McGeoch, D. J., 1981, Identification and mapping of two polypeptides encoded within the herpes simplex virus type 1 thymidine kinase gene sequences, *J. Virol.* **38**:593.

Preston, V. G., 1981, Fine-structure mapping of herpes simplex virus type 1 temperature-sensitive mutations within the short repeat region of the genome, *J. Virol.* **39**:150.

Preston, V. G., Davison, A. J., Marsden, H. S., Timbury, M. C., Subak-Sharpe, J. H., and Wilkie, N. M., 1978, Recombinants between herpes simplex virus types 1 and 2: Analyses of genome structures and expression of immediate early polypeptides, *J. Virol.* **28**:499.

Preston, V. G., Coates, J. A. V., and Rixon, F. J., 1983, Identification and characterization of a herpes simplex virus gene produce required for encapsidation of virus DNA, *J. Virol.* **45**:1056.

Proudfoot, N., and Brownlee, G., 1976, 3' noncoding region sequences in eucaryotic mRNA, *Nature* **263**:211.

Purifoy, D. J. M., and Powell, K. L., 1976, DNA-binding proteins induced by herpes simplex virus type 2 in HEp-2 cells, *J. Virol.* **19**:717.

Purifoy, D. J. M., Lewis, R., and Powell, K., 1977, Identification of the herpes simplex virus DNA polymerase gene, *Nature (London)* **269**:621.

Rafield, L., and Knipe, D. 1984, Characterization of the major mRNAs transcribed from the genes for the glycoprotein gB and DNA-binding protein ICP-8 of HSV-1, *J. Virol.* **49**:960.

Rakusanova, T., Ben-Porat, T., Himeno, M., and Kaplan, A., 1971, Early functions of the genome of herpesvirus. I. Characterization of RNA synthesized in cycloheximide-treated, infected cells, *Virology* **46**:877.

Rapp, F., and Li, J.-L. H., 1974, Demonstration of the oncogenic potential of herpes simplex viruses and human cytomegalovirus, *Cold Spring Harbor Symp. Quant. Biol.* **39**:747.

Rapp, F., and Westmoreland, D., 1976, Cell transformation by DNA-containing viruses, *Biochim. Biophys. Acta* **458**:167.

Read, G. S., and Frenkel, N., 1983, Herpes simplex virus mutants defective in the virion-associated shutoff of host polypeptide synthesis and exhibiting abnormal synthesis of α (immediate early) viral polypeptides, *J. Virol.* **46**:498.

Read, G. S., and Summers, W. C., 1982, In vitro transcription of the thymidine kinase gene of herpes simplex virus, *Proc. Natl. Acad. Sci. USA* **79**:5215.

Reed, C. L., and Rapp, F., 1976, Induction of murine p30 by superinfecting herpesviruses, *J. Virol.* **19**:1028.

Reed, C. L., Cohen, G. H., and Rapp, F., 1976, Detection of a virus-specific antigen on the surface of herpes simplex virus-transformed cells, *J. Virol.* **15**:668.

Reyes, G., LaFemina, R., Hayward, S. D., and Hayward, G. S., 1979, Morphological transformation by DNA fragments of human herpesviruses: Evidence for two distinct transforming regions in herpes simplex virus types 1 and 2, and lack of correlation with biochemical transfer of the thymidine kinase gene, *Cold Spring Harbor Symp. Quant. Biol.* **44**:629.

Reyes, G. R., Jeang, K.-T., and Hayward, G. S., 1982, Transfection with the isolated herpes simplex virus thymidine kinase genes. I. Minimal size of the active fragments from HSV-1 and HSV-2, *J. Gen. Virol.* **62**:191.

Rixon, F. J., and Clements, J. B., 1982, Detailed structural analysis of two spliced HSV-1 immediate-early mRNAs, *Nucleic Acids Res.* **10**:2241.

Rock, D. L., and Fraser, N. W., 1983, Detection of HSV-1 genome in central nervous system of latently infected mice, *Nature* **302**:523.

Ruyechan, W. T., Morse, L. S., Knipe, D. M., and Roizman, B., 1979, Molecular genetics of herpes simplex virus. II. Mapping of the major viral glycoproteins and of the genetic loci specifying the social behavior of infected cells, *J. Virol.* **29**:677.

Sandri-Goldin, R. M., Goldin, A., Holland, L., Glorioso, J., and Levine, M., 1983, Expression of HSV β and γ genes integrated in mammalian cells and their induction by an α gene product, *Mol. Cell. Biol.* **3**:2028.

Sasaki, Y., Sasaki, R., Cohen, G. H., and Pizer, L. I., 1974, RNA polymerase activity and inhibition in herpesvirus-infected cells, *Intervirology* **3**:147.

Sharp, J. A., Wagner, M. J., and Summers, W. C., 1983, Transcription of herpes simplex virus genes in vivo: Overlap of a late promoter with the 3' end of the early thymidine kinase gene, *J. Virol.* **45**:10.

Showalter, S. D., Zweig, M., and Hampar, B., 1981, Monoclonal antibodies to herpes simplex virus type 1 proteins, including the immediate-early protein ICP 4, *Infect. Immun.* **34**:684.

Silverstein, S., Bachenheimer, S., Frenkel, N., and Roizman, B., 1973, Relationship between post-transcriptional adenylation of herpesvirus RNA and mRNA abundance, *Proc. Natl. Acad. Sci. USA* **70**:2101.

Silverstein, S., Millette, R., Jones, P., and Roizman, B., 1976, RNA synthesis in cells infected with herpes simplex virus. XII. Sequence complexity and properties of RNA differing in extent of adenylation, *J. Virol.* **18**:977.

Smiley, J. R., Swan, H., Pater, M. M., Pater, A., and Halpern, M. E., 1983, Positive control of the herpes simplex virus thymidine kinase gene requires upstream DNA sequences, *J. Virol.* **47**:301.

Spang, A. E., Godowski, P. J., and Knipe, D. M., 1983, Characterization of herpes simplex virus 2 temperature-sensitive mutants whose lesions map in or near the coding sequences for the major DNA-binding protein, *J. Virol.* **45**:332.

Spear, P., and Roizman, B., 1980, Herpes simplex virus, in: *Molecular Biology of Tumor Viruses*, Part II, *DNA Tumor Viruses* (J. Tooze, ed.), 2nd ed., pp. 615–746, Cold Spring Harbor Laboratory, New York.

Stenberg, R. M. and Pizer, L. I., 1982, Herpes simplex virus-induced changes in cellular and adenovirus RNA metabolism in an adenovirus type 5-transformed human cell line, *J. Virol.* **42**:474.

Stevens, J. G., 1975, Latent herpes simplex virus and the nervous system, *Curr. Top. Microbiol. Immunol.* **70**:31.

Stevens, J. G., 1980, Herpetic latency and re-activation, in: *Oncogenic Herpesviruses*, Vol. II (F. Rapp, ed.), pp. 1–17, CRC Press, Boca Raton, Fla.

Stevens, J. G., and Cook, M., 1973, Latent herpes simplex virus infection, in: *Virus Research, Second ICN–UCLA Symposium of Molecular Biology* (C. Fox and W. Robinson, eds.), pp. 437–446, Academic Press, New York.

Stinski, M., Thomsen, D., Stenberg, R. M., and Goldstein, L., 1983, Organization and expression of the immediate-early genes of human cytomegalovirus, *J. Virol.* **46**:1.

Stow, N., Subak-Sharpe, J., and Wilkie, N. M., 1978, Physical mapping of HSV-1 mutations by marker rescue, *J. Virol.* **28**:182.

Stringer, J., Holland, L., Swanstrom, R., Pivo, K., and Wagner, E., 1977, Quantitation of herpes simplex virus type 1 RNA in infected HeLa cells, *J. Virol.* **21**:889.

Stringer, J. R., Holland, L. E., and Wagner, E. K., 1978, Mapping early transcripts of herpes simplex virus type 1 by electron microscopy, *J. Virol.* **29**:56.

Strnad, B. C., and Aurelian, L., 1976, Proteins of herpesvirus type 2. II. Studies demonstrating a correlation between a tumor-associated antigen (AG-4), *Virology* **73**:244.

Strobel-Fidler, M., and Franke, B., 1980, Alkaline deoxyribonuclease induced by HSV type 1: Composition and properties of the purified enzyme, *Virology* **103**:493.

Suh, M., 1982, Characterization of a polypeptide present in herpes simplex virus type 2-transformed and -infected hamster embryo cells, *J. Virol.* **41**:1095.

Suh, M., Kessous, A., Poirier, N., and Simard, R., 1980, Immunoprecipitation of polypeptides from hamster embryo cells transformed by herpes simplex virus type 2, *Virology* **104**:303.

Swain, M. A., and Galloway, D. A., 1983, The nucleotide sequence of the HSV-2 thymidine kinase gene, *J. Virol.* **46**:1045.

Swanstrom, R., and Wagner, E., 1974, Regulation of synthesis of herpes simplex type 1 virus mRNA during productive infection, *Virology* **60**:522.

Swanstrom, R., Pivo, K., and Wagner, E., 1975, Restricted transcription of the herpes simplex virus genome occurring early after infection and in the presence of metabolic inhibitors, *Virology* **66**:140.

Sweet, R. W., Chao, M. V., and Axel, R., 1982, The structure of the thymidine kinase gene promoter: Nuclease hypersensitivity correlates with expression, *Cell* **31**:347.

Sydiskis, R. J., and Roizman, B., 1967, The disaggregation of host polyribosomes in productive and abortive infection with herpes simplex virus, *Virology* **32**:678.

van Santen, V., Cheung, A., Hummel, M., and Kieff, E., 1983, RNA encoded by the 1R1-U2 region of Epstein–Barr virus DNA in latently infected, growth transformed cells, *J. Virol.* **46**:424.

Vernon, S. K., Ponce de Leon, M., Cohen, G. H., Eisenberg, R. J., and Rubin, B. A., 1981, Morphological components of herpesvirus. III. Localization of herpes simplex virus type 1 nucleocapsid polypeptides by immune electron microscopy, *J. Gen. Virol.* **54**:39.

Wagner, E., 1972, Evidence of transcriptional control of the herpes simplex virus genome in infected human cells, *Virology* **47**:502.

Wagner, E., 1983, Transcription patterns in HSV infections, in: *Advances in Viral Oncology*, Vol. III (G. Klein, ed.), pp. 239–270, Raven Press, New York.

Wagner, E., and Roizman, B., 1969a, Ribonucleic acid synthesis in cells infected with herpes simplex virus. I. Patterns of ribonucleic acid synthesis in productively infected cells, *J. Virol.* **4**:36.

Wagner, E., and Roizman, B., 1969b, RNA synthesis in cells infected with herpes simplex virus. II. Evidence that a class of viral mRNA is derived from a high molecular weight precursor synthesized in the nucleus, *Proc. Natl. Acad. Sci. USA* **64**:626.

Wagner, E., Swanstrom, R., and Stafford, M., 1972, Transcription of the herpes simplex virus genome in human cells, *J. Virol.* **10**:675.

Wagner, E., Anderson, K., Costa, R., Devi, G., Gaylord, B., Holland, L., Stringer, J., and Tribble, L., 1981, Isolation and characterization of HSV-1 mRNA, in: *Herpesvirus DNA: Developments in Molecular Virology*, Vol. 1 (Y. Becker, ed.), pp. 45–67, Martinus Nijhoff, The Hague.

Wagner, M. J., Sharp, J. A., and Summers, W. C., 1981, Nucleotide sequence of the thymidine kinase gene of herpes simplex virus type 1, *Proc. Natl. Acad. Sci. USA* **78**:1441.

Ward, R. L., and Stevens, J. G., 1975, Effect of cytosine arabinoside on viral-specific protein synthesis in cells infected with herpes simplex virus, *J. Virol.* **15**:71.

Wathem, M., Thomsen, D., and Stinski, M., 1981, Temporal regulation of human cytomegalovirus transcription at immediate early and early times after infections, *J. Virol.* **38**:446.

Watson, R. J., and Clements, J. B., 1978, Characterization of transcription-deficient temperature-sensitive mutants of HSV-1, *Virology* **91**:364.

Watson, R. J., and Clements, J. B., 1980, A herpes simplex virus type 1 function continuously required for early and late virus RNA synthesis, *Nature* **285**:329.

Watson, R. J., Preston, C. M., and Clements, J. B., 1979, Separation and characterization of herpes simplex virus type 1 immediate-early mRNA's, *J. Virol.* **31**:42.

Watson, R. J., Sullivan, M., and Van de Woude, G. F., 1981a, Structures of two spliced herpes simplex virus type 1 immediate-early mRNA's which map at the junctions of the unique and reiterated regions of the virus DNA S component, *J. Virol.* **37**:431.

Watson, R. J., Umene, K., and Enquist, L. W., 1981b, Reiterated sequences within the intron of an immediate-early gene of herpes simplex virus type 1, *Nucleic Acids Res.* 30:4189.

Watson, R. J., Weis, J. H., Salstrom, J. S., and Enquist, L. W., 1982, Herpes simplex virus type-1 glycoprotein D gene: Nucleotide sequence and expression in *Escherichia coli*, *Science* **218**:381.

Watson, R. J., Colberg-Poley, A. M., Marcus-Sekura, C. J., Carter, B. J., and Enquist, L. W., 1983, Characterization of the herpes simplex virus type 1 glycoprotein D mRNA and expression of this protein in *Xenopus* oocytes, *Nucleic Acids Res.* **11**:1507.

Weller, S. K., Lee, K. J., Sabourin, D. J., and Schaffer, P. A., 1983, Genetic analysis of temperature-sensitive mutants which define the gene for the major herpes simplex virus type 1 DNA-binding protein, *J. Virol.* **45**:354.

Wigdahl, B. L., Scheck, A., DeClercq, E., and Rapp, F., 1982, High efficiency latency and activation of herpes simplex virus in human cells, *Science* **217**:1145.

Wigdahl, B. L., Ziegler, R. J., Sneve, M., and Rapp, F., 1983, Herpes simplex virus latency and reactivation in isolated rat sensory neurons, *Virology* **127**:159.

Wigler, M., Silverstein, S., Lee, L.-S., Pellicer, A., Cheng, Y.-C., and Axel, R., 1977, Transfer of purified herpes virus thymidine kinase gene to cultured mouse cells, *Cell* **11**:223.

Wohlrab, F., Garrett, B. K., and Francke, B., 1982, Control of expression of the herpes simplex virus-induced deoxypyrimidine triphosphatase in cells infected with mutants of herpes simplex virus types 1 and 2 and intertypic recombinants, *J. Virol.* **43**:935.

Wright, S., deBoer, E., Grosveld, F. G., and Flavell, R. A., 1983, Regulated expression of the human β-globin gene family in murine erythroleukemia cells, *Nature* **305**:333.

Youssoufian, H., Hammer, S. M., Hirsch, M. S., and Mulder, C., 1982, Methylation of the viral genome in an in vitro model of herpes simplex virus latency, *Proc. Natl. Acad. Sci. USA* **79**:2207.

Zezulak, K. M., and Spear, P. G., 1983, Characterization of a herpes simplex virus type 2 75,000-molecular-weight glycoprotein antigenically related to herpes simplex virus type 1 glycoprotein C, *J. Virol.* **47**:553.

Zezulak, K. M., and Spear, P. G., 1984, Mapping of the structural gene for the HSV-2 counterpart of HSV-1 glycoprotein C and identification of a type-2 mutant which does not express this protein, *J. Virol.* **49**:741.

Zweig, M., Heilman, C. J., Jr., Rabin, H., and Hampar, B., 1980, Shared antigenic determinants between two distinct classes of proteins in cells infected with herpes simplex virus, *J. Virol.* **35**:644.
Zweig, M., Showalter, S. D., Bladen, S. V., Heilman, C. J., and Hampar, B., 1983, Herpes simplex virus type 2 glycoprotein gF and type 1 glycoprotein C have related antigenic determinants, *J. Virol.* **47**:185.

CHAPTER 3

Molecular Biology of Pseudorabies Virus

Tamar Ben-Porat and Albert S. Kaplan

I. INTRODUCTION

Pseudorabies virus (PRV) causes a natural infection in swine similar to that of herpes simplex virus (HSV) in man, and B virus in monkeys. Sabin (1934) was the first to recognize that these viruses are naturally related. As Sabin put it, "Their generic relationship may perhaps be based on the following properties, which, taken together, are not possessed by any other virus: (a) they are pantropic, i.e., they have affinities for cells derived from all embryonic layers; (b) they are neuroinvasive, i.e., they are capable of invading the central nervous system from a peripheral focus; (c) they give rise to a similar intranuclear body; (d) they have a similar, though not identical, range of susceptible hosts; and (e) they appear to have a partial immunological relationship." Later studies established that PRV is indeed a herpesvirus, as determined by a variety of morphological, biological, and biochemical characteristics. To this may be added the fact, established by Shope (1935), that PRV is latent in swine, its natural host, as HSV is in man.

The formal taxonomic name given to PRV is Suid herpesvirus 1 (Matthews, 1982). It belongs to the subfamily Alphaherpesvirinae.

This review will emphasize the biochemical and molecular events occurring during the interaction of PRV with its host. Some aspects of the morphology and growth cycle of the virus, as well as a description of the diseases caused by PRV, are briefly described; reviews on these aspects of the interactions of PRV with its hosts are available, (Kaplan,

TAMAR BEN-PORAT and ALBERT S. KAPLAN • Department of Microbiology, Vanderbilt University School of Medicine, Nashville, Tennessee 37232.

1969, 1973; Baskerville *et al.*, 1973; Gustafson, 1975). For the sake of completeness, we have also included in this chapter some unpublished information that is available in our laboratory dealing with the enumeration and the identification of PRV macromolecules, information that has not been published to date because it is of a descriptive nature only.

II. MORPHOLOGY

PRV, after being thin-sectioned or stained negatively, has a diameter of 150–186 nm as measured by electron microscopy. Negative staining with phosphotungstic acid has provided the most reliable estimate of the size of PRV, which appears to be approximately 180 nm in diameter (Kaplan and Vatter, 1959; Reissig and Kaplan, 1962; Felluga, 1963; Darlington and Moss, 1968).

The general appearance of PRV in thin section is similar to that of HSV, which has been studied more extensively. The electron-dense central body in the particle may vary in shape and appears round, oval, or sometimes like a short-ended rod, depending most likely on the angle of the cut during sectioning. The central body lies within a zone of lower density, which, in turn, is surrounded by a double—sometimes triple—membrane. When examined by negative staining with phosphotungstic acid, the typical herpesvirus particle, a nucleocapsid composed of 162 capsomers surrounded by a floppy envelope, is revealed.

III. STABILITY IN THE PRESENCE OF PHYSICAL AND CHEMICAL AGENTS

Because PRV has a lipid-containing envelope, it is sensitive to inactivation by lipid solvents such as ethyl ether (Kaplan and Vatter, 1959). It is inactivated by compounds that react with nucleic acids, such as nitrous acid (Ivaničová *et al.*, 1963); it is also irreversibly photosensitized by ethidium chloride (Content and Cogniaux-LeClerc, 1968). A variety of other compounds, ranging from metallic ions to a number of organic chemicals affecting proteins such as fluorocarbons (Ivaničová, 1961), dithioerythritol (Gainer *et al.*, 1971), urea, and SDS (Golais and Sabó, 1975), have been found, not unexpectedly, to inactivate PRV infectivity.

PRV is relatively thermostable (Kaplan and Vatter, 1959). Some attenuated strains are more heat-resistant than virulent strains of PRV (Bodon *et al.*, 1968; Bartha *et al.*, 1969) but this does not appear to be the case universally (Golais and Sabó, 1975). Furthermore, it has been reported that different strains of PRV vary in their sensitivity to trypsin and that there is a correlation between this characteristic and virulence (Platt *et al.*, 1980).

As expected, PRV is inactivated exponentially by irradiation with UV light (Kaplan, 1962; Pfefferkorn and Coady, 1968; Zavadova and Zavada, 1968; Dilovski, 1969; Ross et al., 1971) and by γ-rays (Kaplan and Ben-Porat, 1966; Sullivan et al., 1971; Sun et al., 1978). UV-irradiated virus can be photoreactivated after infection of chick embryo but not primary rabbit kidney cells, indicating that the nature of the host cell plays a role in this phenomenon (Pfefferkorn et al., 1965, 1966; Pfefferkorn and Coady, 1968).

To avoid inactivation, PRV is best stored in a medium containing protein and kept at −70°C (Kaplan, 1969). The virus retains infectivity after lyophilization (Popescu et al., 1969). The best media for freeze-drying PRV are those containing glutamate, either alone or in combination with sucrose and either dextran or phosphates (Scott and Woodside, 1976).

IV. VIRUS PROTEINS

A. Structural Protein Components

The number of polypeptides that have been identified in preparations of purified PRV not surprisingly reflects the resolution of the system used to separate them. Originally, the minimal number of major virus polypeptides present in purified virions was estimated to be 10 (Kaplan and Ben-Porat, 1970b; Ben-Porat et al., 1970). Subsequently, using techniques that allowed a better resolution, Stevely (1975) showed that at least 20 polypeptides can be resolved from preparations of purified virions.

Using gels of varying concentrations, 27 bands can be detected by one-dimensional polyacrylamide gel electrophoresis (PAGE) and at least 40 unrelated "spots" (probably generated by unrelated polypeptides) can be resolved by two-dimensional gel electrophoresis (T. Ben-Porat, unpublished results). Four of the virion proteins are phosphopropteins and at least four are glycoproteins (Table I). All the glycoproteins of PRV are sulfated (Kaplan and Ben-Porat, 1976). The glycoproteins appear as shadows on the autoradiograms of the ^{3}H-leucine-labeled virion proteins but can be clearly resolved when the virions are labeled with glucosamine (Fig. 1).

Although the structural glycoproteins of PRV migrate as four main bands (Ben-Porat and Kaplan, 1970), several minor bands can also be distinguished. However, because these minor bands are not present in all virion preparations, they may represent either breakdown products or contaminants. Analysis by two-dimensional gels also reveals that virions contain only four major glycoproteins (Hampl et al., 1984).

The simplest capsids that can be isolated from infected cells (capsid I) are composed of three major proteins [142K, 35K, and 32K, which were previously referred to as proteins 2, 7, and 8, respectively (Ben-Porat et al., 1970; Kaplan and Ben-Porat, 1970b)] and one minor protein (62K). These polypeptides are invariably present in all empty capsids isolated from cells infected with all temperature-sensitive (ts) mutants in which

TABLE I. Protein Composition of Virus Particles as Determined by One-Dimensional PAGE[a]

Apparent MW ($\times 10^{-3}$)		Intact virion	Nucleocapsids[b]	Type I capsid[c]	Type II capsid[d]
>200		+[e]	+		
142		+	+	+	+
125	(glyco)	+			
115		+	+		
112	(phospho)	+	+		+
98	(glyco)	+			
85		+	+		+
81		+	+		
74a		+	+		
74b	(glyco)	+			
67		+			
63	(phospho)	+			
62		+	+	+	+
58	(glyco)	+			
57	(phospho)	+			
56		+			
54		+			
41		+	+		+
40		+	+		
39		+			
38		+	+		+
35				+	+
32		+	+	+	+
30		+	+		+
29	(phospho)	+			
27		+	+		+
24		+	+		
10		+	+		

[a] The molecular weights of the PRV proteins were estimated using commercial standards as well as the proteins of HSV-1 and adenovirus, type 2. The values obtained with the three different standards differed somewhat. The indicated molecular weights therefore represent at best an estimate of the size of the PRV proteins.
[b] The nucleocapsids were obtained after deenveloping virions with nonionic detergent.
[c] Type I capsids can be isolated consistently from mutants in which nucleocapsid formation is blocked. Preparations of pure type I capsids (not contaminated with type II capsids) have also occasionally been obtained from wild-type infected cells.
[d] Type II capsids can be isolated from cells infected with wild-type virus. These preparations probably contain a mixture of true capsids and of nucleocapsids that have lost their DNA. The composition of these capsids is similar to the composition of the nucleocapsids isolated from infected cells by Stevely (1975).
[e] +, presence of protein.

capsids are assembled but which are blocked prior to nucleocapsid assembly (Ladin et al., 1982, and T. Ben-Porat, unpublished results). Empty capsid preparations obtained from wild-type infected cells (type II capsid) usually contain several other proteins as well (see Table I and Fig. 1). In general, the ratio of empty capsids to nucleocapsids that can be isolated from wild-type infected cells by sucrose gradients greatly exceeds the ratio determined by examination of thin sections in the electron microscope.

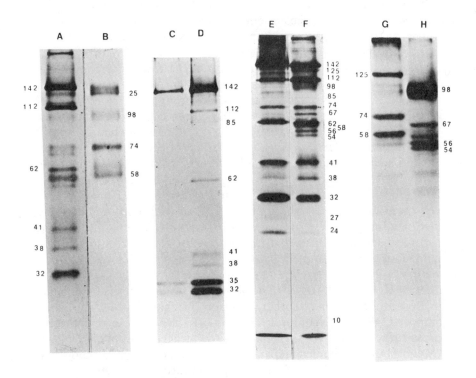

FIGURE 1. SDS-PAGE analysis of the proteins of PRV. Lane A, Complete virions (^{3}H-leucine labeled); lane B, Complete virions (^{3}H-glucosamine labeled); lane C, Capsids I (^{3}H-leucine labeled); lane D, Capsids II (^{3}H-leucine labeled); lane E, Nucleocapsids after de-envelopment of complete virions with Triton X-100 (^{3}H-leucine labeled); lane F, Complete virions (^{3}H-labeled); lane G, Crosslinked complex of glycoproteins (membrane, fraction II) present in the virus membrane (^{3}H-leucine labeled); lane H, Non-crosslinked proteins (membrane, fraction I) present in the virus membrane (^{3}H-leucine labeled). From Ladin *et al.* (1982) and Hampl *et al.* (1984).

It is likely therefore that the empty capsid II preparations (isolated in sucrose gradients) probably consist of a mixture of true empty capsids and of nucleocapsids that have lost their DNA during isolation.

Stevely (1975) examined the protein composition of nucleocapsids isolated from infected cells and showed that these nucleocapsid preparations contain eight proteins. The composition of these nucleocapsids closely resembles that of class II empty capsids isolated from wild-type infected cells (see Table I).

Nucleocapsids can also be recovered after treatment of virions with nonionic detergent (Kaplan and Ben-Porat, 1970b). This treatment removes all the glycoproteins from the virion and the liberated nucleocapsids (which sediment more slowly in sucrose gradients than virions) can be isolated. These nucleocapsids contain a full complement of DNA but have lost all the glycoproteins (Kaplan and Ben-Porat, 1970b), as well as several other proteins (Fig. 1, Table I).

It is of interest that the 35K protein, one of the major components of empty capsids, is greatly underrepresented in mature virions (and in deenveloped nucleocapsids) compared to the other two major capsid proteins (142K, 32K). It can be estimated that empty capsids contain at least 20-fold more 35K protein relative to the 142K protein than do mature virions (Ladin et al., 1982). Thus, either most of the 35K protein is processed and acquires a different mobility during some step leading to the formation of the mature virions or it is degraded (or released) as would be expected if it were a scaffolding protein (Casjens and King, 1975).

As mentioned above, treatment of virions with nonionic detergent removes the membrane proteins from the virions. The solubilized proteins can be separated into two fractions, which differ from each other in their sedimentation and chromatographic behavior. Fraction I consists of one major glycoprotein, 98K, as well as several nonglycosylated virus proteins, 67K, 56K, 54K (Fig. 1). The proteins in this fraction do not appear to be cross-linked to one another. Fraction II behaves as a complex and is composed of three glycoproteins (125K, 74K, 58K). These glycoproteins are cross-linked to one another via disulfide bridges (Hampl et al., 1984).

Monoclonal antibodies prepared against the virus glycoproteins react either with the 98K glycoprotein or coprecipitate, as expected, all three glycoproteins in the cross-linked complex. However, even after treatment of the complex with DTE (which leads to its dissociation) prior to immunoprecipitation, two of the glycoproteins (125K, 74K) are coprecipitated by several independently isolated monoclonal antibodies indicating that these proteins share extensive homologies; furthermore, partial peptide digestion analysis confirms the homology between all three glycoproteins (58K, 74K, 125K) (Hampl et al., 1984).

While most structural proteins in different strains of PRV have similar PAGE migration rates, some differ so that these strains can be distinguished from one another by this procedure. Some differences in the glycoproteins of various field isolates have also been observed (T. Ben-Porat, unpublished results).

It has been reported that like many other enveloped viruses, PRV possesses protein kinase activity (Tan, 1975; W. S. Stevely, M. Katam, and D. P. Leader, personal communication). The activity associated with PRV is, however, relatively low and is only slightly stimulated by the addition of detergents to the virion preparations (Tan, 1975).

B. Virus Neutralization

Monoclonal antibodies that react with the 98K envelope glycoprotein neutralize PRV infectivity. Monoclonal antibodies directed against some of the other membrane glycoproteins have very little neutralizing activity (Hampl *et al.*, 1984).

C. Antigenic Relationship Between PRV and Some Other Herpesviruses

No reciprocal neutralization among antisera against PRV and other herpesviruses has been demonstrated (Kaplan and Vatter, 1959; Watson *et al.*, 1967; Killington *et al.*, 1977). However, Watson *et al.* (1967) showed that antisera against HSV or B virus produced one precipitin line against PRV in agar immunodiffusion tests. This observation was extended by Honess *et al.* (1974) who demonstrated that PRV shares at least two antigens with HSV-1 and HSV-2; Killington *et al.* (1977) showed that PRV shares at least one antigen with bovine mammillitis virus, and that PRV and equine abortion virus share more antigens than either do with the other viruses (Killington *et al.*, 1977). Furthermore, it has been shown that antisera to PRV immune-agglutinate naked capsids but not enveloped particles of HSV, indicating that the common antigens of these viruses are located internally (Honess *et al.*, 1974).

On the basis of cross-reactivity, HSV-1, HSV-2, and bovine mammillitis virus can be placed in one group and PRV and equine abortion virus in another (Yeo *et al.*, 1981). It is of interest that the one antigenic determinant common to all five viruses (which may be considered to be a group-specific antigen) is the major DNA-binding protein.

V. VIRUS GENOME

A. Size and Base Composition

PRV contains 10–11 µg DNA/100 µg protein (Ben-Porat *et al.*, 1975b). The genome of PRV is a linear, duplex DNA molecule; it exhibits hyperchromicity upon heat denaturation and its density is altered after denaturation (Kaplan and Ben-Porat, 1964). The molecular weight of the DNA is approximately 90×10^6, as determined by sedimentation velocity analysis (Ben-Porat *et al.*, 1975b), electron microscopy (Rubenstein and Kaplan, 1975; Stevely, 1977), and restriction enzyme analysis (Ben-Porat and Rixon, 1979). The virus genome has a relatively high G + C content (73 mole%) (Ben-Porat and Kaplan, 1962; Kaplan and Ben-Porat, 1964; Russell and Crawford, 1964; Sydiskis, 1969; Plummer *et al.*, 1969b; Ludwig *et al.*, 1971; Graham *et al.*, 1972; Halliburton, 1972); it is not glycosylated (Russell and Crawford, 1964).

B. Genome Structure

1. General Characteristics

Partial denaturation mapping has shown that the genome of PRV is not circularly permuted (Rubenstein and Kaplan, 1975). Analysis of the genome of PRV by electron microscopy, as well as by restriction enzymes, shows that it contains a sequence of DNA, the short unique (U_S) sequence, with a molecular weight of 6×10^6, that is bracketed by an inverted repeated sequence of a molecular weight of approximately 10×10^6. Another unique sequence, the long unique (U_L) (molecular weight 65×10^6), comprises the remainder of the molecule (Stevely, 1977; Ben-Porat et al., 1979). Thus, examination of self-annealed denatured DNA molecules reveals the presence of a single-stranded loop representing the U_S region, a double-stranded region representing the reannealed inverted repeats, and a long, single-stranded tail representing the U_L region. In addition, a short, single-stranded tail appears on about 20% of the molecules at the junction of the single- and double-stranded regions. The appearance of this tail indicates that in some molecules the terminal sequence of the inverted repeat is not part of the internal inverted repeat (Ben-Porat et al., 1979).

The genome of PRV has been classified as a class 2 herpesvirus DNA molecule (Honess and Watson, 1977) or as a class D molecule (Roizman, 1982). This class of herpesvirus genomes is composed of two sets of unique sequences, U_L and U_S; the latter is bracketed by inverted repeats.

2. Lack of Homology between the Ends of the DNA Molecule

Because the genome of PRV forms circles within the infected cells and replicates in the form of circular and concatemeric molecules (see below), it is noteworthy that no homology between the ends (which would permit circle formation) has been detected.

That terminally redundant sequences, if present, must be rather small was deduced from the fact that the two ends of the virus genome vary in their denaturation patterns (Rubenstein and Kaplan, 1975). Also, no cohesive ends were exposed by exonuclease digestion of the virus DNA (Ben-Porat et al., 1979). Furthermore, no sequence homology between the ends can be detected using probes that should have revealed relatively short stretches of homology (L. Harper and T. Ben-Porat, unpublished results). Sequencing of cloned restriction fragments originating from the ends of the molecules also failed to reveal any sequence homology between the ends (I. Davidson, personal communication).

Interestingly, the genome of PRV is relatively resistant to enzymes that modify the 5′ ends of DNA. Extensive attempts to detect the presence of proteins bound to the ends of the genome proved fruitless (L. Harper and T. Ben-Porat, unpublished results).

3. Heterogeneity of Ends of the DNA Molecule

Some heterogeneity at the end of the U_L sequence has been observed in the genomes of some strains of PRV. Digestion of plaque-purified virus DNA with restriction enzymes generates two main submolar end fragments from the left end of the genome (end of U_L sequence), which differs in length by approximately 150 bp (L. Harper and T. Ben-Porat, unpublished results).

4. The Genome Is Found in Two Isomeric Forms

The U_S sequence of the genome, which is bracketed by the inverted repeats, is found in two orientations relative to the U_L sequence, and the DNA molecules isolated from populations of virions consist of equimolar amounts of these two isomeric forms. Thus, digestion of the DNA with restriction endonucleases that do not cut within the repeated sequences (e.g., *BglII*, see Figs. 2 and 3) produce, in addition to the molar fragments originating entirely from the unique regions of the genome, two half-molar terminal fragments and two half-molar fragments spanning the internal repeat.

5. Minor Repetitions within the Genome

Analysis by electron microscopy of exonuclease-digested, self-annealed PRV DNA reveals different size "lariat" structures, indicating that sequences homologous to the end of the molecules are present within the genome at various distances from the end. Notably, such a sequence is present at approximately 350 nucleotides from the end of the U_L (Ben-Porat *et al.*, 1979).

Fine mapping also reveals the presence of adjacent duplicated sequences present at several locations on the genome. Indeed, the majority of differences in restriction enzyme patterns that can be observed between the genomes of different isolates of PRV may be due to reduplications of restricted regions of the genome (L. Harper and T. Ben-Porat, unpublished results).

6. Single-Stranded Interruptions

PRV sediments heterogeneously in alkaline sucrose gradients, indicating that the DNA contains nicks, gaps, or alkali-sensitive bonds. The size distribution of the fragments generated after denaturation of the DNA in alkali or formamide is the same, thereby eliminating alkali-sensitive bonds as the reason for the fragmentation of the DNA upon denaturation (Ben-Porat *et al.*, 1979).

Analyses of the size distribution of the fragments produced after denaturation of the DNA, as well as of the number and positions on the DNA

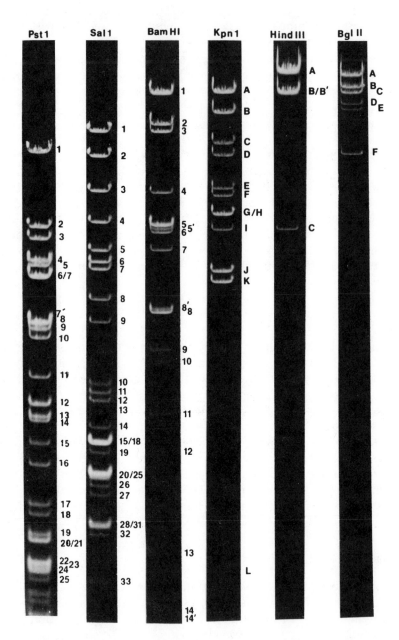

FIGURE 2. Restriction enzyme cleavage patterns of the PRV genome.

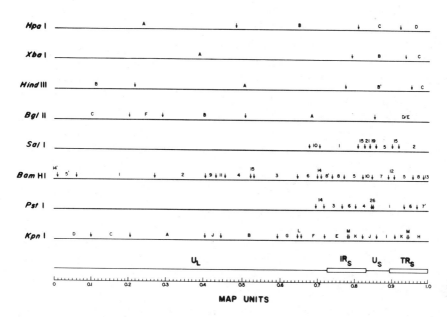

FIGURE 3. Restriction enzyme cleavage maps of PRV DNA. *Bgl*II does not cut the inverted repeats but does cut both the U$_S$ and the U$_L$ sequences. Two half-molar terminal fragments (D and E) and two half-molar fragments spanning the repeats and part of the unique sequences are generated (fragments A); the latter two half-molar fragments comigrate (see Figure 2). The restriction fragments generated from the U$_L$ by *Sal*I and *Pst*I have not been completely mapped; the U$_L$ region has consequently been left blank. The maps for *Hpa*I, *Xba*I, *Hind*III, *Bgl*II, and *Kpn*I were constructed by D. Powell, J. B. Clements and N. N. Wilkie (personal communication).

of the gaps produced by ExoIII digestion, show that the nicks are present at random sites on the DNA and that different molecules have varying numbers of single-stranded interruptions (between 1 and 10) (Ben-Porat *et al.*, 1979). The function of the nicks (or gaps), as well as the factors responsible for their presence, is at present unknown. It is interesting to note, however, that concatemeric DNA molecules accumulating in ts DNA$^+$ mutants defective in encapsidation at the nonpermissive temperature, contain fewer nicks per genome size DNA than do mature DNA molecules (T. Ben-Porat, unpublished results). This finding suggests that the nicks may be introduced into the DNA during the process of maturation and encapsidation.

C. Restriction Enzyme Cleavage Maps

The patterns obtained by digestion of PRV DNA with different restriction enzymes and the available restriction enzyme maps are illustrated in Figs. 2 and 3, respectively. The approximate sizes of the restriction fragments are given in Table II. Only one restriction enzyme (*Bgl*II) that cleaves the genome of PRV in the unique sequences but not in the inverted repeats has been identified. *Eco*R1 does not cleave the genomes of more than 20 different strains of PRV that have been analyzed (T. Ben-Porat, unpublished results).

D. Functional Identity of the Two Isomeric Forms of PRV DNA

Since the discovery that the unique sequences of the genomes of some of the herpesviruses exist in different orientations with respect to one another, a question that has aroused a great deal of interest is whether just one of these isomeric forms of the DNA is infectious. That both isomeric forms of the DNA of PRV are infectious is indicated by the following sets of observations.

Digestion of PRV DNA with the restriction endonuclease *Bgl*II, which does not cut in the repeat region, gives rise to two half-molar terminal fragments, each of which originates from a different isomeric form of the DNA. The presence of one or the other of these half-molar fragments is thus diagnostic of the presence of either one of the two isomeric forms of PRV DNA. Using this type of analysis, it was found that virions containing either isomeric form of the DNA adsorb equally well to susceptible cells and that both isomeric forms of the DNA reach the nucleus and form circles with equal efficiency (Ben-Porat *et al.*, 1980).

There is no exclusion of one of the isomeric forms of the DNA from replication because in cells infected at high multiplicities (when the virions may provide each other with "helper" functions), approximately 70% of the parental virus DNA replicates (Ben-Porat *et al.*, 1976a). Furthermore, virions containing either of the isomeric forms of the DNA are infectious and will replicate when cells are infected at low multiplicity (in the absence of possible helper functions provided by one of the isomers). Analysis of the DNA replicating in these cells during the first round of replication showed that while most of the parental DNA replicated in the form of circles, some molecules with free ends that had replicated could be detected. These free ends originated from either of the isomeric forms, indicating that both isomers of the DNA replicate. Because each infected cell that synthesizes early virus antigen (which is presumably essential for viral DNA synthesis) proceeds to produce an infective center (DeMarchi *et al.*, 1979), one may conclude that viruses containing either isomeric form of PRV DNA are infectious. Thus, inversion of the U_S region of the PRV genome does not affect its infectivity.

E. Strain Variability

The restriction patterns of the DNA of various field isolates (as well as of vaccine strains) have been examined in several laboratories (Heppner

TABLE II. Estimation of the Molecular Weights of Fragments Generated by Digestion of PRV(Ka) DNA with Restriction Enzymes

BamHI		KpnI		SalI		PstI		BglII		XbaI		HindIII		HindIII	
1[a]	19.5[b]	A[a]	18.3[b]	1[a]	10.0[b]	1[a]	9.0[b]	A[a]	33.0[b]	A[a]	72.00[b]	A[a]	51.0[b]	A[a]	42.0[b]
2	12.5	B	13.5	2	7.5	2	4.8	B	20.5	B	14.00	B	19.5	B	32.0
3	11.0	C	9.5	3	5.5	3	4.5	C	18.5	C	6.0	B'	17.0	C	11.0
4	6.0	D	8.3	4	4.6	4	4.0	D	13.7 (0.5M)			C	4.6	D	7.1
5	5.1	E	6.3	5	4.5	5	3.8	E	12.3 (0.5M)						
5'	5.0	F	6.1	6	4.0	6	3.6	F	8.0						
6	4.9	G	5.4	7	3.6	7–8	2.8								
7	4.4	H	5.3	8	3.0	9	2.6								
8'	3.0	I	4.8	9	2.6	10	2.5								
8	2.9	J	3.8	10	2.0	11	1.9								
9	2.3	K	3.5	11	1.95	12–13	1.85								
10	2.2	L	0.6	12	1.85	14	1.55								
11	1.8	M	0.4	13	1.8	15	1.5								
12	1.4			14	1.5	16	1.4								
13	1.1			15–18	1.4	17	1.3								
14	0.8			19	1.2	18	1.25								
14'	0.75			20–24	0.95	19	1.2								
15	0.5			25–50	0.9–0.1[c]	20	1.15								
						21–23	1.1								
						24–60	1.0–0.1[c]								

[a] Fragment designation
[b] Molecular weight (× 10⁻⁶)
[c] The exact molecular weights of these fragments have not been determined.

et al., 1981; Gielkens and Berns, 1982; Ludwig *et al.*, 1982; Ben-Porat *et al.*, 1984a; Lomniczi *et al.*, 1984). The DNA of different PRV strains isolated from different, as well as from the same, geographic areas can be distinguished from one another (Paul *et al.*, 1982; Gielkens and Berns, 1982; Ludwig *et al.*, 1982; Ben-Porat *et al.*, 1984a). Not surprisingly, the greatest differences in restriction fragment patterns are observed when one compares the patterns published by different investigators who have analyzed field isolates from different geographic areas.

Variation in the sizes of restriction fragments generated from most regions of the genome can be identified in different isolates. However, some regions of the genome appear to be more variable than others. These regions include the inverted repeats, as well as regions close to the left end of the genome (Gielken and Berns, 1982; Ben-Porat *et al.*, 1984a).

A variation in the restriction patterns of the genomes of mutants, which behave genetically as being altered in a single gene (Ihara *et al.*, 1982), has also been observed (Ben-Porat *et al.*, 1984a). Interestingly, the regions of the genome that had a modified restriction pattern were unrelated to the regions of the genome in which the mutations had been located by marker rescue. Such variations in the restriction patterns of the mutant genomes were observed in TK⁻ mutants that had been obtained by exposure of the infected cells to Ara-T (a procedure that is probably mutagenic), as well as in ts mutants obtained after mutagenesis by either UV light irradiation, treatments with nitrosoguanidine, or substitutions in the DNA of BUdR for thymidine. Despite the low doses of mutagen, which had been used to avoid isolation of mutants with multiple lesions, secondary nonlethal mutations nevertheless must have been introduced into many of the genomes. Interestingly, the changes in the restriction patterns of the genomes of the mutants mimicked those most prevalently observed between field isolates. Repeated plaque purification of the same isolates, as well as passage of the virus in animals (Ben-Porat *et al.*, 1984a), has not, however, resulted in changes in the restriction pattern of the genome to date.

The changes in the restriction patterns of different field isolates (or different mutants) observed are due mainly to the deletion (or acquisition) of sequences that do not affect virus viability (Ben-Porat *et al.*, 1984a, and L. Harper and T. Ben-Porat, unpublished results).

F. Genomes of Vaccine Strains

Several attenuated strains of PRV have been isolated by repeated passage in culture (Baskersville *et al.*, 1973); some of these are currently used as vaccines. The genomes of these vaccine strains have been analyzed by restriction endonucleases (Geck *et al.*, 1982; Gielkens and Berns, 1982; Paul *et al.*, 1982). Of interest is the fact that two independently isolated attenuated strains of PRV (the Bartha and the Norden vaccine strains) both have a deletion in the U_S between 0.855 and 0.882 map units (MU) (Lomniczi et al., 1984a). This deletion appears to be a characteristic of these attenuated strains and has not been observed in the genomes of any other strain of PRV analyzed to date. Therefore, it appears likely that one of the functions specified by the region deleted from the two vaccine strains is somehow involved in virulence.

Of special interest is the fact that the Norden vaccine strain contains a genome that is a class 3 herpesvirus DNA molecule; i.e., in contrast to other strains of PRV, in which only the U_S inverts itself relative to the U_L, in the Norden strain both the U_S and U_L sequences invert themselves relative to one another (Lomniczi et al., 1984). In the genome of the Norden strain, a sequence of nucleotides, normally present in all other PRV strains at the end of the U_L only, has been translocated to a region adjacent to the junction of the U_L and the internal inverted repeat. As a consequence, both the U_L and the U_S are bracketed by inverted repeats. It is probable that this translocation provides the basis for the inversion of the U_L region of the genome (Lomniczi et al., 1984).

VI. VIRUS REPLICATION IN CELL CULTURE

A. Virus Growth Cycle

As is the case for all bacterial and animal viruses, adsorption of PRV is an electrostatic process. It is not affected by temperature nor is it an energy-requiring process (Kaplan, 1962). It does, however, require the presence of some electrolytes (Farnham and Newton, 1959). The rate of adsorption of PRV to rabbit kidney cells is relatively rapid [50% in 30 min (Kaplan and Vatter, 1959)] but is much slower to chick embryo cells (Béládi, 1962).

In contrast to adsorption, the penetration of virus into cells is an energy-requiring process (Kaplan, 1969). Following penetration, the virions lose their outer envelope rapidly. The first steps leading to the uncoating of the nucleocapsids probably occur in the cytoplasm. Thus, in cells infected at the nonpermissive temperature with a ts mutant defective in uncoating, the virus loses its outer envelope and nucleocapsids accumulate in the cytoplasm (Feldman et al., 1981), indicating that the intact nucleocapsid does not penetrate the nucleus. On the other hand, during the normal course of infection, many of the structural proteins of the virus become associated with the cell nucleus (L. Feldman and T. Ben-Porat, unpublished results). The identity of the proteins that accompany the viral genome into the nucleus has not been studied in detail.

The growth characteristics of the virus depend to some extent on the virus strain, as well as upon the type of cells and the type of cultures (monolayer, cell suspension, or single cell) being infected. Thus, the length of the eclipse period, as well as of the latent period, varies in different cells. For example, the eclipse period for PRV-infected rabbit kidney and chick embryo cells is 3 hr (Kaplan and Vatter, 1959; Rada and Zemla, 1973; Lomniczi, 1974), while it is 5 hr for BHK21/13 cells (Darlington and Moss, 1968). Similarly, the latent phase in PRV-infected rabbit kidney cells and chick embryo cells is 4 hr (Kaplan and Vatter, 1959; Rada and Zemla, 1973; Lomniczi, 1974; 5 hr in BHK21/13 cells (Darlington and Moss, 1968), and 12 hr in monkey kidney cells (von Kerékjártó and Rhode, 1957). Furthermore, during the course of studies in which the growth characteristics of various PRV strains were followed, considerable differences in the length of time required for different virus strains to complete the growth cycle in a particular cell type were found, possibly reflecting the degree to which a particular strain is "adapted" to a particular host cell (T. Ben-Porat, unpublished results). The final yield of virus produced by infected cells also depends upon a number of factors, such as the virus strain, the type of cell infected, and the thermostability of the virus strain.

The ratio of infectious to noninfectious particles in a population of virions has been determined for one laboratory strain to be approximately 1:20 (Ben-Porat et al., 1974). Most of the virions in this population can adsorb to and penetrate into the cells, and the large majority of the virus genomes reach the nucleus (Ben-Porat et al., 1980). Furthermore, each particle that can initiate infection, i.e., induce the expression of early antigens in the cells, will go on to form a plaque (DeMarchi et al., 1979). The low ratio of infectious to noninfectious particles thus probably reflects the fact that most of the infecting virions do not come in contact with the appropriate cell targets; it does not necessarily reflect the presence of "defective" virions.

B. Cytopathic Effects on Cells in Culture

Infection with PRV engenders drastic changes in their host cells, causing them to undergo degeneration and, eventually, death. A series of characteristic cytopathic alterations occurs that may vary, depending on the virus strain and the host cell. One of the first, most obvious effects of PRV infection is the inhibition of mitosis; this inhibition is a rapid event, occurring within a short period of time after virus attachment (Stoker and Newton, 1959; Reissig and Kaplan, 1960). Another striking cytological feature of infection, probably common to all PRV strains, is the formation of intranuclear inclusions; their characteristics have been reviewed (Kaplan, 1969).

Two other types of characteristic changes are induced in monolayers of cells by infection with PRV: (1) The infected cells may fuse and form syncytia (also known as polykaryocytes), or (2) the infected cells may become rounded, frequently ballooned, with the formation of only an occasional small polykaryocyte. These two types of degenerative changes

TABLE III. Host Cell Range

Cells	References
Mouse fibroblasts	Scherer (1953)
HeLa	Scherer and Syverton (1954)
Rabbit kidney	Barski et al. (1955), Kaplan and Vatter (1959), Smith et al. (1966)
Pig kidney	Singh (1962), McFerran and Dow (1962), Stewart et al. (1974)
Lamb kidney	Ceccarelli and Del Mazza (1958)
Dog kidney	Aurelian and Roizman (1965)
Monkey kidney	von Kerékjártó and Rhode (1957). Follett et al. (1975)
Chick embryo fibroblasts	Cserey-Pechany et al. (1951), Béládi and Ivánovics (1954), Albrecht et al. (1963)
Baby hamster kidney	Sharma et al. (1962), Chantler and Stevely (1973), Dubovi and Youngner (1976)
Human fibroblasts	Plummer et al. (1969a)

were described many years ago by Lauda and Rezek (1926). Since that time there have been many descriptions of these two kinds of cytopathologic changes in PRV-infected cells (Kaplan, 1969; Darlington and Granoff, 1973).

The kind of effect (rounding or syncytia formation) one observes is dependent upon the genetic constitution of the infecting virus. Thus, variants that produce one or the other type of cellular degeneration have been repeatedly isolated (Tokumaru, 1957; Kaplan and Vatter, 1959; Golais and Sabó, 1976; Bitsch, 1980). The type of cell infected, however, also seems to affect the cytopathology of the virus–cell complex. The same virus isolates may produce different kinds of cytopathic effects depending on the cell culture they infect (Škoda et al., 1968; Ludwig et al., 1974; Golais and Sabó, 1976). Thus, in addition to the genotype of the infecting virus, its phenotypic expression, as modified by the host cells, will also influence the type of cytopathic effect produced.

The formation of mature virus is not essential for the development of cytopathic changes; they will occur when the formation of mature virions is inhibited by a variety of drugs (reviewed by Ben-Porat and Kaplan, 1973). Cell fusion is inhibited by Con A and 2-deoxy-D-glucose; the factor responsible for fusion is therefore likely a glycoprotein (Ludwig et al., 1974).

C. Host Range

In vitro: PRV will multiply in a wide variety of cells cultivated *in vitro*, some of which are listed in Table III. The different types of cells will, however, vary in the yield of infectious virus they produce and different strains of virus will differ in the relative yield of virus produced in different cell lines.

In vivo: The natural host of PRV is swine. Under natural conditions, the virus causes fatal infection of cattle, as well as of sheep, dogs, and cats (Gustafson, 1981). Wild animals, such as foxes, skunks, rats, mice, cottontail rabbits, muskrats, raccoons, badgers, woodchucks, opossums, and deer, are also susceptible to infection with PRV (Trainer and Karstad, 1963; Kirkpatrick *et al.*, 1980).

Experimentally, PRV can be transmitted by the inoculation of the following animals: chickens, rabbits, guinea pigs, cats, embryonated eggs, pigeons, sheep, hamsters, marmosets, rhesus monkeys, rats, mink, and ferrets (Faria and Braga, 1934; Remlinger and Bailly, 1934; Hurst, 1936; Galloway, 1938; Bailly, 1938; Glover, 1939; Burnet *et al.*, 1939; Toneva, 1961; Tokumaru and Scott, 1964; Goto *et al.*, 1968; McFerran and Dow, 1970; Goto *et al.*, 1971). There are some species of animals that appear to be refractory to infection: certain primates (Barbary apes and chimpanzees) and poikilotherms (Gustafson, 1981).

VII. RNA SYNTHESIS IN PRV-INFECTED CELLS

A. Patterns of Transcription of the Virus Genome

It is generally assumed that the unmodified host cell polymerases can transcribe the incoming herpesvirus genome, because naked virus DNA is infectious (Sheldrick *et al.*, 1973). Furthermore, because there is no evidence to show that herpes virions possess an RNA polymerase, the assumption is made that during the normal course of infection the first virus transcripts are also synthesized by an unmodified host RNA polymerase. Indeed, studies using inhibitors, as well as cell lines resistant to α-amanitin, have implicated RNA polymerase II in the transcription of the HSV genome (Alwine *et al.*, 1974; Ben-Zeev and Becker, 1977). As expected, PRV DNA is also infectious (see below) and therefore at least part of the virus genome can be transcribed by an unmodified cellular polymerase II.

On the basis of the differential transcription patterns of the virus genome at various stages of infection, the transcriptional program of the genome of PRV may be subdivided into at least the following three classes, which are transcribed successively and which exhibit increasing complexity: (1) "immediate–early" (IE) RNA, which is synthesized in the absence of protein synthesis; (2) "early" RNA, which is synthesized before virus DNA synthesis is initiated; (3) "late" RNA, which is synthesized after virus DNA synthesis has taken place in the infected cells.

The complexity of the transcripts accumulating within the infected cells at various stages of infection varies greatly. Only approximately 7% of the total virus DNA sequences are transcribed as IE RNA (Rakusanova *et al.*, 1971; Jean *et al.*, 1974; Feldman *et al.*, 1979). Early RNA is complementary to only half as many sequences as is late RNA which hy-

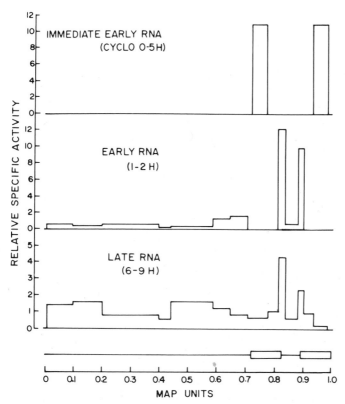

FIGURE 4. Relative abundance of transcripts synthesized at various stages of infection originating from different regions of the PRV genome. Modified from Feldman *et al.* (1979).

bridizes to about 50% of the virus DNA sequences (Rakusanova *et al.*, 1971; Jean *et al.*, 1974; Feldman *et al.*, 1979).

The relative abundance of transcripts originating from various regions of the virus genome that accumulate under different conditions within the infected cells is illustrated in Fig. 4, which summarizes the cumulative data obtained with restriction fragments generated by several enzymes (Feldman *et al.*, 1979). The difference in the relative abundance of the transcripts is mainly due to differential transcription rates; the patterns obtained by labeling the RNA (in nuclear monolayers) during short labeling periods were similar to the patterns obtained by labeling intact cells for longer periods (Feldman *et al.*, 1979, 1982).Thus, selective degradation of parts of the transcribed sequences is not the main reason for the differences in the abundances of different transcripts at different stages of infection and it is clear that controls operate at the level of transcription of the PRV genome. The factors that modulate the transcriptional program of the PrV genome have not been identified to date.

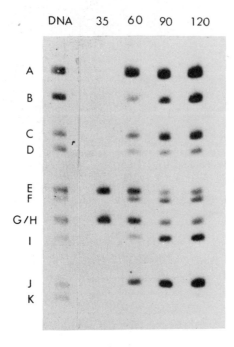

FIGURE 5. Sequential accumulation of virus transcripts at early stages of infection. Rabbit kidney cells were labeled with $^{32}PO_4$ from the time of infection. At the indicated time periods (min), they were harvested and the RNA extracted and hybridized to KpnI-digested PRV DNA fixed to nitrocellulose filters. Modified from Feldman et al. (1979).

B. Transition from IE RNA to Early RNA Transcription

The sequential transcription of different regions of the genome and the transition from the synthesis of IE RNA to that of early RNA during the earliest stages of the normal infective process are illustrated in Fig. 5. Scanning of appropriately exposed strips shows that IE RNA (hybridizing to KpnI fragments E and H only) is transcribed for the first 40 minutes of the infective process only (Feldman et al., 1982). With the cessation of transcription of IE RNA, there is an increase in the transcription of other virus sequences (early RNA).

The sequences from which IE RNA originates are not detectably transcribed during the early phase of infection; pulse labeling of nuclear monolayers obtained from infected cells shows that between 50 and 90 min postinfection the IE region of the genome is not transcribed (Feldman et al., 1982). However, by 3 hr postinfection, this region is again transcribed, but in contrast to the IE transcripts synthesized at the earliest stages of infection, the transcripts originating from that region at later times after infection are not transported efficiently to the cytoplasm. Furthermore, these transcripts have higher sedimentation characteristics than do the "true" IE transcripts (Feldman et al., 1982).

The observation that IE RNA only is transcribed in cycloheximide-treated cells infected with PRV (Rakusanova et al., 1971; Jean et al., 1974) implicates an IE function either directly or indirectly in the transition to the transcription of early RNA. The evidence obtained with *ts* mutants defective in IE functions also shows that the IE proteins are continuously required for the orderly transcriptional program of PRV (Ihara *et al.*, 1983). Furthermore, in addition to being required for the transition of IE to early RNA synthesis, expression of the IE protein also leads to the simultaneous loss of recognition of the IE promoter (Jean *et al.*, 1974; Ihara *et al.*, 1983).

The mechanism by which the IE protein regulates transcription is presently unknown. It is possible that it modifies the RNA polymerase, or another cellular factor that modulates transcription, or that it modifies the DNA template.

C. Transition from Early to Late RNA Transcription

By definition, transition from early to late transcription occurs only after virus DNA synthesis has been initiated. The situation appears, however, to be much more complex and many different DNA$^-$ *ts* mutants have patterns of transcription intermediate between those seen at early and late stages of the normal course of infection (L. Feldman and T. Ben-Porat unpublished results). This is observed by analysis of Southern blots (Southern, 1975) of RNA accumulating in the mutant infected cells during relatively long labeling periods, as well as after a short labeling period, in nuclear monolayers.

The factors affecting the switch from early to late RNA synthesis in PRV-infected cells are largely unknown. One can postulate, as has been done for other systems, that a modification of the DNA template (which is achieved when the DNA is in replicative form) is directly involved. Alternatively, the switch from early to late RNA synthesis may be mediated, at least in part, by other factors, the presence of which in the cells is dependent on replicating virus DNA.

D. Characteristics of Virus Transcripts

1. The IE Transcripts

The IE transcripts have been mapped to a 4.1×10^6-dalton segment of DNA located internally within the inverted repeat (approximate map positions 0.95–0.99). The transcript is approximately 6 kb in size; it is normally synthesized up to approximately 40–50 min postinfection. While this transcript is not synthesized at later stages of infection, larger transcripts that include these sequences are synthesized beginning at approximately 3 hr postinfection (Rakusanova *et al.*, 1971; Feldman *et al.*, 1979, 1982; Ihara *et al.*, 1983). These transcripts are not precursors of mRNA (Feldman *et al.*, 1982).

Analysis of nuclear and cytoplasmic IE RNA by the Northern technique, using nick-translated restriction fragments of PRV DNA as probes, shows that only one IE transcript is synthesized and that the sizes of the nuclear and the cytoplasmic transcripts are similar (Ihara *et al.*, 1983).

IE RNA synthesized at the earliest times in the normal course of infection is rapidly transported to the cytoplasm (Feldman *et al.*, 1982).

2. Early Transcripts

As mentioned above, early transcripts are complementary to most sequences of the genome. Two exceptions have been detected: (1) the region from which IE RNA originates and (2) a region within the middle of the inverted repeats (about 5 kbp; see Fig. 4).

Analysis of the size distribution of the early transcripts synthesized during a short pulse shows that most are approximately of mRNA size and that they are rapidly transported to the cytoplasm (Deatly *et al.*, 1984). The majority of these transcripts are relatively stable and decay with a half time of several hours.

3. Late Transcripts

Late transcripts may be divided into two groups. One group is composed of transcripts of approximately mRNA size that make up only a fraction of the total "late" virus RNA synthesized by the cells. This RNA is transported rapidly to the cytoplasm. The other group of transcripts consists of very large molecules that turn over rapidly (50% of the transcripts turn over within 30 min) and are not precursors to mRNA (Deatly *et al.*, 1984). While some differential turnover of transcripts originating from specific regions of the genome can be detected, as a whole, the unstable sequences appear to be distributed fairly evenly throughout the genome (Feldman, 1980). Whether these large transcripts have a function in the biology of the virus or whether they represent aberrant transcripts, possibly reflecting weak termination sites, is not known.

E. Accumulation of Virus mRNAs in the Infected Cells

As mentioned above, the rate of transcription of different regions of the genome differs both at early and at late stages of infection; some regions are more actively transcribed than others. Similarly, the degree of accumulation of different mRNA species varies greatly. In general, it appears that the regions that are most actively transcribed are also those from which mRNAs accumulate most abundantly.

The number of different major mRNA species that accumulate in the cells at various stages of infection has been determined by the Northern technique, using nick-translated, cloned virus DNA restriction fragments as probes.

On the basis of the sequential appearance of the various mRNA species, as well as on the basis of their level of abundance at various stages of the infective process (as determined by the relative intensity of the bands), the virus mRNAs may be divided into four groups.

1. IE mRNA

Only a single mRNA, approximately 6 kb in size, belongs to this class. Thus, the IE functions of PRV represent a much more restricted set of functions than do those of HSV. Two other major IE transcripts of HSV that originate from the junction of the U_S and the repeat regions appear to have equivalents in two early transcripts of PRV (Feldman *et al.*, 1979). Thus, it appears that some of the functions that are included in the IE class of HSV are part of the early class of functions in PRV.

2. Early mRNA

mRNA species belonging to this class are synthesized at the early stage of infection and accumulate up to about 3–4 hr postinfection. Thereafter, the level of most of these mRNA species remains constant while the level of some species declines. In the absence of DNA synthesis, continuous accumulation of some of these mRNA molecules is observed. Different DNA $^-$ ts mutants, however, vary somewhat in this respect (L. Feldman and T. Ben-Porat unpublished results).

3. Early-Late mRNA

mRNA species belonging to this class start being synthesized at early times (1 hr) after infection, but accumulate abundantly between 3 and 9 hr postinfection (by 9 hr postinfection, transcription of the virus genome declines in the system studied).

4. Late mRNA

These mRNA species start appearing by 2.5 hr postinfection (after virus DNA synthesis has started) and accumulate abundantly thereafter up to the end of the infectious cycle.

The map positions and the sizes of the abundant mRNA species belonging to these four classes are given in Table IV.

Analysis of the sizes of the abundant mRNAs originating from some regions of the genome showed that their cumulative sizes exceed the coding capacity of those regions. For example, the genomic region in the U_S that encompasses 84.3–89.1 MU (6.6 kbp) specifies at least seven major mRNAs with a cumulative size of 19.2 kb. (The sizes of these mRNAs differ from those detected with probes consisting of the sequences adjacent to that region of the genome.) Selection of these mRNA species followed by translation of the selected RNA *in vitro* results in the synthesis of at least six (Lomniezi *et al.*, 1984b) distinct polypeptides and it

TABLE IV. Estimation of Sizes of the Abundant mRNA Species Belonging to Different Classes that Originate from Different Regions of the PRV Genome[a]

Fragment	Map unit (in hundredths)	IE (kb)	Early (kb)	Early–late (kb)	Late (kb)
BamHI 14'	0.0–0.9	—	—	1.8	—
BamHI 5'	0.9–6.0	—	1.5, 1.7	1.8, 2.8	—
BamHI 1	6.0–27.1	—	4.3, 1.6, 1.7, 2.1	4.8, 3.8, 2.8, 2.3	9.4, 6.4, 1.8, 1.0
SalI 12	8.5–10.6	—	—	4.8	6.4
SalI 3	14.8–20.8	—	4.3	3.8	—
SalI 9	20.8–23.6	—	1.6, 1.7	—	1.0, 9.4
SalI 23	23.6–24.6	—	—	—	9.4
SalI 11	24.6–26.7	—	2.1	2.8	1.8
BamHI 2	27.1–40.6	—	4.1	1.6, 1.7, 3.0	1.9, 3.8, 4.2
BamHI 9	40.6–43.5	—	—	1.3, 2.0, 4.0	4.4
BamHI 11	43.5–45.5	—	1.7, 2.2, 2.6	4.0	—
BamHI 4	45.5–51.9	—	—	1.5, 2.1, 4.4	5.5
BamHI 15	51.9–52.4	—	—	—	0.5
BamHI 3	52.4–64.1	—	2.4, 2.8, 3.6, 3.9	0.8, 1.7, 1.8	—
BamHI 6	64.1–69.5	—	1.9, 3.0, 3.5, 3.7	1.2	—
BamHI 14	69.5–70.6	—	1.9	—	—
BamHI 8'	70.6–73.7	—	—	—	—
BamHI 8	73.7–76.7	6.0	—	—	—
BamHI 5	76.7–82.0	—	1.7, 2.0	—	—
BamHI 10	82.0–84.5	—	1.6, 3.5	2.2	4.3
BamHI 7	84.5–89.1	—	1.3, 1.6, 3.5	2.0, 2.8	3.9, 4.1
BamHI 12	89.1–91.0	—	1.3	2.2	4.3

[a] The estimate of the sizes of the mRNA represents an approximation only.

appears therefore that the various size mRNAs code for different size proteins. Transcription of both strands of the DNA would give rise to only 13.2-kb RNA (instead of the 19.2 kb mRNA seen). Thus, as is the case for HSV, as well as for most other eukaryotic DNA viruses, there appears to be efficient use of the coding capacity of the PRV genome. Whether any of the messages share a common 5' initiation site or common 3' termination site has not as yet been determined.

VIII. PROTEIN SYNTHESIS IN THE INFECTED CELLS

A. Patterns of Virus Protein Synthesis

Infection of cells with PRV inhibits cellular protein synthesis (Hamada and Kaplan, 1965); during the normal course of infection, this inhibition occurs gradually. Thus, when infected cells are labeled during early stages of infection, the virus protein bands appear on the autoradiograms against a background of cellular protein bands. At later times after infection, most of the bands on the autoradiograms of the gels have migration characteristics that distinguish them from those of the cellular protein bands.

Approximately 40 new virus protein bands can be identified by 5 hr postinfection in one-dimensional gels (using various gel concentrations). However, many more new "spots" can be distinguished on two-dimensional gels. Typical two-dimensional gels of proteins synthesized by cells at early stages of infection and by uninfected cells are illustrated in Fig. 6. Two points emerge from an analysis of the autoradiograms of such gels: (1) Even at early stages of infection, a relatively large number of polypeptides (approximately 30 new major "spots") can be identified that are not present in the uninfected cell lysate. (2) While by 3 hr postinfection the synthesis of most cellular polypeptides has been inhibited to some degree, the synthesis of some has been inhibited to a much greater degree. Thus, infection appears to affect the synthesis (or processing) of the various cellular polypeptides differentially.

An analysis of scans of the one-dimensional PAGE autoradiograms of the proteins synthesized by infected cells at various stages of infection shows that the virus proteins (i.e., protein bands not observed in uninfected cells) can be subdivided into five classes, depending on the stage of infection during which they are synthesized most abundantly (Fig. 7).

1. Class I Proteins (IE Proteins)

These include the proteins synthesized at very early stages of infection (up to 2½ hr postinfection only) (Ben-Porat et al., 1975a). This class of virus proteins is overproduced when IE mRNA is allowed to accumulate by treatment of the infected cells with inhibitors of protein syn-

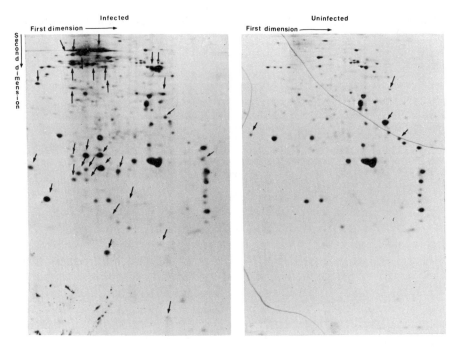

FIGURE 6. Two-dimensional gel electrophoresis of proteins synthesized by infected and uninfected cells. The cells were labeled with [³H]leucine between 2.0 and 3.0 hr postinfection. Arrows on the autoradiogram of infected cell lysates indicate virus proteins; arrows on the autoradiogram of unifected cell lysates indicate cellular proteins, the synthesis of which has been inhibited by infection to a greater extent than that of others. The autoradiogram of infected cell proteins was exposed 20 days; that of uninfected proteins was exposed 10 days (T. Ben-Porat, unpublished results).

thesis and then allowed to be expressed. These proteins are also over-produced in cells infected at the nonpermissive temperature with mutants defective in the IE protein (Ihara *et al.*, 1983). Only one IE protein (180K) is specified by the PRV genome (Ihara *et al.*, 1983).

Control of abundance of IE proteins in the infected cells operates at at least two levels: (1) self-regulation at the level of transcription (see above) and (2) regulation of message stability in which an early function appears to play a role (Feldman *et al.*, 1982). While the IE protein is not synthesized beyond 2.5 hr postinfection (Ben-Porat *et al.*, 1975a), its continued presence in the infected cells is essential to the maintenance of the normal transcriptional program (Ihara *et al.*, 1983), as is the 175K IE protein of HSV (Watson and Clements, 1980).

2. Class II Proteins (Early Proteins)

These include the proteins synthesized most abundantly between 1 and 4 hr postinfection, i.e., prior to and during the early stages of DNA

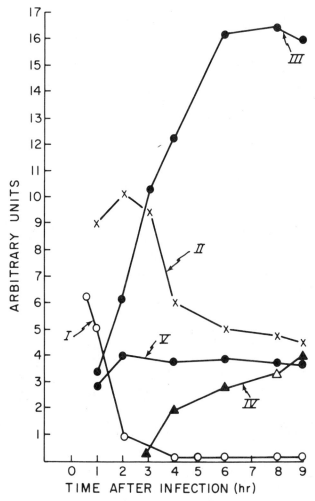

FIGURE 7. Synthesis of various classes of virus proteins at different times after infection. Infected cells were labeled with [³H]leucine for 1-hr intervals at various times after infection. The proteins were electrophoresed and autoradiograms prepared. The autoradiograms were scanned, the area under the peaks was determined, and the relative abundance of the various proteins that had been synthesized at different times after infection was deduced. The Roman numerals refer to the protein class. (T. Ben-Porat, unpublished results).

synthesis. Thereafter, the rate of their synthesis declines. The rate of synthesis of these proteins continues to increase with time after infection when virus DNA synthesis is inhibited or in cells infected at the nonpermissive temperature with some DNA⁻ ts mutants (A. Deatley and T. Ben-Porat, unpublished results). The phenomenon may be related to the overproduction under similar conditions of early virus enzymes, which probably belong to this class of proteins (Kaplan and Ben-Porat, 1968).

3. Class III Proteins (Early–Late Proteins)

These include proteins whose synthesis is detectable as early as 90 min postinfection but which are synthesized at a maximal rate at later times after infection (between 4 and 9 hr postinfection). When the synthesis of virus DNA is prevented (either in cells infected with ts mutants or by treatment of the infected cells with inhibitors), the synthesis of class III proteins continues in some cases at approximately the same relatively low rate between $1\frac{1}{2}$ and 9 hr postinfection. The major capsid proteins belong to this class.

4. Class IV Proteins (Late Proteins)

The synthesis of these proteins does not start in cells in which virus DNA synthesis is inhibited (DNA$^-$ ts mutants at the nonpermissive temperature or in the presence of inhibitors of DNA synthesis). They also are not detectably synthesized before 2.5–3 hr postinfection. Some of the virus envelope proteins belong to this class.

5. Class V Proteins (Unvarying Proteins)

These are synthesized in relatively large quantities throughout the infectious cycle; no significant change in the rate of their synthesis is detected between 1 and 9 hr postinfection (i.e., between the beginning and the end of the growth cycle). These proteins are not detectably synthesized by uninfected cells and do not appear to be cellular proteins.

B. Mapping of Virus Polypeptides

Selection of mRNA species specified by defined regions of the virus genome, followed by *in vitro* translation of these mRNAs, allows the determination of the map position coding for the different virus polypeptides. These polypeptides can be identified by their apparent molecular weight, as determined by PAGE and by immunoprecipitation with specific antibodies, thereby mapping on the genome the genes coding for these antigenically identified proteins.

Fewer than 40 polypeptides have been identified to date by selection and translation of mRNAs extracted from infected cells (S. Watanabe and T. Ben-Porat, unpublished results). This is a considerably smaller number than expected on the basis of the number of "unrelated" virus polypeptides that can be resolved by two-dimensional gel electrophoresis. It is likely that only those polypeptides for which abundant mRNA is present in the cells were identified and that the polypeptides specified by minor species of mRNAs were not detected.

The salient points to emerge from the available information are as follows: (1) The IE 180K protein, as well as the 136K DNA-binding and the 142K major capsid proteins, originate from regions of the genome to which they had previously been assigned on the basis of marker rescue experiments of ts mutants defective in these proteins (Ladin *et al.*, 1982; Ihara *et al.*, 1983; Ben-Porat *et al.*, 1983a). (2) The genes specifying for three capsid proteins (142K, 62K, 32K) are clustered in the same region of the genome (0.445–0.519 MU).

C. Control of Virus Protein Synthesis

As mentioned above, inspection and scanning of the PAGE profiles of lysates of infected cells show that the types of proteins synthesized, as well as their relative abundance, vary at different stages of infection (Fig. 7). Evidence has been obtained implicating controls of the expression of the virus genome at the levels of transcription and of mRNA stability (see above). In addition, it appears that controls are also exerted at the level of translation of the mRNAs. This is indicated by the fact that under certain conditions of infection, the polysomal mRNA species are only a subset of the mRNA species present in the cytoplasm of the infected cells (A. Deatly and T. Ben-Porat, unpublished results).

McCrath and Stevely (1980) have studied the translation of virus mRNA *in vitro* and have found that the conditions for optimum stimulation of amino acid uptake in rabbit reticulocyte lysates are similar for mRNA obtained from infected and from uninfected cells. At least part of the polypeptides that are synthesized under these conditions are viral. These findings have been confirmed in our laboratory (A. Deatley and T. Ben-Porat unpublished results). Several additional points emerge from an analysis of the types of polypeptides that are synthesized when rabbit reticulocyte lysates are primed with mRNA obtained from PRV-infected cells at various stages of the infective process (A. Deatly and T. Ben-Porat, 1984).

(1) Despite the fact that cellular proteins are synthesized at only low levels (or not at all). Under some conditions of infection, mRNA preparations isolated from the cytoplasm of these cells contain relatively large amounts of cellular messages that are translationally competent (Ihara *et al.*, 1983). [The presence of these cellular messages in the RNA preparations can also be detected by the Northern or dot-blot hybridization procedures using specific cellular cloned genes as probes (S. Watanabe and T. Ben-Porat, unpublished results)].

(2) While IE mRNA, as well as many early–late mRNAs, are efficiently translated in reticulocyte lysate systems, some early virus mRNAs are translated poorly or not at all. It is probable that some factor(s) essential for the translation of these early mRNA species, but not of other virus mRNA species, is limiting in the system. This factor(s) has not been identified.

D. Functions of Some Nonstructural Virus Proteins

The enumeration of the virus proteins synthesized by the infected cells discussed above indicates that, as expected, PRV codes for many nonstructural proteins. While the functions of most of these proteins remain obscure, some information concerning a few of them is available and is discussed below.

1. The IE Protein

PRV specifies only one (180K) IE protein (Ihara *et al.*, 1983). This protein probably corresponds to the 175K IE protein of HSV (Morse *et al.*, 1978; Preston *et al.*, 1978; Dixon and Schaffer, 1980; Watson *et al.*, 1979). Indeed, the regions of the genome coding for these proteins in HSV and PRV show a relatively high degree of sequence complementarity (Ben-Porat *et al.*, 1983b).

The IE protein of PRV appears to be multifunctional and to regulate several processes in the infected cells.

a. Role of the IE Protein in the Transition from IE to Early Phase of Transcription

The observation that only IE RNA is transcribed in cycloheximide-treated infected cells (see section VIII.D.1.) implicates an IE function either directly or indirectly in the transition from transcription of IE RNA to transcription of early RNA. Indeed, evidence obtained with ts mutants deficient in IE proteins suggests that functional IE protein is continuously required for the orderly transcriptional program of PRV (Ihara *et al.*, 1983), as it is of HSV-1 (Preston, 1979; Watson and Clements, 1980).

b. Role of the IE Protein in the Repression of Synthesis of IE mRNA

During the normal course of infection, PRV transcripts accumulate up to 50 min postinfection; between approximately 50 min and 2.5–3 hr postinfection the IE region of the genome is not transcribed (Feldman *et al.*, 1982). Similarly, after shiftdown from the nonpermissive to the permissive temperature of cells infected with a ts IE mutant, transcription of the IE region of the genome, which had been continuous at the nonpermissive temperature, is arrested (Ihara *et al.*, 1983). Furthermore, in cells incubated from the time of infection with cycloheximide, IE transcripts accumulate continuously; however, after the removal of the inhibitor the transcription of that region ceases as the result of the synthesis of the IE protein (Feldman *et al.*, 1979). Thus, the expression of IE proteins leads to a loss of recognition of the IE promoter and appears to be directly involved in the repression of IE mRNA.

c. Role of the IE Protein in Inhibition of Cellular Protein Synthesis

The IE protein is involved in the inhibition of cellular protein synthesis, as deduced from the following observations: When cycloheximide-treated infected cells are allowed to synthesize RNA (but not proteins) during the first few hours after infection, most of the cell-specific polysomes within the infected cells remain intact. Furthermore, for the first 3 min after resumption of protein synthesis is allowed in these cells, reformation of cellular polysomes occurs; several minutes thereafter, however, cell-specific polysomes disappear and only virus mRNA remains polysome-associated. It was concluded from these observations that the virus RNA that had accumulated in the infected cells in the absence of protein synthesis (IE RNA) was translated after protein synthesis was allowed to proceed, and that the IE proteins are probably responsible for the subsequent disaggregation of cellular polysomes (Ben-Porat et al., 1971; Rakusanova et al., 1971). This conclusion was reinforced by the finding that expression of early function is blocked under these conditions (Jean et al., 1974). The IE protein must therefore be responsible for the disappearance of the cellular polysomes. Also, in cells infected at the nonpermissive temperature with a ts mutant in which only the IE protein is expressed, inhibition of the synthesis of cellular proteins occurs. Because the incoming virions do not inhibit cellular protein synthesis, it appears that the IE protein synthesized at the nonpermissive temperature in the mutant infected cells, though unable to induce the transition from IE to early transcription, still appears able to inhibit (at least partially) the synthesis of cellular proteins (Ihara et al., 1983).

2. DNA-Binding Proteins

Virus-specified chromatin-associated, arginine-rich, acid-extractable proteins have been identified by several investigators in PRV-infected cells (Shimono and Kaplan, 1968; Stevens et al., 1969; Chantler and Stevely, 1973). Some of these proteins appear to be structural and can be isolated from preparations of purified virions (Chantler and Stevely, 1973). As expected, a relatively large number of virus proteins (at least 16) (T. Ben-Porat, unpublished results) that bind to both single- and double-stranded DNA can be isolated from PRV-infected cells. While the functions of most of these proteins remain obscure, a few have been studied in some detail.

a. The 136K DNA-Binding Protein

This is the major DNA-binding protein synthesized by PRV; it displays the characteristic kinetics of synthesis of a class II (early) protein. This protein has been isolated and characterized by Littler et al. (1981)

and Yeo *et al.* (1981) and has been found to be antigenically related to the major DNA-binding protein of HSV.

Some of the functions of this protein have been established using cells infected at the nonpermissive temperature with mutants defective in this protein. The 136K DNA-binding protein is required for the following: (1) the initiation of the first round of virus DNA synthesis; (2) later rounds of DNA replication; (3) the stabilization of progeny virus DNA (Ben-Porat *et al.*, 1983a). In the absence of a functional 136K protein, progeny virus DNA that has accumulated within the infected cells is degraded and it appears that the 136K protein plays a role in the protection of the progeny DNA from nucleolytic attack.

b. The 10K and 15K DNA-Binding Proteins

Two small, nuclear, acid-soluble proteins, 10K and 15K, are closely associated with the fast-sedimenting structures of concatemeric DNA that are present within the infected cells at late stages of infection. Part of these proteins, while associated with concatemers, is protected from the action of proteases, probably because the proteins are present at the center of the concatemeric tangles of DNA. Removal of the proteins from the concatemeric DNA by treatment with high salt concentrations (e.g., isopycnic centrifugation in CsCl) results in the introduction of double-stranded breaks in the virus DNA. Whether this reflects physical breakage or is the result of an enzymatic action (similar to that, for example, of a topoisomerase) is not known. Furthermore, it is also not clear whether the effect is due directly to the small DNA-binding proteins or whether some other protein(s) present in amounts too small to be detected by the methods that have been used may be responsible for the breaks.

Both the 10K and the 15K DNA-binding proteins are late proteins; i.e., they are not detectable in lysates of infected cells during early stages of infection, as well as in lysates of cells infected at the nonpermissive temperature with most DNA⁻ ts mutants. The 10K is a structural virus protein, but the 15K is nonstructural (Ben-Porat *et al.*, 1984d).

3. Excreted Glycoproteins

The envelope of PRV is composed of four major (and possibly several minor) glycoproteins. In addition, one nonstructural glycoprotein (90K) is produced in relatively large amounts by cells infected with the virus (Erickson and Kaplan, 1973); this protein is also efficiently excreted from the cells. While all virus glycoproteins are sulfated to varying degrees, the 90K glycoprotein is sulfated to a much higher extent than the others (Kaplan and Ben-Porat, 1976).

The sulfated moiety of the 90K excreted glycoprotein is not linked to carbohydrates, as is characteristic of mammalian mucopolysaccharides

(Erickson, 1976). Indeed, Pennington and McCrae (1977) have shown that while this protein is usually first glycosylated and then sulfated, the protein is sulfated even when glycosylation is inhibited. This supports the finding of Erickson (1976) mentioned above.

The 90K excreted glycoprotein does not react with any one of 10 different monoclonal antibodies directed against the virus structural glycoprotein isolated to date (H. Hampl and T. Ben-Porat, unpublished results). Furthermore, antisera raised against partially purified preparations of this protein react poorly with the virus structural glycoproteins. Therefore, while it appears unlikely that this protein is a modified structural glycoprotein, this possibility has not, as yet, been completely eliminated. The function of this protein is unknown.

4. Virus Enzymes

PRV probably specifies many different enzymes; only a few such enzymes have been identified to date.

In rabbit kidney cells infected with PRV there is an increase in the activity of TK, beginning about 1.5–2 hr postinfection (Nohara and Kaplan, 1963; Kamiya et al., 1964; Kaplan et al., 1965). The TKs present in infected and uninfected cells are antigenically distinct (Hamada et al., 1966), indicating that the newly synthesized enzyme is probably specified by PRV. That this is indeed the case is clearly shown by the fact that TK$^-$ mutants of the virus can be isolated (Ben-Porat et al., 1983b).

The TK activity specified by PRV differs in several aspects from that specified by HSV. Buchan and Watson (1969) analyzed the antigenic determinants of the TKs produced by BHK21 cells infected with either PRV or HSV and found the enzymes to be serologically distinct. Furthermore, the TK produced by PRV-infected cells differs also in other properties from the HSV-induced enzyme. The HSV-induced enzyme can phosphorylate deoxyuridine (Kit et al., 1967) whereas the TK induced by PRV cannot (Kaplan and Ben-Porat, 1970a). Furthermore, whereas the TKs specified by HSV-1 and HSV-2 are able to phosphorylate both thymidine and deoxycytidine, the enzyme specified by PRV will phosphorylate only thymidine (Kaplan et al., 1967; Jamieson et al., 1974).

PRV also codes for a DNA polymerase. The PRV-induced DNA polymerase differs antigenically from cellular DNA polymerase, as well as from the DNA polymerase induced by HSV-1 and HSV-2; it also differs in its requirement for KCl in in vitro assays (Halliburton and Andrew, 1976). Temperature-sensitive mutants of PRV deficient in the DNA polymerase have been isolated (Ben-Porat et al., 1982b, 1983b).

As high concentrations of thymidine in the growth medium will inhibit DNA synthesis of uninfected cells as a result of feedback inhibition of the synthesis of deoxycytidylic acid (Reichard et al., 1961; Morris et al., 1963), but will not inhibit DNA synthesis in cells infected with PRV (Kaplan, 1964), it was concluded that PRV probably specifies a ribonucleotide reductase that is resistant to feedback inhibition by dTTP (Kaplan and Ben-Porat, 1968). Recent experiments by Lankinen *et al.* (1982) have shown this to be the case. Infection of mouse L cells with PRV results in a 10-fold increase in the activity of ribonucleotide reductase. The properties of the induced enzyme differ from those of the enzyme present in uninfected cells; it has no requirement for ATP as a positive effector and is not feedback-inhibited by dTTP or dATP.

DNase is another enzyme induced in cells infected with PRV. The activity of this enzyme is inhibited by homologous antiserum but not by antiserum against HSV. The PRV-induced DNase is an exonuclease that catalyzes the release of deoxyribonucleoside-5'-monophosphates sequentially from the 3' terminus of the DNA (Keir, 1968).

After infection of rabbit kidney cells with PRV, there is also an increase in the level of dTMP kinase activity (Nohara and Kaplan, 1963). The dTMP kinase activity present in PRV-infected cells exhibits a greater thermostability than the activity present in uninfected cells and, in contrast to the latter, does not require substrate for stabilization (Nohara and Kaplan, 1963). However, the increased level of dTMP kinase activity in the PRV-infected cells, although prevented by inhibitors of protein synthesis, is not the result of the synthesis of new enzyme protein; instead, the enzyme in uninfected cells, which is normally unstable *in vitro*, appears to acquire new properties after infection (Kaplan *et al.*, 1967).

IX. DNA SYNTHESIS IN PRV-INFECTED CELLS

A. Rate of Virus DNA Synthesis

The rate of PRV DNA replication has been estimated to be approximately 1 μm (2×10^6 daltons)/min at 37°C; part of the DNA, however, replicates at twice that rate. The time required to replicate a unit size genome is thus for the most part approximately 45 min (Ben-Porat *et al.*, 1977).

PRV DNA replicates semiconservatively (Kaplan and Ben-Porat, 1964) and at early stages of DNA synthesis the virus DNA appears to replicate in a geometric fashion (Ben-Porat *et al.*, 1976a). However, at later times, only part of the newly replicated DNA molecules replicate again (Kaplan, 1964), even when encapsidation of the DNA is prevented (A. S. Kaplan and T. Ben-Porat unpublished results). Whether this is due to a limiting supply of some factors required for DNA replication or whether it reflects the mode of PRV DNA replication remains to be ascertained.

B. Properties of Intracellular Virus DNA

On the basis of a combination of results obtained from sedimentation studies, analysis of the structure of the DNA by electron microscopy, and analysis of the DNA with restriction endonucleases, the intracellular structures of virus DNA may be subdivided into three types: (1) structures of parental virus DNA before the initiation of virus DNA synthesis; (2) structures of replicating DNA during the first round(s) of replication; (3) structures of replicating DNA during late phases of replication.

1. Structure of Parental Virus DNA before Initiation of Virus DNA Synthesis

Despite the relatively high ratio of physical to infectious particles in populations of PRV, a large proportion of the virions will participate in the initial phases of the infective process; most of the virions can adsorb to the cells and 30 to 90% (depending on the quality of the virus and on the host cells) of the DNA in the adsorbed virions becomes associated with the nucleus of the cell. A significant proportion of the nicks present in the parental virus DNA is ligated and practically all the PRV DNA that penetrates the nucleus will replicate (Ben-Porat et al., 1976a).

Virus DNA molecules with single-stranded ends, as well as a small number of circular molecules, have been observed in the electron microscope at very early stages of infection, indicating that the parental genomes form circles (Jean and Ben-Porat, 1976). Analysis by means of restriction endonuclease cleavage of parental PRV DNA present in the infected cells before DNA synthesis has been initiated also indicated that most of the parental virus DNA molecules lose their free ends; instead of fragments representing the free ends, a new fragment consisting of the two end fragments joined together appears, indicating that the parental virus DNA forms circles (Ben-Porat et al., 1980).

Because many of the parental virus DNA molecules that could be isolated from the nuclei of the infected cells had single-stranded ends, it was postulated that upon entering the cell nucleus, the virus DNA is digested by an exonuclease and that circle formation occurred by annealing of cohesive ends. However, because no sequence homology between the two ends of the molecule exists (I. Davidson, personal communication; L. Harper and T. Ben-Porat, in preparation), this is probably not correct and parental DNA most likely circularizes by blunt end ligation.

It has been established that expression of virus functions is not required for circularization of the genome; it occurs efficiently in cycloheximide-treated cells (Jean and Ben-Porat, 1976; L. Harper and T. Ben-Porat, in preparation). Although it has been established that eukaryotic cells can efficiently blunt end ligate DNA, it is tempting to postulate that a structural virus protein may play a role in this process.

2. Structures of Replicating Virus DNA during the First Round of Replication

Examination by electron microscopy of PRV DNA molecules present within the infected cells soon after virus DNA synthesis has begun revealed the presence of circular or linear unit-size molecules containing large replicative loops; branched molecules of varying sizes were also observed. Because branched molecules, as well as molecules with "eyes," were not observed when DNA synthesis had been inhibited, it was concluded that these structures most likely represent replicative intermediates (Jean *et al.*, 1977). The information derived from electron microscopic analyses of replicating PRV DNA must be viewed with caution because of the inherent fragility of this DNA and the consequent generation of artifactual structures. The interpretation of the results obtained from electron microscopic analysis is, however, facilitated by the results obtained from restriction enzyme analysis. Analysis by this method of PRV DNA synthesized during the first round of replication reveals that most of the DNA is "endless." The end fragments normally present in mature DNA are underrepresented or missing completely and instead, a new fragment composed of the two joined end fragments appears (Ben-Porat and Veach, 1980). As most of the DNA synthesized during the first round of replication sediments with a maximum S value approximately twice that of mature DNA (Ben-Porat and Tokazewski, 1977), it is unlikely that the DNA is in the form of concatemers of a size compatible with the relatively small number of free ends observed. One is thus led to conclude on the basis of these results as well as those obtained from analysis of the DNA by electron microscopy that replication is initiated mainly on circular molecules by a mechanism that gives rise to "Cairns-type" structures. Although most of the PRV DNA is "endless" during the first round of replication, some replication of molecules with free ends can also be detected (Ben-Porat and Veach, 1980).

3. Structures of Replicating Virus DNA during the Late Phase of
   Replication

Only linear unit-size and longer-than-unit-size molecules, as well as concatemeric tangles, were observed by electron microscopy at this stage of virus DNA synthesis. Molecules with "eyes" and branches observed during the first round of replication were not present during later phases of DNA replication (Ben-Porat *et al.*, 1976b; Jean *et al.*, 1977).

Analysis of the fragments generated by digestion with restriction endonucleases of DNA synthesized during the late phase of replication shows that the free ends normally present in mature PRV DNA are underrepresented, and a fragment composed of the two joined end fragments appears (Ben-Porat and Rixon, 1979). The newly synthesized DNA is composed of tandem arrays of virus genomes, as indicated by the following. (1) Electron microscopic analysis of the DNA revealed the presence of longer-than-unit-size linear molecules. (2) Longer-than-unit-size DNA can be delineated within the concatemeric tangles. (3) Circular virus DNA molecules are not observed during the late phase of replication. (4) Some longer-than-unit-size single strands of virus DNA can be isolated after denaturation of the virus DNA. (5) In cells infected with DNA[+] mutants in which cleavage of the DNA to unit-size linear molecules is defective, a large proportion of the DNA that accumulates at the non-permissive temperature sediments in alkaline sucrose gradients with an S value indicating that it is at least two unit-sizes in length; electron microscopic examination of this DNA confirms the presence of longer-than-unit-size single-strands (T. Ben-Porat and M. Blankenship, unpublished results). Taken together, these data indicate that the virus DNA is in the form of linear concatemers rather than in the form of interlocked circular molecules.

The concatemeric tangles of PRV DNA resemble those of some bacteriophages. Thus, the tangles of virus DNA have dense cores similar to those observed in the replicating DNA of T4 or T7 bacteriophages (Huberman, 1968; Bernstein and Bernstein, 1974). Also, concatemeric PRV DNA sediments faster than can be accounted for by its molecular weight (i.e., degree of underrepresentation of free ends), indicating that the concatemers have an unusually compact conformation, a characteristic, for example, exhibited by the concatemers of T4 DNA (Frankel, 1968; Altman and Lerman, 1970; Curtis and Alberts, 1976). Furthermore, the high S value of the concatemeric structures is not affected by treatment with detergents, RNase, or proteases (Ben-Porat et al., 1976b).

Some of the characteristics of the concatemeric DNA could be due to the presence within the center of the tangles of virus proteins that are protected from proteolytic digestion by the DNA. Exposure of the concatemeric structures to high salt (isopycnic centrifugation in CsCl or KI) results in the loss of their high S values with the concomitant removal of the proteins that had been associated with the structures (Ben-Porat et al., 1984d). These results suggest that the rapidly sedimenting structure with which the concatemeric DNA is associated represents virus concatemeric DNA molecules folded around and held together by proteins.

X. ORIGINS OF DNA REPLICATION

Information concerning the origins of replication of PRV DNA is available from several different lines of evidence.

Originally, evidence was obtained by electron microscopy. The positions and the sizes of "eyes," as well as the lengths of the branches, on virus DNA extracted from infected cells suggested that replication of the DNA molecules may originate at at least two locations, one near or at the end of the molecule and one at a site approximately 20 μm from one of the ends (Jean et al., 1977). As it has become increasingly clear that replicating DNA is fragile, some of the replicating molecules that have been analyzed may have been broken circular molecules. This casts some doubt on the validity of the conclusions drawn from electron microscopic data regarding the location of the origins of replication.

Because the DNA of herpesviruses replicates as concatemers and because replication bears no temporal relationship to DNA maturation, it is not possible to locate the origin of replication in a Dintzis-type experiment (Dintzis, 1961), as has been done for the DNA of some other viruses, such as, for example, SV40 or adenovirus. The origins of replication of PRV DNA have, therefore, been located by labeling the DNA at the beginning of the first round of DNA replication. Analysis of this DNA by blot hybridization or by direct digestion revealed a gradient of radioactivity along the genome, with the highest specific activity being in the fragments originating from the inverted repeats (Ben-Porat and Veach, 1980). As replication proceeded, the specific activity of the different regions of the genome equalized, the gradient of radioactivity moving from the end of the molecule bearing the inverted repeat to the other end. The labeling pattern of the DNA indicated that the main origin of replication is located in the inverted repeats but that replication is also initiated on some molecules at other positions along the genome, in particular at a location around the middle of the U_L sequence. Furthermore, the results indicated that replication proceeded mainly unidirectionally.

Some evidence concerning the origins of DNA replication may also be obtained from the analyses of defective DNA. By analogy with other systems, it appears likely that origins of replication are essential to the replication of the defective DNA and that a segment of DNA containing the origin must therefore be present in this DNA. Indeed, analysis of defective PRV DNA indicates that different populations of defective DNA become enriched either in sequences present in the middle of the U_L or in sequences derived from the inverted repeats (Rixon and Ben-Porat, 1979). Furthermore, the sequences for which defective DNA is enriched replicate first within cells coinfected with standard and defective virions, indicating that an origin of replication must be present in the defective DNA (C. Wu and T. Ben-Porat, unpublished results). Some cloned restriction fragments of defective DNA will also, when cotransfected into susceptible cells with wild-type PRV DNA, allow replication of the plasmid sequences (C. Wu and T. Ben-Porat, unpublished results), indicating that they bear origins of replication. Results obtained from these types of experiments corroborate the conclusion drawn from other types of evidence (see above) concerning the locations of origins of replication on the PRV genome.

The fact that at least two origins of replication have been identified in the PRV genome, one in the inverted repeat and one at about the middle of the U_L sequence, is of interest and requires comment. Because DNA replication occurs in two phases (i.e., at early stages on circular unit-size molecules as Cairns-type structures and at later times presumably via a rolling circle mechanism in association with large concatemers), it is conceivable that the two origins are used with different efficiencies during each phase. It has been shown that during the first round of replication the origin present in the inverted repeat is used preferentially (Ben-Porat and Veach, 1980). On the other hand, studies with defective interfering particles, which are composed of tandem arrays of sequences comprising the origin in the U_L, show that in cells coinfected with these defective particles, the U_L origin is used preferentially at early stages of infection; this defective DNA replicates almost exclusively up to about 6 hr postinfection (C. Wu and T. Ben-Porat, unpublished results). Thus, both origins can function at early stages of infection.

The findings that during the normal course of infection the origin in the repeat is used preferentially for the first round(s) of replication, while in cells coinfected with defective particles, which are enriched for the origin in the U_L, the latter is used preferentially at early stages of infection, do not necessarily conflict. Thus, the origin in the repeat could be used preferentially at early stages of infection but the origin in the U_L could also be recognized. Because the origin in the U_L is represented in cells coinfected with the defective particles in vast excess over that present in the repeats, it might bind to and exhaust a limited pool of proteins essential for the initiation of DNA replication; initiation at the origins present in the DNA of wild type helper virus (including the origin in the repeats, which is normally used preferentially during the early stages of DNA replication) might thereby be effectively inhibited. At later times postinfection, the essential proteins might no longer be limiting and initiation at the origins of replication present in the standard virus DNA will also occur.

XI. MODE OF PRV DNA REPLICATION

Because no terminal redundancy is found on the genome of PRV, it is probable that upon entering the cells the genome termini are ligated directly. As circularization of the genome can occur in the absence of expression of virus functions, ligation of the ends must be mediated either by cellular proteins or by virus proteins that accompany the virus genome into the cell nucleus. The fact that purified virus DNA is infectious indicates that cellular proteins may be competent to mediate the circularization. However, the process of virus DNA replication may be somewhat different in cells transfected with virus DNA, and in cells infected with intact virions. The possibility that some virion proteins facilitate the circularization of the virus genome needs to be explored further.

After the DNA has circularized [and recombination as well as probably isomerization has occurred (see below)], DNA replication is initiated at a preferred origin of replication. The first round of replication occurs on a Cairns-type structure, which has been visualized by electron microscopy (Jean and Ben-Porat, 1976). Thereafter, the DNA is in a concatemeric form in which unit-size DNA molecules are in head-to-tail

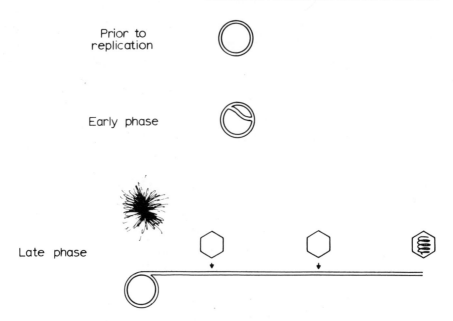

FIGURE 8. Diagrammatic representations of the stages of PRV DNA replication.

alignment (Ben-Porat and Rixon, 1979); it is likely that replication occurs by a rolling circle mechanism. Figure 8 summarizes the mode of replication of the PRV DNA.

XII. MAINTENANCE OF SEQUENCE IDENTITY OF THE INVERTED REPEATS

Analysis of the genome of isolates of PRV with restriction enzymes and examination of their denatured, self-annealed DNA in the electron microscope indicate that they contain genomes in which the homology between the repeats bracketing the unique sequences appears to be complete. This may be so because processes that ensure the identity of the repeats against genetic drift operate as if, for example, one of the repeats were to act as a template for the replication (or repair) of the other (Skare and Summers, 1977; Roizman, 1979). On the other hand, equalization of the sequences within each of the two inverted repeats may occur as a result of intermolecular recombination. Thus, the progeny of a mutant with unequal repeats may, by intermolecular recombination, give rise to virions with genomes with equal repeats that will then segregate.

Mutant viruses containing genomes in which the two repeat regions differ can be used to ascertain the processes leading to the equalization of these repeats. In principle, starting with a virion containing DNA with unequal repeats, one would expect, after extensive intermolecular exchange of the repeat region, 50% of the DNA molecules to have unequal repeats and 50% to have equal repeats (25% of each kind). This was found to be the case when the DNA in single virions present within one plaque produced by a virus containing two different sizes of inverted repeats was analyzed (Ben-Porat et al., 1984b). A similar result was also obtained when the virions from this plaque were further passaged several times at high multiplicity and individual plaques produced by the progeny virions were analyzed. Thus, it appears that propagation of virions containing genomes with unequal repeats at high multiplicity results in a population of virions in which molecules with equal and unequal repeats are present at the frequency predicted on the basis of random reassortment of the repeats between the DNA molecules (probably reflecting the high degree of intermolecular recombination between viral genomes). Equalization of the two repeats within a DNA molecule can therefore occur by intermolecular exchange of repeat sequences between DNA molecules with unequal repeats. Furthermore, intermolecular exchange of repeats is a rapid event; it appears to be complete by the time a virion has produced a small plaque. Thus, within the limitations of our ability to measure the two types of events, intermolecular exchange of repeats appears to be as rapid as is isomerization of the U_S sequence of the molecule (Ben-Porat et al., 1980, 1984b).

When the population of virions in a plaque formed by a virion containing unequal repeats was passaged several times at low multiplicity and the individual plaques produced by the progeny were analyzed, the frequency of the molecules with unequal repeats was very low (Ben-Porat et al., 1984b). Thus, passage of virions with unequal repeats at low multiplicity results in segregation of virions containing equal repeats that have been generated by intermolecular recombination and that are propagated as such thereafter. Because passage of the virions at low multiplicity probably mimics conditions of virus infection in nature, equalization of the inverted repeats in the genomes of virions in nature can be attributed to intermolecular recombination between the progeny of a virion with a mutation in one of the repeats, resulting in virions with equal repeats, and subsequent segregation of the recombinant progeny virions. A special mechanism ensuring equalization of the repeats (such as one repeat acting as a template for the synthesis or repair of the other) need not be invoked.

Figure 9 illustrates how intermolecular recombination leading to equalization of the repeats may occur. For the sake of simplicity, the molecules are represented in linear form, even though recombination probably occurs mainly between circular molecules (Ben-Porat, 1982). According to model I, intermolecular recombination generates molecules in which the internal repeat becomes terminal; recombination between a molecule in the original orientation with a molecule that has inverted,

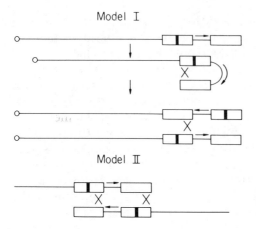

FIGURE 9. Diagrams illustrating how equalization of the inverted repeats might be achieved. The blocks with or without a black stripe represent the two unequal inverted repeats. The crosses indicate sites of recombination. The arrows above the region of the genome bracketed by the inverted repeats indicate the orientation of the U_S sequence.

generates two molecules, each of which has equal repeats. Two recombinational events, one intramolecular (which also generates an inversion of the U_S) and one intermolecular, must occur according to this model to result in equalization of the repeats within the molecule.

Model II illustrates how intermolecular recombination between the two molecules, which are aligned so that the unique sequences are in different orientations, can result in the equalization of the repeats. The model postulates that an even number of recombinational events occurs because molecules that have experienced an even number of recombinational events may have an advantage in replication (Ihara *et al.*, 1982). According to this model, equalization of the repeats (as well as inversion of the U_S) occurs via intermolecular recombination. On the basis of the available information, we cannot conclude which of the models is correct.

XIII. ISOMERIZATION OF THE VIRUS GENOME

When Sheldrick and Berthelot (1974) first reported the presence of inverted repeated sequences in the DNA of HSV, they suggested that these repeats might promote isomerization of the unique sequences they bracket. This prediction has proved to be correct.

Isomerization of the U_S region of the PRV genome (which is bracketed by inverted repeats) is a rapid process; not only is it completed by the time the progeny of a single plaque has been amplified, but analysis of the virus DNA within a single small plaque also revealed the presence of equimolar amounts of the two isomeric forms of the PRV DNA (Ben-Porat *et al.*, 1980).

Evidence indicating that the processes that lead to the inversion of the unique sequence in PRV DNA occur prior to DNA replication was obtained by analyzing the fate of parental virus DNA strands that are transferred to progeny virions in density shift experiments (Ben-Porat *et al.*, 1982a). The most salient finding from these studies was that few recombinational events between parental DNA strands and progeny DNA strands accumulating in the cells could be detected. Indeed, both genetic and biochemical evidence indicate that recombination occurs mainly between parental DNA molecules prior to or during the earliest stages of DNA replication (Ben-Porat *et al.*, 1982a; Ihara *et al.*, 1982). The fact that parental virus DNA strands that have replicated and are transferred to progeny virions have not undergone recombination with progeny DNA strands shows that isomerization of the molecules does not occur as a result of autonomous replication of the U_S and the U_L regions of the genome, followed by their reassociation, nor does it occur by a mechanism that is linked to DNA replication. The available evidence indicates that both isomerization of the molecules (either by intramolecular or by intermolecular recombination) and general recombination (intermolecular) are early events that occur prior to DNA replication.

In several prokaryotic systems the inversion of unique sequences bracketed by inverted repeats is sequence specific and mediated by specific proteins. It was thus tempting to postulate that the inversions of the unique sequences in the herpesviruses are achieved in a similar fashion. An interesting observation, germane to the problem concerning the mechanism by which inversion occurs, deserves mention. PRV DNA is classified as a class 2 herpesvirus DNA molecule; i.e., although it contains both a U_S and a U_L sequence, only the U_S sequence is bracketed by inverted repeats and only the U_S sequence inverts. Thus, only two isomeric forms of the DNA exist. The Norden (Bucharest) vaccine strain of PRV, derived from a virulent strain that had been attenuated by more than 800 passages in embryonated eggs and in chick embryo fibroblasts, contains a genome that can be classified as a class 3 herpesvirus DNA molecule (Lomniczi *et al.*, 1984); both the U_S and the U_L sequences are bracketed by inverted repeats and both invert themselves with respect to one another. Thus, four isomeric forms of the genome of the Norden vaccine strain are found. In the Norden strain, a sequence of nucleotides (estimated to be 0.3×10^6 daltons in size) normally present in all other PRV strains at the end of the U_L only, has been translocated in inverted form to the other end of the U_L (adjacent to the internal inverted repeat). It is likely that as a result of this translocation and the consequent presence of an inverted repeat bracketing the U_L sequence of the Norden strain, the U_L sequence inverts itself.

As discussed above, the two ends of the genome of wild-type PRV do not share any homology; neither has homology between the two ends of the Norden strain been detected (L. Harper and T. Ben-Porat, unpublished results). If a sequence-specific protein were responsible for the inversion of the U_S sequence in PRV, one would not expect it to recognize the sequences bracketing the U_L of the Norden strain. Because inversion of the U_L occurs at high frequency in the Norden strain, it seems unlikely that the inversion of unique sequences bracketed by inverted repeats is mediated by a sequence-specific protein.

The finding that genetic markers on the PRV genome behave as though they were linked in a circular fashion (see below), even though the molecule has unique ends, may also have some bearing on the events that may lead to the isomerization of the molecules. This observation is best explained by invoking the possibility that molecules that have undergone an even number of recombinational events have an advantage in replication. It is interesting to note that a double crossover between two DNA molecules, one in each repeat, would give rise to the inversion of the unique segment of DNA bracketed by these repeats. This is shown by the model for isomerization depicted in Fig. 10. This model is in essence similar to that proposed previously by other investigators (Sheldrick and Berthelot, 1974; Skare and Summers, 1977), but is modified to include the findings that (1) the DNA molecules are in a circular form during recombination and (2) molecules that have experienced an even number of crossover events are favored.

XIV. PROCESS OF VIRION ASSEMBLY

A. Capsid Assembly

The capsids of PRV are assembled in the nucleus of the infected cell, as are the capsids of HSV (Morgan et al., 1954; Reissig and Melnick, 1955), but the virus proteins are synthesized in the cytoplasm (Fujiwara and Kaplan, 1967; Ben-Porat et al., 1969).

The processes leading to the accumulation of the capsid proteins in the nucleus, as well as the details of the process of capsid assembly, are not well understood. However, some information concerning these processes is available and indicates that the migration of the capsid proteins to the nucleus is linked to capsid assembly and is dependent upon the synthesis of several functional proteins and the processing of at least one.

Under conditions of arginine deprivation, capsid proteins are synthesized but do not migrate to the nucleus (Mark and Kaplan, 1971), and capsids are not assembled. While most virus proteins are synthesized by arginine-starved cells, three arginine-rich acid-soluble proteins are not, raising the possibility that these acid-soluble proteins may play a role in capsid assembly (Chantler and Stevely, 1974).

Studies with DNA$^+$ ts mutants have yielded some information concerning the processes leading to the formation of the empty capsids. In DNA$^+$ mutant-infected cells, in which capsids are not assembled, essentially all the virus-induced polypeptides are synthesized with the following two notable exceptions. (1) In cells infected with some ts mutants, the major capsid protein (142K) is present in diminished amounts because a thermosensitive 142K protein is synthesized. (2) In cells infected with

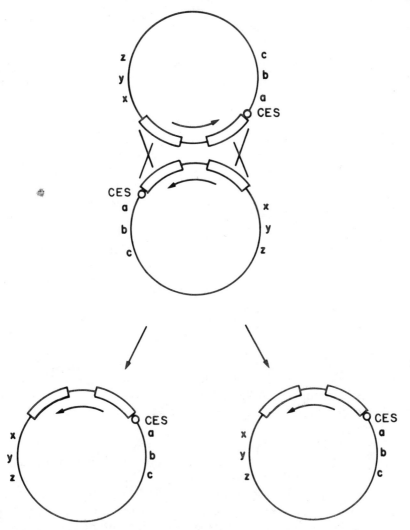

FIGURE 10. Model of isomerization. A double crossover event involving both of the inverted repeats between two molecules in the same isomeric form gives rise to two molecules in which the U_S sequence has been inverted. The blocks indicate the inverted repeats. The arrows above the region of the genome bracketed by the repeats indicate the orientation of the U_S sequence, CES denotes the cleavage–encapsidation site. Modified from Ben-Porat *et al.* (1982).

all DNA$^+$ capsid$^-$ mutants studied, the 35K capsid protein is not detectable. The 35K protein is also not detectable after a relatively short labeling period in wild-type infected cells but becomes prominent after longer labeling periods because it is processed from a precursor protein. In the cells infected with the capsid$^-$ mutants, a 35K precursor protein (antigenically related to the 35K protein) is detectable but processing to

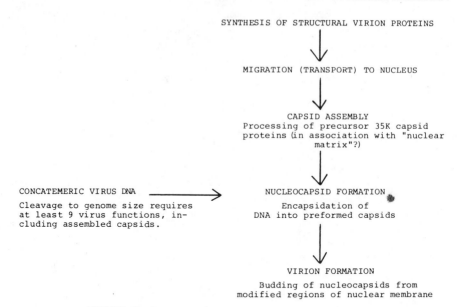

SYNTHESIS OF STRUCTURAL VIRION PROTEINS

MIGRATION (TRANSPORT) TO NUCLEUS

CAPSID ASSEMBLY
Processing of precursor 35K capsid
proteins (in association with "nuclear
matrix"?)

CONCATEMERIC VIRUS DNA ————————→   NUCLEOCAPSID FORMATION

Cleavage to genome size requires   Encapsidation of
at least 9 virus functions, in-   DNA into preformed capsids
cluding assembled capsids.

VIRION FORMATION
Budding of nucleocapsids from
modified regions of nuclear membrane

FIGURE 11. Sequence of events leading to virion assembly.

the 35K does not take place. Because the lack of processing of the 35K
protein occurs in cells infected with mutants belonging to different complementation groups, it appears that the processing of the 35K protein is
dependent on several gene products and is related to capsid assembly
(Ladin et al., 1982).

Analysis of the distribution of the capsid proteins between nuclear
and cytoplasmic fractions of infected cells revealed that in the capsid⁻
mutant-infected cells, the major capsid proteins do not become associated
with the nuclear fraction (Ladin et al., 1982). Thus, the accumulation of
capsid proteins in the nucleus, the assembly of virus capsids, and the
processing of the p35K capsid proteins appear to be interdependent events.

It appears likely that the following sequence of events occurs leading
to the assembly of capsids in the nuclei (Fig. 11). Structural virus proteins
are transported (or diffuse) into the nucleus of the infected cells. Processing of the precursor 35K protein to the 35K capsid protein and concomitant capsid assembly occurs. Assembly promotes further transport
of diffusion of the structural proteins by creating a "sink" favoring continued accumulation of capsid proteins (and of capsids) in the nuclei. A
defect in any one of the structural capsid proteins or in any other protein
that may be required for capsid assembly, results in a lack of capsid assembly and also leads both to a lack of processing of the 35K precursor
protein and to a lack of accumulation of virus capsid proteins in the nuclei
of the infected cells.

B. Precursor Relationship between Empty Capsids and Nucleocapsids

Three types of virus particles may be distinguished by examination of thin sections of infected cells in the electron microscope, as well as by isolation of particles from the cells (see above): empty capsids (devoid of DNA), nucleocapsids (containing DNA), and virions (enveloped nucleocapsids). The three different types of structure represent different stages in the process of virion assembly, and precursor relationships between these structures have been established. Temperature-sensitive mutants of PRV able to assemble capsids but not nucleocapsids at the nonpermissive temperature have been isolated and have proved useful in establishing these precursor relationships (Ladin *et al.*, 1980). It was established that after shiftdown of cells infected with these mutants (in the presence of an inhibitor of protein synthesis) to the permissive temperature, the empty capsids that had accumulated at the nonpermissive temperature disappeared and, instead, full capsids appeared, indicating that the empty capsids that had assembled in the cells at the nonpermissive temperature are precursors of the nucleocapsids. Similarly, the nucleocapsids eventually disappeared and mature enveloped virions appeared, again indicating a precursor relationship between the two.

C. Cleavage of Concatemeric DNA and Nucleocapsid Formation

It has been known for some time that virus DNA synthesized at various times after infection has an equal chance to be packaged into virions and that in the infected cells, there is a pool from which virus DNA is withdrawn at random to be encapsidated (Ben-Porat and Kaplan, 1963). This conclusion has been verified recently (Ben-Porat *et al.*, 1982a).

Because replicating DNA of PRV is "endless," cleavage must occur at specific sites along the molecules to generate the mature PRV genomes, i.e., linear DNA molecules with unique ends. It has been established that in wild-type infected cells, "maturation" of virus DNA to unit length does not require DNA synthesis but does require concomitant protein synthesis, probably because the proteins necessary for virus DNA maturation are present in limiting amounts (Ben-Porat *et al.*, 1976b; Ladin *et al.*, 1980).

To elucidate the mechanism by which virus DNA matures from the concatemeric forms into unit-length genomes, DNA$^+$ ts mutants of PRV defective in the processes of cleavage of concatemeric DNA to genome size were used. Mutants belonging to nine different complementation groups behave in this manner, indicating that maturation of the DNA is a complex process. In cells infected with the mutants belonging to six complementation groups, capsids with electronluscent cores (empty capsids) accumulated. However, in cells infected with mutants belonging to three complementation groups, empty capsids failed to accumulate; these mutants are defective in both capsid assembly and virus DNA maturation, indicating that a single function can affect both processes. Thus, it appears that cleavage of concatemeric PRV DNA to unit size is dependent upon capsid assembly. However, in addition, several other factors (not directly associated with capsid assembly) are also required (Ladin et al., 1980). At present little is known about these factors.

Available evidence seems to indicate that encapsidation occurs by a process that may be related to "headful" encapsidation but that specific cleavage sites are also recognized. While the capsids seem to accommodate DNA molecules that are both larger as well as smaller than the standard size (Ben-Porat et al., 1975b; Rubenstein and Kaplan, 1975; Ben-Porat et al., 1984), it has been found generally that the DNA molecules that are encapsidated are approximately the size of standard DNA. For example defective virions contain DNA molecules in which many of the virus genes are missing but due to reiterations of restricted regions of the genome the overall sizes of the defective genomes are similar to those of the standard virus. Thus, something akin to the "headful" mechanism of encapsidation may occur in this system. Because the genome of PRV has a unique set of end sequences, site-specific cleavage of concatemeric DNA must also occur.

The sequence of events leading to virion assembly thus is as follows. (1) The capsid proteins are assembled into empty shells; (2) concatemeric virus DNA is cleaved at specific sites and is packaged into these capsids, the capsids themselves, as well as other noncapsid virus proteins, being responsible for cleavage of virus concatemeric DNA to linear mature-genome-length DNA. The sequence of events leading to assembly of the virus particles is outlined in Fig. 11.

D. Biogenesis of the Virus Membrane

Relatively little information regarding membrane changes induced by infection with PRV is available.

The nuclear membrane in cells infected with PRV becomes convoluted as it does in cells infected with other herpesviruses (Darlington and Moss, 1968). In all cell types infected with different virus strains examined to date, the virus acquires its envelope by budding from the nuclear membranes. The proteins normally associated with the cellular membranes are not, however, present in the areas of these membranes from which the virus buds because purified virion preparations are free of cellular proteins (Ben-Porat et al., 1970).

Infection of cells with PRV stimulates phospholipid metabolism; there is an increase after infection in the incorporation of $[^{32}P]$-, $[^{3}H]$choline, and $[^{3}H]$myoinositol into phospholipids (Ben-Porat and Kaplan, 1971). While the increase in incorporation of these labeled precursors into cytoplasmic membranes is significant, it is especially marked into the inner nuclear membrane. Infection also changes the relative degree of incorporation of labeled precursors into different phospholipids. Approximately three times more of the total label derived from [^{32}P]- or [^{3}H]choline incorporated into phospholipids appears in sphingomyelin in infected than in uninfected cells. This relative increase parallels the relative richness of the inner nuclear membrane of the cells in sphingomyelin (Ben-Porat and Kaplan, 1971, 1972).

Despite the stimulation in phospholipid metabolism after infection, most of the phospholipids that become part of the virus membrane preexist in the cell at the time of infection. The virus envelope is, however, not derived from random areas of nuclear membrane into which virus-specific proteins have been inserted but from areas that have been metabolically more active than the remainder of the nuclear membrane (Ben-Porat and Kaplan, 1971, 1972). Thus, areas of the nuclear membrane destined to become virus envelopes appear to be assembled *de novo* mainly from phospholipids present in the cells at the time of infection and from virus-specific glycoproteins synthesized by the cells.

The factors controlling recognition by the nucleocapsids of the areas of the nuclear membrane that are virus-specific are not known. It is likely that the nucleocapsids bear on their surface a "recognition component" because only nucleocapsids acquire an outer envelope and envelopment of empty capsids is a rare event. Thus, it is likely that empty and full capsids differ in their surface structure with respect to the presence of this component.

XV. ORGANIZATION OF THE GENOME OF PRV

A. Phenotypic Characterization of Mutants

1. Complementation Analysis

Mutants of PRV have been isolated in several laboratories, classified into functionally different groups by means of complementation tests, and partially characterized phenotypically.

Pringle *et al.* (1973) reported the isolation of 10 *ts* mutants belonging to 9 complementation groups. Vo-Doung *et al.* (1977) have isolated 9 ts mutants belonging to four complementation groups. Ben-Porat *et al.* (1982b) have placed 31 ts mutants into 19 complementation groups. Unfortunately, a collaborative study in which the mutants isolated in different laboratories are compared to determine whether they are mutated in different genes has not been performed.

In addition to these mutants, one mutant (*ts*L) that does not complement any of the other *ts* mutants and that is defective in uncoating at the nonpermissive temperature has also been identified (Feldman *et al.*, 1981). Whether this mutant is altered in a gene that is different from any of the genes that have been mutated in the other ts mutants is unknown. Mutants defective in the TK gene have also been isolated (Ben-Porat *et al.*, 1983b).

The genome of PRV is sufficiently large to code for at least 100 proteins (with an approximate molecular weight of 50K). It is clear, therefore, that further isolation of additional mutants is required if the virus genome is to be saturated with mutations.

The preliminary phenotypic characterization of most of the mutants of PRV published to date deals mainly with determining whether cells infected with these mutants at the nonpermissive temperature synthesize virus DNA, whether they assemble virus particles, whether they specify defective enzymes, such as TK, polymerase, or DNase, and whether they inhibit cellular DNA synthesis.

In none of the cells infected with either DNA$^+$ or DNA$^-$ ts mutants of PRV described to date are nucleocapsids or enveloped particles formed at the nonpermissive temperature; no mutant of PRV defective in a virus glycoprotein or in the process of nucleocapsid envelopment has been described. The characteristics of some mutants belonging to different complementation groups isolated in our laboratory are given in Table V.

2. The DNA$^-$ Mutants

Two laboratories have reported that the mutants belonging to about half of the complementation groups identified were DNA$^-$ at the nonpermissive temperature. In the study of Pringle *et al.* (1973), 4 of the 9 complementation groups were DNA$^-$. Similarly, mutants in 10 of the 19 complementation groups identified by Ben-Porat *et al.* (1982b) were defective in their ability to synthesize DNA despite the fact that the mutants were isolated after mutagenesis of the virus stocks by three different methods. Mutants in only two complementation groups did not induce the synthesis of a functional DNA polymerase. While one of the polymerase$^-$ mutants appears to be truly defective in the DNA polymerase gene [it induces the synthesis of a thermolabile DNA polymerase (Ben-Porat *et al.*, 1983b)], the other mutant is an IE regulatory mutant defective in the expression of all early functions (Ihara *et al.*, 1983). The function defective in mutants belonging to another DNA$^-$ complementation group has also been identified. These mutants are defective in the major (136K) DNA-binding protein (Ben-Porat *et al.*, 1983a). The other DNA$^-$ mutants (which fall into 7 complementation groups) must therefore be defective in other virus proteins necessary for DNA synthesis. Thus, as previously demonstrated for HSV (Schaffer, 1975), a relatively large proportion of the functions specified by the PRV genome is necessary for the process of virus DNA synthesis. Interestingly, some of the DNA$^-$ mutants isolated by Pringle *et al.*, (1973) were less effective in

TABLE V. Characteristics of Mutants in Different Complementation Groups[a]

Complementation group	Mutant class[b]	Rescuing fragment MU	DNA synthesis	Polymerase activity	TK activity	Assembly of capsids	Assembly of nucleocapsids	Margination of chromatin	Inhibition of cell DNA synthesis	Inhibition of cell protein synthesis	Affected function
1	IE	0.95–0.99	−	−	−	−	−	−	±	±	180K IE protein
2	E	0.09–0.18	−	+	+	+	−	+	+	+	ND[c]
3	E	0.14–0.18	−	+	+	+	−	+	+	+	136K DBP
4	E	0.09–0.18	±	+	+	+	−	+	+	+	ND
5	E	0.09–0.18	−	−	+	+	−	+	+	+	DNA polymerase
6	E	0.18–0.27	±	+	+	+	−	+	+	+	ND
7	E	0.09–0.18	−	+	+	−	−	+	+	+	ND
8	E	0.27–0.41	−	+	+	+	−	+	+	+	ND
9	E	0.64–0.70	−	+	+	+	−	+	+	+	ND
10	E	0.01–0.06	−	+	+	+	−	+	+	+	ND
11	L	0.45–0.52	+	+	+	+	−	+	+	+	ND
12	L	0.45–0.52	+	+	+	−	−	+	+	+	ND
13	L	0.45–0.52	+	+	+	−	−	+	+	+	142K capsid protein
14	L	0.18–0.27	+	+	+	+	−	+	+	+	ND
15	L	0.18–0.27	+	+	+	+	−	+	+	+	ND
16	L	0.09–0.18	+	+	+	+	−	+	+	+	ND
17	L	0.09–0.18	+	+	+	+	−	+	+	+	ND
18	L	0.52–0.54	+	+	+	+	+	+	+	+	ND
19	TK−	0.43–0.45	+	+	−	+	−	+	+	+	Thymidine kinase
	Uncoating	ND	−	−	−	−	−	−	−	−	Capsid protein?

[a] From Ihara et al., (1982), Feldman et al., (1982), Ben-Porat et al., 1983a,b and unpublished results.
[b] IE, immediate-early; E, early (DNA− or DNA±); L, late (DNA+).
[c] ND, not determined.

inhibiting cellular DNA synthesis than were the wild-type or the DNA $^+$ mutants.

The fact that cells infected with most but not all of the DNA $^-$ mutants assemble naked capsids (see Table V) deserves comment. It has been known for some time that virus particles devoid of electron-dense cores (i.e., of DNA) are assembled in cells infected with PRV under conditions that do not allow virus DNA synthesis to occur (Reissig and Kaplan, 1962) and that therefore all the proteins necessary for capsid assembly can probably be classified as early–late proteins (see above). Of interest is the fact that in cells infected with one DNA $^-$ mutant (complementation group 8) empty capsids are not assembled. The major virus proteins are, however, synthesized in cells infected with this virus (T. Ben-Porat, unpublished results). If this mutant is indeed defective in a single gene, as is suggested by the results of the complementation tests, as well as by the fact that this mutant is rescued by a single specific restriction fragment (Ben-Porat et al., 1982b; Ihara et al., 1982), one would be led to conclude that capsid proteins are somehow involved in DNA synthesis.

Analysis by PAGE of the proteins synthesized by cells infected at the nonpermissive temperature with DNA $^-$ ts mutants revealed the presence of early and early–late virus proteins (Ben-Porat et al., 1983b, and T. Ben-Porat unpublished results). Late proteins are not detectably synthesized in cells infected with DNA $^-$ mutants. However, the ratio of different early to early–late proteins that are synthesized varies somewhat in the different mutant-infected cells. This might reflect a different degree of "leakiness" of the mutant but might also be directly affected by the genetic lesion in the mutants.

3. The DNA $^+$ Mutants

All DNA $^+$ mutants that have been studied (belonging to nine complementation groups) are defective in their ability to cleave concatemeric DNA, and nucleocapsids are not assembled in cells infected with these mutants at the nonpermissive temperature. The PAGE profile of the proteins synthesized by cells infected at the nonpermissive temperature with DNA $^+$ mutants or with wild type virus are similar with the exception that processing of the p35K protein to the 35K protein is not detectable in the capsid $^-$ mutants (Ladin et al., 1982).

B. The Genetic Map

Intermolecular recombination between the genomes of PRV is a rather frequent event in cells infected with this virus. The recombinational frequency has been estimated to be 3% per kilobase (Ihara et al., 1982).

Pringle et al. (1973) isolated 10 ts mutants belonging to nine different complementation groups; genetic crosses between these mutants showed that they could be arranged in a linear array on a genetic map. In another study, it was found that when relatively closely linked markers were analyzed, clear-cut linear linkage maps between the mutants studied could be constructed (Ihara *et al.*, 1982); each group of closely linked markers gave internally consistent results but when these partial linear linage maps were used to construct a map that included all markers, the genetic map appeared to have no termini and the same markers appeared at each end. Furthermore, markers that mapped physically at opposite ends of the genome (as determined by marker rescue) were found to be closely linked on the genetic map (Ihara *et al.*, 1982). Thus, the genetic map of PRV is topologically equivalent to a circle.

The finding that the genetic map of PRV is circular was surprising because even though the virus genomes may recombine while in the form of circles or head-to-tail concatemers (the intracellular form in which the DNA is found), one would not expect the genetic map to be circular because the mature form of the virus genome has unique ends and cleavage of the DNA prior to or during packaging should obscure any circular linkage between the genetic markers. One explanation for the apparent circularity of the genetic map is that there is a prevalence of even-numbered crossover events between recombining molecules or alternatively that molecules that have experienced an even number of recombinational events have a replicative advantage. As mentioned above, recombination is an event that occurs during the early stages of infection, a time when most of the genomes are in a circular form. In order for these molecules to retain the form of unit-size circles after recombination, an even number of crossover events must take place. If one postulates that genomes that have retained the form of a unit-size circle prior to or during the first round of replication have a selective advantage in replication, the genetic map would appear to be circular.

XVI. CELLULAR MACROMOLECULAR SYNTHESIS IN THE INFECTED CELLS

A. Protein Synthesis

Infection of cells with PRV results in an inhibition of cell-specific protein synthesis; this can be deduced from the reduction in the rate of incorporation of essential amino acids into the proteins synthesized by the infected cells. Interestingly, there is a difference in the magnitude of this reduction depending upon which amino acid is used. Thus, while the rate of leucine incorporation during the early stages of infection is relatively stable, as a result of a smooth switchover from the synthesis of cell-specific proteins to the synthesis of virus-specific proteins, the rate of lysine incorporation decreases rapidly after infection. Furthermore, the rate of arginine incorporation increases during the early stages of infection (Kaplan *et al.*, 1970). These changes in the rate of incorporation of amino acids after infection are not due to changes in intracellular pools but appear to reflect a decrease in the rate of synthesis of cellular proteins and a concomitant increase in the synthesis of virus proteins (Kaplan *et al.*, 1970; Ben-Porat *et al.*, 1971; Saxton and Stevens, 1972).

The incoming PRV virions do not inhibit cellular protein synthesis. Thus, actinomycin-treated, infected and uninfected cells synthesize approximately the same amounts of protein. Furthermore, infection of cells with a mutant unable to express any virus functions (an uncoating mutant) does not interfere with cellular protein synthesis (Feldman *et al.*, 1981; Ihara *et al.*, 1983).

Kennedy *et al.* (1981) have reported that the ribosomal S6 protein is more highly phosphorylated in PRV-infected cells than in uninfected cells. Furthermore, while the S16 or S18 ribosomal proteins are not phosphorylated in uninfected cells, it becomes phosphorylated after infection. On the basis of these findings the authors concluded that the increased phosphorylation of the ribosomal proteins in the infected cells could possibly be related to the inhibition of translation of cellular mRNA.

The inhibition of cell-specific protein synthesis is due, at least in part, to the synthesis within the infected cells of factors that prevent the association of cellular messages with ribosomes. The inhibition of the synthesis of cellular proteins by the IE proteins occurs at the level of translation. Thus, few cellular proteins are synthesized *in vivo* by 30 min after the removal of cycloheximide from infected cells which had been exposed to the drug from the time of infection because of the expression of the IE proteins (see Section VIII.C.1c). However, translationally competent cellular mRNA can be isolated from these cells (Ihara *et al.*, 1983).

It is interesting to note that, as mentioned above, at relatively early stages of the normal course of infection, the inhibition of synthesis of some cellular polypeptides is more pronounced than that of others (Fig. 6). Furthermore, while the level of most cellular mRNA within the infected cells remains constant for several hours, as determined by the Northern or dot-blot hybridization procedures using cloned cellular genes as probes, as well as by *in vitro* translation of the mRNA isolated from the infected cells, a reduction in the amounts of others is observed (S. Watanabe, A. Deatly, and T. Ben-Porat, in preparation). The available data thus show that the inhibition of cellular protein is achieved at the level of cellular mRNA translation as well as at the level of message stability. This latter effect is differential, some mRNA species being affected more rapidly than others. Thus, the process of inhibition of cellular protein synthesis is complex and does not affect all cellular mRNA species in an identical fashion.

B. RNA Synthesis

1. Ribosomal RNA

The inhibition of the synthesis of mature rRNA is relatively rapid; it is inhibited by more than 95% by 4 hr postinfection. The synthesis of precursor 45S RNA, although reduced, is inhibited less extensively. The formation of RNA is thus affected by infection both at the level of the transcription of precursor rRNA, as well as at the level of its processing (Rakusanova and Kaplan, 1970; Kyriakidis and Stevely, 1982). The 45S precursor rRNA synthesized in the infected cells is also undermethylated (Furlong et al., 1982).

2. Messenger RNA

Transcription of cellular DNA occurs in the infected cells at a reduced rate. The turnover of the cellular transcripts normally observed in uninfected cells also is less pronounced in the infected cells (Rakusanova et al., 1972).

The synthesis of cellular mRNA is inhibited more rapidly by infection with PRV than is the overall synthesis of total cellular RNA. This can be deduced from the fact that at early stages of infection (when about 80% as much polysome-associated cellular mRNA is present in the infected as in the uninfected cells), few of cellular transcripts that are synthesized by the infected cells becomes associated with polysomes (Rakusanova et al., 1972). If one assumes that newly synthesized mRNA can compete successfully with existing polysomal mRNA for association with ribosomes, one is led to conclude that functional cellular mRNA synthesis is inhibited more rapidly than transcription, probably reflecting an interference by infection with the orderly processing of the cellular transcripts.

C. DNA Synthesis

Cellular DNA synthesis is inhibited gradually in PRV-infected cells (Kaplan and Ben-Porat, 1963). This inhibition is most likely related to the inhibition of cellular protein synthesis (see above).

XVII. SEQUENCE HOMOLOGY BETWEEN THE GENOMES OF PRV AND HSV

The sequence homology between PRV and HSV-1 and HSV-2 has been shown by both DNA–DNA hybridization (Ludwig et al., 1972) and

DNA–RNA hybridization (Bronson *et al.*, 1972) to be a maximum of 8%. Analysis of the distribution of related sequences on these genomes by the Southern technique using whole heterologous genomes as probes indicated that the sequences that are homologous are distributed throughout most of the genomes. Homology is most pronounced in sequences in the U_L regions, as well as in those regions of the inverted repeat that code for the IE functions. Some regions in which no homology between PRV and HSV-1 or HSV-2 exists were also found. It is noteworthy that these regions were more commonly found in the genome of HSV-1 than that of HSV-2, suggesting that HSV-2 is more closely related to PRV than is HSV-1 (Rand and Ben-Porat, 1980).

The finding that PRV and HSV share relatively few homologous sequences but that these are distributed throughout most of their genomes is compatible with the accepted view that although there has been considerable evolutionary divergence, these viruses have evolved from a common ancestor. Despite this divergence, a degree of colinearity between the genomes of PRV and of HSV has been retained. This partial colinearity can be detected in cross-hybridization experiments (Ben-Porat *et al.*, 1983b; Davison and Wilkie, 1983), as well as by a comparison of the positions on the physical map of HSV-1 and PRV of genes coding for similar functions (Ben-Porat *et al.*, 1983b). The colinearity between PRV and HSV-1 is with the I_L configuration of HSV, not the prototype configuration in which the genome of HSV is usually displayed.

A noteworthy exception to colinearity is found in a region of the genome that includes the genes coding for the 136K major DNA binding protein and for the DNA polymerase. Thus, despite the fact that the DNA-binding proteins of PRV and HSV are closely related antigenically (Littler *et al.*, 1981; Ben-Porat *et al.*, 1983a), it appears that the region of the genome containing this gene has been either translocated or inverted on the genome of one of these viruses. The available information concerning the colinearity between the genomes of HSV-1 and PRV is summarized in Fig. 12.

XVIII. AUJESZKY'S DISEASE

Aujeszky's disease, a disease caused by PRV, has been recognized for a long time. Descriptions of the disease in cattle appeared in the United States in the early part of the 19th century (Hanson, 1954). This disease, which has been called "Mad Itch" in the United States because of the vigorous rubbing of their affected parts by infected cattle, was also prevalent in other parts of the world; it was first demonstrated by Aujeszky (1902) in Hungary to be the result of a virus infection. In 1931, Shope showed that the "mad itch" appearing in cattle in Iowa was the same

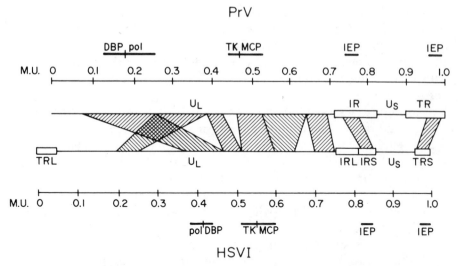

FIGURE 12. Cross-hybridization and position of genes coding for known virus functions on the physical maps of the PRV and HSV-1 genomes. The genome of HSV-1 is in the I_L arrangement. DBP, DNA-binding protein; pol, DNA polymerase; TK, thymidine kinase; MCP, major capsid protein; IEP, immediate–early protein. Hatched areas indicate areas of cross-homology.

disease as that described by Aujeszky (Shope, 1931, 1932). The disease is prevalent in most European countries. Obel (1965) reported its first appearance in Sweden. Pseudorabies also has been reported to occur in China (Lin, 1947).

Not long after the virus etiology of Aujeszky's disease was discovered, Marek (1904) called the disease "infectious bulbar paralysis" because it affected the medullary centers in rabbits and under certain conditions of infection, only the medulla appeared to be affected. However, it turned out that in other animals (monkeys and swine), there is no particular susceptibility of the medulla to PRV and the name "infectious bulbar paralysis" is therefore not a particularly apt description of this disease. Because of certain clinical aspects of the disease in some animals, there was at first some confusion about the relation of this virus to rabies, whence the name "pseudorabies," but it was soon found that the two viruses (and the diseases that they cause) are totally unrelated.

The clinical features of PRV infection in swine depend on the age of the animal. The most acute and fatal form of the disease occurs in newborn pigs; usually 100% of these animals die. Pigs that are 1 month old are not as severely affected and the mortality rate is lower. The disease is even less severe in 5- to 6-month-old pigs. Adult pigs are rarely fatally infected; usually they suffer a subclinical or very mild infection (Kaplan, 1969; McKercher, 1973; Lee and Wilson, 1979; Gustafson, 1981). However, some strains of virus may cause severe symptoms even in adult swine (B. Lomniczi, personal communication).

Pregnant sows frequently abort after infection. Infection very early during pregnancy usually results in the resorption of the embryos; later during gestation, infection will result in fetus fatality and expulsion (Gustafson, 1981).

In some cases, skin lesions resembling those of herpes labialis in man have been observed on the snouts of young pigs infected with PRV (Gustafson and Kanitz, 1975). In most other susceptible species, notably cattle, the course of the disease is rapid and usually fatal. Epizootics of pseudorabies occur in those countries of the world where pigs are produced. The question of transmission of the virus was puzzling for some time. It is now generally agreed that the main reservoir is the latently infected pig (Gustafson, 1981).

PRV persists in the latent state in infected pigs that have recovered from the disease; this latency plays an important role in the transmission of virus to susceptible animals. Latent virus in pigs can be reactivated by the stress of adverse climatic conditions (Howarth, 1969), as well as by experimental means (Mock et al., 1981; Lee and Shiffy, Crandell et al., both quoted by Davies and Beran, 1980; H.-J. Rhiza, personal communication). PRV or its genome has been detected in various tissues of latently infected swine (Sabó and Rajčáni, 1976; Gutekunst, 1979). After virus can no longer be detected or recovered from most organs, it is still present in the trigeminal ganglia (Beran et al., 1980; Ben-Porat et al., 1984a; H.-J. Rhiza, personal communication; B. Easterday and S. McGregor, personal communication) and can be recovered by cocultivation with susceptible cells. Proof of shedding from a persistently infected pig in isolation that had recovered from the disease has been obtained (Davies and Beran, 1980).

The natural route of infection in swine is via susceptible cells in the nasopharyngeal and respiratory tracts. Virus reaches these areas by droplet infection or by ingestion (Gustafson, 1975). Following multiplication in these areas, the virus penetrates the CNS by way of the trigeminal and olfactory nerves, resulting in a diffuse, nonsuppurative, meningoencephalomyelitis and ganglioneuritis (Saunders and Gustafson, 1964; McFerran and Dow, 1965; Masič et al., 1965; Sabó et al., 1969; Blaškovič et al., 1970; Wittmann et al., 1980). Virus may also be isolated from white blood cells, indicating that virus may spread to various organs, including the CNS via hematogenous and lymphatic pathways (Wittmann et al., 1980).

ACKNOWLEDGMENTS. The authors' research reported herein was supported by NIH Grant AI-10947.

REFERENCES

Albrecht, P., Blaškovič, D., Jakubik, J., and Leššo, J., 1963, Demonstration of pseudorabies virus in chick embryo cell cultures and infected animals by the fluorescent antibody technique, *Acta Virol.* **7:**289–296.

Altman, S., and Lerman, L., S., 1970, Kinetics of intermediates in the intracellular synthesis of bacteriophage T4 deoxyribonucleic acid, *J. Mol. Biol.* **50:**235–261.

Alwine, J. C., Steinhart, W. L., and Hill, C. W., 1974, Transcription of herpes simplex type 1 DNA in nuclei isolated from infected HEp-2 and KB cells, *Virology* **60:**302–307.

Aujeszky, A., 1902, Über eine neue Infektionskrankheit bei Haustieren, *Zentralbl. Bakteriol. Parasitenkd. Infektionskr. Hyg. Abt. 1 Orig.* **32:**353–373.

Aurelian, L., and Roizman, B., 1965, Abortive infection of canine cells by herpes simplex virus. II. Alternative suppression of synthesis of interferon and viral constituents, *J. Mol. Biol.* **11:**539–548.

Bailly, J., 1938, Maladie d'Aujeszky, in: *Les Ultravirus des Maladies Animales* (C. Levaditi, P. Lepine, and J. Verge, eds.), pp. 703–727, Libraire Maloine, Paris.

Barski, G., Lamy, M., and Lepine, P., 1955, Culture de cellules trypsinées de rein de lapin et leur application a l'etude des virus du groupe herpetique, *Ann. Inst. Pasteur* **89:**415–427.

Bartha, A., Belák, S., and Benyeda, J., 1969, Trypsin and heat resistance of some strains of the virus group, *Acta Vet. Hung.* **19:**97–99.

Baskerville, A., McFerran, J. B., and Dow, C., 1973, Aujeszky's disease in pigs, *Vet. Bull.* **43:**465–480.

Béládi, I., 1962, Study on the plaque formation and some properties of the Aujeszky disease virus on chicken embryo cells, *Acta Vet. Acad. Sci. Hung.* **12:**417–422.

Béládi, I., and Ivánovics, G., 1954, Immunisierung von Laboratoriumstieren mit dem Virus der Aujeszky'schen Krankheit nach dessen Inactivierung mit Ultraviolett-strahlen, *Acta Microbiol. Acad. Sci. Hung.* **11:**151–160.

Ben-Porat, T., 1982, Organization and replication of herpesvirus DNA, in: *Organization and Replication of Viral DNA* (A. S. Kaplan, ed.), pp. 147–172, CRC Press, Boca Raton, Fla.

Ben-Porat, T., and Kaplan, A. S., 1962, The chemical composition of herpes simplex and pseudorabies viruses, *Virology* **16:**261–266.

Ben-Porat, T., and Kaplan, A. S., 1963, The synthesis and fate of pseudorabies virus DNA in infected mammalian cells in the stationary phase of growth, *Virology* **20:**310–317.

Ben-Porat, T., and Kaplan, A. S., 1970, Synthesis of proteins in cells infected with herpesviruses. V. Viral glycoproteins, *Virology* **41:**265–273.

Ben-Porat, T., and Kaplan, A. S., 1971, Phospholipid metabolism of herpesvirus-infected and uninfected rabbit kidney cells, *Virology* **45:**252–264.

Ben-Porat, T., and Kaplan, A. S., 1972, Studies on the biogenesis of the herpesvirus envelope, *Nature (London)* **235:**164–166.

Ben-Porat, T., and Kaplan, A. S., 1973, Replication—Biochemical aspects, in: *The Herpesviruses* (A. S. Kaplan, ed.), pp. 163–220, Academic Press, New York.

Ben-Porat, T., and Rixon, F. J., 1979, Replication of herpesvirus DNA. IV. Analysis of concatemers, *Virology* **94:**61–70.

Ben-Porat, T., and Tokazewski, S., 1977, Replication of herpesvirus DNA. II. Sedimentation characteristics of newly-synthesized DNA, *Virology* **79:**292–301.

Ben-Porat, T., and Veach, R. A., 1980, Origin of replication of the DNA of a herpesvirus (pseudorabies), *Proc. Natl. Acad. Sci. USA* **77:**172–175.

Ben-Porat, T., Shimono, H., and Kaplan, A. S., 1969, Synthesis of proteins in cells infected with herpesvirus. 2. Flow of structural viral proteins from cytoplasm to nucleus, *Virology* **37:**56–61.

Ben-Porat, T., Shimono, H., and Kaplan, A. S., 1970, Synthesis of proteins in cells infected with herpesvirus. IV. Analysis of the proteins in viral particles isolated from the cytoplasm and the nucleus, *Virology* **41:**256–264.

Ben-Porat, T., Rakusanova, T., and Kaplan, A. S., 1971, Early functions of the genome of herpesvirus. II. Inhibition of the formation of cell-specific polysomes, *Virology* **46**:890–899.

Ben-Porat, T., DeMarchi, J. M., and Kaplan, A. S., 1974, Characterization of defective, interfering viral particles present in a population of pseudorabies virions, *Virology* **60**:29–37.

Ben-Porat, T., Kervina, M., and Kaplan, A. S., 1975a, Early functions of the genome of herpesviruses. V. Serological analysis of "immediate-early" proteins, *Virology* **65**:355–362.

Ben-Porat, T., Lonis, B., and Kaplan, A. S. 1975b, Further characterization of a population of defective, interfering pseudorabies virions, *Virology* **65**:179–186.

Ben-Porat, T., Stehn, B., and Kaplan, A. S., 1976a, Fate of parental herpesvirus DNA, *Virology* **71**:412–422.

Ben-Porat, T., Kaplan, A. S., Stehn, B., and Rubenstein, A. S., 1976b, Concatemeric forms of intracellular herpesvirus DNA, *Virology* **69**:547–560.

Ben-Porat, T., Blankenship, M. L., DeMarchi, J. M., and Kaplan, A. S., 1977, Replication of herpesvirus DNA. III. Rate of DNA elongation, *J. Virol.* **22**:734–741.

Ben-Porat, T., Rixon, F. J., and Blankenship, M. L., 1979, Analysis of the structure of the genome of pseudorabies virus, *Virology* **95**:285–294.

Ben-Porat, T., Veach, R. A., and Ladin, B. F., 1980, Replication of herpesvirus DNA. VI. Virions containing either isomer of pseudorabies virus DNA are infectious, *Virology* **102**:370–380.

Ben-Porat, T., Brown, L., and Veach, R. A., 1982a, Recombination occurs mainly between parental genomes and precedes DNA replication in pseudorabies virus-infected cells, *J. Virol.* **44**:134–143.

Ben-Porat, T., Hoffmann, P., Brown, L., Feldman, L., and Blankenship, M. L., 1982b, Partial characterization of temperature-sensitive mutants of pseudorabies virus, *Virology* **122**:251–267.

Ben-Porat, T., Veach, R. A., and Hampl, H., 1983a, Functions of the major nonstructural DNA binding proteins of a herpesvirus (pseudorabies), *Virology* **124**:411–424.

Ben-Porat, T., Veach, R. A., and Ihara, S., 1983b, Localization of the regions of homology between the genomes of herpes simplex virus, type 1, and pseudorabies virus, *Virology* **127**:194–204.

Ben-Porat, T., Deatly, A. M., Easterday, B. C., Galloway, D., Kaplan, A. S. and McGregor, S., 1984a, Latency of pseudorabies virus, in: Latent herpes virus infections in veterinary medicine, *Current Topics in Veterinary Medicine and Animal Science* (G., Wittman, R. M., Gaskell, and H.-J., Rziha, eds.), pp. 365–383, Martinus Nijhoff Publishers, Boston.

Ben-Porat, T., Deatly, A., Veach, R. A., and Blankenship, M. L., 1984b, Equalization of the inverted repeat sequences of the pseudorabies virus genome by intermolecular recombination, *Virology* **132**:303–314.

Ben-Porat, T., Wu, C., Harper, L., and Lomniczi, B., 1984c, Biological significance of the alterations in restriction patterns of the genomes of different isolates of pseudorabies virus, in: *Herpesvirus, UCLA Symposia on Molecular and Cellular Biology*, New Series, Vol. 21 (Rapp, F., ed.), Alan R. Liss, Inc. New York, NY, (in press).

Ben-Porat, T., Veach, R. A., Blankenship, M. L., and Kaplan, A. S., 1984d, Differential association with cellular substructures of pseudorabies virus DNA during early and late phases of replication, *Virology*, in press.

Ben-Zeev, A., and Becker, Y., 1977, Requirements of host cell RNA polymerase II in the replication of herpes simplex virus in α-amanitin-sensitive and resistant cell lines, *Virology* **76**:246–253.

Beran, G. W., Davis, E. B., Arambulo, P. V. 3rd, Will, L. A., Hill, H. D., and Rock, D. L., 1980, Persistence of pseudorabies virus in infected swine, *J. Am. Vet. Med. Assoc.* **176**:998–1000.

Bernstein, C., and Bernstein, H., 1974, Coiled rings of DNA released from cells infected with bacteriophage T7 and T4 or from uninfected *E. coli*, *J. Virol.* **13**:1346–1355.

Bitsch, V., 1980, Correlation between the pathogenicity of field strains of Aujeszky's disease virus and their ability to cause cell fusion—syncytia formation—in cell cultures, *Acta Vet. Scand.* **21**:708–710.

Blaškovič, D., Sabó, A., and Rajćani, J., 1970, Experimentelle Pathogenese der Aujeszkyschen Krankheit beim Ferkel, *Arch. Exp. Vet. Med.* **24**:9–27.

Bodon, L., Meszaros, J., Papp-Vid, G., and Romvary, J., 1968, Properties of Aujeszky's disease virus strains isolated from swine pneumonia cases, *Acta Vet. Hung.* **18**:107–109.

Bronson, D. L., Graham, B. J., Ludwig, H., Benyesh-Melnick, M., and Biswal, N., 1972, Studies on the relatedness of herpesviruses through DNA–RNA hybridization, *Biochim. Biophys. Acta* **259**:24–34.

Buchan, A., and Watson, D. H., 1969, The immunological specificity of thymidine kinase in cells infected by viruses of the herpes group, *J. Gen. Virol.* **4**:461–463.

Burnet, F. M., Lush, D., and Jackson, A. V., 1939, The relationship of herpes and B viruses: Immunological and epidemiological considerations, *Aust. J. Exp. Biol. Med. Sci.* **17**:41–51.

Casjens, S., and King, J., 1975, Virus assembly, *Annu. Rev. Biochem.* **44**:555–611.

Ceccarelli, A., and Del Mazza I., 1958, Coltura del virus d'Aujeszky su cellule renali di agnello, *Zooprofilassi* **13**:159–167.

Chantler, J. K., and Stevely, W. S., 1973, Virus-induced proteins in pseudorabies-infected cells. I. Acid-extractable proteins of the nucleus, *J. Virol.* **11**:815–822.

Content, J., and Cogniaux-LeClerc, J., 1968, Comparison of the *in vitro* action of ethidium chloride on animal viruses with that of other photodyes, *J. Gen. Virol.* **3**:63–75.

Cserey-Pechani, E., Béládi, I., and Ivánovics, G., 1951, Züchtung und Wertmessung des Virus der Aujeszky'schen Krankheit in Gewebekulturen, *Acta Physiol. Acad. Sci. Hung.* **2**:229.

Curtis, J. M., and Alberts, B., 1976, Studies on the structure of intracellular bacteriophage T4 DNA, *J. Mol. Biol.* **102**:793–816.

Darlington, R. W., and Granoff, A., 1973, Replication—Biological aspects, in: *The Herpesviruses* (A. S. Kaplan, ed.), pp. 93–132, Academic Press, New York.

Darlington, R. W., and Moss, L. H., III, 1968, Herpesvirus envelopment, *J. Virol.* **2**:48–55.

Davies, E. B., and Beran, G. W., 1980, Spontaneous shedding of pseudorabies virus from a clinically recovered postparturient sow, *J. Am. Vet. Med. Assoc.* **176**:1345–1347.

Davison, A. J., and Wilkie, N., 1983, Location and origin of homologous sequences in the genomes of five herpesviruses, *J. Gen. Virol.* **64**:1927–1942.

Deatly, A. M., Feldman, L. T., and Ben-Porat, T., 1984, The large late virus transcripts synthesized in *Herpesvirus suis* (pseudorabies) virus-infected cells are not precursors of mRNA, *Virology* **135**:452–465.

DeMarchi, J. M., Ben-Porat, T., and Kaplan, A. S., 1979, Expression of the genome of noninfectious particles in stocks of standard and defective interfering pseudorabies virions, *Virology* **97**:457–463.

Dilovski, M., 1969, Inactivation of Aujeszky virus by ultraviolet irradiation, *Vet. Bull.* **39**:774.

Dintzis, H. M., 1961, Assembly of the peptide chain of hemoglobin, *Proc. Natl. Acad. Sci. USA* **47**:247–261.

Dixon, R. A. F., and Schaffer, P. A., 1980, Fine structure mapping and functional analysis of ts mutants of the gene encoding the HSV-1 immediate-early protein VP175, *J. Virol.* **36**:189–203.

Dubovi, E. J., and Youngner, J. S., 1976, Inhibition of pseudorabies virus replication by vesicular stomatitis virus. II. Activity of defective interfering particles, *J. Virol.* **18**:534–541.

Erickson, J. S., 1976, A pseudorabies virus-specific sulfated moiety, *Virology* **71**:598–601.

Erickson, J. S., and Kaplan, A. S., 1973, Synthesis of proteins in cells infected with herpesvirus. IX. Sulfated proteins, *Virology* **55**:94–102.

Faria, A., and Braga, A., 1934, Neuro-histo-patologia da cinomose, *Rev. Dep. Nac. Prod. Anim.* (*Braz.*) **1**:23–25.

Farnham, A. E., and Newton, A. A., 1959, The effect of some environmental factors on herpes virus growth in HeLa cells, *Virology* **7**:449–461.

Feldman, L. T., 1980, Herpesvirus RNA metabolism: Operational controls at the transcriptional and posttranscriptional levels in pseudorabies virus-infected cells, Ph.D. thesis, Vanderbilt University.

Feldman, L., Rixon, F. J., Jean, J.-H., Ben-Porat, T., and Kaplan, A. S., 1979, Transcription of the genome of pseudorabies virus (a herpesvirus) is strictly controlled, *Virology* **97**:316–327.

Feldman, L., Blankenship, M. L., and Ben-Porat, T., 1981, Isolation and characterization of a temperature-sensitive uncoating mutant of pseudorabies virus, *J. Gen. Virol.* **54**:333–342.

Feldman, L. T., DeMarchi, J. M., Ben-Porat, T., and Kaplan, A. S., 1982, Control of abundance of immediate-early mRNA in herpesvirus (pseudorabies)-infected cells, *Virology* **116**:250–263.

Felluga, B., 1963, Electron microscopic observations on pseudorabies virus development in a line of pig kidney cells, *Ann. Sclavo* **5**:412–424.

Follett, E. A. C., Pringle, C. R., and Pennington, T. H., 1975, Virus development in enucleate cells: Echovirus, poliovirus, pseudorabies virus, reovirus, respiratory syncytial virus, and Semliki Forest virus, *J. Gen. Virol.* **26**:183–196.

Frankel, F. R., 1968, DNA replication after T4 infection, *Cold Spring Harbor Symp. Quant. Biol.* **33**:485–493.

Fujiwara, S., and Kaplan, A. S., 1967, Site of protein synthesis in cells infected with pseudorabies virus, *Virology* **40**:60–68.

Furlong, C. J., Kyriakidis, S., and Stevely, W. S., 1982, Synthesis and methylation of ribosomal RNA in HeLa cells infected with the herpesvirus pseudorabies virus, *Arch. Virol.* **73**:329–335.

Gainer, J. H., Long, J., Jr., Hill, P., and Capps, W. J., 1971, Inactivation of the pseudorabies virus by dithiothreitol, *Virology* **45**:91–100.

Galloway, I. A., 1938, Aujeszky's disease, *Vet. Rec.* **50**:745–762.

Geck, P., Jr., Nagy, E., and Lomniczi, B., 1982, Differentiation between Aujeszky's disease virus strains of different virulence by restriction enzyme analysis of the DNA, *Mag. Allatorv. Lapja* **37**:651–656.

Gielkens, A. L. J., and Berns, A. J. M., 1982, Differentiation of Aujeszky's disease virus strains by restriction endonuclease analysis of the viral DNAs, in: *Aujeszky's Disease* (G. Wittmann and S. A. Hall, eds.), pp. 3–13, Nijhoff, The Hague.

Glover, R. E., 1939, Cultivation of the virus of Aujeszky's disease on the chorioallantoic membrane of the developing egg, *Br. J. Exp. Pathol.* **20**:150–158.

Golais, F., and Sabó, A., 1975, The effect of temperature and urea on virulent and attenuated strains of pseudorabies virus, *Acta Virol.* **19**:387–392.

Golais, F., and Sabó, A., 1976, Susceptibility of various cell lines to virulent and attenuated strains of pseudorabies virus, *Acta Virol.* **20**:70–72.

Goto, H., Gorham, J. R., and Hagen, K. W., 1968, Clinical observations of experimental pseudorabies in mink and ferrets, *Jpn. J. Vet. Sci.* **30**:257–263.

Goto, H., Burger, D., and Gorham, J. R., 1971, Quantitative studies of pseudorabies virus in mink, ferrets, rabbits, and mice, *Jpn. J. Vet. Sci.* **33**:145–153.

Graham, B. J., Ludwig, H., Bronson, D. L., Benyesh-Melnick, M., and Biswal, N., 1972, Physicochemical properties of the DNA of herpesviruses, *Biochim. Biophys. Acta* **259**:13–23.

Gustafson, D. P., 1975, Pseudorabies, in: *Diseases of Swine* (H. W. Dunne and A. D. Leman, eds.), 4th ed., pp. 391–410, State University Press, Ames, Iowa.

Gustafson, D. P., 1981, Herpesvirus diseases of mammals and birds: Comparative aspects and diagnosis, in: *Comparative Diagnosis of Viral Diseases* (E. Kurstak and C. Kurstak, eds.), Vol. III, pp. 205–263, Academic Press, New York.

Gustafson, D. P., and Kanitz, C. L., 1975, Pseudorabies in perspective, *Proc. U.S. Anim. Health Assoc.* **79**:287–293.

Gutekunst, D. E., 1979, Latent pseudorabies virus infection in swine detected by RNA–DNA hybridization, *Am. J. Vet. Res.* **40:**1568–1572.

Halliburton, I. W., 1972, Biochemical comparisons of type 1 and type 2 herpes simplex viruses, in: *Oncogenesis and Herpesviruses* (P. M. Biggs, G. de Thé, and L. N. Payne, eds.), pp. 432–438, IARC, Lyon, France.

Halliburton, I. W., and Andrew, J. C., 1976, DNA polymerase in pseudorabies virus-infected cells, *J. Gen. Virol.* **30:**145–148.

Hamada, C., and Kaplan, A. S., 1965, Kinetics of synthesis of various types of antigenic proteins in cells infected with pseudorabies virus, *J. Bacteriol.* **89:**1328–1334.

Hamada, C., Kamiya, T., and Kaplan, A. S., 1966, Serological analysis of some enzymes present in pseudorabies virus-infected and noninfected cells, *Virology* **28:**261–281.

Hampl, H., Ben-Porat, T., Ehrlicher, L., Habermehl, K.-O., and Kaplan, A. S., 1984, Characterization of the proteins of the envelope of pseudorabies virus, *J. Virol.*, in press.

Hanson, R. P., 1954, The history of pseudorabies in the United States, *J. Am. Vet. Med. Assoc.* **124:**259–261.

Heppner, B., Darai, G., Podesta, B., Pauli, G., and Ludwig, H., 1981, Strain differences in pseudorabies viruses, *Med. Microbiol. Immunol.* **169:**128.

Honess, R. W., and Watson, D. H., 1977, Unity and diversity in the herpes viruses, *J. Gen. Virol.* **37:**15–37.

Honess, R. W., Powell, K. L., Robinson, D. J., Sim, C., and Watson, D. H., 1974, Type-specific and type-common antigens in cells infected with herpes simplex virus type 1 and on the surface and naked particles of the virus, *J. Gen. Virol.* **22:**159–169.

Howarth, J. A., 1969, A serologic study of pseudorabies in swine, *J. Am. Vet. Med. Assoc.* **154:**1583–1589.

Huberman, J. A., 1968, Replicating mammalian and T4 bacteriophage DNA, *Cold Spring Harbor Symp. Quant. Biol.* **33:**509–524.

Hurst, E. W., 1936, Studies on pseudorabies (infectious bulbar paralysis, mad itch). III. The disease in the rhesus monkey, *Macaca mulatta, J. Exp. Med.* **63:**449–463.

Ihara, S., Ladin, B. F., and Ben-Porat, T., 1982, Comparison of the physical and genetic maps of pseudorabies virus shows that the genetic map is circular, *Virology* **122:**268–278.

Ihara, S., Feldman, L., Watanabe, S., and Ben-Porat, T., 1983, Characterization of the immediate-early functions of pseudorabies virus, *Virology* **131:**437–454.

Ivaničová, S., 1961, Inactivation of Aujeszky's disease (pseudorabies) by fluorocarbon, *Acta Virol.* **5:**328.

Ivaničová, S., Škoda, R., Mayer, V., and Sokol, F., 1963, Inactivation of Aujeszky's disease (pseudorabies) virus by nitrous acid, *Acta Virol.* **7:**7–15.

Jamieson, A. T., Gentry, G. A., and Subak-Sharpe, J. H., 1974, Induction of both thymidine and deoxycytidine kinase activity by herpesviruses, *J. Gen. Virol.* **24:**465–480.

Jean, J.-H., and Ben-Porat, T., 1976, Appearance *in vivo* of single-stranded complementary ends on parental herpesvirus DNA, *Proc. Natl. Acad. Sci. USA* **73:**2674–2678.

Jean, J.-H., Ben-Porat, T., and Kaplan, A. S., 1974, Early functions of the genome of herpesvirus. III. Inhibition of transcription of the viral genome in cells treated with cycloheximide early during the infective process, *Virology* **59:**516–523.

Jean, J.-H., Blankenship, M. L., and Ben-Porat, T., 1977, Replication of herpesvirus DNA. I. Electron microscopic analysis of replicative structures, *Virology* **79:**281–291.

Kamiya, T., Ben-Porat, T., and Kaplan, A. S., 1964, The role of progeny viral DNA in the regulation of enzyme and DNA synthesis, *Biochem. Biophys. Res. Commun.* **16:**410–415.

Kaplan, A. S., 1962, Analysis of the intracellular development of a DNA-containing mammalian virus (pseudorabies) by means of ultraviolet light irradiation, *Virology* **16:**305–313.

Kaplan, A. S., 1964, Studies on the replicating pool of viral DNA in cells infected with pseudorabies virus, *Virology* **24:**19–25.

Kaplan, A. S., 1969, Herpes simplex and pseudorabies viruses, in: *Virology Monographs* (S. Gard, C. Hallauer, and K. F. Meyer, eds.), Vol. 5, pp. 1–115, Springer-Verlag, Berlin.

Kaplan, A. S. (ed.), 1973, *The Herpesviruses*, Academic Press, New York.

Kaplan, A. S., and Ben-Porat, T., 1963, The pattern of viral and cellular DNA synthesis in pseudorabies virus-infected cells in the logarithmic phase of growth, *Virology* **19**:205–214.

Kaplan, A. S., and Ben-Porat, T., 1964, Mode of replication of pseudorabies virus DNA, *Virology* **23**:90–95.

Kaplan, A. S., and Ben-Porat, T., 1966, Mode of antiviral action of 5-iodouracil deoxyriboside, *J. Mol. Biol.* **19**:320–332.

Kaplan, A. S., and Ben-Porat, T., 1968, Effect of nucleoside analogues on cells infected with pseudorabies virus, in: *Medical and Applied Virology* (M. Sanders and E. H. Lennette, eds.), pp. 56–76, Green, St. Louis.

Kaplan, A. S., and Ben-Porat, T., 1970a, Nucleoside analogues as chemotherapeutic antiviral agents, *Ann. N.Y. Acad. Sci.* **173**:346–361.

Kaplan, A. S., and Ben-Porat, T., 1970b, Synthesis of proteins in cells infected with herpesvirus. VI. Characterization of the proteins of the viral membrane, *Proc. Natl. Acad. Sci. USA* **66**:799–806.

Kaplan, A. S., and Ben-Porat, T., 1976, Synthesis of proteins in cells infected with herpesvirus. XI. Sulfated, structural glycoproteins, *Virology* **70**:561–563.

Kaplan, A. S., and Vatter, A. E., 1959, A comparison of herpes simplex and pseudorabies viruses, *Virology* **7**:394–407.

Kaplan, A. S., Ben-Porat, T., and Kamiya, T., 1965, Incorporation of 5-bromodeoxyuridine and 5-iododeoxyuridine into viral DNA and its effect on the infective process, *Ann. N.Y. Acad. Sci.* **130**:226–239.

Kaplan, A. S., Ben-Porat, T., and Coto, C., 1967, Studies on the control of the infective process in cells infected with pseudorabies virus, in: *Molecular Biology of Viruses* (S. J. Colter and W. Paranchych, eds.), pp. 527–545, Academic Press, New York.

Kaplan, A. S., Shimono, H., and Ben-Porat, T., 1970, Synthesis of proteins in cells infected with herpesviruses. III. Relative amino acid content of various proteins formed after infection, *Virology* **40**:90–101.

Keir, H. M., 1968, Virus-induced enzymes in mammalian cells infected with DNA viruses, in: *The Molecular Biology of Viruses* (L. V. Crawford and M. G. P. Stoker, eds.), pp. 67–99, 18th Symp. Soc. Gen. Microbiol., Cambridge, England.

Kennedy, I. M., Stevely, W. S., and Leader, D. P., 1981, Phosphorylation of ribosomal proteins in hamster fibroblasts infected with pseudorabies virus or herpes simplex virus, *J. Virol.* **39**:359–366.

Killington, R. A., Yeo, J., Honess, R. W., Watson, D. H., Duncan, B. E., Halliburton, I. W., and Mumford, J., 1977, Comparative analysis of the proteins and antigens of five herpesviruses, *J. Gen. Virol.* **37**:297–310.

Kirkpatrick, C. M., Kanitz, C. L., and McCrocklin, S. M., 1980, Possible role of wild animals in transmission of pseudorabies to swine, *J. Wildl. Dis.* **16**:601–614.

Kit, S., Dubbs, D. R., and Anken, M., 1967, Altered properties of thymidine kinase after infection of mouse fibroblast cells with herpes simplex virus, *J. Virol.* **1**:238–240.

Kyriakidis, S., and Stevely, W. S., 1982, The effect of herpesvirus infection on ribosomal RNA synthesis and on nucleolar size and number in HeLa cells, *Arch. Virol.* **71**:79–83.

Ladin, B. F., Blankenship, M. L., and Ben-Porat, T., 1980, Replication of herpesvirus DNA. V. Maturation of concatemeric DNA of pseudorabies virus to genome length is related to capsid formation, *J. Virol.* **33**:1151–1164.

Ladin, B. F., Ihara, S., Hampl, H., and Ben-Porat, T., 1982, Pathway of assembly of herpesvirus capsids: An analysis using DNA$^+$ temperature-sensitive mutants of pseudorabies virus, *Virology* **116**:554–561.

Lankinen, H., Gräslund, A., and Thelander, L., 1982, Induction of a new ribonucleotide reductase after infection of mouse L cells with pseudorabies virus, *J. Virol.* **41**:893–900.

Lauda, E., and Rezek, P., 1926, Zur Histopathologie des Herpes Simplex, *Virchows Arch. Path. Anat.* **262**:827–837.

Lee, J. Y. S., and Wilson, M. R., 1979, A review of pseudorabies (Aujeszky's disease) in pigs, *Can. Vet. J.* **20**:65–69.

Lin, Y. C., 1947, Aujeszky's disease, the first case reported in China, *Chin. J. Anim. Husb.* **6**:6–7 (*Abstr. Vet. Bull.* **18**:200).

Littler, E., Yeo, Y., Killington, R. A., Purifoy, D. J. M., and Powell, K. L., 1981, Antigenic and structural conservation of herpesvirus DNA binding proteins, *J. Gen. Virol.* **56**:409–419.

Lomniczi, B., 1974, Biological properties of Aujeszky's disease (pseudorabies) virus strains with special regard to interferon production and interferon sensitivity, *Arch. Gesamte Virusforsch.* **44**:205–214.

Lomniczi, B., Blankenship, M. L., and Ben-Porat, T., 1984a, Deletions in the genomes of pseudorabies virus vaccine strains and existence of four isomers of the genomes. *J. Virol.* **49**:970–979.

Lomniczi, B., Watanabe, S., Ben-Porat, T., and Kaplan, A. S., 1984b, Genetic basis of the neurovirulence of pseudorabies virus, *J. Virol.*, in press.

Ludwig, H., Biswal, N., Bryan, J. T., and McCombs, R. M., 1971, Some properties of the DNA from a new equine herpesvirus, *Virology* **45**:534–537.

Ludwig, H. O., Biswal, N., and Benyesh-Melnick, M., 1972, Studies on the relatedness of herpesviruses through DNA–DNA hybridization, *Virology* **49**:95–101.

Ludwig, H., Becht, H., and Rott, R., 1974, Inhibition of herpes virus-induced cell fusion by concanavalin-A, antisera, and 2-deoxy-D-glucose, *J. Virol.* **14**:307–314.

Ludwig, H., Heppner, B., and Herrman, S., 1982, The genomes of different field isolates of Aujeszky's disease virus, in: *Aujeszky's Disease* (G. Wittmann and S. A. Hall, eds.), pp. 15–22, Nijhoff, The Hague.

McFerran, J. B., and Dow, C., 1962, Growth of Aujeszky's virus in rabbits and tissue cultures, *Br. Vet. J.* **118**:386–389.

McFerran, J. B., and Dow, C., 1965, The distribution of the virus of Aujeszky's disease (pseudorabies) in experimentally infected swine, *Am. J. Vet. Res.* **26**:631–635.

McFerran, J. B., and Dow, C., 1970, Experimental Aujeszky's disease, *Br. Vet. J.* **126**:173–179.

McGrath, B. M., and Stevely, W. S., 1980, The characteristics of the cell-free translation of mRNA from cells infected with the herpes virus pseudorabies virus, *J. Gen. Virol.* **49**:323–332.

McKercher, D. G., 1973, Viruses of other vertebrates, in: *The Herpesviruses* (A. S. Kaplan, ed.), pp. 456–464, Academic Press, New York.

Marek, J., 1904, Klinische Mitteilungen, *Z. Tiermed.* **8**:389–392.

Mark, G. E., and Kaplan, A. S., 1971, Synthesis of proteins in cells infected with herpesvirus. VII. Lack of migration of structural viral proteins to the nucleus of arginine-deprived cells, *Virology* **45**:53–60.

Masič, M., Ercegan, M., and Petrovič, M., 1965, Die Bedeutung der Tonsillen, für die Pathogenese und Diagnose der Aujeszkyschen Krankheit bei Schweinen, *Zentralbl. Veterinaermed. Reihe B* **12**:398–405.

Matthews, R. E. F., 1982, Classification and nomenclature of viruses, *Intervirology* **17**:1–200.

Mock, R. E., Crandell, R. A., and Mesfin, G. M., 1981, Induced latency in pseudorabies vaccinated pigs, *Can. J. Comp. Med.* **45**:56–59.

Morgan, C., Ellison, S. A., Rose, H. M., and Moore, D. H., 1954, Structure and development of viruses as observed in the electron microscope, *J. Exp. Med.* **100**:195–202.

Morris, N. R., Reichard, P., and Fischer, G. A., 1963, Studies concerning the inhibition of cellular reproduction by deoxyribonucleosides. II. Inhibition of the synthesis of deoxycytidine by thymine, deoxyadenosine, and deoxyguanosine, *Biochim. Biophys. Acta* **68**:93–99.

Morse, L. S., Pereira, L., Roizman, B., and Schaffer, P. A., 1978, Anatomy of herpes simplex virus DNA. X. Mapping of viral genes by analysis of polypeptides and functions specified by HSV-1 × HSV-2 recombinants, *J. Virol.* **26**:389–410.

Nohara, H., and Kaplan, A. S., 1963, Induction of a new enzyme in rabbit kidney cells by pseudorabies virus, *Biochem. Biophys. Res. Commun.* **12**:189–193.

Obel, A. L., 1965, Aujeszky's disease, *Sven. Vet. Tidskr.* **17**:214–215.

Paul, P. S., Mengeling, W. L., and Pirtle, E. C., 1982, Differentiation of pseudorabies (Aujeszky's disease) virus strains by restriction endonuclease analysis, *Arch. Virol.* **73**:193–198.

Pennington, T. M., and McCrae, M. A., 1977, Processing of the pseudorabies virus-induced protein which is glycosylated, sulfated and excreted, *J. Gen. Virol.* **34**:155–165.

Pfefferkorn, E. R., and Coady, H. M., 1968, Mechanism of photoreactivation of pseudorabies virus, *J. Virol.* **2**:474–479.

Pfefferkorn, E. R., Rutstein, G., and Burge, B. W., 1965, Photoreactivation of pseudorabies virus, *Virology* **27**:457–459.

Pfefferkorn, E. R., Burge, B. W., and Coady, H. M., 1966, Characteristics of the photoreactivation of pseudorabies virus, *J. Bacteriol.* **92**:856–861.

Platt, K. B., Maré, C. S., and Hinz, P. N., 1980, Differentiation of vaccine strains and field isolates of pseudorabies (Aujeszky's disease) virus: Trypsin sensitivity and mouse virulence markers, *Arch. Virol.* **63**:107–114.

Plummer, G., Bowling, C. P., and Goodheart, C. R., 1969a, Comparison of four horse herpesviruses, *J. Virol.* **4**:738–746.

Plummer, G., Goodheart, C. R., Henson, D., and Bowling, C. P., 1969b, A comparison study of the DNA density and behavior in tissue culture of fourteen different herpesviruses, *Virology* **39**:134–137.

Popescu, A., Aderca, I., and Bern, S., 1969, Immunobiological properties and behavior in foal kidney cells of freeze-dried pseudorabies virus stored for 3½–4 years, *Lucr. Inst. Cercet. Vet. Bioprep. Pasteur* **6**:455–466.

Preston, C. M., 1979, Control of herpes simplex virus type 1 mRNA synthesis in cells infected with wild-type virus or the temperature sensitive mutant tsK, *J. Virol.* **29**:275–284.

Preston, V. G., Davison, A. J., Marsden, M. S., Timbury, M. C., Subak-Sharpe, J. M., and Wilkie, N. M., 1978, Recombinants between herpes simplex types I and II: Analysis of genome structures and expression of immediate-early polypeptides, *J. Virol.* **28**:499–517.

Pringle, H., Howard, D. K., and Hay, J., 1973, Temperature-sensitive mutants of pseudorabies virus with differential effects on viral and host DNA synthesis, *Virology* **55**:495–505.

Rada, B., and Zemla, J., 1973, Uridine kinase level in cells infected with western equine encephalomyelitis, vesicular stomatitis, pseudorabies and polyoma viruses, *Acta Virol.* **17**:111–115.

Rakusanova, T., and Kaplan, A. S., 1970, Synthesis of cell-specific RNA in cells infected with a herpesvirus, *Fed. Proc.* **29**:309.

Rakusanova, T., Ben-Porat, T., Himeno, M., and Kaplan, A. S., 1971, Early functions of the genome of herpesvirus. I. Characterization of the RNA synthesized in cycloheximide-treated, infected cells, *Virology* **46**:877–889.

Rakusanova, T., Ben-Porat, T., and Kaplan, A. S., 1972, Effect of herpesvirus infection on the synthesis of cell-specific RNA, *Virology* **49**:537–548.

Rand, T. H., and Ben-Porat, T., 1980, Distribution of sequences homologous to the DNA of herpes simplex virus, types 1 and 2, in the genome of pseudorabies virus, *Intervirology* **13**:48–53.

Reichard, P., Canellakis, Z. N., and Canellakis, E. S., 1961, Studies on a possible regulatory mechanism for the biosynthesis of deoxyribonucleic acid, *J. Biol. Chem.* **236**:2514–2519.

Reissig, M., and Kaplan, A. S., 1960, The induction of amitotic nuclear division by pseudorabies virus multiplying in single rabbit kidney cells, *Virology* **11**:1–11.

Reissig, M., and Kaplan, A. S., 1962, The morphology of noninfective pseudorabies virus produced by cells treated with 5-fluorouracil, *Virology* **16**:1–8.

Reissig, M., and Melnick, J. L., 1955, The cellular changes produced in tissue cultures by herpes B virus correlated with the concurrent multiplication of the virus, *J. Exp. Med.* **101**:341–351.

Remlinger, P., and Bailly, J., 1934, A l'etude du virus de la "Maladie d'Aujeszky," *Ann. Inst. Pasteur* **52**:361.

Rixon, F. J., and Ben-Porat, T., 1979, Structural evolution of the DNA of pseudorabies defective viral particles, *Virology* **97**:151–163.

Roizman, B., 1979, The structure and isomerization of herpes simplex virus genomes, *Cell* **16**:481–494.

Roizman, B., 1982, The family Herpesviridae: General description, taxonomy, and classification, in: *The Herpesviruses* (B. Roizman, ed.), Vol. 1, pp. 1–23, Plenum Press, New York.

Ross, L. J.., Wildy, P., and Cameron, K. R., 1971, Formation of small plaques by herpesviruses irradiated with ultraviolet light, *Virology* **45**:808–812.

Rubenstein, A. S., and Kaplan, A. S., 1975, Electron microscopic studies of the DNA of defective and standard pseudorabies virions, *Virology* **66**:385–392.

Russell, W. C., and Crawford, L. V., 1964, Some properties of the nucleic acids from some herpes group viruses, *Virology* **22**:288–292.

Sabin, A. B., 1934, The immunological relationships of pseudorabies (infectious bulbar paralysis, mad itch), *Br. J. Exp. Pathol.* **15**:372–380.

Sabó, A., and Rajčáni, J., 1976, Latent pseudorabies virus in pigs, *Acta Virol.* **20**:208–214.

Sabó, A., Rajčáni, J., and Blaškovič, D., 1969, Studies on the pathogenesis of Aujeszky's disease. III. The distribution of virulent virus in piglets after intranasal infection, *Acta Virol.* **13**:407–414.

Saunders, J. R., and Gustafson, D. P., 1964, Serological and experimental studies of pseudorabies in swine, *Proc. 68th Annu. Meet. U.S. Livestock Sanit. Assoc.* p. 256.

Saxton, R. E., and Stevens, J. G., 1972, Restriction of herpes simplex virus replication by poliovirus: A selective inhibition of viral translation, *Virology* **48**:207–220.

Schaffer, P. A., 1975, Temperature-sensitive mutants of herpesviruses, *Curr. Top. Microbiol. Immunol.* **70**:50–100.

Scherer, W. F., 1953, The utilization of a pure strain of mammalian cells (Earle) for the cultivation of viruses *in vitro*. I. Multiplication of pseudorabies and herpes simplex viruses, *Am. J. Pathol.* **29**:113–137.

Scherer, W. F., and Syverton, J. T., 1954, The viral range *in vitro* of a malignant human epithelial cell (strain HeLa, Gey). I. Multiplication of herpes simplex, pseudorabies, and vaccinia viruses, *Am. J. Pathol.* **30**:1057–1073.

Scott, E. M., and Woodside, W., 1976, Stability of pseudorabies virus during freeze-drying and storage: Effect of suspending media, *J. Clin. Microbiol.* **4**:1–5.

Sharma, J. M., Burger, D., and Kenzy, S. G., 1962, Serological relationships among herpesviruses: Cross-reaction between Marek's disease virus and pseudorabies virus as detected by immunofluorescence, *Infect. Immun.* **5**:406–411.

Sheldrick, P., and Berthelot, N., 1974, Inverted repetitions in the chromosomes of herpes simplex virus, *Cold Spring Harbor Symp. Quant. Biol.* **39**:667–678.

Sheldrick, P., Laithier, M., Lando, D., and Rhyner, M., 1973, Infectious DNA from herpes simplex virus: Infectivity of double-stranded and single-stranded molecules, *Proc. Natl. Acad. Sci. USA* **70**:3621–3625.

Shimono, H., and Kaplan, A. S., 1968, Histone synthesis in cells infected with a virulent or an oncogenic virus, *Fed. Proc.* **27**:615.

Shope, R. E., 1931, An experimental study of mad itch with special reference to its relationship to pseudorabies, *J. Exp. Med.* **45**:233–248.

Shope, R. E., 1932, Identity of the viruses causing "mad itch" and pseudorabies, *Proc. Soc. Exp. Biol. Med.* **30**:308–309.

Shope, R. E., 1935, Experiments on the epidemiology of pseudorabies. I. Mode of transmission of the disease in swine and their possible role in its spread to cattle, *J. Exp. Med.* **62**:85–89.

Singh, K. V., 1962, A plaque assay of pseudorabies virus in monolayers of porcine kidney cells, *Cornell Vet.* **52:**237–246.

Skare, J., and Summers, W. C., 1977, Structure and function of herpesvirus genomes. II. Eco RI, Xba I and Hind III endonuclease cleavage sites on herpes simplex virus type 1 DNA, *Virology* **76:**581–596.

Škoda, R., Becker, C. H., and Jamrichová, O., 1968, Zur Morphologie der Gewebekulturen nach Infektion mit dem Herpes-Suis-Virus (Aujeszkysche Krankheit), *Arch. Exp. Vet. Med.* **22:**1051–1082.

Smith, R. D., Henson, D., Gehrke, J., and Barton, J. R., 1966, Reversible inhibition of DNA virus replication with mithramycin, *Proc. Soc. Exp. Biol. Med.* **121:**209–211.

Southern, E. M., 1975, Detection of specific sequences among DNA fragments separated by gel electrophoresis, *J. Mol. Biol.* **98:**503–517.

Stevely, W. S., 1975, Virus-induced proteins in pseudorabies-infected cells. II. Proteins of the virion and nucleocapsid, *J. Virol.* **16:**944–950.

Stevely, W. S., 1977, Inverted repetition in the chromosome of pseudorabies virus, *J. Virol.* **22:**232–234.

Stevens, J. G., Kado-Boll, G. J., and Haven, C. B., 1969, Changes in nuclear basic proteins during pseudorabies virus infection, *J. Virol.* **3:**490–497.

Stewart, W. C., Cabrey, E. A., Kresse, J. I., and Snyder, M. L., 1974, Infections of swine with pseudorabies virus and enteroviruses: Laboratory confirmation, clinical and epizootiological features, *J. Am. Vet. Med. Assoc.* **165:**440–442.

Stoker, M. G. P., and Newton, A., 1959, The effect of herpes virus on HeLa cells dividing parasynchronously, *Virology* **7:**438–448.

Sullivan, R., Fassolitis, A. C., Larkin, E. P., Read, R. B., and Peeler, J. T., 1971, Inactivation of thirty viruses by gamma radiation, *Appl. Microbiol.* **22:**61–65.

Sun, I. L., Gustafson, D. P., and Scherba, G., 1978, Comparison of pseudorabies virus inactivated by bromoethylene-imine, ^{60}Co irradiation, and acridine dye in immune assay systems, *J. Clin. Microbiol.* **8:**604–611.

Sydiskis, R. J., 1969, Precursor products found in formaldehyde-fixed lysates of BHK-21 cells infected with pseudorabies virus, *J. Virol.* **4:**283–291.

Tan, K. B., 1975, Comparative study of the protein kinase associated with animal viruses, *Virology* **64:**566–570.

Tokumaru, T., 1957, Pseudorabies virus in tissue culture: Differentiation of two distinct strains of virus by cytopathogenic pattern induced, *Proc. Soc. Exp. Biol. Med.* **96:**55–60.

Tokumaru, T., and Scott, T. F. M. 1964, The herpesvirus group: Herpesvirus hominis, herpesvirus simiae, herpesvirus suis, in: *Diagnostic Procedures for Viral and Rickettsial Diseases* (E. Lennette and N. Schmidt, eds.), 3rd ed., pp. 381–433, American Public Health Association, New York.

Toneva, V., 1961, Obtention d'une souche non-virulente du virus de la maladie d'Aujeszky au moyen de passages et de l'adaption des pigeons, *C.R. Acad. Bulg. Sci.* **14:**187–199.

Trainer, D. O., and Karstad, L., 1963, Experimental pseudorabies in some wild North American mammals, *Zoonoses Res.* **2:**135–150.

Vo-Duong, H., Stäber, H., Waschke, K., and Rosenthal, H. A., 1977, Isolation and characterization of temperature-sensitive mutants of pseudorabies virus, *Acta Virol.* **21:**397–404.

von Kerékjártó, B., and Rhode, B., 1957, Über die Vermehrung des Aujeszky-virus auf Affennieren-Epithelkulturen, *Z. Naturforsch.* **12b:**292–298.

Watson, D. H., Wildy, P., Harvey, B. A. M., and Shedden, W. I. H., 1967, Serological relationships among viruses of the herpes group, *J. Gen. Virol.* **1:**139–141.

Watson, R. J., and Clements, J. B., 1980, Identification of a HSV-1 function continuously required for synthesis of early and late virus RNA, *Nature (London)* **285:**329–330.

Watson, R. J., Preston, C. M., and Clements, J. B., 1979, Separation and characterization of herpes simplex virus type 1 immediate-early mRNAs, *J. Virol.* **31:**42–52.

Wittmann, G., Jakubik, J., and Ahl, R., 1980, Multiplication and distribution of Aujeszky's disease (pseudorabies) virus in vaccinated and nonvaccinated pigs after intranasal infection, *Arch. Virol.* **66:**227–240.

Yeo, J., Killington, R. A., Watson,D. H., and Powell, K. L., 1981, Studies on cross-reactive antigens in the herpesviruses, *Virology* **108:**256–266.

Zavadova, Z., and Zavada, J., 1968, Host-cell repair of ultraviolet-irradiated pseudorabies virus in chick embryo cells, *Acta Virol.* **12:**507–514.

CHAPTER 4

Herpes Simplex Virus Latency

Terence J. Hill

I. INTRODUCTION

A. Development of the Concept of Latent Infection

The epidemiology of herpes simplex virus (HSV) infections in man has a number of unusual features (see Whitley, this volume). In particular, the recurrent lesions (herpes labialis, herpes genitalis, and herpes keratitis) often appear at the same peripheral site, the recurrences are often precipitated by particular stimuli, e.g., fever, excess exposure to sunlight, etcetera, and serum antibodies to the virus remain constant throughout life irrespective of the frequency of recurrent disease. Moreover, during intervals of clinical normality it has proved difficult to isolate virus from peripheral tissues such as the skin where lesions are prone to occur (see Section II.B). Such observations led to the concept that recurrent herpetic disease arises from an endogenous, lifelong, "hidden" or latent infection rather than from frequent exogenous reinfection with the virus (Burnet and Williams, 1939). This "natural history" of the virus infection poses a number of fascinating questions: What tissue or tissues harbor the latent infection? How is the latent infection established and maintained? What mechanisms underlie the reactivation of latency and the production of recurrent disease? Such questions have intrigued virologists since the early part of this century but definitive answers only began to appear during the last decade. In particular, the experimental confirmation of the role of the nervous system in latent infections with HSV has prompted a number of recent excellent reviews (Stevens and Cook, 1973a,b; Docherty and Chopan, 1974; Baringer, 1975; Stevens, 1975, 1978, 1980; Klein,

TERENCE J. HILL • Department of Microbiology, University of Bristol Medical School, Bristol BS8 1TD, England.

1976; Longson, 1978; Wildy *et al.*, 1982). The present review will pay particular attention to recent developments in our understanding of those functions of the virus and the host that seem to be important in the establishment, maintenance, and reactivation of the latent infection.

B. Neurotropism of HSV and Animal Models

The neurotropism of HSV that is often manifest following injection of the virus into a peripheral site, e.g., the eye of a laboratory animal such as the rabbit, was among the most striking observations recorded in the early days of animal virology (Doer, 1920; Doer and Vöchting, 1920; Friedenwald, 1923). Such observations led to a series of elegant studies, by Goodpasture in the 1920s, on the neuropathology of HSV infection in the rabbit (Goodpasture and Teague, 1923; Goodpasture 1925a,b, 1929). These studies, primarily on the ability of the virus to spread to the CNS and cause lesions there, engendered the view that the nervous system might also be involved in the pathogenesis of herpetic disease in man. Indeed, Goodpasture (1929) made what subsequently proved to be the farsighted prediction that HSV may establish a latent infection in the neurons of the sensory ganglion of the fifth cranial nerve (the trigeminal or Gasserian ganglion). This ganglion supplies sensory nerves to the ophthalmic and orofacial areas in which the recurrent herpetic lesions are often seen.

The ability of HSV to spread to the CNS and there produce severe destructive and often fatal infection was subsequently demonstrated in other animals beside the rabbit, in particular the guinea pig and mouse (reviewed by Baringer, 1975). Thus, virologists interested in HSV are fortunate in having available a number of laboratory animals that are susceptible to infection with this human pathogen. Indeed, many of the advances referred to in this chapter are derived from the use of such models. However, although severe CNS disease is a common sequel to infection with HSV in laboratory animals, it is a rare event in man (Longson and Bailey, 1977; Longson *et al.*, 1980). This fact alone suggests important differences between the controls that limit the infection in man and experimental animals. Such differences should prompt a degree of caution in extrapolating from animal models to the human disease.

II. THE SITE OF LATENT INFECTION

A. Neural Sites

1. Peripheral Nervous System

Observations made at the beginning of this century on the coincidence of pathological lesions in the human trigeminal ganglion with pneumonitis and herpes labialis (Howard, 1903, 1905) and also the development of cutaneous herpetic lesions after trigeminal nerve surgery (Cushing, 1905) gave a hint of the possible involvement of the trigeminal nerve in the pathogenesis of herpetic disease in man. These early clues and the prediction of Goodpasture mentioned above, were supported by more careful documentation of the cutaneous herpetic lesions that frequently follow surgery of the trigeminal tract (Carton and Kilbourne, 1952; Carton, 1953; Pazin et al., 1978). Carton and Kilbourne noted that 2–4 days after section of the root of the fifth cranial nerve, oral or facial herpetic lesions (but not ophthalmic) occurred in about 90% of individuals. The lesions developed provided that the ganglion was not destroyed and the peripheral nerve remained intact (Carton, 1953). Carton and Kilbourne (1952) and Ellison et al. (1959) interpreted their observations as indicating that changes induced in the ganglion by surgery caused reactivation of latent HSV in the skin. Paine (1964) in a later review gave an excellent summary of all the neurosurgical and pathological observations linking the trigeminal ganglion with herpetic disease in man. He was led to favor the original idea of Goodpasture of latent infection in the ganglion and suggested the following outline for the pathogenesis of HSV infection in man: (1) during primary infection the virus grows in epithelial cells and enters nerve endings, (2) the virus spreads, probably via the nerve, to the trigeminal ganglion, (3) in the ganglion a latent infection is established in which the virus may be in a noninfectious form [at that time several workers had reported failure to isolate infectious HSV from human trigeminal ganglia: Burnet and Lush (1939), Richter (1944), Carton and Kilbourne (1952), Ellison et al. (1959)], and (4) the latent infection is activated and infectious virus spreads, again via the nerve, to the skin where the clinical lesion is produced.

Experimental proof that HSV can indeed establish a long-term "silent" infection in the nervous system began to appear in the 1970s. Plummer et al. (1970) found that 9–11 months after intramuscular injection of HSV-2 into rabbits, virus could be recovered from their CNS and sensory ganglia by inoculation of trypsinized suspensions of these neural tissues onto cell monolayers. Such observations were confirmed and much extended by Stevens and his colleagues in the mouse (Stevens and Cook, 1971) and rabbit (Stevens et al., 1972). In their first study in mice, Stevens and Cook found that infectious virus was demonstrable in the lumbosacral sensory ganglia for only a short time (usually not later than day 7) after inoculation of HSV into a rear footpad. Isolation of virus was attempted by the conventional procedure of inoculating cell cultures with homogenates of the tissue. However, if the sensory ganglia (associated with the sciatic nerve of the originally inoculated limb) were cultured in vitro as explants, infectious virus could be isolated from the culture medium at 7 or 14 days after explantation. Using this method, virus could not be grown from the sciatic nerve trunk, spinal cord, or medulla oblongata. By this explantation method, virus was recovered from the lumbosacral ganglia of the majority of mice even 4 months after they had recovered from the primary infection. In subsequent experiments the appearance of virus was usually detected by culturing the ganglion explants, sometimes for many weeks, in the presence of a cell monolayer susceptible to HSV—the so-called "cocultivation" method (Stevens et al., 1972; Knotts et al., 1973). A further modification of the explantation method that is now often used involves culture of the ganglion for a short period (usually 4–6 days). The presence of virus is then detected by inoculating a homogenate of the ganglion onto cell monolayers (Wohlenberg et al., 1979; Harbour et al., 1981). This technique gives similar results to the cocultivation method (T. Hill et al., unpublished results) and has the advantage of producing the results in a shorter time. Furthermore the titer of virus in the ganglion homogenate may give some crude quantification of latent infection (Klein, 1980). Other attempts to quantify latency in the ganglion have involved disaggregation of the tissue and detection of the number of cells that give rise to infectious centers (Walz et al., 1976), assay of virus produced at different times after culture of the ganglia (Klein, 1980), and observation of the time of appearance of CPE in the indicator cells when the cocultivation method is used (Klein, 1980).

The availability of meaningful and convenient methods for quantifying the extent of latent infection in sensory ganglia would be extremely valuable in many experimental situations, e.g., more detailed assessment of the effects of antiviral drugs or vaccines.

It has been reported that the sensitivity of the explant method can be significantly enhanced by dispersion of the ganglion with trypsin/EDTA before cocultivation of the cell suspension (Nesburn et al., 1980). Lewis et al. (1982) have also described methods for optimizing the recovery of HSV from human trigeminal ganglia. Whichever of the explantation methods is used to detect latent infection in ganglia, or any other tissue, none of them reveal the state of the virus during latency. Virus isolated from ganglia by such methods could arise from either of two states: (1) from small foci of normal productive infection yielding insufficient virus to be detected by inoculation of ganglion homogenates into cell cultures (during the culture of the ganglion in vitro, controls active in the host animal would be absent, thus allowing the infection in the ganglion tissue to spread and produce detectable levels of infectious virus); or (2) from a viral genome that in vivo is completely or partially repressed but during culture of the ganglion in vitro is reactivated to produce infectious virus. Hence, it should be emphasized that the definition of latent infection according to the results of these explantation methods is entirely operational. As discussed later, there may be situations in which the complete viral genome is demonstrable in tissues by methods such as nucleic acid hybridization but no reactivation of virus can be induced. By the results of explantation methods, such tissues would be considered incorrectly to be free from latent infection.

Using explant methods similar to those developed by Stevens and his colleagues, HSV was isolated from 9 to 86% of human trigeminal ganglia taken at postmortem (Bastian *et al.*, 1972; Baringer and Swoveland, 1973; Rodda *et al.*, 1973; Plummer, 1973; Warren *et al.*, 1977). The picture was completed when Baringer (1974) demonstrated latent infection with HSV-2 in human lumbosacral ganglia. Similar results were obtained for the analogous ganglia in mice (Walz *et al.*, 1977) and monkeys (Reeves *et al.*, 1976) that had recovered from vaginal infection with HSV-2.

From subsequent studies it became apparent that, depending on the site of primary infection, HSV could establish a latent infection in any type of nerve ganglion, including autonomic ganglia in mice (Price *et al.*, 1975a), rabbits (Martin *et al.*, 1977), and humans (Warren *et al.*, 1978).

More recently, Tullo *et al.* (1983b) have shown that within the human trigeminal ganglion, latent infection with HSV is most common in the maxillary and mandibular parts.

The predilection of the virus for establishing latency in nervous tissue was clearly demonstrated in experiments by Cook and Stevens (1976) in which mice were injected intravenously with HSV. Despite widespread exposure of different tissues to the virus, latent infection was demonstrated only in central and peripheral nervous tissues (including the adrenal medulla).

2. Central Nervous System

By the culture of tissue explants, latent infection with HSV has been demonstrated in the CNS of rabbits (Plummer *et al.*, 1970), mice and rabbits (Knotts *et al.*, 1973), mice (Cook and Stevens, 1976; Cabrera *et al.*, 1980), and guinea pigs (Tenser and Hsiung, 1977). However, the incidence of detectable latency in such tissues was much lower than in sensory ganglia and in two reports no latent infection was demonstrated in the brain stem after ocular infection in mice (Knotts *et al.*, 1974; Tullo *et al.*, 1982a). It seems unlikely that this apparent difference between the CNS and the PNS is due to the inability of the virus to reach the CNS. In spite of the presence of high titers of HSV in many areas of the brains of mice that had survived the acute infection, latent infection was not demonstrated in their brain stem (Tullo *et al.*, 1982a; Hill *et al.*, 1983a).

By culture of tissue explants, Openshaw (1983) detected latent HSV in the retinas of mice that had recovered from corneal infection (embryologically and anatomically the retina forms part of the CNS).

As yet there have been no reports of the demonstration, by explantation methods, of latent infection with HSV in the human CNS. However, the preliminary report of DNA homologous to HSV DNA in human brain (Sequiera *et al.*, 1979) has now been confirmed in more detail by Fraser *et al.* (1981) who found such DNA in 7 of 11 human brains (of the 7 three were from normal individuals and the remainder from patients with multiple sclerosis). In most cases the DNA present appeared to represent the complete viral genome. Moreover, Cabrera et al. (1980) demonstrated DNA homologous to viral DNA in the brains of 30% of mice in which the incidence of latent infection (demonstrated by an explantation method) in the brain was 5% and in the trigeminal ganglia 95%. Hence, the apparently low incidence of latency in the CNS may merely reflect difficulties in demonstrating its presence by the usual explantation techniques. Such difficulties may arise from the fact that the latent infection may be less easily reactivated in neurons of the CNS than in those of the PNS. This will be discussed further in Section IV. However, the possibility of these differences calls for caution in extrapolating from experiments on the establishment of latent infection by HSV mutants in the mouse brain (Lofgren et al., 1977; Watson et al., 1980) to what might happen in sensory ganglia.

The observations of Cabrera et al. (1980) also raise the issue of whether latent infection is best defined on the basis of results of explantation methods alone, a problem already referred to. Some tissues like the brain may rarely, if ever, yield infectious virus by such methods. Nevertheless, these tissues may need to be considered as latently infected if they can be shown to contain complete copies of the viral DNA.

B. Extraneural Sites

As the recurrent lesions caused by HSV normally occur in peripheral sites such as skin, it was natural that these tissues should be suspected as sites of latent infection (Carton and Kilbourne, 1952; Carton, 1953; Ellison et al., 1959). However, attempts in man to detect virus in homogenates or explant cultures of such tissues, particularly skin from areas where lesions were prone to develop, have proved unsuccessful (Findlay and MacCallum, 1940; Anderson and Hamilton, 1949; Coriell, 1963; Rustigian et al., 1966; Smith and McLaren, 1977). Moreover, facial skin taken from areas prone to show recurrent disease, did not develop lesions when transplanted to other sites of the body (Nicolau and Poincloux, 1928; Stalder and Zurukzoglu, 1936). However, Hoyt and Billson (1976) reported recurrent herpes simplex in patients 7–10 days after "blowout" fractures had severed the nerve supply to the area where the lesions developed. This raises the possibility that the lesions arose from virus already present in the skin at the time of injury.

Shedding of virus in body secretions in the absence of clinical disease has been described in man and rabbits: in human saliva (Douglas and Couch, 1970), secretions of the human genitourinary system (Deardourff et al., 1974; Orsi et al., 1978) human tears (Kaufman et al., 1967), and rabbit eye secretions (Kaufman et al., 1967; Nesburn et al., 1967; Laibson and Kibrick, 1969). Such shedding has been interpreted as indicating latent infection with HSV in salivary or lacrimal glands (Kaufman *et al.*, 1967, 1968; Brown and Kaufman, 1969).

However, the presence of infectious virus in peripheral tissues in the absence of clinical disease may not indicate a latent infection at that site but rather the presence of microfoci of infection resulting from an input of virus from the primary site of latency, namely the sensory ganglion. This interpretation was favored by Hill *et al.* (1980) to account for the presence of HSV in the skin of about 10% of clinically normal, latently infected mice.

Some internal organs, e.g., parts of the genital tract, have autonomic neurons closely associated with them and latent infection has been demonstrated in such neurons (Price *et al.*, 1975a). This may explain the latent infection found in the vaginouterine tissue of more than 25% of mice after vaginal inoculation with HSV-1 or HSV-2 (Walz *et al.*, 1977).

However, the work of Scriba (1977) on HSV infection in the guinea pig suggests that the possibility of latency in extraneural tissues cannot be entirely excluded. By culture of skin explants, Scriba has shown that HSV-2 can be present in the footpad of the majority of latently infected guinea pigs in the absence of clinical lesions and in the absence of any nerve supply from the lumbosacral ganglia (Scriba, 1981a). In the case of HSV-1 it seems that the virus persists almost entirely in the peripheral tissues of the guinea pig, at the site of inoculation, with little or no latent infection in the lumbosacral ganglia (Scriba, 1977; Scriba and Tatzber, 1981). Moreover, in mice latently infected with a strain of HSV-2 or various of its ts mutants, virus has been recovered from explant cultures of the footpad (site of inoculation) and/or the sensory ganglia (Al–Saadi *et al.*, 1983).

Whether such extraneural latent infection can occur in man remains to be established. In this respect a situation that would merit further investigation is the presence of virus in the stromal tissues of the eye during chronic herpetic keratitis (Dawson and Togni, 1976; Meyers-Elliott *et al.*, 1980). Indeed, although in the past it has proved difficult to culture virus from such lesions, HSV has recently been isolated from two of three corneal discs (from patients with chronic stromal keratitis) that had been removed during corneal grafting and cultured as explants *in vitro* (Shimeld *et al.*, 1982).

The detection of RNA complementary to HSV in mononuclear cells from the peripheral blood of patients with Behcet's syndrome and recurrent oral ulcers also suggests that in some circumstances the virus may be latent in extraneural sites (Eglin *et al.*, 1982).

Despite these possibilities of extraneural latency, the weight of evidence at present favors the nervous system, in particular sensory ganglia, as the primary site of latent infection and thereby the source of virus responsible for recurrent disease (Paine, 1964). Indeed, the observations of Carton (1953) mentioned previously, suggest that in man the integrity of the trigeminal nerve and ganglion is essential for continued recurrence of cutaneous lesions.

III. ESTABLISHMENT OF LATENT INFECTION IN GANGLIA

A. Introduction

The establishment of latency in ganglia does not merely involve the interaction of the virus with the "target" cell (which, as will become clear, is almost certainly the ganglionic neuron). Before this final phase in the process, there are other important and distinct but interrelated phases, namely growth of virus in peripheral tissues such as skin, entry of virus into nerve endings, and spread of virus (usually via nerves) to the ganglion. At all of these phases there may be subtle interaction between particular viral and host functions. In what follows an attempt is made to dissect these functions, but with regard to the final phase (the interaction of the virus with the "target" cell in the ganglion) the discussion at this stage will be limited to functions of the virus. Because changes in certain host functions (particularly those concerning the metabolism of neurons) appear to be involved in the reactivation of latent infection, the role of such functions in the establishment and maintenance of latency in neurons will be discussed in a later section dealing with reactivation.

The main anatomical structures concerned in the different phases of the infection are shown diagrammatically in Fig. 1. An excellent account of the peripheral nerve and its related structures is given by Landon (1976).

B. Spread of Virus from the Periphery

It is now clear that latent infection with HSV can be established in the sensory ganglia of laboratory animals after infection of a peripheral site such as the mouse footpad (Stevens and Cook, 1971) or the mouse ear (Hill *et al.*, 1975). This raises the question of how the virus spreads from such peripheral sites to the sensory ganglia. Over the years a number of workers have shown that although there may be a transient viremia following peripheral inoculation of virus (particularly in newborn mice), there is no significant spread of virus to the CNS from the bloodstream in adult animals (Goodpasture, 1925a; Johnson, 1964; Wildy, 1967; Cook and Stevens, 1973; Knotts *et al.*, 1974; Renis *et al.*, 1976). Moreover, administration of anti-HSV serum to mice did not prevent the spread of virus to the CNS (Wildy, 1967; Cook and Stevens, 1973). The accumulated work of these same authors shows, from the sequence with which infectious virus appears in different tissues and the distribution of the pathological lesions, that the peripheral nerve is involved in the spread of

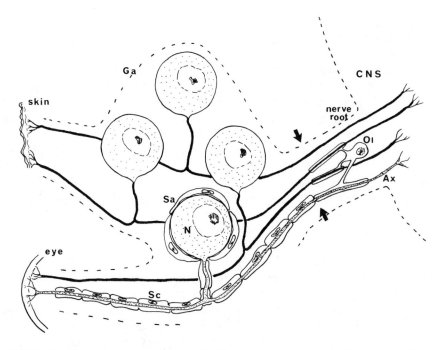

FIGURE 1. Diagram of the main anatomical features of a sensory ganglion and peripheral nerve. Only one neuron is shown in full detail. Ga, ganglion; Sa, satellite cell; Sc, Schwann cell; N, neuron; Ol, oligodendroglial cell; Ax, axon; arrows, junction of PNS and CNS. The diagram is particularly illustrative of the ophthalmic part of the trigeminal (fifth cranial nerve) ganglion, which supplies sensory nerves to the eye and some areas of the skin of the head.

virus to the ganglia and CNS. Indeed, Wildy (1967) found that cutting the peripheral nerve that supplied the inoculation site could completely prevent the spread of virus to the CNS.

Spread of virus to the ganglion in the endoneural fluid of the nerve sheath seems unlikely as the net movement of such fluid is away from the CNS (Weiss et al., 1945; Mellick and Cavanagh, 1967). A number of workers have suggested that virus may spread along peripheral nerves by sequential infection of contiguous Schwann cells (Fig. 1). This proposal arose from the observation of virus particles in Schwann cells by electron microscopy (Dawson et al., 1966; Rabin et al., 1968; Severin and White, 1968; Rajčáni and Conen, 1972) and of viral antigens in Schwann cells by immunofluorescence (Johnson, 1964; Yamamoto et al., 1965; Rajčáni et al., 1969) or the immunoperoxidase technique (Lascano and Berria, 1980). However, two points need to be considered in the interpretation of such observations: (1) the detection of viral antigens in Schwann cells by immunofluorescence or the immunoperoxidase method does not necessarily indicate the production of infectious virus; and (2) apart from Dawson et al., who used rabbits, all of the other work quoted above was done with neonatal or suckling mice. In such animals the peripheral nerves are largely unmyelinated and therefore the metabolism of their Schwann cells will be very different from that in older animals in which myelination is complete. Moreover, in the sciatic nerve of the 2-day-old mouse, 27% of Schwann cells are still dividing (Asbury, 1967); in the adult this figure is less than 0.1% (Bradley and Asbury, 1970). Such physiological differences between the Schwann cells of neonatal and adult mice may have marked effects on their susceptibility to HSV. Indeed, from all other studies that have made careful electron microscopic observations of HSV infection in Schwann cells of older mice (3 weeks of age or more), it is clear that these cells are either infected abortively or are totally resistant to infection (Dillard *et al.*, 1972; Cook and Stevens, 1973; Knotts *et al.*, 1974). These observations suggest that at least in the adult mouse, spread of virus via Schwann cells is unlikely to be of significance.

In contrast to the PNS, the glial cells of the CNS, in particular the oligodendrocytes and astrocytes, are susceptible to infection with HSV (Townsend and Baringer, 1976, 1978; Townsend, 1981a; Hill, 1983). Hence, in the CNS, cell-to-cell spread of virus might occur although its speed would be relatively slow, the limiting factors being the number of cells per unit length of nerve fiber and the growth rate of the virus. Evidence for such slow spread (about 1 mm/day) comes from the morphological observations of Narang (1977) and Narang and Codd (1978) on the movement of infection along the optic nerve of young rabbits. (The optic nerve embryologically and structurally is part of the CNS.) This very slow cell-to-cell spread of virus in the optic nerve contrasts markedly with the rate at which herpesviruses (HSV and pseudorabies) apparently travel along peripheral nerves (2–10 mm/hr). Such speeds have been determined by observing the time at which infectious virus can first be detected in sensory ganglia after peripheral inoculation of virus [for HSV, Kristensson *et al.* (1971), Cook and Stevens (1973); for pseudorabies, McCracken *et al.* (1973), Field and Hill (1974, 1975)]. This relatively rapid spread of herpesviruses along peripheral nerves is very similar to that reported for normal retrograde (toward the CNS) flow of proteins in axons—2–10 mm/hr (Ochs, 1974; Kristensson, 1978). Axonal microtubules are thought to be involved in the mechanism of normal retrograde axonal transport (Ochs, 1974; Kristensson, 1978). Therefore, the interruption of the spread of HSV along a peripheral nerve by local treatment of the nerve with colchicine (a drug that disrupts microtubules) adds weight to the idea that HSV can travel by intraaxonal transport (Kristensson *et al.*, 1971). Further support for this concept comes from two reports concerning experiments with neuronal cultures *in vitro*. In the first, cultures of rat dorsal root ganglia were inoculated with HSV onto the tip of the axonal outgrowth and virus was shown (by immunofluorescence) to reach the neural cell bodies without involvement of Schwann cells (Ziegler and Herman, 1980). In the second, Price *et al.* (1982) used cultures of autonomic neurons from the superior cervical ganglia of young rats. The cultures contained very few glial cells and the neuronal cell bodies were connected by a network of axonal fibers. After infection of the cultures with HSV, viral antigen was shown by immunoperoxidase staining to spread along such fibers. Moreover, the presence of antiviral serum in the culture medium slowed down but did not prevent the spread of virus.

Direct observation of HSV in axons by electron microscopy has proved difficult. This probably reflects the large sampling errors inherent in the technique. Nevertheless, intraaxonal particles have been observed both for HSV (Hill *et al.*, 1972; Cook and Stevens, 1973; Baringer and Swoveland, 1974; Kristensson *et al.*, 1974) and for pseudorabies virus (McCracken *et al.*, 1973; Field and Hill, 1974). However, in all these cases the particles observed may represent virus arising from productively infected ganglionic neurons or CNS tissue rather than virus traveling to the ganglion from the peripheral site of inoculation.

In summary, it seems most likely that HSV makes use of a normal neuronal function (retrograde axonal flow) in order to travel in peripheral nerves to the ganglion and CNS. At present there is little information on whether particular viral components, e.g., envelope glycoproteins, are of importance in the entry of virions into nerve endings in peripheral tissues. In this respect two observations may be of relevance: Vahlne *et al.* (1978, 1980) have shown that there may be HSV type-specific receptors in mouse brain synaptosomes and Ziegler and Pozos (1981) have found that N-acetylneuraminic acid may form part of receptors for HSV on axonal processes in culture. In this respect, differences between strains of HSV in their ability to enter nerve endings might be relevant to differences in neurovirulence.

A further and striking consequence of the intraneural spread of HSV, which can occur frequently in experimental animals and occasionally in man, is the zosteriform spread of herpes simplex lesions (reviewed by Hill, 1983; Blyth *et al.*, 1984). Such spread indicates that once the virus has infected the sensory ganglion and/or the CNS, it may then travel centrifugally (again via axonal flow) to reach parts of the dermatome remote from the site of inoculation. The possible mechanisms underlying zosteriform spread are discussed by Hill (1983) and similar mechanisms may be involved in the more widespread establishment of latent infection by the "backdoor route" (see below).

C. Events in the Peripheral Tissue during Primary Infection

In theory it is possible that after gaining access to tissues such as skin, HSV could pass directly into peripheral nerve endings. By using large doses of virus this has been demonstrated experimentally with the herpesvirus of pseudorabies (Field and Hill, 1975) and it is likely that the same could be achieved with HSV. However, under circumstances of natural infection where the virus inoculum may be small, it seems more probable that some virus growth in peripheral tissues will precede entry of particles into the nerve. The extent of such growth would clearly affect the amount of virus in the tissue and the time for which it was present. This in turn may greatly influence the chance of virions entering the nervous system and thereby may affect the incidence of latent infection in the ganglia. Hence, strains of virus or virus mutants that grow poorly in tissues such as skin may be rendered "latency negative" simply because there is little chance of them entering the peripheral nerve endings.

In the normal course of events the growth of virus in peripheral tissues is of limited duration; e.g., after inoculation of the mouse footpad (Stevens and Cook, 1971) or mouse ear (Hill et al., 1975), virus titers reach a peak at 3–4 days after inoculation and the virus is cleared from the tissues 3–7 days later. However, as mentioned above, zosteriform spread of lesions can frequently occur in experimental animals. Hence, an oversimplistic picture of the growth of HSV in the periphery should be avoided, for such spread can occur over the whole dermatome (including the inoculation site) and can begin as soon as 2–3 days after inoculation (T. Hill, W. Blyth, and D. Harbour, unpublished observations).

Many of the immunological mechanisms known to be involved in the control of HSV infections (reviewed by Nahmias and Ashman, 1978; Babiuk and Rouse, 1979; Wildy et al., 1982; Rouse and Lopez, 1984) are almost certainly involved in the clearance of virus and the termination of the primary infection in the peripheral tissue. Hence, the integrity of such mechanisms is likely to be of prime importance in limiting the amount of virus available to enter the nervous system and thereby to establish latency.

It follows from these arguments that from the point of view of treating the primary infection, this stage in the pathogenesis of the disease is an important "Achilles' heel" with respect to the establishment of latency. Hence, chemotherapeutic or immunological treatments may not only directly affect the infection in the nervous system but also, and perhaps more importantly, may indirectly reduce the incidence of latency by decreasing the amount of virus available to enter nerve endings. Indirect evidence for such effects comes from the observation of a decreased incidence of latent infection in animals treated with antiviral drugs at or very near the time of primary infection (Wohlenberg et al., 1976; Klein et al., 1977, 1978a,b, 1979). Similar effects are reported in animals vaccinated before infection or after secondary infection (Price et al., 1975b; McKendall, 1977; Klein et al., 1978a; Price and Schmitz, 1979; Tullo et al., 1982a, 1983a). Walz et al. (1976) have also shown that after infection of immunized animals the number of latently infected cells in ganglia is lower than in controls (infected but nonimmunized). Moreover, treatment of mice with immune serum at the time of primary infection may reduce the incidence of latent infection (McKendall et al., 1979) and/or reduce the extent of the latent infection in the ganglion (Klein, 1980).

Conversely, factors that disable defense mechanisms and thereby allow increased virus growth in peripheral tissues might be expected to increase the chance of establishing latency. Thus, treatment of adult mice with silica [which disables macrophages (Zisman *et al.*, 1970; Hirsch *et al.*, 1970)] before infection with HSV (Blyth *et al.*, 1980a) and treatment of guinea pigs with immunosuppressive steroids (Tenser and Hsiung, 1977) produced a higher incidence of latent infection than in control animals.

While the extent of virus growth in the periphery is probably of importance in affecting the chance of establishing latency in ganglia, the *site* of primary infection may to a large extent determine which ganglia or parts of ganglia are involved. This is of particular importance in man with respect to the trigeminal ganglion. Embryological observations show that this large sensory ganglion is formed by fusion of two separate nerves, the opthalmic and maxillomandibular branches (Kappers *et al.*, 1960). Moreover, in the adult animal the neurons of the three parts of the ganglion (ophthalmic, maxillary, and mandibular) are within separate regions and the nerves from the three parts each supply a different peripheral area (Gregg and Dixon, 1973; Arvidson, 1979). In the majority of cases, primary infection with HSV-1 occurs in the oropharynx (Buddingh *et al.*, 1953), an area supplied by the maxillary and mandibular nerves. Hence, it might be expected that, at least initially, latent infection would be restricted to the maxillary/mandibular part of the ganglion. This is borne out by the observation that section of the trigeminal sensory root of man, a process that would affect the whole ganglion, was followed by herpetic lesions in the maxillary/mandibular and not the ophthalmic region (Carton and Kilbourne, 1952). Indeed, in a small series of human trigeminal ganglia, Tullo *et al.* (1983b) found that latent infection was limited to maxillary and mandibular parts. However, if recurrent herpetic eye disease is associated with a latent infection in the ophthalmic part of the trigeminal ganglion, the question arises as to how such latency is established. The following possibilities have been considered by Tullo *et al.* (1982a,b). (1) An individual already carrying latent infection in the maxillary/mandibular parts of the ganglion may be reinfected exogenously or by autoinoculation in the eye and thereby latent infection may be established in the ophthalmic part of the ganglion. The difficulty of establishing a latent infection in immune animals (Tullo *et al.*, 1982a) argues against such an event. (2) Arguments against hematogenous spread of virus have already been mentioned. (3) During primary infection, spread of virus from the maxillary/mandibular parts of the ganglion to the ophthalmic is unlikely as there are no interneuronal connections in sensory ganglia. Moreover, the glial cells that surround the neurons and axons are relatively resistant to infection by HSV (Dillard *et al.*, 1972; Cook and Stevens, 1973) and are likely to form a barrier to such spread. (4) Latent infection could be established in the ophthalmic part of the trigeminal ganglion by centripetal spread of virus to the brain stem (via maxillary/mandibular neurons) and then by centrifugal spread to the ophthalmic division [the so-called "backdoor" route (Tullo et al., 1982b)]. The sequence with which virus can be isolated from different sites of the nervous system of mice after ocular infection (Tullo et al., 1982a; Hill et al., 1983a) or lip infection (Tullo et al., 1982b) suggests that this is a likely course of events. Hence, by outflow of virus from the CNS during the acute disease, the virus may establish latency in neurons that do not supply the site of primary infection. The mechanisms underlying this outflow may be similar to those involved in zosteriform spread of lesions (Hill, 1983). The occurrence of latent infection in contralateral sensory ganglia may be further evidence of such spread (Knotts et al., 1973; Tenser and Hsiung, 1977; McKendall et al., 1979; Tullo et al., 1982a). The results of Lonsdale et al. (1979) and Gerdes et al. (1981) provide indirect evidence of similar spread of HSV in man. By analysis of viral DNA with restriction enzymes, they showed that in a single individual, isolates from different ganglia (Lonsdale et al.) or peripheral sites (Gerdes et al.) were of the same strain of HSV. Moreover, the possibility of HSV DNA in different parts of the human brain (Fraser et al., 1981) suggests the virus can reach many parts of the nervous system.

Thus, the single exposure to oropharyngeal infection by HSV-1 in childhood may, by spread of virus in the nervous system, result in latent infection in several neural sites. In a few individuals, such sites may include the ophthalmic as well as the maxillary/mandibular parts of the trigeminal ganglion (Tullo et al., 1982a,b).

At least in experimental studies the route of inoculation is a further factor that can markedly affect the incidence of subsequent latent infection. Thus, after subcutaneous inoculation of the skin of the pinna of mice with 10^5 PFU HSV-1, the incidence of latent infection was 40%. In contrast, scarification of the pinna with this dose gave an incidence of latent infection of nearly 100% (Blyth et al., 1984). Presumably, scarification facilitates access of the virus to the sensory nerve endings in the skin.

D. The Latently Infected Cell

The release of infectious virus from ganglia of latently infected animals and man implies the presence of the whole genome of HSV in these tissues. Which cell type in the ganglion harbors the viral genome?

Cook et al. (1974) found that ganglia from latently infected mice were induced to produce virus not only by explantation in vitro but also by transplantation to the peritoneum of syngeneic animals. By immunofluorescent and electron microscopy, viral products were identified first in the neurons of such transplanted ganglia. This was confirmed by labeling of neuronal nuclei in autoradiographs of sections from explanted ganglia that had been incubated in the presence of [³H]thymidine. Moreover, by using radiolabeled virus-specific cRNA as the probe, viral DNA was detected in the nuclei of neurons in ganglion explants by *in situ* hybridization. This technique was also used to show viral DNA in neurons of ganglia from latently infected rabbits (zur Hausen and Schulte-Holthausen, 1975) and viral mRNA in neurons of human sensory ganglia (Galloway *et al.*, 1979). In dissociated cultures of latently infected mouse ganglia, HSV antigens were first detected on day 3 after dissociation in neurons (Kennedy *et al.*, 1983). The neurons were identified by a specific antineuronal monoclonal antibody.

Unequivocal evidence for the neuron as the site of latent infection was provided by McLennan and Darby (1980) in mice latently infected with ts mutants of HSV-1. The latent infection with the mutants was activated *in vitro* by the usual explantation method but at the nonpermissive temperature. Reactivation *in vivo* was by section of the peripheral nerve associated with the ganglion. In the latter case the body temperature of the mouse was relied upon to provide the nonpermissive conditions. At the nonpermissive temperature the spread of infection from the primary site of activation to adjacent nonlatently infected cells could not occur. Under such conditions, either *in vivo* or *in vitro*, viral antigens were detected by immunofluorescence only in neurons.

Further evidence for latent infection in neurons is provided by the demonstration, again by immunofluorescence, of the α polypeptide ICP4 in the nuclei of ganglionic neurons of latently infected rabbits (Green *et al.*, 1981b).

At present there is little information on the proportion of ganglionic neurons that may harbor the latent infection at any one time. Walz *et al.* (1976) have attempted to investigate this by enzymatically dispersing latently infected ganglia from mice. By cocultivating such cell suspensions they determined that latent infection was reactivated in 0.1% of ganglionic cells, i.e., about 1% of neurons. Using similar methods, but with more positive means of identifying neurons, Kennedy *et al.* (1983) estimated that 0.4% of neurons produced reactivated virus. If, as suggested by these experiments, only a proportion of ganglionic neurons may be latently infected, the question arises as to whether this proportion can be affected once latency is established. An increase in latently infected neurons might occur following secondary infection with the same of a different strain of virus, or as a consequence of the "round-trip" mechanism (Klein, 1976). In the "round-trip" hypothesis it is proposed that during episodes of recurrent peripheral disease, e.g., in the skin, virus is reseeded into the ganglion. This was originally conceived as a possible mechanism for maintenance of the latent infection in the ganglion in the presence of the neuronal loss that might follow the reactivation of latency. If no such loss occurs, the "round trip" of virus might result in an increase in the number of latently infected neurons. However, it may be that the presence of the viral genome within a latently infected neuron renders that cell "immune" to the establishment of latency by a second virus. Moreover, by the production of a diffusible repressorlike molecule from latently infected cells, such immunity might also be conferred on adjacent uninfected neurons. By the use of viruses with particular markers it should be possible to determine, at the least, whether more than one strain of HSV can establish latency in the same ganglion.

E. Viral Functions Associated with the Establishment and Maintenance of Latency in Neurons

Once the virus has reached the ganglion it seems likely that the successful establishment of latency will depend on particular interactions between the virus and the "target" cell, i.e., the neuron.

Particular characteristics of the neuron may play an important part in these interactions but these will be considered later in the section dealing with reactivation of latency. In the discussion that follows, a number of viral functions are considered with respect to whether they are necessary for the establishment of latency in neurons.

1. Productive Infection

During intraaxonal transport, virus replication would be impossible, but once the virus arrives in the ganglion, replication could theoretically occur in the nucleated cell bodies of the neurons, the glial cells, and fibroblasts. Following the inoculation of a peripheral site, such as the mouse footpad, infectious virus can be detected in the relevant sensory ganglia from about the 1st to the 8th day after infection; peak titers occur at about the 4th day (Stevens and Cook, 1971; Cook and Stevens, 1973). Theoretically, this appearance of virus in the ganglion might merely represent virions accumulating there as a result of an input from the periphery. However, there is clearly viral replication in some ganglion cells, as productive infection has been observed in ganglionic neurons by electron microscopy (Dillard et al., 1972; Schwartz and Elizan, 1973; Cook and Stevens, 1973; Knotts et al., 1974). In contrast to the productive infection in neurons, electron microscopic observations by these same authors suggest that the surrounding satellite cells are abortively infected. Confirmation of this differing response of neurons and peripheral glia has been obtained in vitro by using explant cultures of sensory ganglia (Hill and Field, 1973).

What is the relevance of these events in the ganglion, particularly the productive infection in neurons, to the establishment of latency? Certain ts mutants of HSV (both DNA$^+$ and DNA$^-$) are able to establish latency in the mouse brain (Lofgren et al., 1977; Watson et al., 1980) and sensory ganglion (McLennan and Darby, 1980) although they are unable to replicate at the inner body temperature of the mouse. This suggests that neither DNA replication nor productive infection is essential for the establishment of a latent infection in neurons. Further indirect evidence for this conclusion is provided by the observation that passive immunization, which reduces the amount of productive replication in the ganglion during primary infection, results in an increased incidence of latent infection (Price and Schmitz, 1979). In contrast, neurectomy (Price and Schmitz, 1979), treatment with cyclophosphamide (Price and Schmitz, 1979; Openshaw et al., 1981), or treatment with 5-hydroxydopamine (Price, 1979) all increase productive infection in ganglia at the time of primary infection and lead to a decreased incidence of latent infection. From such observations, Sekizawa et al. (1980) and Openshaw et al. (1981) have postulated that ganglionic neurons may respond to HSV infection in either of two ways. Some neurons may be permissive for infection and perhaps thereby are killed while others may be nonpermissive and thereby latency can be established. They further postulated that the permissivity of some neurons may be induced by irritation of their nerve endings. Such irritation might result from the inflammation in the peripheral tissues at the time of primary infection.

The administration of HSV antiserum soon after infection apparently allows latency to be established more rapidly—2–4 days after infection rather than the usual 10–14 days (Sekizawa et al., 1980). An alternative interpretation of such observations might be that even during the course of a normal primary infection, latency is established in nonpermissive neurons as soon as the virus reaches the ganglion (1–2 days after infection). However, this rapid establishment of latency would not normally be demonstrable because for several days it would be masked by the productive infection in other permissive cells. Hence, the administration of HSV antiserum might ablate such productive infection and thereby allow the demonstration of latency at an early stage.

In summary, it seems that the productive infection seen in the ganglia of laboratory animals during primary infection with HSV is unnecessary for the establishment of latency. Indeed, there may be an inverse relationship between productive and latent infection.

2. Structure of the Viral Genome

By analogy with certain DNA tumor viruses it might be proposed that the establishment of latency involves integration of the viral genome into that of the host and/or persistence of the viral genome in an extrachromosomal plasmidlike state. In this respect, Becker (1978, 1979) has postulated that by virtue of its unique structure, HSV DNA could undergo intramolecular recombination to form figure-8-shaped molecules. Such molecules might then exist in latently infected cells as plasmids or become integrated into the host cell DNA. Yanagi et al. (1979) and, more extensively, Ritchie and Timbury (1980) have drawn attention to the similarities between the arrangement of the HSV genome and that of transposons (units of DNA that can readily "move" from one site to another within the DNA of prokaryotic and eukaryotic organisms). Indeed, they suggest that the HSV genome resembles two transposable elements (the long and short unique regions, each bounded by inverted repeat sequences) that might integrate independently into the host cell genome. While there are superficial similarities between the HSV genome and transposons, there is at least one notable difference. The terminal repeat sequences of transposons are often as small as 40 base pairs and can be as large as 1.5 kb (Kopecko, 1980) whereas the repeat regions of HSV are much larger (the ab region being 9.6 kb and the ca region 6.9). Indeed, the repeat regions of HSV DNA are themselves equivalent in size to some bacterial transposons, many of which are 3–10 kb in length (see also Kopecko, 1980). Therefore, a modification of the Ritchie and Timbury proposals would be that the HSV repeat regions themselves may behave like transposons. By so doing [and by analogy with compound transposons in bacteria (Kleckner, 1981)] they may be able to mobilize the unique regions with which they are linked and insert them into the genome of the host cell.

If during establishment of latency the viral genome is integrated into that of the host, it will be important to determine whether the process is site specific (as in the case of λ phage and the *E. coli* chromosome) or more random (as in the case of μ phage) (see also Kopecko, 1980). Moreover, it remains to be established whether such integration may be limited to certain host cells, e.g., neurons. Observations on the growth of HSV in BSC-1 cells or BHK21 cells (Biegeleisen and Rush, 1976; Yanagi *et al.*, 1979) suggest that integration of viral DNA into that of the host cell may occur during a normal productive infection. Hence, while such integration may occur in many cell types, the establishment of latency may require the infection of a nonpermissive cell in which late functions of the viral genome are not expressed and thereby cell death is avoided. If this is so, then latency and transformation by tumor viruses such as SV40 may indeed have much in common, as transformation also requires the infection of a nonpermissive cell (Tooze, 1973). This concept would also conform with the conclusion (see above) that the establishment of latency does not require virus replication.

It will be apparent that information on the precise state of the viral genome in the latently infected neuron is almost nonexistent. Because, in some cases, the viral DNA found in the human brain contained terminal fragments, Fraser *et al.* (1981) suggested that the genome may be present in a linear, nonintegrated form. More recent molecular evidence is conflicting. Rock and Fraser (1983) demonstrated the presence of the HSV genome in the brains of latently infected mice, but the viral DNA lacked the terminal fragments, suggesting that the viral genome was in a nonlinear, episomelike state. The terminal repeat fragments were detected by Puga *et al.* (1984) in the viral DNA from the trigeminal ganglia of latently infected mice. However, these fragments had structural changes that were suggestive of integration into cellular DNA.

3. Specific Viral Genes

Studies with ts mutants of HSV suggest that specific viral genes are necessary for the establishment of latent infection in the mouse brain (Lofgren *et al.*, 1977; Watson *et al.*, 1980). One such latency-negative mutant (*ts*K) is defective in the production of the immediate–early (α) polypeptide ICP4 (otherwise known as VMW 175) (Preston, 1979a). The significance of this observation has been greatly increased by the report, previously mentioned, that this same polypeptide, appears to be continuously expressed in the nuclei of latently infected ganglionic neurons (Green *et al.*, 1981b). During the course of normal virus replication the polypeptide ICP4 appears to have a regulatory function. In particular, it may control the transition from immediate–early to early and late protein synthesis (Preston, 1979b). Hence, this immediate early function may also be involved in the regulation of the genome that may be necessary for both establishment and maintenance of latency.

Other mutants such as *ts*A examined by Lofgren *et al.* (1977) and Watson *et al.* (1980), which are defective in genes beyond the immediate–early region of the genome, were also unable to establish latency in the CNS. This suggests that later functions of the genome may also be involved.

It is noteworthy that in all these experiments with *ts* mutants the viruses were assessed for their ability to establish latency in the mouse brain. Therefore, in the interpretation of the results it is worth reemphasizing the note of caution made earlier that there may be differences between the PNS and the CNS in the ability of HSV to establish latency and/or be reactivated from the latent state.

As suggested by Marsden (1980) the study of specific deletion mutants of HSV should greatly facilitate clarification of which viral genes may be important in the establishment of latency.

Another function of the virus that has been considered of possible relevance to latency is the enzyme thymidine kinase (TK). It is clear that TK expression is unnecessary for replication of the virus in rapidly growing cells in culture. However, mutants lacking this function (TK⁻) grow less well in serum-starved (slowly growing) cells (Jamieson *et al.*, 1974). Hence, *in vivo* the viral TK may be necessary for virus replication in nondividing cells (e.g., neurons) or cells that are slowly growing (e.g., in the epidermis). Certainly, TK⁻ mutants of HSV are usually extremely avirulent (Field and Wildy, 1978) and grow poorly in skin and CNS (Field and Wildy, 1978; Field and Darby, 1980), in sensory ganglia (Tenser and Dunstan, 1979; Tenser *et al.*, 1979), and in autonomic ganglia (Price and Kahn, 1981). Moreover, such mutants produce a low incidence of latent infection in sensory ganglia (Tenser *et al.*, 1979; Tenser and Dunstan, 1979) and autonomic ganglia (Price and Kahn, 1981). Such observations have led to the view that viral TK may be necessary for the establishment of latency (Tenser and Dunstan, 1979). However, the fact that many TK⁻

viruses are apparently "latency-negative" may be related to their poor growth in tissues such as skin and thereby their reduced chance of entry into nerve endings. This suggestion is supported by the observation that inoculation of a very large dose of TK$^-$ virus (strain B2006) into the skin of the mouse pinna, produced latent infection in the cervical ganglia of 50% of animals (Field and Wildy, 1978). Using the same virus strain for ocular infection but at a lower dose, Price and Kahn (1981) were unable to demonstrate latency in the superior cervical ganglia of mice.

In interpreting experiments on the significance of the TK gene in pathogenesis and latency it is noteworthy that different strains and mutants of HSV produce different amounts of TK (Field and Darby, 1980). In this respect a mutant TK$^{1/4}$ (producing 25% of the level of the parent strain) was found to establish latency in mice but was avirulent unless inoculated at high doses (Gordon et al., 1983).

As virus replication in neurons appears to be unnecessary for establishment of latency (see above), it seems unlikely that the inability of TK$^-$ viruses to replicate in neurons would be the reason for their apparent inability to establish latency (Price and Kahn, 1981). Indeed, Price and Kahn postulated that TK$^-$ viruses may be able to establish latency in neurons but because such viruses replicate poorly in these cells the latent infection cannot be reactivated. If this is so, TK$^-$ viruses will appear to be "latency-negative" because the usual ganglion explantation methods depend upon reactivation as the means of showing a virus to be "latency-positive." Confirmation of this hypothesis will require the application of other techniques, such as nucleic acid hybridization, to detect the putative TK$^-$ genome in neurons.

In summary, the relevance of the viral TK gene to latency is still far from clear. However, it seems that under some circumstances TK$^-$ mutants can establish a latent infection and therefore this viral function does not appear to be an absolute requirement for latency to occur.

4. Fc Receptors

Cells productively infected with HSV are known to express receptors for the Fc portion of immunoglobulins (Watkins, 1964). Such receptors may be part of the viral glycoprotein gCE, which is inserted into both the host cell membrane and the viral envelope (Baucke and Spear, 1979; Para et al., 1980). On an entirely hypothetical basis it was suggested that the binding of IgG to HSV Fc receptors on cell membranes (Costa and Rabson, 1975) or the double binding of antiviral IgG to such receptors and viral antigens (Lehner et al., 1975) might modulate virus replication and thereby might be involved in the establishment and maintenance of latency. The possible role of antibody in the establishment and maintenance of latency will be considered later. The discussion here will be limited to the likelihood of Fc receptors being present on the surface of latently infected neurons. As yet there have been no reports of Fc receptors on such cells and although their presence cannot be excluded it seems unlikely for a number of theoretical reasons: (1) the production of gCE may be a late function of the genome (one of the γ polypeptides). Hence, unless there were partial transcription of the genome, the expression of such a function would be more likely to occur during a productive infection, an event that invariably leads to cell death. (2) It has been demonstrated that HEp-2 cells in culture can acquire Fc receptors from the envelope of the input (infecting) virions (Para *et al.*, 1980). Therefore, in attempting to overcome the difficulties mentioned in (1), it has been suggested that neurons might acquire receptors in a similar manner (Spear *et al.*, 1981). However, the cell membrane is a highly dynamic structure. It is therefore difficult to envisage how such passively acquired molecules could be maintained for long periods on the surface of the latently infected neuron.

5. Defective Interfering Particles

The occurrence of defective interfering (DI) particles has been described for a number of RNA viruses, and the ability of DI particles to interfere with the replication of nondefective (standard) virus led to the proposal that they may play a role in the modulation of infections *in vivo* (Huang and Baltimore, 1970). DI particles have also been described for herpesviruses, including HSV, and much is known about their molecular biology (reviewed by Frenkel, 1981). It seems that in the mouse the presence of HSV DI particles can decrease the severity of the infection in the skin and CNS (Zenda and Murray, 1981). However, nothing is known at present of the possible role of HSV DI particles in latency. It is conceivable that the presence of DI particles in the inoculum or their production in the skin during primary infection, might interfere with standard virus and reduce its entry into the nervous system. It is also possible that DI particles might interfere more directly in the establishment of latency by affecting the interaction between the neuron and the standard virus. In this respect it will be of interest to determine whether despite their defective genomes, DI viruses can themselves produce a latent infection.

F. The State of Latency: Presence and Expression of Viral Functions in the Latently Infected Ganglion

The precise state of the virus in latently infected ganglia has been a matter for speculation for some time. Roizman (1965, 1974) proposed two possible states of latency. One is the *dynamic state* in which the presence of virus is maintained by a low level of productive infection. Such a state would resemble persistent infections described *in vivo* and *in vitro* for other viruses, e.g., the paramyxoviruses (ter Meulen and Carter, 1982). The other is the *static state* in which the viral genome is totally or partially repressed. In this state the genome might either be integrated into the host cell chromosome (in a manner analogous to a lysogenic bacteriophage) or be present as an extrachromosomal plasmidlike structure [like EBV in some lymphocytes (Adams and Lindahl, 1975)]. The weight of evidence in favor of the dynamic or static state of latency is best assessed by considering the presence and expression of various viral functions in the latently infected ganglion.

1. Infectious Virus

Most workers report that after primary infection in experimental animals, infectious virus can be isolated from sensory ganglia for only a short time, usually not later than 7–10 days after inoculation (Stevens and Cook, 1971; Cook and Stevens, 1973). However, in one report (Schwartz et al., 1978) it is claimed that by inoculating homogenates of mouse ganglia onto explant cultures of fetal mouse dorsal root ganglia, infectious virus could be detected for up to 8 months after infection. The incidence of virus isolation from ganglia fell from 80% at 1 month after infection to 20% at 6 months.

By electron microscopy of serial sections of trigeminal ganglia from latently infected rabbits, Baringer and Swoveland (1974) found a very small number of productively infected cells (probably neurons). However, such cells could represent either a low level of persistent infection (dynamic state) or the spontaneous reactivation of virus from a truly latent (i.e., static) state. The latter may be more likely in view of the known propensity of the latently infected rabbit to shed virus spontaneously in its tears (Nesburn et al., 1967).

Treatment of latently infected mice or rabbits with various potent antiviral drugs did not affect the incidence of latent infection in their ganglia (Field et al., 1979; Klein et al., 1979, 1981; Blyth et al., 1980b; Svennerholm et al., 1981; Field and De Clercq, 1981; Nesburn et al., 1983). This provides indirect evidence for the lack of a dynamic state of latency, as such drugs only affect replicating virus.

2. Viral Genome

As evidenced by the release of infectious virus after explantation in vitro, most latently infected ganglia must contain the complete HSV genome. However, Brown et al. (1979) found that by superinfection with ts mutants of HSV, viral genomes could be rescued from human ganglion cultures that had failed to yield infectious virus. This suggests that some neural tissues that are "latency-negative" by the usual cultural tests may contain defective or noninducible viral genomes.

The presence of viral DNA in latently infected mouse ganglia (0.1 genome equivalent/cell) was demonstrated by a liquid-phase hybridization technique (Puga et al., 1978).

3. Viral Antigens

Stevens and Cook (1971) were unable to detect viral antigens in latently infected mouse ganglia by immunofluorescence. Using a general antiserum to HSV, Green et al. (1981b) were similarly unsuccessful with latently infected ganglia from rabbits. However, by using a monospecific antiserum they demonstrated the presence of the α polypeptide ICP4 in the nuclei of ganglionic neurons.

4. Viral Enzymes

During the normal productive replication of HSV a number of specific viral enzymes are produced. In one report (Yamamoto et al., 1977) it was claimed that one such enzyme (TK) was present in the ganglia of mice up to 60 days after infection, a time normally regarded as being in the latent phase. This observation was not confirmed by Fong and Scriba (1980) who found in guinea pigs that TK could be detected in skin and ganglia during primary infection but not in the ganglia when latency was established (at 2 weeks and at 1 year after infection).

5. Viral mRNA

Even in the absence of productive infection, viral mRNA should be present in ganglia if viral polypeptides such as ICP4 (Green et al., 1981b) or viral enzymes such as TK (Yamamoto et al., 1977) are expressed during the latent phase. However, no viral transcripts were detected in latently infected mouse ganglia (either soon after establishment of latency or at later times) by a liquid-phase hybridization method (Puga et al., 1978). It was estimated that this method had a lower limit of sensitivity of one full complement of viral mRNA per 2000 cells. Hence, a transcript equivalent to about 5% of the viral genome would have been undetected. Therefore, more sensitive methods may be required to demonstrate the presence of a putative transcript equivalent to a single viral gene.

By using in situ hybridization, viral transcripts have been detected in human thoracic, lumbar, and sacral ganglia (Galloway et al., 1979, 1982). However, there is need for caution in interpreting these observations. Unlike the situation with experimental animals some delay is inevitable between the death of the individual and the removal of tissues for processing in the laboratory. Therefore, the degenerative changes that occur rapidly after death, particularly in neurons, might initiate reactivation of the latent infection in a manner analogous to that caused by explantation of the ganglia in vitro. Such complete or partial reactivation would lead to the production of viral mRNA and thereby give a false impression of the true state of latency in vivo. However, this criticism is less readily applicable to experiments involving guinea pigs in which HSV mRNA was detected in latently infected trigeminal ganglion neurons by in situ hybridization (Tenser et al., 1982).

From the information reviewed above and summarized in Table I there is clearly little evidence for the presence of normal productive replication of HSV in the latently infected ganglion. Therefore, at present the weight of evidence favors the static state of latency in which perhaps only some early functions of the viral genome are expressed. The extent of such expression and the factors involved in the apparent block in transcription of the viral genome remain to be determined. Cultures of infected neurons *in vitro* that mimic latency *in vivo* (e.g., Wigdahl *et al.*, 1983) may be helpful in this respect.

Some observations on latently infected ganglia already referred to [the loss of TK (Yamamoto *et al.*, 1977), the gradual decline in incidence of infectious virus (Schwartz *et al.*, 1978)] suggest that over a period of time there may be gradual changes in the latent state. Confirmation of such changes and an investigation of their nature are urgently required.

IV. REACTIVATION OF THE LATENT INFECTION AND THE PRODUCTION OF RECURRENT DISEASE

A. Introduction

1. Terminology

During latency, infectious HSV can rarely be isolated from affected tissues such as sensory ganglia but, as already discussed, the appearance of virus can be induced by culture of the tissues *in vitro*. It is now known (Table II. and discussed in detail below) that a variety of procedures will also induce the appearance of infectious virus in latently infected ganglia *in vivo*. The demonstration of such virus by inoculation of cell cultures with homogenates of the latently infected tissue (without intact, viable cells) is taken to indicate *reactivation* of the virus. Hence, as in the case of latency, reactivation is also defined in operational terms. Some authors have extended the use of *reactivation* to include the production of recurrent clinical disease. However, to aid clarity of thought and discussion it would seem desirable to adopt a more precise terminology.

For such reasons Wildy *et al.* (1982) proposed the following definitions:

- *Reactivation*: the "reawakening" of virus from the latent state, either spontaneously or as a result of external stimuli, so that infectious virus may be isolated
- *Recrudescence*: the initiation of a clinical lesion in a peripheral tissue as a result of reactivation in the associated sensory ganglion
- *Recurrence*: the presence of infectious virus in a peripheral tissue (probably resulting from reactivation in the ganglion) in the absence of a clinical lesion

TABLE I. Presence and Expression of Viral Functions in Latently Infected Ganglia

			Host	Reference
Viral genome	Complete	+	Mouse	Stevens and Cook (1971, 1973a,b)
	Complete	+	Human	Reviewed by Baringer (1975)
	Defective or noninducible	+	Human	Brown et al. (1979)
Viral DNA		+	Mouse	Puga et al. (1978)
Viral polypeptides	Antigens	−	Mouse	Stevens and Cook (1971, 1973a,b)
		+ ICP4	Rabbit	Green et al. (1981b)
	Thymidine kinase	+	Mouse	Yamamoto et al. (1977)
		−	Guinea pig	Fong and Scriba (1980)
		−	Mouse	Puga et al. (1978)
Viral mRNA		+	Human	Galloway et al. (1979, 1982)
		+	Guinea pig	Tenser et al. (1982)
Productive infection		−	Mouse	Stevens and Cook (1971, 1973a,b)
		+	Mouse	Schwartz et al. (1978)

TABLE II. Reactivation of Latent Infection in Ganglia of Mice *in Vivo*

Site of primary infection[a]	Treatment	Reactivation of virus in:	Reference
Footpad	Section of peripheral nerve	Dorsal root ganglia	Walz et al. (1974)
	Intratracheal injection of pneumococci or mucin	Dorsal root ganglia	Stevens et al. (1975)
Intraocular	Postganglionic neurectomy	Superior cervical autonomic ganglion	Price and Schmitz (1978)
Cornea	Cyclophosphamide or X-rays	Trigeminal ganglion	Openshaw et al. (1979a)
	Cyclophosphamide, prednisolone, antithymocyte serum, or trauma to ganglion	Trigeminal ganglion	Hill et al. (1981)
Cornea or lip	Dry ice on lip	Trigeminal ganglion (Reactivation shown by seroconversion of "antibody-negative" mice)	Openshaw et al. (1979b)
Cornea or lip + passive immunization	Dry ice on lip		Sekizawa et al. (1980)
Skin of ear pinna	Cellophane tape stripping, application of DMSO, xylene, or retinoic acid to ear	Cervical dorsal root ganglia	Harbour et al. (1983b)

[a] All models used HSV-1.

While this attempt at a more careful use of terminology is to be welcomed, some confusion may still arise between *recurrence* and *recrudescence* and therefore I shall not adopt these two definitions. Nevertheless, the stricter definition of *reactivation* as proposed by Wildy *et al.* (1982) will be used, particularly as it relates to the appearance of infectious virus at the site of latency (Hill, 1981).

2. The Consequences of Latent Infection and Its Reactivation

There has been a general assumption that while the virus remains in the latent state, in the sensory neuron or elsewhere, the vital functions of the host cell are unaffected. This is perhaps more likely to be the case if, as previously concluded, latency is of the static type in which few if any viral functions are expressed. However, it would be of interest to determine whether, even in such a static state, the presence of the virus might alter some of the *nonvital* functions of the host cell. For example, the presence of the latent HSV genome in a neuron might produce subtle changes in the ability of the cell to synthesize neuropeptides. However, so far as is known at present the effects of the latent infection on the host cell only become manifest when the latent state is disturbed by endogenous or exogenous factors and viral infectivity is reactivated. On occasions this reactivation may lead to signs of disease in the host and the consequences of latent infection then become obvious to the patient and clinician. Hence, a consideration of reactivation and the pathogenesis of recurrent herpetic disease is essential to complete the picture of latency. Moreover, an understanding of how the latent infection is reactivated should shed further light on the factors that are involved in its maintenance and control.

B. Mechanisms Underlying the Reactivation from Latency

1. A Paradox

The relationship that must exist between HSV and the host cell during latency presents an intriguing paradox. While latency is maintained the neuron behaves as a nonpermissive cell with respect to virus replication. However, when reactivation occurs the host cell–virus relationship is altered so that the neuron becomes permissive and virus replication is initiated. Such an alteration might involve the removal or modification of controls that operate to maintain latency. In this section an attempt is made to unravel the extent to which host and viral factors may play a role in these control processes.

2. Host Factors and Reactivation

The reactivation of the latent infection that follows explantation of sensory ganglia *in vitro* was the means by which latency was first demonstrated experimentally (see above). Subsequently, reactivation in ganglia was also demonstrated *in vivo* in a number of model systems, particularly in mice, after a variety of procedures (Table II). Most of these procedures fall into the following categories: (1) nerve damage (trauma to the ganglion or nerve section), (2) trauma (physical or chemical) to the peripheral site receiving innervation from the latently infected ganglion, (3) immunosuppressive treatments. All these *in vivo* procedures and the processes involved in removal and explantation of ganglia *in vitro* are very likely to induce changes in cell and tissue functions. It is therefore important to consider which, if any, of these changes might affect the virus–host balance so as to induce reactivation. Two possibilities will be discussed: (1) that latency is maintained by "inhibitors" of host origin, in particular immune factors, and that reactivation is caused by a depression of their effects; (2) that latency is maintained in the neuron as a result of particular features of its physiology and that reactivation follows physiological changes in the cell that allow increased expression of the viral genome.

a. Immune Mechanisms

Mention has already been made of the postulate that latency in the neuron may be maintained by the double binding of antiviral IgG to viral antigens and Fc receptors on the cell surface (Lehner *et al.*, 1975). In support of this idea, Costa *et al.* (1977) found that high concentrations of normal rabbit IgG inhibited the growth of HSV in Vero and Y-79 cells. However, the unlikely occurrence of Fc receptors on the surface of latently infected neurons has already been discussed.

Whether or not Fc receptors are involved in maintenance of latency, Stevens and Cook (1974) produced evidence for the possible role of anti-HSV IgG. Latently infected ganglia were transplanted in chambers covered with Millipore filters (pore size 0.22 μm) into the peritoneal cavity of immune and nonimmune mice. After 4 days the ganglia were removed and sections examined for foci of viral antigens and DNA. Ganglia from the immune mice showed fewer foci of reactivated virus than those from nonimmune animals. Further experiments showed this effect was due to antiviral IgG. These experiments led Stevens and Cook to suggest that antiviral IgG plays a part in the maintenance of latency. However, it seems likely that the processes involved in transplantation of the ganglia would initiate reactivation. Hence, an alternative interpretation of their results would be that the antiviral IgG, perhaps in combination with the lymphocytes present in the ganglion, can limit the spread of the reactivated virus. Hence, the apparent effects of antiviral IgG may indicate the ability of immune mechanisms to limit the spread of reactivated HSV infection in ganglia rather than an involvement in the maintenance of latency.

Further doubt has been cast on the role of immunoglobulins in the control of latency by the observation that latency can occur for long periods in sensory ganglia of mice in the absence of detectable serum neutralizing antibody (Sekizawa et al., 1980). Moreover, reactivation of virus from latently infected neurons in vitro took place in the presence of neutralizing antibody (Kennedy et al., 1983). This does not of course rule out the possibility that other kinds of antibody might be involved in maintaining the latent state (Nash, 1981).

The induction of reactivation in the ganglion by various immuno-suppressive treatments (Table II) might be taken to indicate a more ob-vious role for immune mechanisms in the maintenance of latency. How-ever, it is quite clear, particularly from the drastic levels of treatment that are necessary, that such procedures could induce reactivation by a variety of nonimmunological means (Hill et al., 1981). For example, an-tithymocyte serum could produce membrane changes in neurons by bind-ing to the theta antigen, which neurons and lymphocytes have in com-mon. Such changes might then lead to reactivation. Similarly, steroids might induce nonspecific changes in neuronal membranes. With regard to cyclophosphamide, Openshaw et al. (1979a) have pointed out that its ability to induce reactivation may be related to the known ability of the drug to damage DNA rather than its immunosuppressive activity.

In summary, there is at present no strong evidence for the involve-ment of immune functions in the maintenance of latency in neurons and therefore in its reactivation. Moreover, it is not clear how some of the stimuli that induce reactivation in ganglia, in particular damage or trauma to peripheral tissues, might act by affecting immune mechanisms in the ganglion.

However, there are indications that immune functions, in particular the production of antibody (Stevens and Cook, 1973; Kino et al., 1982; Cook and Stevens, 1983) and the production of interferon (Sokawa et al., 1980), may be involved in clearance of infectious virus from ganglia dur-ing the acute disease.

b. Physiology of the Neuron

At least two features of the neuron's physiology may be relevant to latency. First, in the adult animal the neuron is a relatively long-lived, nondividing cell in which cellular DNA synthesis only occurs as a con-sequence of repair mechanisms (Sanes and Okun, 1972; Ishikawa et al., 1978). If during latency the viral and host genomes were closely associ-ated, the quiescent nature of the host's genome might allow the virus to remain undisturbed for long periods of time. Second, the neuron is a fully differentiated cell with very specialized functions. In such a cell large, areas of the genome would remain untranscribed. As HSV uses the host cell RNA polymerase B to transcribe its DNA (Constanzo et al., 1977), an association of the viral DNA with such "silent" areas of the neuronal

DNA might allow the viral genome to remain unexpressed during latency. Under such circumstances full expression of the viral genome (i.e., reactivation) might be caused when these "silent" areas of the neuronal DNA were transcribed. This might occur when the cell was called on to perform "out of the ordinary" functions, e.g., the repair of damage to its membrane or cytoplasmic processes. Therefore, it is noteworthy that such damage (nerve section or trauma to the ganglion) is a well-recognized cause of reactivation of latency in the ganglion *in vivo* (Table II). Moreover, the cutting of nerves involved in removal of ganglia for explantation may provide the main stimulus for the reactivation observed during culture *in vitro*.

The normal repair processes initiated by nerve section include an increase in protein synthesis in the neuronal cell body and an increase in transcription of neuronal DNA (reviewed by Grafstein, 1975). It is not known whether other inducers of reactivation in the ganglion, in particular physical or chemical trauma to the skin, might induce similar changes in neuronal metabolism. However, it seems likely that such stimuli would induce minor damage to peripheral nerve endings and that this would be followed up by repair.

Hence, the general concept emerges, as discussed by Price and Schmitz (1978), that changes in neuronal metabolism may lead to the reactivation of latent infection with HSV, i.e., the neuron becomes permissive for virus replication. Further support for this concept comes from observations on the effect of neuronal metabolic changes on productive infection with HSV in the superior cervical ganglion of mice during the acute disease. Thus, postganglionic neurectomy (Price and Schmitz, 1979) and treatment with 6-hydroxydopamine, a drug that selectively injures adrenergic nerve terminals (Price, 1979), both increased the titers of virus produced in the ganglion.

It is well recognized that in comparison with the neurons of the PNS, those in the CNS have a lesser capacity to regrow and repair their nerve processes (Grafstein, 1975). Recent evidence suggests that PNS and CNS neurons may not be intrinsically different in this respect but that the ability of peripheral neurons to regrow damaged axons depends on the environment provided by peripheral glial cells (Benfey and Aguayo, 1982). Nevertheless, the apparent lack of reparative ability of CNS neurons may be of relevance to the previously mentioned apparent low incidence of latent infection with HSV in the CNS as compared with the PNS. If, as argued above, reactivation is in some way related to metabolic changes in the neuron associated with repair processes, the relative lack of such processes in CNS neurons might render reactivation more difficult in these cells. This in turn might lead to a falsely low measure of the true incidence of latency in CNS tissues.

The molecular mechanisms underlying the response of the ganglionic neuronal cell body to axotomy, in particular the increase in transcription (Watson, 1974), are not fully understood. However, in other cells it appears that the extent of methylation of the cytosine bases in DNA is of great significance in the control of gene expression (reviewed by Doerfler, 1981). In general, it seems that such methylation leads to the silencing of genes whereas for expression the gene needs to be under-methylated. This same generalization also appears to be true for the few viral systems that have been examined. In particular, it has been shown with the following viruses that viral sequences integrated into the host cell genome are extensively methylated whereas during productive in-fection this is usually not so: herpes saimiri in lymphoid tumor cells (Desrosiers et al., 1979), HSV in human lymphoblastoid cells of T-cell origin (Youssoufian et al., 1982), adenovirus type 12 in transformed ham-ster cells (Sutter and Doerfler, 1980). It remains to be established whether methylation of the viral DNA, perhaps via host cell functions, plays any part in controlling the expression of the HSV genome during latency. However, it is noteworthy that dimethylsulfoxide [a chemical known to cause hypomethylation of cellular DNA (Christman et al., 1977)] is a potent inducer of reactivation of HSV in ganglia (Hill et al., 1983a; Har-bour et al., 1983b). Moreover, in mouse cells transformed by HSV and containing the inactive viral TK gene, the gene can be activated by treat-ment with 5-azacytidine (Clough et al., 1982), a drug known to cause hypomethylation of DNA. It was also shown that the inactive TK gene was methylated whereas the active gene was unmethylated.

In summary, many of the stimuli that cause reactivation in gangli-onic neurons may do so by affecting the transcription of the host cell genome and thereby, pari passu, by affecting that of the virus, a possibility first mentioned by Roizman (1974).

Viewed in this light, the establishment and maintenance of latency may depend heavily on the peculiar properties of the neuron as a host cell. Hence, in many circumstances latency may be the "natural" rela-tionship between HSV and the neuron (Blyth and Hill, 1984) rather than a situation in which "external" factors (such as immune mechanisms) are necessary to "force" the virus into latency and/or to maintain the latent state.

There may indeed be other cells that possess some of the features of neurons discussed above and thereby they too may have the capacity to harbor latent HSV. In this respect it is of interest that a model of latent infection with HSV can be established in a lymphoblastoid T-cell line of human origin (Hammer et al., 1981). Moreover, the infection was reac-tivated by exposure of the cells to phytohemagglutinin, a mitogen that dramatically alters the DNA metabolism of T lymphocytes.

3. Viral Factors and Reactivation

Just as particular viral functions may be involved in the establish-ment of latency in neurons, so the same and/or different functions may be involved in its maintenance and reactivation. As yet there is little information in this respect but, as already mentioned, Price and Kahn (1981) have made the suggestion that because TK⁻ viruses grow poorly in neurons they may be unable to undergo reactivation.

It may also be possible that the maintenance of latency depends on the production of a virally coded "repressor" in which case reactivation might result from negation of its effects. As yet there is no evidence for a function of this type although the presence of ICP4 in the nuclei of latently infected neurons (Green *et al.*, 1981b) raises the possibility that latency is maintained by the continued expression of a particular viral gene or genes. If this proves to be so, it may be that reactivation results from an interaction between the kind of host factors discussed above, particularly those concerned with transcription of the host genome, and the expression of those viral genes that maintain the latent state.

4. Fate of the Neuron after Reactivation

Whatever the mechanism of reactivation, its consequence is a productive infection with HSV at the site of latent infection. In the ganglion the cell in which infectious virus is produced is probably the neuron, for viral antigens appear there after explantation *in vitro* (Cook *et al.*, 1974; Rajčáni *et al.*, 1976; McLennan and Darby, 1980; Kennedy *et al.*, 1983) and after nerve section *in vivo* (McLennan and Darby, 1980). The consequence of reactivation for the neuron may be, as for productive infection with HSV in other cells, death of the cell. McLennan and Darby (1980) found that 2 weeks after section of the peripheral nerve close to latently infected cervical ganglia in mice (a procedure known to cause reactivation, Table II) the incidence of latent infection in the ganglia was 25% compared with 100% in control animals. Their interpretation of this result was that reactivation led to neuronal death and thereby a reduction in the incidence of latent infection.

Unfortunately, these observations are also explicable in terms of the nonspecific neuronal death that is known to follow section of a peripheral nerve adjacent to a ganglion. In this respect, Ranson (1906) showed a 50% loss of neurons in the cervical ganglia of rats after a nerve section procedure similar to that used by McLennan and Darby (1980).

If sensory neurons in man do indeed succumb to reactivation of latency, there might eventually be a noticeable loss of sensation in the relevant peripheral tissue. So far such loss has not been reported except in herpetic keratitis (Duke-Elder, 1965; Tullo *et al.*, 1983c). However, the putative deficit in sensation might be masked by reorganization of function within the CNS.

Klein (1976) has pointed out that if reactivation is an "end event" for the neuron, there should be a gradual depletion of latent infection within the ganglion. To overcome this he has proposed the following series of events as the "round-trip" hypothesis: (1) reactivation results in

death of the neuron and translocation of the virus via the axon to epidermal cells; (2) the virus infects epidermal cells from where it passes to
the terminals of neighboring axons; (3) the virus passes by retrograde
axonal transport to establish latency in fresh ganglionic neurons. Confirmation of this interesting idea will require methods for precise quantification of latency in ganglia and the demonstration that latency can
be "reestablished" in ganglia that already carry latently infected cells.

C. Events That May Result from Reactivation of Latency

As already discussed, reactivation of latent infection with HSV has
been shown with human sensory and autonomic ganglia *in vitro*. However, in terms of the definition outlined above (the appearance of infectious virus in the ganglion), reactivation in human ganglia *in vivo* has
not been demonstrated; indeed, for obvious technical reasons it may prove
impossible to do so. Nevertheless, circumstantial evidence, e.g., the occurrence of herpetic lesions after surgery to the trigeminal ganglion (Carton and Kilbourne, 1952), suggests that some kind of reactivation in the
ganglion does occur in man. In what follows, the possible consequences,
both actual and hypothetical, of such reactivation are considered. While
the discussion is directed particularly to events in man, extensive reference is also made to evidence from a variety of animal models.

1. Shedding of Virus at the Periphery without Associated Clinical Disease

In the earlier discussion on the site of latent infection, it was pointed
out that infectious virus can occasionally be isolated from peripheral
tissues or secretions such as tears of latently infected humans and animals
[a situation termed *recurrence* by Wildy *et al.* (1982)]. With the possible
exception of the guinea pig (Scriba, 1977, 1981), observations in man and
rabbit suggest that the latently infected sensory ganglion is the source of
such virus shed at the periphery. Hence, in man, besides producing overt
clinical lesions, surgical manipulation of the trigeminal ganglion also
causes asymptomatic shedding of virus in the saliva (Pazin *et al.*, 1978).
Similarly, dental extraction procedures in man produce either clinical
herpes labialis or shedding of virus without disease (Openshaw and Bennett, 1982). Shedding of virus without clinical disease has also been reported in rabbits after electrical or mechanical stimulation of the latently
infected trigeminal ganglion (Nesburn *et al.*, 1976a,b, 1977; Green *et al.*,
1981a) and after iontophoresis of epinephrine into the cornea (Kwon *et
al.*, 1981). These observations raise the important question, considered
in detail later, of why the presence of virus in peripheral tissues is not
necessarily associated with clinical disease?

2. Disorders Associated with the Autonomic Nervous System

The fact that HSV can establish a latent infection in autonomic ganglia of man and animals led Price and Notkins (1977) to suggest that such infection might be associated with disease. They postulated that disease might result from dysfunction of the ganglion itself or from the production of recurrent lesions in organs such as the gut that may receive their nerve supply from a latently infected autonomic ganglion. As yet there is no direct evidence to support these interesting ideas although Vestergaard and Rune (1980) and Van der Merwe and Alexander (1982) have reported a higher level of HSV-1 type-specific antibodies in patients with recurrent duodenal ulcers.

Wellings (1968) has speculated that dysfunction of the parasympathetic ciliary ganglion (producing various ocular abnormalities) may be associated with infection with HSV.

3. Psychiatric Disorders or Chronic Neurological Disease

There are reports that patients with some psychiatric disorders have an increased incidence of antibodies to HSV (Cleobury et al., 1971; Lycke et al., 1974). Similarly, patients with Bell's palsy (facial paralysis) may have a higher incidence and titers of antibodies to HSV than controls (Vahlne et al., 1981). The observation that HSV DNA may be present in many parts of the human brain (Sequiera et al., 1979; Fraser et al., 1981) may be of relevance to these serological findings. Clearly, it is most important to determine to what extent HSV or fragments of its genome can be demonstrated in the CNS of the population at large. When information from studies of this kind is available, it may be possible to get a clearer picture of the relevance of HSV to chronic nervous diseases.

The concept of a possible relationship between HSV and chronic neurological disease in man has received some support from the observation that demyelination of CNS axons can occur during acute infection with the virus in experimental animals (Townsend and Baringer, 1976, 1978, 1979; Kristensson et al., 1978; Townsend, 1981a,b; Martin, 1982; Martin et al., 1982). A mouse model has recently been described for studying the relationship between such demyelination and the function of the seventh (facial) cranial nerve (Hill et al., 1983a; Hill, 1983).

Fujinami et al. (1983) have demonstrated that a monoclonal antibody to an HSV-1 phosphoprotein (ICP6) cross-reacts with an intermediate filament protein (probably vimentin) found in many normal cells. This raises the possibility of the occurrence of autoimmune reactions during the acute and/or chronic phases of HSV infections.

4. Encephalitis

Encephalitis is a rare but serious consequence of infection with HSV in man (Longson and Bailey, 1977; Longson et al., 1980). In theory the disease might occur under three possible circumstances: (1) after primary infection, perhaps occurring later in life than usual; (2) after centrifugal spread of virus to the brain following reactivation of latent infection in the trigeminal ganglion (Davis and Johnson, 1979) [the distribution of viral antigens in the brain of cases of encephalitis does not support this suggestion (Esiri, 1982)]; (3) after reactivation of latent infection in the brain itself. As yet there are no clear indications of which of these possibilities is most likely to occur in man although epidemiological observations suggest that many cases may be due to reactivation of a latent infection (Leider *et al.*, 1965; Longson and Bailey, 1977; Longson *et al.*, 1980). More recent studies, using the precision of typing HSV with restriction endonucleases, suggest that some cases of encephalitis do represent a primary infection whereas others undoubtedly arise from an existing latent infection (Whitley *et al.*, 1982).

As already discussed it is known that in mice, rabbits, and guinea pigs, HSV can establish latency in the CNS. Moreover, in latently infected rabbits, encephalitis can be precipitated by injection of epinephrine (Schmidt and Rasmussen, 1960) or by anaphylactic shock (Good and Campbell, 1948). In the latently infected mouse there is some evidence that drastic immunosuppression (by X-rays and antithymocyte serum) may induce signs of encephalitis (Kastrukoff *et al.*, 1981). However, such treatment would produce many effects in addition to immunosuppression. Latent infection in the CNS and its reactivation are discussed further by Hill (1983).

5. Recurrence of Lesions at the Periphery

a. Clinical Problem

The most common and troublesome manifestation of the latent infection with HSV is recurrence of lesions at peripheral sites such as the orofacial area, eye, and genitalia. Hence, this consequence of latent infection will be considered in more detail. In many individuals, recurrence of lesions is often precipitated by particular stimuli, which may be exogenous (e.g., sunlight, trauma) or endogenous (e.g., stress, fever, menstruation) (Wheeler, 1975). The appearance of orofacial herpetic lesions after trigeminal surgery in man (see above) provided the first clear indication that such recurrences resulted from an endogenous input of virus to the periphery from the latent infection in the ganglion. The gathering of further observations on the natural history of the latent infection in man will continue to be useful (e.g., Spruance *et al.*, 1977). However, it seems likely that much information about the pathogenesis of recurrent disease and the reactivation of latency will continue to come from animal models.

TABLE III. Animal Models Showing Spontaneous Recurrent Disease

Site of primary infection (animal)	Type of HSV	Recurrent disease	Reference
Cornea (rabbit)	Type 1	Keratitis	Nesburn et al. (1967)
Skin of ear pinna (mouse)	Type 1	Cutaneous lesions	Hill et al. (1975)
Footpad (guinea pig)	Type 2	Cutaneous lesions	Scriba (1975)
Vagina (guinea pig)	Type 2	Genital lesions	Scriba (1976), Stanberry et al. (1982)
Vagina (Cebus monkey)	Type 2	Genital lesions	Reeves et al. (1981)
Heart (mouse)	Type 2	Heart infection	Grodums et al. (1981)

b. Recurrent Disease in Animal Models

Animals models in which recurrent clinical herpetic lesions (spontaneous and/or induced) have been described are listed in Tables III and IV. Because of their unpredictability the occurrence of spontaneous lesions is of limited value for experimental studies. Such studies require models in which the recurrent lesions occur within a well-defined time after an inducing stimulus. In the rabbit eye model this has been partially achieved in that shedding of virus in the tears of about 80% of animals can be induced within 48 hr of direct stimulation of the latently infected trigeminal ganglion (Nesburn et al., 1977; Green et al., 1981a) or within 3 days of iontophoresis of epinephrine into the cornea (Kwon et al., 1981). However, in neither case was clinical ocular disease associated with shedding of virus.

At present the only animal model in which recurrent clinical lesions, both spontaneous (Hill et al., 1975) and induced (Hill et al., 1978), have been described is the "mouse ear model" developed in our laboratory. Some of the important features of this model are described in Fig. 2 and Table V.

The mouse ear model has a number of important advantages. (1) The ear is a relatively flat, hairless area of skin that is easily injected and in which lesions are visible without depilation. (2) In albino mice the development of erythema is a clearly visible sign of recurrent disease (Hill et al., 1975; 1978). Under a dissecting microscope, herpetic vesicles and/or pustules can usually be seen in the erythematous area (Hill et al., 1982). (3) Recurrent clinical lesions can be induced by stimuli that are also effective in man (e.g., UV light and trauma). (4) It is economical to use large groups of animals and, when necessary, inbred strains are available (Har-

TABLE IV. Animal Models in Which Recurrent Skin Disease or Virus Shedding in the Eye Can Be Induced

Site of primary infection[a] (animal)	Treatment	Observation	Reference
Cornea (rabbit)	Simulation of trigeminal ganglion	Shedding of virus in tears	Nesburn et al. (1976a,b), Green et al. (1981a)
	Iontophoresis of epinephrine into cornea	Shedding of virus in tears[b]	Kwon et al. (1981)
	UV irradiation of cornea and systemic and topical treatment with steroids	Shedding of virus in tears and keratitis	Spurney and Rosenthal (1972)
Skin of flank (hairless mouse)	Systemic prednisone	Skin lesions in 17%	Underwood and Weed (1974)
Skin of flank (mouse)	Plucking hair	Skin lesions in 50–80%	Hurd and Robinson (1977)
Skin of ear pinna (mouse)	UV irradiation of pinna or injection of prostaglandins or saline into pinna	Appearance of virus in skin	Blyth et al. (1976)
	Cellophane tape stripping of pinna	Recurrent erythema and skin lesions in 30–50%	Hill et al. (1978, 1982)
	Intraperitoneal amphetamine sulfate	Recurrent erythema	Blue et al. (1981)

[a] HSV-1 used in all cases.
[b] Also induces reactivation in trigeminal ganglion (J. Hill et al., 1983).

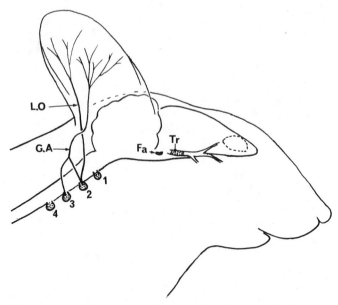

FIGURE 2. Mouse ear model: anatomical and virological aspects (Hill *et al.*, 1975, 1978). L.O, lesser occipital nerve; G.A, great auricular nerve; 1–4, cervical ganglia; Tr, trigeminal ganglion (fifth cranial); Fa, root of facial nerve (seventh cranial, supplies motor nerves to ear muscles). *Mice*: female, 4-week outbred, Swiss white injected intradermally in right ear with 6×10^4 or 3×10^5 PFU. *Virus*: HSV-1, strain SC16. *Incidence of latent infection in ganglia*: C2 82%, C3 57%, C4 32%; trigeminal 26% vagal 10% superior cervical 6%, facial 0%. *Spontaneous recurrence of clinical disease*: approximately 3.5%; *induced recurrent disease*: 30–50% after cellophane tape stripping.

bour *et al.*, 1981). The availability of the mouse ear model has already been of value not only in studies on latency and recurrent disease but also in tests on the efficacy of antiviral drugs for treatment of primary (Field *et al.*, 1979) and recurrent herpetic disease (Blyth *et al.*, 1980b; Hill *et al.*, 1982). Moreover, the model has been of value in recent studies on the immunology of HSV infection (Darville and Blyth, 1982; reviewed by Wildy *et al.*, 1982).

c. Where Do Inducers of Recurrent Disease Act?

With the growing realization of the sensory ganglion as the site of latent infection with HSV, the idea developed that stimuli that induce recurrent disease act solely in such sites. After reactivation it was envisaged that virus would spread via the nerve to the peripheral tissue and there cause a lesion (Paine, 1964; Stevens, 1975). In a theoretical consideration of the mechanism underlying the production of recurrent disease, Hill and Blyth (1976) described this chain of events as the "ganglion trigger" theory. However, they felt the theory did not satisfactorily explain a number of observations. (1) After some stimuli, recurrent lesions appear

TABLE V. Some Immunological Aspects of the Mouse Ear Model

Immune function	Time of appearance after primary infection (days)	Duration	Effects of induced recurrent disease	Reference
Serum neutralizing antibody	8	Levels constant throughout life	None	Darville and Blyth (1982)
Cytotoxic T cells in draining lymph node	4 (peak 6–9)	Not detectable after day 12	Unknown but reappear rapidly in draining lymph node after reinoculation of virus into ear	Nash et al. (1980a)
Delayed-type hypersensitivity	4–5	Throughout life	Unknown but almost certainly involved in recurrent lesion	Nash et al. (1980b)

so rapidly [e.g., 24–48 hr after fever therapy (Warren et al., 1940; Keddie et al., 1941)] that is seems difficult to accommodate all the events required by the "ganglion trigger" theory (reactivation in the ganglion, spread of virus to the periphery, development of lesion). (2) Reactivation in the ganglion is not automatically followed by peripheral disease (Stevens et al., 1975). (3) Although some stimuli (e.g., UV light, chemical or physical trauma) may by indirect means affect the ganglion, they have more obvious effects (often lasting several days) on the peripheral tissue. Such effects usually involve the release of mediators of inflammation and repair of minor tissue damage. (4) As already discussed, infectious virus can be present in peripheral tissues of man and animals in the absence of overt clinical disease. This suggests that a supply of virus from the ganglion is not the only factor required to produce disease. (5) The "ganglion trigger" theory implies that each reactivation in the ganglion is followed by a recurrent lesion. Hence, an individual who rarely experiences such lesions would rarely have the antigenic stimulus provided by the release and growth of virus. However, in latently infected humans and mice the levels of anti-HSV antibodies in the serum remain constant throughout life and are unaffected by recurrence of lesions (Douglas and Couch, 1970; Darville and Blyth, 1982). This observation suggests a frequent antigenic stimulus that is irrespective of the frequency of recurrent disease.

In an attempt to accommodate these observations, Hill and Blyth (1976, 1977) argued that a supply of virus from the latently infected ganglion is not the only factor in recurrence of disease. In particular, they proposed that some of the changes produced in peripheral tissues by the stimuli of recurrent disease may produce conditions favorable for virus growth (Harbour et al., 1977). Such changes would thereby facilitate the production of a clinical lesion. This idea was central to Hill and Blyth's "skin trigger" theory. It was envisaged that without the effect of the "skin trigger" the virus would be eliminated by normal defense mechanisms. The processes underlying the induction of recurrent disease may differ slightly in different peripheral tissues such as the skin and cornea (Blyth et al., 1981). However, in the discussion that follows, it will be assumed for the moment, despite the use of the term "skin trigger," that "triggers" act in a similar manner in sites other than skin.

A requirement of the "skin trigger" theory is some degree of coincidence between the presence of virus in the peripheral tissue and the effect of the "trigger." The frequent presence of microfoci of infection in the peripheral tissue would increase the likelihood of this coincidence. Evidence for the presence of such virus in the absence of clinical disease has already been discussed. Moreover, in the mouse ear model it was shown that HSV could be isolated from explant cultures of the skin from 10% of clinically normal, latently infected mice; the skin was taken from the original site of primary infection (Hill et al., 1980).

Alternatively, virus might arrive in the skin from the ganglion at a time when, due to the recent occurrence of a "skin trigger," the conditions were still favorable for virus growth.

As originally proposed in the "skin trigger" theory it was envisaged that the supply of virus to the skin from the ganglion, and thereby the presence of microfoci of infection in the skin, would occur frequently. Such a supply of virus could arise either from a dynamic state of latency in the ganglion or by frequent, spontaneous breakdown of a static state. However, since the "skin trigger" idea was proposed, experiments with the mouse ear model have shown that stimuli to the skin (at the original site of primary infection) that induce recurrent disease, also reactivate latency in the ganglion (see previous discussion and Table II). After such a stimulus, e.g., stripping the ear with cellophane tape, recurrent lesions (vesicles that develop into pustules) appear in the skin by day 3–6 (Hill et al., 1982) and these lesions are associated with the presence of infectious virus in the skin. However, soon after stripping the skin (1–2 days) and usually before virus appears there, *infectious* virus can be isolated from the latently infected cervical ganglia that supply sensory nerves to the skin of the stripped ear (Hill et al., 1983a; Harbour et al., 1983b). Moreover, if these nerves are cut immediately after stripping the skin, recurrent cutaneous lesions do not appear (Hill et al., 1983b). These observations suggest that the reactivated infection in the ganglia (caused by the peripheral stimulus) is the source of virus for recurrent disease and that such virus reaches the skin via the peripheral nerve.

However, we have also shown in the mouse ear model that some peripheral stimuli, in particular application of DMSO to the skin of the ear, are efficient at inducing reactivation in the ganglia but not at producing significant recurrence of disease (Hill, 1981; Hill et al., 1983a; Harbour et al., 1983b). This result is the more surprising because the ganglionic reactivation induced by DMSO is followed by appearance of virus in the skin. Such observations led to the suggestion that stimuli that induce recurrent clinical disease may need to act as both ganglion *and* skin triggers (Hill, 1981; Harbour et al., 1983b) (Fig. 3). Hence, three possible "scenarios" emerge from this discussion:

i. Recurrence of Lesions after a "Skin Trigger." A peripheral stimulus produces local changes in the skin (or other peripheral tissue) that lead to an increase in susceptibility of the tissue to HSV infection and/or a depression of local defense mechanisms. Such changes allow microfoci of HSV infection already present (either as a latent infection in the skin or derived from latent infection in the ganglion) to grow and produce a clinical lesion. This sequence might explain the very rapid development of lesions seen after stimuli such as fever therapy.

ii. "Ganglion and Skin Trigger." The peripheral stimulus first induces reactivation of the latent infection in the ganglion. The virus

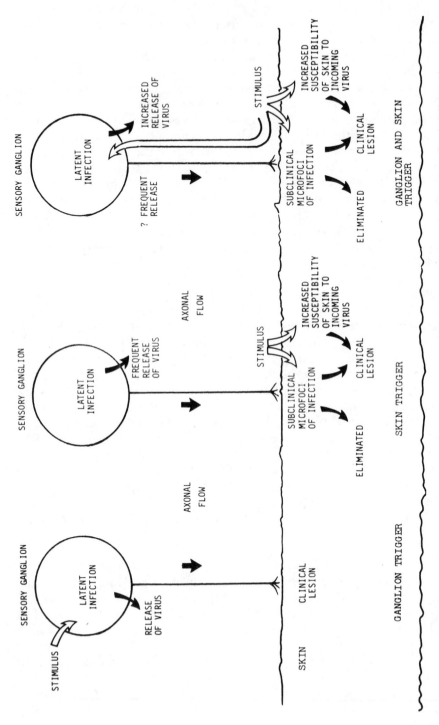

FIGURE 3. Induction of recurrent herpes simplex: possible modes of action. Reprinted with permission from Hill (1981).

spreads, probably intraaxonally, to the skin where the stimulus has also produced the kind of changes described in (i). By virtue of these changes the virus produces a clinical lesion. Because the reactivation in the ganglion results in production of infectious virus, it is possible that this is the form in which the virus is carried in the nerve to the skin. However, it is conceivable that the virus may travel intraaxonally as a subviral particle, even as infectious DNA. In whatever state the virus is transported, it is likely that it will be delivered, via the fine, naked nerve endings in the epidermis, close to the plasma membrane of an epidermal cell.

  iii. Recurrence of Lesions after a "Ganglion Trigger." Some peripheral stimuli, e.g., DMSO in the mouse ear model, may act largely as "ganglion triggers" and therefore produce virus shedding but no recurrent disease [a "recurrence" in the terminology of Wildy *et al.* (1982)]. However, a sufficiently severe stimulus to the ganglion might produce large amounts of reactivated virus that when supplied to the skin, might overwhelm the defense mechanisms and thereby produce lesions in the absence of a "skin trigger." This sequence of events may underlie the means by which direct trauma to the human trigeminal ganglion produces cutaneous herpetic lesions (Carton and Kilbourne, 1952). Alternatively, it has been suggested that patients who suffer from recurrent herpes may have long-term minor defects in their cell-mediated immunity (an idea discussed in detail below). In the presence of such defects a "ganglion trigger" alone might be sufficient to produce recurrent disease.

  The evidence accumulating from the mouse ear model suggests that of the "scenarios" outlined above, the second (a combined "ganglion and skin trigger") may provide the most accurate description of induced recurrent disease. Further work will be needed to determine whether in some circumstances the other "scenarios" may also occur in experimental animals and man.

  Having discussed the sequence of events that may occur in the pathogenesis of recurrent disease and the sites at which inducers of the disease might act, it remains to consider in more detail the mechanisms underlying this pathogenesis. As for discussions on other aspects of latency these mechanisms will be considered from the point of view of host and viral factors.

d. Host Factors and Recurrent Disease

  The host factors, in particular the changes in neuronal physiology, that may be involved in reactivation of latency in ganglia have already been considered. This section will deal with other aspects of the host that may play a role in the development of recurrent disease. Particular attention will be given to the mechanisms by which inducers of ganglionic reactivation *in vivo* may also affect other sites, in particular the peripheral tissues in which recurrent lesions develop; i.e., what are the mechanisms underlying the action of "skin triggers"?

The immunological response is one of the most obvious host functions that may be of importance in the pathogenesis of recurrent herpetic disease. A number of observations suggest that immune mechanisms themselves may contribute to the pathology of the recurrent lesion: virus particles in vesicular fluid appear to be complexed with antibody (Daniels et al., 1975); latently infected humans and animals have a cutaneous delayed-type hypersensitivity response to HSV (reviewed by Nash et al., 1980b); the vesicular lesion rapidly develops into a pustule as a result of invasion by polymorphonuclear leukocytes and mononuclear cells; studies in nude mice suggest that the activity of T cells may significantly increase the severity of the pathology in herpetic keratitis (Metcalf et al., 1979).

However, herpetic lesions that occur during immunosuppression in man are often abnormally severe and long-lasting (Montgomerie et al., 1969; Rand et al., 1976; Pass et al., 1979). Increased duration of recurrent lesions after immunosuppression was also observed in the mouse ear model (Blyth et al., 1980a) and in another mouse model (Hurd and Robinson, 1977). Hence, as might be expected, immune mechanisms [of the type reviewed by Nahmias and Ashman (1978), Babiuk and Rouse (1979), Wildy et al. (1982), Rouse and Lopez (1984)] appear to be involved in the healing of the recurrent lesion.

After immunosuppression in man some authors have reported no increased incidence of recurrent herpes simplex (Spencer and Anderson, 1970) while many others (reviewed by Ho, 1977) observed more frequent recurrent disease. In experiments with latently infected mice, immunosuppression either failed to produce recurrent disease (Stevens and Cook, 1973a,b; Blyth et al., 1976; Hurd and Robinson, 1977) or produced a low incidence [17% in hairless mice treated with corticosteroids (Underwood and Weed, 1974)]. Further studies in the mouse ear model with a variety of immunosuppressive agents did not show a significant increase in the incidence of recurrent disease above the spontaneous background level (Blyth et al., 1980a). These observations on latently infected mice are of interest in view of the reports that in such animals immunosuppressive treatments appear to act as "ganglion triggers" (Table II).

The apparent difference in the effect of immunosuppression on the incidence of recurrent disease in man and experimental mice may arise from the much more long-term treatments used in man. However, in the mouse ear model (Blyth et al., 1980a) it was shown that immunosuppression did not increase the incidence of recurrent disease induced by mild trauma to the skin (stripping with cellophane tape).

There has been considerable interest in the possible existence of long-term minor immunological defects in patients suffering from recurrent herpes simplex. So far there is no evidence in these patients of such defects in their humoral response to HSV. Recurrent disease occurs in the face of high levels of circulating neutralizing antibody (Douglas and Couch, 1970), an observation confirmed in the mouse ear model (Darville and Blyth, 1982). Moreover, although vaccination or hyperimmunization of latently infected guinea pigs produced a significant increase in titers of serum neutralizing antibodies, there was no effect on the incidence of spontaneous recurrent disease in guinea pigs (Scriba, 1981b) and only a small reduction in induced recurrence in mice (Darville and Blyth, 1982). Furthermore, in patients with frequent recurrence of lesions there appears to be no qualitative or quantitative defects in the production of antibodies to particular viral antigens (Zweerink and Stanton, 1981).

Tokumaru (1966) suggested that a depression in serum IgA levels might be related to recurrence of herpes labialis. However, using a sensitive radioimmunoassay for HSV IgA, comparable titers were found in seropositive adults with and without recurrent herpes labialis (Friedman and Kimmel, 1982).

In patients prone to recurrent herpes simplex the evidence for minor defects in various cell-mediated immune functions is conflicting. Several reports show that such patients have an active virus-specific lympho-proliferative response to HSV (Wilton et al., 1972; Russell, 1973; Rasmussen et al., 1974; Steele et al., 1975; Russell et al., 1976). Some authors have shown these responses are depressed during active recurrent disease (Rosenberg et al., 1974; Lopez and O'Reilly, 1977; El Araby et al., 1978).

Russell et al. (1975), Reichman et al. (1977), and Fujimiya et al. (1978) found no defects in natural killer cell-mediated cytotoxicity in patients suffering from recurrent disease, whereas Steele et al. (1975) and Thong et al. (1975) observed decreased lymphocyte cytotoxicity responses in such patients.

There is one report of a slight decrease in HSV-induced interferon from mononuclear cells in vitro from patients with more frequent recurrence of lesions (Rasmussen et al., 1974). More recently, Cunningham and Merigan (1983) observed that within 3 weeks of an episode of herpes labialis the patients' peripheral blood lymphocytes spontaneously secreted IFN-γ in vitro. Moreover, there was a strong correlation between the peak interferon level and the time to the next recurrence: the lower the level, the shorter the time. They suggested that this production of IFN-γ is either a direct determinant of the frequency of recurrence or a marker of other immune events determining the frequency.

Some authors have reported minor defects in T-cell-mediated effect or responses to HSV. Thus, O'Reilly et al. (1977) found a decreased production in lymphocyte migration inhibition factor (LIF) immediately before and at the time of recurrent disease. Similar observations were made by Sheridan et al. (1982) at the time of recrudescence of HSV-2 lesions. Moreover, these authors observed a significant increase in suppressor T cells in peripheral blood during recrudescent disease and 24–48 hr before its onset. Furthermore, Wilton et al. (1972) and Shillitoe et al. (1977) found that T lymphocytes from patients who had frequent recurrent lesions, failed to produce macrophage migration inhibition factor (MIF) in response to HSV. A similar observation was made by Gange et al. (1975) in patients at the time of active recurrent lesions. However, no such defect in MIF production was reported by Russell et al. (1976).

A deficiency in LIF and lymphocyte proliferative response has also been observed in spleen cells of guinea pigs prone to develop recurrent HSV-2 lesions (Donnenberg et al., 1980). However, unlike humans, there is evidence that a large proportion of such animals may be persistently infected with the virus in peripheral tissues (Scriba, 1977). It will be of interest to determine whether this chronic extraneural infection is the cause or the result of the observed immunological defects.

Hence, in man there is as yet no clear-cut evidence for the involvement of "naturally" occurring minor impairments in the immune system (particularly in cell-mediated function) in the causation of recurrent herpes simplex. Indeed, even if these defects do exist it is unclear how they might arise. Could they be the result rather than the cause of recurrent disease or could they arise from the specific immunological suppression that can occur during primary infection with the virus (Nash et al., 1981; Schrier et al., 1983)? Moreover, it should be stressed that all of the observations on human patients mentioned above were made on cells from peripheral blood. It is not clear to what extent the behavior of these cells gives an accurate measure of immune functions in peripheral tissues such as skin. In this respect it is noteworthy that in the experiments on recurrent HSV-2 in guinea pigs (Donnenberg et al., 1980), spleen cells showed impairment of LIF and lymphocyte transformation to HSV, whereas in circulating lymphocytes, LIF was impaired but the transformation response was intact. Further, it is clear from experiments in rabbits (Jacobs et al., 1976) and mice (Nash et al., 1980a) that immune responses to HSV are concentrated heavily in the draining lymph nodes at the site of primary infection. Moreover, the experiments already mentioned with the mouse ear model (Blyth et al., 1980a) indicate that systemic treatment with powerful immunosuppressive drugs is much less effective than mild trauma to the skin in inducing recurrent lesions. Such observations argue strongly for the importance of local factors in the pathogenesis of recurrent disease.

These local factors might include, as first suggested by Hill and Blyth (1976), the production of prostaglandins. These are a family of hormone-like substances with a wide variety of activities in vivo. They are derived from fatty acids (particularly arachidonic acid, a constituent of membrane phospholipids) and according to their structure fall into several main series such as PGE and PGF. Prostaglandins are produced locally in tissues in response to many kinds of damage (Goldyne, 1975) including those stimuli (e.g., UV light and trauma) that induce recurrent herpes. Prostaglandins in the skin are among the mediators of the inflammatory response (Goldyne, 1975) and they may also be concerned with the mitotic changes of epidermal cells following stimuli such as UV irradiation (Eaglstein and Weinstein, 1975). Therefore, with respect to the possible action of prostaglandins as "skin triggers" it is noteworthy that after stripping the skin of mouse ear with cellophane tape (as discussed previously this induces both reactivation in the ganglion and an incidence of recurrent disease of 30–50%) the levels of PGE_2 in the skin are raised for at least 4 days (Harbour et al., 1983b). In contrast, after application of DMSO to the ear (a treatment that reactivates ganglionic latency but produces little recurrent disease) there is no such increase in PGE_2.

With respect to effects on the immune system there is growing evidence that prostaglandins (in particular those of the E series) can have a suppressive effect on the activity of lymphocytes (Goodwin et al., 1977; Fulton and Levy, 1980; Rogers et al., 1980; see also references in these papers). Indeed, it has been shown that prostaglandins can reduce antibody-dependent cell cytotoxicity against HSV-infected cells (Russell and Miller, 1978; Trofatter and Daniels, 1979).

Another feature of the local immune system that may be affected by "skin triggers" is the Langerhans cell (a macrophagelike cell within the epidermis). There is now extensive evidence that these cells are in the "front line" of the immune response in the skin and are concerned with initial presentation of antigens to lymphoid cells (reviewed by Friedman, 1981). In particular, it has been shown that human Langerhans cells are able to present HSV in an immunogenic state to T lymphocytes (Braathen et al., 1980). Therefore, it is of interest that at least two inducers of recurrent disease in man and mouse (Table IV) cause a transient depletion of Langerhans cells in the epidermis, namely UV light (Bergstresser et al., 1980) and cellophane tape stripping (Lessard et al., 1968). Indeed, it has been postulated that the absence of Langerhans cells from the central cornea may be involved in the pathogenesis of chronic herpetic keratitis (Streilen et al., 1979).

Besides depressing local defense mechanisms, "skin triggers" might also act, as suggested by Hill and Blyth (1976), by increasing the susceptibility of skin cells to infection with HSV. Such changes in susceptibility may also be mediated through the production of prostaglandins, as Harbour et al. (1978) found that prostaglandins enhanced the spread of HSV in Vero cell cultures. This enhancement may be related to the fact that prostaglandins increase the intercellular adhesion between uninfected Vero cells and cells infected with HSV (Harbour et al., 1983a). Of more relevance to the skin, we have also found that PGE_2 enhances the spread of HSV in cultures of mouse keratinocytes (A. Whitby, W. Blyth, and T. Hill, unpublished observations).

Besides causing release of inflammatory mediators, inducers of recurrent disease (e.g., UV irradiation and mild trauma) also produce minor damage to epithelial cells. This damage is followed by repair processes that involve changes in the growth and differentiation of the epithelium. It is not clear whether such changes might also be involved in the action of "skin triggers." However, as early as 1923, Teague and Goodpasture observed that zosteriform spread of herpetic lesions could be produced in rabbits and guinea pigs by painting the skin with coal tar in areas adjacent to the point of primary inoculation with HSV. The coal tar induced hyperplasia of the epidermis with many young, undifferentiated cells (it should also be noted that this inflammatory agent would almost certainly cause increased prostaglandin production in the skin). They postulated that such cells might be highly susceptible to HSV and thereby the zosteriform spread of the infection would be facilitated. More recently, we have shown *in vitro* that undifferentiated cultures of mouse keratinocytes are highly susceptible to HSV infection but with increased differentiation this susceptibility is markedly decreased (A. Whitby, W. Blyth, and T. Hill, unpublished observations).

Hence, by entirely local effects, in part perhaps mediated through prostaglandins, "skin triggers" may produce transient immunosuppression and/or increased susceptibility of epidermal cells to infection with HSV. Clearly, such effects might be even more likely to result in a clinical lesion if they were to occur against a background of existing immunological defects of the kind discussed previously.

The genotype of the host is a further factor of possible importance in herpetic disease. There is clear evidence that in the mouse, resistance to primary infection with HSV is genetically controlled (reviewed by Lopez, 1981); this resistance may be mediated by NK cells. With respect to recurrent disease, we have shown in the mouse ear model a much lower incidence of recurrent disease (induced by stripping with cellophane tape) in BALB/c mice than in any of the other four albino inbred strains tested (Harbour *et al.*, 1981). Further work is needed to determine whether there is any relationship between resistance to primary infection with HSV and resistance to development of recurrent disease.

In humans there is one report of a higher incidence of the HLA antigen Al in patients with recurrent herpes labialis (Russell and Schlaut, 1975). There are also a number of conflicting reports of a positive association of different HLA antigens and recurrent herpetic keratitis (see several papers in the symposium edited by Sundmacher, 1981). However, as pointed out by Volker-Dieben *et al.* (1981), some of these associations, in particular that between keratitis and HLA-B5 (Zimmerman *et al.*, 1977), may prove to be insignificant after careful correction in the statistical analysis for the number of antigens studied. Even when such associations are clarified, they are unlikely to reveal much of the mechanisms underlying the pathogenesis of recurrent disease.

Spruance and Chow (1980) have explored the possibility that genetically controlled variations in the susceptibility of epidermal cells may play a role in recurrent herpes labialis. However, they found no significant difference in the growth of HSV in cultures of epidermal cells from patients with frequent episodes of herpes labialis and similar cultures from control patients (seropositive but with no history of recurrent disease).

A further observation, as yet unexplained, that indicates a role for host factors in herpetic disease is the twofold higher incidence of herpetic keratitis in males as compared with females (Wilhelmus et al., 1981; Volker-Dieben et al., 1981).

Thus, there are hints that genetic and/or hormonal host factors may also play some part in the development of recurrent herpetic disease but more detailed investigations are needed.

A final point for consideration is the possible influence of one episode of recurrent disease on the likely occurrence of subsequent episodes. With respect to induced recurrent disease there are indications that repeated treatment with an inducer may be associated with a fall in the incidence of disease at a later as compared with the initial time of induction. Thus, in humans a high incidence (46–70%) of herpes simplex occurred after the first occasion of fever therapy but after further occasions this incidence fell to 5–20% (Warren et al., 1940; Keddie et al., 1941). In the mouse ear model, Harbour et al. (1981) found that recurrent disease could be induced repeatedly by stripping the ears of mice with cellophane tape at monthly intervals; in two groups of mice the stripping was started at 6 or 22 weeks after infection. However, the incidence of such disease fell from 30–40% on the first three occasions of stripping to 0–5% on the ninth occasion (for mice first stripped 6 weeks after infection) or on the fourth occasion (for mice first stripped 22 weeks after infection). The reasons for such decreased incidence of disease are unknown. Frequent attempts to induce recurrent disease may repeatedly "trigger" virus in the ganglion and/or skin. This might lead to an increase in virus available to stimulate the immune system and thereby to the low incidence of recurrent disease. Alternatively, because stimuli such as cellophane stripping are known to act as "ganglion triggers" (Hill et al., 1983a; Harbour et al., 1983b), there may be a loss of latently infected neurons (McLennan and Darby, 1980). Hence, repeated attempts to induce recurrent disease might produce a depletion of the reservoir of latent infection in the ganglion and thereby a decreased likelihood of recurrent disease.

e. Viral Factors and Recurrent Disease

There are clear indications that the viral genotype can affect the pattern of disease after primary infection in laboratory animals. Thus, isolates and strains of HSV differ markedly in their neurovirulence (Hill et al., 1975; Lopez, 1975). Wander et al. (1980) have shown that HSV strains differ in their ability to produce stromal and epithelial disease after inoculation into the rabbit eye. Moreover, this ability is related to specific regions of the viral genome (Centifanto-Fitzgerald et al., 1982). With respect to recurrent herpetic infection in man there is considerable variability in its expression, e.g., in the frequency and severity of recurrences in different individuals. Although, as previously discussed, host factors may play an important part in the expression of recurrent disease, it is also important to consider whether the viral genotype may be involved. Differences between viruses in this respect could arise at the level of reactivation in the ganglion and/or at the level of the recurrent infection in the peripheral tissues.

V. EVOLUTIONARY ASPECTS OF LATENCY

It seems likely that the host's immune response has been one of the major selection pressures operating on HSV during its evolution. In response to this immunological pressure the virus appears to have developed at least two major strategies. First, in the latently infected neuron the expression of viral antigens is severely restricted. Second, even when infectious virus is produced it is protected by remaining, for much of the time, within the neuron and its processes. Thus, during the early phases of the establishment of latency the virus is transported intraaxonally, i.e., within a cytoplasmic process of a neuron, to the ganglion. When the virus reaches the neuronal cell bodies in the ganglion, productive virus replication does not appear to be necessary for the virus–cell interaction that leads to latency. Therefore, few, if any, viral antigens may be produced in the neuron at this stage. Even at a later time when reactivation in the ganglion occurs and infectious virus is formed, the virus that will result in recurrent disease probably remains within the neuron and is delivered, again by axonal flow, to arrive close to the surface of the "target" cell in the epithelium. Moreover, during both primary and recurrent infection in peripheral tissues, HSV can, to some degree, avoid both the extracellular environment and immune attack by the formation of polykaryocytes and direct cell-to-cell spread (Wheeler, 1960; Roizman, 1965). The virus may also have evolved methods for moderating the immune system [e.g., by the induction of specific suppressor lymphocytes (reviewed by Wildy et al., 1982; Schrier et al., 1983)] that enable the virus to persist in the host.

It also appears that during the course of evolution the virus has shown considerable opportunism in "using" a number of normal host functions for different phases of the host–virus interaction. Examples of such functions that have been mentioned include various aspects of neuronal physiology, namely axonal flow (involved in transport of virus to and from the ganglion), the highly differentiated, nondividing state of the neuron (possibly important in the establishment and maintenance of latency), changes in neuronal metabolism associated with damage and repair (probably concerned with reactivation of latency). Moreover, it is possible that the production of host cell factors such as prostaglandins may be involved in the development of recurrent disease.

By the development of such subtle interactions between virus and host functions, the latent state has evolved and thereby the viral genome is preserved for the lifetime of the host. Maintenance of the genome within the host population requires the infection of new susceptible individuals; this appears to be the role of reactivation, virus shedding, and recurrent disease.

The success of these viral strategies is evidenced by the widespread incidence of infection with HSV. Thus, depending on socioeconomic conditions, the genome of HSV-1 can be present in up to 90% of humans (see Whitley, this volume). Hence, it seems likely that man will be troubled by the virus for many years to come. However, within the next decade the evolution of human knowledge should lead to a greater understanding of this successful parasite, and thereby more effective methods of chemotherapy and immunological protection should become available. Indeed, when we know more of the biology, both of the virus and host, we may learn eventually how to eliminate HSV from its "hiding place" in the nervous system.

ACKNOWLEDGMENTS. The development of many of the ideas incorporated in this review has been helped over the years by discussions with many colleagues. In particular, I should like to thank Drs. Bill Blyth, David Harbour, Andrew Tullo, Andrew Whitby, Hugh Field, David Easty, and Peter Bennett. I am also grateful to my wife, Mary, for her encouragement during the writing of this review.

Work done in the Laboratory has been supported by grants from the Medical Research Council, the Wellcome Foundation and Wellcome Trust.

REFERENCES

Adams, A., and Lindahl, T., 1975, Epstein–Barr virus genomes with properties of circular DNA molecules in carrier cells, *Proc. Natl. Acad. Sci. USA* **72**:1477.

Al-Saadi, S. A., Clements, G. B., and Subak-Sharpe, J. H., 1983, Viral genes modify herpes simplex virus latency both in mouse footpad and sensory ganglia, *J. Gen. Virol.* **64**:1175.

Anderson, S. G., and Hamilton, J., 1949, The epidemiology of primary herpes simplex infection, *Med. J. Aust.* **1**:308.

Arvidson, B., 1979, Retrograde transport of horseradish peroxidase in sensory and adrenergic neurons following injection into the anterior chamber, *J. Neurol.* **8**:751.

Asbury, A. K., 1967, Schwann cell proliferation in developing mouse sciatic nerve, *J. Cell Biol.* **34**:735.

Babiuk, L. A., and Rouse, B. T., 1979, Immune control of herpes virus latency, *Can. J. Microbiol.* **25**:267.

Baringer, J. R., 1974, Recovery of herpes simplex virus from human sacral ganglions, *New Engl. J. Med.* **291**:828.

Baringer, J. R., 1975, Herpes simplex virus infection of nervous tissue in animals and man. *Prog. Med. Virol.* **20**:1.

Baringer, J. R., and Swoveland, P., 1973, Recovery of herpes simplex virus from human trigeminal ganglia, *New Engl. J. Med.* **288**:648.

Baringer, J. R., and Swoveland, P., 1974, Persistent herpes simplex virus infection in rabbit trigeminal ganglia, *Lab. Invest.* **30**:230.

Bastian, F. O., Rabson, A. S., Yee, C. L., and Tralka, T. S., 1972, Herpes virus hominis: Isolation from human trigeminal ganglia, *Science* **178**:306.

Baucke, R. B., and Spear, P. G., 1979, Membrane proteins specified by herpes simplex viruses. V. Identification of an Fc-binding glycoprotein, *J. Virol.* **32:**779.

Becker, Y., 1978, A model for intramolecular recombination of herpes virus DNA leading to the formation of circular, circular-linear and fragmented genomes, *J. Theor. Biol.* **75:**339.

Becker, Y., 1979, A model for herpes virus (class 3) DNA during infection, latency and transformation, *Perspect. Biol. Med.* **23:**92.

Benfey, M., and Aguayo, A. J., 1982, Extensive elongation of axons from rat brain into peripheral nerve grafts, *Nature* **296:**150.

Bergstresser, P., Toews, G. B., and Streilein, J. W., 1980, Natural and perturbed distributions of Langerhans cells: Response to ultraviolet light, heterotopic skin grafts and dinitrofluorobenzene sensitisation, *J. Invest. Dermatol.* **75:**73.

Biegeleisen, K., and Rush, M. G., 1976, Association of herpes simplex virus type 1 DNA with host chromosomal DNA during productive infection, *Virology* **69:**246.

Blue, W. T., Winland, R. D., Stobbs, D. G., Kirksey, D. F., and Savage, R. E., 1981, Effects of adenosine monophosphate on the reactivation of latent herpes simplex virus type 1 infections of mice, *Antimicrob. Agents Chemother.* **20:**547.

Blyth, W. A., and Hill, T. J., 1984, Establishment, maintenance and control of herpes simplex virus latency, in: *Immunobiology of Herpes Simplex Virus Infections* (B. T. Rouse and C. Lopez, eds.), p. 9., CRC Press, Boca Raton, Florida.

Blyth, W. A., Hill, T. J., Field, H. J., and Harbour, D. A., 1976, Reactivation of herpes simplex virus infection by ultraviolet light and possible involvement of prostaglandins, *J. Gen. Virol.* **33:**547.

Blyth, W. A., Harbour, D. A., and Hill, T. J., 1980a, Effect of immunosuppression on recurrent herpes simplex in mice, *Infect. Immun.* **29:**902.

Blyth, W. A., Harbour, D. A., and Hill, T. J., 1980b, Effect of acyclovir on recurrence of herpes simplex skin lesions in mice, *J. Gen. Virol.* **48:**417.

Blyth, W. A., Hill, T. J., and Harbour, D. A., 1981, Mechanisms of control of latent herpes simplex virus infection in the skin, and eye, in: *Herpetic Eye Disease* (R. Sundmacher, ed.), p. 43, Springer-Verlag (Bergmann).

Blyth, W. A. Harbour, D. A., and Hill, T. J., 1984, Pathogenesis of zosteriform spread of herpes simplex virus in the mouse, *J. Gen. Virol.* (in press).

Braathen, L. R., Berle, E., Mobech-Hanssen, V., and Thorsby, E., 1980, Studies on human epidermal Langerhans cells. II. Activation of human T lymphocytes to herpes simplex virus. *Acta Derm. Venereol.* **60:**381.

Bradley, W. G., and Asbury, A., 1970, Duration of synthesis phase in neurolemma cells in mouse sciatic nerve during degeneration, *Exp. Neurol.* **26:**275.

Brown, D. C., and Kaufman, H. E., 1969, Chronic herpes simplex infection of the ocular adnexa, *Arch. Ophthalmol.* **81:**837.

Brown, S. M., Subak-Sharpe, J. H., Warren, K. G., Wroblewska, Z., and Koprowski, H., 1979, Detection by complementation of defective or uninducible (herpes simplex type 1) virus genomes latent in human ganglia, *Proc. Natl. Acad. Sci. USA* **76:**2364.

Buddingh, G. J., Shrum, D. I., Lannier, J. C., and Guidry, D. J., 1953, Studies of the natural history of herpes simplex infections, *Pediatrics* **11:**595.

Burnet, F. M., and Lush, D., 1939, Herpes simplex studies on the antibody content of human sera, *Lancet* **1:**629.

Burnet, F. M., and Williams, S. W., 1939, Herpes simplex: A new point of view, *Med. J. Aust.* **1:**637.

Cabrera, C. V., Wohlenberg, C., Openshaw, H., Rey-Mendez, M., Puga, A., and Notkins, A. L., 1980, Herpes simplex virus DNA sequences in the CNS of latently infected mice, *Nature (London)* **288:**288.

Carton, C. A., 1953, Effect of previous sensory loss on the appearance of herpes simplex, *J. Neurosurg.* **10:**463.

Carton, C. A., and Kilbourne, E. D., 1952, Activation of latent herpes simplex by trigeminal sensory-root section, *New Engl. J. Med.* **246:**172.

Centifanto-Fitzgerald, Y. M., Yamaguchi, T., Kaufman, H. E., Tognon, M., and Roizman, B., 1982, Ocular disease pattern induced by herpes simplex virus is genetically determined by specific region of viral DNA, *J. Exp. Med.* **155**:475.

Christman, J. K., Price, P., Pedrinan, L., and Acs, G., 1977, Correlation between hypomethylation of DNA and expression of globin genes in Friend erythroleukaemia cells, *Eur. J. Biochem.* **81**:53.

Cleobury, J. R., Skinner, G. R. B., Thouless, M. E., and Wildy, P., 1971, Association between psychopathic disorder and serum antibody to herpes simplex virus (type 1), *Br. Med. J.* **1**:438.

Clough, D. W., Kunkel, L. M., and Davidson, R. L., 1982, 5-Azacytidine induced reactivation of a herpes simplex thymidine kinase gene, *Science* **216**:70.

Constanzo, B. F., Campadelli-Fiume, G., Foa-Tomasi, L., and Cassai, E., 1977, Evidence that herpes simplex virus DNA is transcribed by cellular RNA polymerase B, *J. Virol.* **21**:996.

Cook, M. L., and Stevens, J. G., 1973, Pathogenesis of herpetic neuritis and ganglionitis in mice: Evidence of intra-axonal transport of infection, *Infect. Immun.* **7**:272.

Cook, M. L., and Stevens, J. G., 1976, Latent herpetic infections following experimental viraemia, *J. Gen. Virol.* **31**:75.

Cook, M. L., and Stevens, J. G., 1983, Restricted replication of herpes simplex virus in spinal ganglia of resistant mice is accompanies by an early infiltration of immunoglobulin G-bearing cells, *Infect. Immun.* **40**:752.

Cook, M. L., Bastone, V. B., and Stevens, J. G., 1974, Evidence that neurons harbor latent herpes simplex virus, *Infect. Immun.* **9**:946.

Coriell, L. L., 1963, in: *Virus Nucleic Acids and Cancer* (B. Roizman, ed.), Williams & Wilkins, Baltimore.

Costa, J. C., and Rabson, A. S., 1975, Role of Fc receptors in herpes simplex virus infection, *Lancet* **1**:77.

Costa, J., Rabson, A. S., Lee, C., and Tralka, T. S., 1977, Immunoglobulin binding to herpes virus-induced Fc receptors inhibits virus growth, *Nature (London)* **269**:251.

Cunningham, A. L., and Merigan, T. C., 1983, Gamma interferon production appears to predict time of recurrence of herpes labialis, *J. Immunol.* **130**:2397.

Cushing, H., 1905, Surgical aspects of major neuralgia of trigeminal nerve: Report of 20 cases of operation upon the Gasserian ganglion with anatomic and physiologic notes on the consequence of its removal, *J. Am. Med. Assoc.* **44**:1002.

Daniels, C. A., Legoff, S. G., and Notkins, A. L., 1975, Shedding of infectious virus–antibody complexes from vesicular lesions of patients with recurrent herpes labialis, *Lancet* **2**:524.

Darville, J. M., and Blyth, W. A., 1982, Neutralizing antibody in mice with primary and recurrent herpes simplex virus infection, *Arch. Virol.* **71**:303.

Davis, L. E., and Johnson, R. T., 1979, An explanation for the localization of herpes simplex encephalitis, *Ann. Neurol.* **5**:2.

Dawson, C. R., and Togni, B., 1976, Herpes simplex eye infections: Clinical manifestations, pathogenesis and management, *Surv. Ophthalmol.* **21**:121.

Dawson, C. R., Togni, B., and Thygeson, P., 1966, Herpes simplex virus particles in the nerves of rabbit corneas after epithelial inoculation, *Nature*, **211**:316.

Deardourff, S. L., De Ture, F. A., Drylie, D. M., Centifanto, Y., and Kaufman, H., 1974, Association between herpes hominis type 2 and the male genitourinary tract, *J. Urol.* **112**:126.

Desrosiers, R. C., Mulder, C., and Fleckenstein, B., 1979, Methylation of Herpesvirus Saimiri DNA in lymphoid tumour cell lines, *Proc. Natl. Acad. Sci. USA* **76**:3839.

Dillard, S. H., Cheatham, W. J., and Moses, H. L., 1972, Electron microscopy of zosteriform herpes simplex infection in the mouse, *Lab. Invest.* **26**:391.

Docherty, J. J., and Chopan, M., 1974, The latent herpes simplex virus, *Bacteriol. Rev.* **38**:337.

Doer, R., 1920, Sitzungsberichte der gesellschaft der schweizerischen augenartzte diskussion, *Klin. Monatsbl. Augenheilkd.* **65**:104.

Doer, R., and Vöchting, K., 1920, Etudes sur le virus de l'herpes febrile, *Rev. Gen. Ophthal- mol. Paris* **34:**409.

Doerfler, W., 1981, DNA methylation—A regulatory signal in eukaryotic gene expression, *J. Gen. Virol.* **57:**1.

Donnenberg, A. D., Chaikof, E., and Aurelian, L., 1980, Immunity to herpes simplex virus type 2: Cell mediated immunity in latently infected guinea pigs, *Infect. Immun.* **30:**90.

Douglas, R. G., and Couch, R. B., 1970, A prospective study of chronic herpes virus infection and recurrent herpes labialis in humans, *J. Immunol.* **104:**289.

Duke-Elder, S., 1965, *System of Ophthalmology*, Vol. VIII, *Diseases of the Outer Eye*, p. 317, Mosby, St. Louis.

Eaglstein, W. H., and Weinstein, G. D., 1975, Prostaglandins and DNA synthesis in human skin: Possible relationship to ultraviolet light, *J. Invest. Dermatol.* **64:**386.

Eglin, R. P., Lehner, T., and Subak-Sharpe, J., 1982, Detection of RNA complementary to herpes simplex virus in mononuclear cells from patients with Behcet's syndrome and recurrent ulcers, *Lancet* **2:**1356.

El Araby, I. I., Chernesky, M. A., Rawls, W. E., and Dent, P. B., 1978, Depressed herpes simplex virus induced lymphocyte blastogenesis in individuals with severe recurrent herpes infections, *Clin. Immunol. Immunopathol.* **9:**253.

Ellison, S. A., Carton, C. A., and Rose, H. M., 1959, Studies of recurrent herpes simplex infections following section of the trigeminal nerve, *J. Infect. Dis.* **105:**161.

Esiri, M. M., 1982, Herpes simplex encephalitis: An immunohistological study of the dis- tribution of viral antigen within the brain, *J. Neurol. Sci.* **54:**209.

Field, H. J., and Darby, G., 1980, Pathogenicity in mice of strains of herpes simplex virus which are resistant to acyclovir *in vitro* and *in vivo*, *Antimicrob. Agents Chemother.* **17:**209.

Field, H. J., and De Clercq, E., 1981, Effects of oral treatment with acyclovir and bromo- vinyldeoxyuridine on the establishment and maintenance of latent herpes simplex virus infection in mice, *J. Gen. Virol.* **56:**259.

Field, H. J., and Hill, T. J., 1974, The pathogenesis of pseudorabies in mice following pe- ripheral inoculation, *J. Gen. Virol.* **23:**145.

Field, H. J., and Hill, T. J., 1975, The pathogenesis of pseudorabies in mice: Virus replication at the inoculation site and axonal uptake, *J. Gen. Virol.* **26:**145.

Field, H. J., and Wildy, P., 1978, The pathogenicity of thymidine kinase-deficient mutants of herpes simplex virus in mice, *J. Hyg.* **81:**267.

Field, H. J., Bell, S. E., Elion, G. B., Nash, A. A., and Wildy, P., 1979, Effect of acycloguanosine treatment on acute and latent herplex simplex infections in mice, *Antimicrob. Agents Chemother.* **15:**554.

Findlay, G. M., and MacCallum, F. O., 1940, Recurrent traumatic herpes, *Lancet* **1:**259.

Fong, B. S., and Scriba, M., 1980, Use of ^{125}I-deoxycytidine to detect herpes simplex virus specific thymidine kinase in tissues of latently infected guinea pigs, *J. Virol.* **34:**644.

Fraser, N. W., Lawrence, W. C., Wroblewska, Z., Gilden, D. H., and Koprowski, H., 1981, Herpes simplex type 1 DNA in human brain tissue, *Proc. Natl. Acad. Sci. USA* **78:**6461.

Frenkel, N., 1981, Defective interfering herpesviruses, in: *The Human Herpesviruses: An Interdisciplinary Perspective* (A. J. Nahmias, W. R. Dowdle, and R. F. Schinazi, eds.), p. 91, Elsevier/North-Holland, Amsterdam.

Friedenwald, J. S., 1923, Studies on the virus of herpes simplex, *Arch. Ophthalmol.* **52:**105.

Friedman, M. G., and Kimmel, N., 1982, Herpes simplex virus-specific serum immunog- lobulin A: Detection in patients with primary or recurrent herpes infections and in healthy adults, *Infect. Immun.* **37:**374.

Friedman, P. S., 1981, The immunobiology of Langerhans cells, *Immunol. Today* **2:**124.

Fujimiya, Y., Babiuk, L. A., and Rouse, B. T., 1978, Direct lymphocytotoxicity against herpes simplex virus infected cells, *Can. J. Microbiol.* **24:**1076.

Fujinami, R. S., Oldstone, M. B., Wroblewska, Z., Frankel, M., and Koprowski, H., 1983, Molecular mimicry in virus infection: Cross reaction of measles virus phosphoprotein or of herpes simplex virus protein with human intermediate filaments, *Proc. Natl. Acad. Sci. USA* **80:**2346.

Fulton, A. M., and Levy, J. G., 1980, The possible role of prostaglandins in mediating immune suppression by nonspecific T suppressor cells, *Cell. Immunol.* **52:**29.

Galloway, D. A., Fenoglio, C. M., Shevchuk, M., and McDougall, J. K., 1979, Detection of herpes simplex RNA in sensory ganglia, *Virology* **95:**265.

Galloway, D. A., Fenoglio, C. M., and McDougall, J. K., 1982, Limited transcription of the herpes simplex virus genome when latent in human sensory ganglia, *J. Virol.* **41:**686.

Gange, R. W., Bats, A. D., Bradstreet, C. M., and Rhodes, E. L., 1975, Cellular immunity in subjects with recurrent herpes simplex lesions and controls, *Br. J. Dermatol.* **97:**539.

Gerdes, J. O., Smith, D. S., and Forstot, S. L., 1981, Restriction endonuclease cleavage of DNA obtained from herpes simplex isolates of two patients with bilateral disease, *Curr. Eye Res.* **1:**357.

Goldyne, M. E., 1975, Prostaglandins and cutaneous inflammation, *J. Invest. Dermatol.* **64:**377.

Good, R. A., and Campbell, B., 1948, The precipitation of latent herpes simplex encephalitis by anaphylactic shock, *Proc. Soc. Exp. Biol. Med.* **68:**82.

Goodpasture, E. W., 1925a, The axis cylinders of peripheral nerves as portals of entry to the central nervous system for the virus of herpes simplex in experimentally infected rabbits, *Am. J. Pathol.* **1:**11.

Goodpasture, E. W., 1925b, The pathways of infection of the central nervous system in herpetic encephalitis of rabbits contracted by contact with a comparative comment on medullary lesions in a case of human poliomyelitis, *Am. J. Pathol.* **1:**29.

Goodpasture, E. W., 1929, Herpetic infections with special reference to involvement of the nervous system, *Medicine (Baltimore)* **7:**223.

Goodpasture, E. W., and Teague, O., 1923, Transmission of the virus of herpes febrilis along nerves in experimentally infected rabbits, *J. Med. Res.* **44:**121.

Goodwin, J. S., Bankhurst, A. D., and Messner, R. P., 1977, Suppression of human T cell mitogenesis by prostaglandin: Existence of a prostaglandin producing suppressor cell, *J. Exp. Med.* **146:**1719.

Gordon, Y., Gilden, G. Y., Shtram, Y., Asher, Y., Tabor, E., Wellish, M., Devlin, M., Snipper, D., Hadar, J., and Becker, Y., 1983, A low thymidine kinase producing mutant of herpes simplex virus type 1 causes latent trigeminal ganglia infections in mice, *Arch. Virol.* **76:**39.

Grafstein, B., 1975, The nerve cell body response to axotomy, *Exp. Neurol.* **48:**32.

Green, M. T., Rosborough, J. P., and Jones, D. B., 1981a, *In vivo* reactivation of herpes simplex virus in rabbit trigeminal ganglia: Electrode model, *Infect. Immun.* **34:**69.

Green, M. T., Courtney, R. J., and Dunkel, E. G., 1981b, Detection of an immediate early herpes simplex virus type 1 polypeptide in trigeminal ganglia from latently infected animals, *Infect. Immun.* **34:**987.

Gregg, J. M., and Dixon, A. D., 1973, Somatotopic organisation of the trigeminal ganglion of the rat, *Arch. Biol.* **18:**487.

Grodums, E. I., Kuling, P. J., and Zbitnew, A., 1981, Experimental studies of acute and recurrent herpes simplex virus infections in the murine heart and dorsal root ganglia, *J. Med. Virol.* **7:**163.

Hammer, S. M., Richter, B. S., and Hirsch, M. S., 1981, Activation and suppression of herpes simplex virus in a human T lymphoid cell line, *J. Immunol.* **127:**144.

Harbour, D. A., Hill, T. J., and Blyth, W. A., 1977, The effect of ultraviolet light on primary herpes virus infection in the mouse, *Arch. Virol.* **54:**367.

Harbour, D. A., Blyth, W. A., and Hill, T. J., 1978, Prostaglandins enhance spread of herpes simplex virus in cell cultures, *J. Gen. Virol.* **41:**87.

Harbour, D. A., Hill, T. J., and Blyth, W. A., 1981, Acute and recurrent herpes simplex in several strains of mice, *J. Gen. Virol.* **55:**31.

Harbour, D. A., Hill, T. J., and Blyth, W. A., 1983a, Prostaglandins enhance intercellular adhesion of Vero cells infected with herpes simplex virus, *J. Gen. Virol.* **64:**507.

Harbour, D. A., Hill, T. J., and Blyth, W. A., 1983b, Recurrent herpes simplex in the mouse: Inflammation in the skin and activation of virus in the ganglia following peripheral stimuli, *J. Gen. Virol.* **64:**1491.

Hill, J., Kwon, B. S., Shimomura, Y., Colborn, G. L., Yaghmai, F., and Gangarosa, L., 1983, Herpes simplex virus recovery in neural tissue after ocular herpes simplex virus shedding induced by epinephrine iontophoresis to the rabbit cornea, *Invest. Ophthalmol. Vis. Sci.* **24:**243.

Hill, T. J., 1981, Mechanisms involved in recurrent herpex simplex, in: *The Human Herpesviruses: An Interdisciplinary Perspective* (A. J. Nahmias, W. R. Dowdle, and R. F. Schinazi, eds.), p. 241, Elsevier/North-Holland, Amsterdam.

Hill, T. J., 1983, Herpesviruses in the central nervous system, in: *Viruses and Demyelinating Diseases* (C. Mims, ed.), p. 29, Academic Press, New York, in press.

Hill, T. J., and Blyth, W. A., 1976, An alternative theory of herpes simplex recurrence and a possible role for prostaglandins, *Lancet* **1:**397.

Hill, T. J., and Blyth, W. A., 1977, Recurrent herpes: Is reactivation controlled in the skin?, *Int. J. Dermatol.* **16:**274.

Hill, T. J., and Field, H. J., 1973, The interaction of herpes simplex virus with cultures of peripheral nervous tissue: An electron microscopic study, *J. Gen. Virol.* **21:**123.

Hill, T. J., Field, H. J., and Roome, A. P. C., 1972, Intraaxonal location of herpex simplex virus particles, *J. Gen. Virol.* **15:**253.

Hill, T. J., Field, H. J., and Blyth, W. A., 1975, Acute and recurrent infection with herpes simplex virus in the mouse: A model for studying latency and recurrent disease, *J. Gen. Virol.* **28:**341.

Hill, T. J., Blyth, W. A., and Harbour, D. A., 1978, Trauma to the skin causes recurrence of herpes simplex in the mouse, *J. Gen. Virol.* **39:**21.

Hill, T. J., Harbour, D. A., and Blyth, W. A., 1980, Isolation of herpes simplex virus from the skin of clinically normal mice during latent infection, *J. Gen. Virol.* **47:**205.

Hill, T. J., Ahluwalia, K., and Blyth, W. A., 1981, Infection with herpes simplex virus in the eye and trigeminal ganglion of mice, in: *Herpetic Ocular Diseases* (R. Sundmacher, ed.), p. 37, Springer-Verlag (Bergmann), Berlin.

Hill, T. J., Blyth, W. A., and Harbour, D. A., 1982, Recurrent herpes simplex in mice: Topical treatment with acyclovir cream, *Antiviral Res.* **2:**135.

Hill, T. J., Blyth, W. A., Harbour, D. A., Berrie, E. L., and Tullo, A. B., 1983a, Latency and other consequences of infection of the nervous system with herpes simplex virus, *Brain Res.* **59:**173.

Hill, T. J., Blyth, W. A., and Harbour, D. A., 1983b, Recurrence of herpes simplex in the mouse requires an intact nerve supply to the skin, *J. Gen. Virol.* **64:**2763.

Hirsch, M. S., Zisman, B., and Allison, A. C., 1970, Macrophage and age dependent resistance to herpes simplex in mice, *J. Immunol.* **104:**1160.

Ho, M., 1977, Viral infections after transplantation in man, *Arch. Virol.* **55:**1.

Howard, W. T., 1903, The pathology of labial and nasal herpes and of herpes of the body occurring in acute croupous pneumonia and their relation to the so-called herpes zoster, *Am. J. Med. Sci.* **125:**256.

Howard, W. T., 1905, Further observations on the relation of lesions of the Gasserian and posterior root ganglia to herpes occurring in pneumonia and cerebrospinal meningitis, *Am. J. Med. Sci.* **130:**1012.

Hoyt, C. S., and Billson, F. A., 1976, Herpes simplex infection after blowout fractures, *Lancet* **2:**1364.

Huang, A. S., and Baltimore, D., 1970, Defective viral particles and viral disease processes, *Nature* **226:**325.

Hurd, J., and Robinson, T. W. E., 1977, Herpes simplex: Aspects of reactivation in a mouse model, *J. Antimicrob. Chemother.* **3:**99.

Ishikawa, T., Takajama, S., and Kitagawa, T., 1978, DNA repair synthesis in rat retinal ganglion cells treated with chemical carcinogen or UV light, *J. Natl. Cancer Inst.* **61:**1101.

Jacobs, R. D., Aurelian, L., and Cole, G., 1976, Cell mediated immune response to herpes simplex virus: Type specific lymphoproliferative responses in lymph nodes draining the site of primary infection, *J. Immunol.* **116**:1520.

Jamieson, A. T., Gentry, G. A., and Subak-Sharpe, J. H., 1974, Induction of both thymidine and deoxycytidine kinase activity by herpes viruses, *J. Gen. Virol.* **24**:465.

Johnson, R. T., 1964, The pathogenesis of herpes virus encephalitis. I. Virus pathways to the nervous system of suckling mice demonstrated by fluorescent antibody staining, *J. Exp. Med.* **119**:343.

Kappers, C. U. A., Huber, G. D., and Crosby, E. C., 1960, *The Comparative Anatomy of the Nervous System of Vertebrates Including Man*, Vol. 1, Hafner, New York.

Kastrukoff, L. F., Long, C., Doherty, P. C., Wroblewska, Z., and Koprowski, H., 1981, Isolation of virus from brain after immunosuppression of mice with latent herpes simplex, *Nature* **291**:432.

Kaufman, H. E., Brown, D. C., and Ellison, E. M., 1967, Recurrent herpes in rabbit and man, *Science* **156**:1628.

Kaufman, H. E., Brown, D. C., and Ellison, E. M., 1968, Herpes virus in the lacrimal gland and conjunctiva, *Am. J. Ophthalmol.* **65**:32.

Keddie, F. M., Rees, R. B., and Epstein, N. N., 1941, Herpes simplex following artificial fever therapy, *J. Am. Med. Assoc.* **117**:1327.

Kennedy, P. G. E., Al-Saadi, S. A. and Clements, G. B., 1983, Reactivation of latent herpes simplex virus from dissociated identified dorsal root ganglion cells in culture, *J. Gen. Virol.* **64**:1629.

Kino, Y., Hayashi, Y., Hayishida, I., and Mori, R., 1982, Dissemination of herpes simplex virus in nude mice after intracutaneous inoculation and effect of antibody on the course of infection, *J. Gen. Virol.* **63**:475.

Kleckner, N., 1981, Transposable elements in prokaryotes, *Annu. Rev. Genet.* **15**:341.

Klein, R. J., 1976, Pathogenetic mechanisms of recurrent herpes simplex viral infections, *Arch. Virol.* **51**:1.

Klein, R. J., 1980, Effect of immune serum on the establishment of herpes simplex virus infection in trigeminal ganglia of hairless mice, *J. Gen. Virol.* **49**:401.

Klein, R. J., Friedman-Kien, A. E., Fondak, A. A., and Buimovici-Klein, E., 1977, Immune response and latent infection after topical treatment of herpes simplex virus infection in hairless mice, *Infect. Immun.* **16**:842.

Klein, R. J., Friedman-Kien, A. E., and Brady, E., 1978a, Latent herpes simplex virus infection in ganglia of mice after primary infection and reinoculation at a distant site, *Arch. Virol.* **57**:161.

Klein, R. J., Friedman-Kien, A. E., and Yellin, P. B., 1978b, Orofacial herpes simplex virus infection in hairless mice: Latent virus in trigeminal ganglia after topical antiviral treatment, *Infect. Immun.* **20**:130.

Klein, R. J., Friedman-Kien, A. E., and De Stefano, E., 1979, Latent herpes simplex virus infections in sensory ganglia of hairless mice prevented by acycloguanosine, *Antimicrob. Agents Chemother.* **15**:723.

Klein, R. J., De Stefano, E., Friedman-Kien, A. E., and Brady, E., 1981, Effect of acyclovir on latent herpes simplex virus infections in trigeminal ganglia of mice, *Antimicrob. Agents Chemother.* **19**:937.

Knotts, F. B., Cook, M. L., and Stevens, J. G., 1973, Latent herpes simplex virus in the central nervous system of rabbits and mice, *J. Exp. Med.* **138**:740.

Knotts, F. B., Cook, M. L., and Stevens, J. G., 1974, Pathogenesis of herpetic encephalitis in mice following ophthalmic inoculation, *J. Infect. Dis.* **130**:16.

Kopecko, D. J., 1980, Specialized genetic recombination systems in bacteria: Their involvement in gene expression and evolution, *Prog. Mol. Subcell. Biol.* **7**:135.

Kristensson, K., 1978, Retrograde transport of macromolecules in axons. *Annu. Rev. Pharmacol. Toxicol.* **18**:97.

Kristensson, K., Lycke, E., and Sjöstrand, J., 1971, Spread of herpes simplex virus in peripheral nerves, *Acta Neuropathol.* **17**:44.

Kristensson, K., Gheth, B., and Wisniewski, H. M., 1974, Study on the propagation of herpes simplex virus (type 2) into the brain after intraocular injection, *Brain Res.* **69**:189.

Kristensson, K. A., Vahlne, A. L., Persson, L. A., and Lycke, E., 1978, Neural spread of herpes simplex virus types 1 and 2 in mice after corneal or subcutaneous (footpad) inoculation, *J. Neurol. Sci.* **35**:331.

Kwon, B. S., Gangarosa, L. P., Burch, K. D., de Baeck, J., and Hill, J. M., 1981, Induction of ocular herpes simplex virus shedding by iontophoresis of epinephrine into rabbit cornea, *Invest. Ophthalmol. Vis. Sci.* **21**:442.

Laibson, P. R., and Kibrick, S., 1969, Recurrence of herpes simplex virus in rabbit eyes: Results of a three year study, *Invest. Ophthalmol.* **8**:346.

Landon, D. N. (ed.), 1976, *The Peripheral Nerve*, Chapman & Hall, London.

Lascano, E. F., and Berria, M. I., 1980, Histological study of the progression of herpes simplex virus in mice, *Arch. Virol.* **64**:67.

Lehner, T., Wilton, J. M. A., and Shillitoe, E. J., 1975, Immunological basis for latency, recurrence and putative oncogenicity of herpes simplex virus, *Lancet* **2**:60.

Lessard, R. J., Wolff, K., and Winkelmann, R. K., 1968, The disappearance and regeneration of Langerhans cells following epidermal injury, *J. Invest. Dermatol.* **50**:171.

Lewis, M. E., Warren, K. G., and Jeffery, V. M., 1982, Factors affecting recovery of latent herpes simplex virus from human trigeminal ganglia, *Can. J. Microbiol.* **28**:123.

Leider, W., Magoffin, R. L., Lennett, E. H., and Leonards, L. N. R., 1965, Herpes simplex encephalitis—Its possible association with reactivated infection, *New Engl. J. Med.* **273**:341.

Lofgren, K. W., Stevens, J. G., Marsden, H. S., and Subak-Sharpe, J. H., 1977, Temperature sensitive mutants of herpes simplex virus differ in the capacity to establish latent infections in mice, *Virology* **76**:440.

Longson, M., 1978, Persistence of herpes simplex virus infections, *Postgrad. Med. J.* **54**:603.

Longson, M., and Bailey, A. S., 1977, *Herpes encephalitis*, in:*Recent Advances in Clinical Virology* (A. P. Waterson, ed.), Vol. 1, p. 1, Churchill & Livingstone, Edinburgh.

Longson, M., Bailey, A. S., and Klapper, P., 1980, *Herpes encephalitis*, in: *Recent Advances in Clinical Virology* (A. P. Waterson, ed.), Vol. 2, p. 147, Churchill & Livingstone, Edinburgh.

Lonsdale, D. M., Brown, M. S., Subak-Sharpe, J. H., Warren, K. G., and Kropowski, H., 1979, The polypeptide and DNA restriction enzyme profiles of spontaneous isolates of herpes simplex virus type 1 from explants of human trigeminal, superior cervical and vagus ganglia, *J. Gen. Virol.* **43**:151.

Lopez, C., 1975, Genetics of natural resistance to herpes virus infections in mice, *Nature (London)* **258**:152.

Lopez, C., 1981, Resistance to herpes simplex virus type 1 (HSV1), *Curr. Top. Microbiol. Immunol.* **92**:15.

Lopez, C., and O'Reilly, R. J., 1977, Cell mediated immune responses in recurrent herpesvirus infections, I. Lymphocyte proliferation assay, *J. Immunol.* **118**:895.

Lycke, E., Norrby, R., and Roos, B., 1974, A serological study on mentally ill patients, *Br. J. Psychiatry* **124**:273.

McCracken, R. M., McFerran, J. B., and Dow, C., 1973, The neural spread of pseudorabies virus in calves, *J. Gen. Virol.* **20**:17.

McKendall, R. R., 1977, Efficacy of herpes simplex virus type 1 immunisation in protecting against acute and latent infection by herpes simplex virus type 2 in mice, *Infect. Immun.* **16**:717.

McKendall, R. R., Klassen, T., and Baringer, J. R., 1979, Host defenses in herpes simplex infections of the nervous system: Effect of antibody on disease and viral spread, *Infect. Immun.* **23**:305.

McLennan, J. L., and Darby, G., 1980, Herpes simplex virus latency: The cellular location of virus in dorsal root ganglia and the fate of the infected cell following virus activation, *J. Gen. Virol.* **51**:233.

Marsden, H., 1980, Herpes simplex virus in latent infections, *Nature* **288**:212.

Martin, J. R., 1982, Spinal cord and optic nerve demyelination in experimental herpes simplex virus type 2 infection, *J. Neuropathol. Exp. Neurol.* **41**:253.

Martin, J. R., Stoner, G. L., and Webster, H. D., 1982, Lethal encephalitis and non lethal multifocal central nervous system demyelination in herpes virus type 2 infections in mice, *Br. J. Exp. Pathol.* **63**:651.

Martin, R. G., Dawson, C. R., Jones, P., Togni, B., Lyons, C., and Oh, J. O., 1977, Herpes virus in sensory and autonomic ganglia after eye infection, *Arch. Ophthalmol.* **95**:2053.

Mellick, R. C., and Cavanagh, J. B., 1967, Longitudinal movement of radioiodinated albumin within extravascular spaces of peripheral nerves following three sorts of experimental trauma, *J. Neurol. Neurosurg. Psychiatry* **30**:458.

Metcalf, J. F., Hamilton, D. S., and Reichart, R. W., 1979, Herpetic keratitis in athymic (nude) mice, *Infect. Immun.* **26**:1161.

Meyers-Elliott, R. H., Pettit, T. H., and Maxwell, W. A., 1980, Viral antigens in the immune ring of herpes simplex stromal keratitis, *Arch. Ophthalmol.* **98**:897.

Montgomerie, J. J., Becroft, D. M. O., Croxson, M. C., Doak, P. B., and North, J. D., 1969, Herpes simplex virus infection after renal transplantation, *Lancet* **2**:867.

Nahmias, A. J., and Ashman, R. B., 1978, The immunology of primary and recurrent herpes virus infection: An overview, *IARC Sci. Publ.* **24**:659.

Narang, H. K., 1977, The pathway into the central nervous system after intraocular injection of herpes simplex (type 1) in rabbits, *Neuropathol. Appl. Neurobiol.* **3**:490.

Narang, H. K., and Codd, A. A., 1978, The pathogenesis and pathway into the central nervous system after intraocular infection of herpes simplex virus type 1 in rabbits, *Neuropathol. Appl. Neurobiol.* **4**:137.

Nash, A. A., 1981, Antibody and latent herpes simplex virus infection, *Immunol. Today* **2**:19.

Nash, A. A., Quartey-Papafio, R., and Wildy, P., 1980a, Cell-mediated immunity in herpes simplex virus-infected mice: Functional analysis of lymph node cells during periods of acute and latent infection, with reference to cytotoxic and memory cells, *J. Gen. Virol.* **49**:309.

Nash, A. A., Field, H. J., and Quartey-Papafio, R., 1980b, Cell-mediated immunity in herpes simplex virus-infected mice: Induction, characterization and antiviral effects of delayed type hypersensitivity, *J. Gen. Virol.* **48**:351.

Nash, A. A., Gell, P. G. H., and Wildy, P., 1981, Tolerance and immunity in mice infected with herpes simplex virus: Simultaneous induction of protective immunity and tolerance to delayed-type hypersensitivity, *Immunology* **43**:153.

Nesburn, A. B., Elliott, J. M., and Leibowitz, H. M., 1967, Spontaneous reactivation of experimental herpes simplex keratitis in rabbits, *Arch. Ophthalmol.* **78**:523.

Nesburn, A. B., Dickinson, R., and Radnoti, M., 1976a, The effect of trigeminal nerve and ganglion manipulation on recurrence of ocular herpes simplex in rabbits, *Invest. Ophthalmol.* **15**:726.

Nesburn, A. B., Dickinson, R., and Radnoti, M., 1976b, Experimental reactivation of ocular herpes simplex in rabbits, *Surv. Ophthalmol.* **21**:185.

Nesburn, A. B., Green, M. T., Radnoti, M., and Walker, B., 1977, Reliable *in vivo* model for latent herpes simplex virus reactivation with peripheral virus shedding, *Infect. Immun.* **15**:772.

Nesburn, A. B., Dunkel, E. C., and Trousdale, M. D., 1980, Enhanced HSV recovery from neuronal tissues of latently infected rabbits, *Proc. Soc. Exp. Biol. Med.* **163**:398.

Nesburn, A. B., Willey, D. E., and Trousdale, M. D., 1983, Effect of intensive acyclovir therapy during artificial reactivation of latent herpes simplex virus, *Proc. Soc. Exp. Biol. Med.* **172**:316.

Nicolau, S., and Poincloux, P., 1928, Etudes cliniques et expérimentales d'un cas d'herpès récidivant du doigt, *Ann. Inst. Pasteur Paris* **38**:977.

Ochs, S., 1974, Systems of material transport in nerve fibres (axoplasmic transport) related to nerve function and trophic control, *Ann. N.Y. Acad. Sci.* **228**:202.

Openshaw, H., 1983, Latency of herpes simplex virus in ocular tissue of mice, *Infect. Immun.* **39**:960.

Openshaw, H., and Bennett, H.E., 1982, Recurrence of herpes simplex virus after dental extraction, *J. Infect. Dis.* **146**:707.

Openshaw, H., Shavrina Asher, L. V., Wohlenberg, C., Sekizawa, T., and Notkins, A. L., 1979a, Acute and latent infection in sensory ganglia with herpes simplex virus: Immune control and virus reactivation, *J. Gen. Virol.* **44**:205.

Openshaw, H., Puga, A., and Notkins, A. L., 1979b, Herpes simplex virus infection in sensory ganglia, immune control, latency and reactivation, *Fed. Proc.* **38**:2660.

Openshaw, H., Sekizawa, T., Wohlenberg, C., and Notkins, A. L., 1981, The role of immunity in latency and reactivation of herpes simplex viruses, in: *The Human Herpesviruses: An Interdisciplinary Perspective* (A. J., Nahmias, W. R. Dowdle, and R. F. Schinazi, eds.), p. 288, Elsevier/North-Holland, Amsterdam.

O'Reilly, R. J., Chibbaro, A., Anger, E., and Lopez, C., 1977, Cell mediated responses in patients with recurrent herpes simplex infections. II. Infection associated deficiency of lymphokine production in patients with recurrent herpes labialis or genitalis, *J. Immunol.* **118**:1095.

Orsi, E. V., Howard, J. L., Baturay, N., Ende, N., Ribot, S., and Eslami, H., 1978, High incidence of virus isolation from donor and recipient tissues associated with renal transplantation, *Nature (London)* **272**:372.

Paine, T. F., 1964, Latent herpes simplex infection in man, *Bacteriol. Rev.* **28**:472.

Para, M. F., Baucke, R. B., and Spear, P. G., 1980, Immunoglobulin G (Fc)-binding receptors on virions of herpes simplex virus type 1 and transfer of these receptors to the cell surface by infection, *J. Virol.* **34**:512.

Pass, R. F., Whitley, R. J., Whelchel, J. D., Diethelm, A. G., Reynolds, D. W., and Alford, C. A., 1979, Identification of patients with increased risk of infection with herpes simplex virus after renal transplantation, *J. Infect. Dis.* **140**:487.

Pazin, G. J., Ho, M., and Jannetta, P. J., 1978, Reactivation of herpes simplex virus after decompression of the trigeminal nerve root, *J. Infect. Dis.* **138**:405.

Plummer, G., 1973, Isolation of herpesviruses from the trigeminal ganglia of man, monkeys and cats, *J. Infect. Dis.* **128**:345.

Plummer, G., Hollingsworth, D. C., Phuangsab, A., and Bowling, C. P., 1970, Chronic infections by herpes simplex viruses and by the horse and cat herpesviruses, *Infect. Immun.* **1**:351.

Preston, C. M., 1979a, Abnormal properties of an immediate early polypeptide in cells infected with herpes simplex virus type 1 mutant tsk, *J. Virol.* **32**:357.

Preston, C. M., 1979b, Control of herpes simplex virus type 1 mRNA synthesis in cells with wild type virus on the temperature sensitive mutant tsk, *J. Virol.* **29**:275.

Price, R. W., 1979, 6-Hydroxydopamine potentiates acute herpes simplex virus infection of the superior cervical ganglion in mice, *Science* **205**:518.

Price, R. W., and Kahn, A., 1981, Resistance of peripheral autonomic neurons to *in vivo* productive infection by herpes simplex virus mutants deficient in thymidine kinase activity, *Infect. Immun.* **43**:571.

Price, R. W., and Notkins, A. L., 1977, Viral infections of the autonomic nervous system and its target organs: Pathogenetic mechanisms, *Med. Hypoth.* **3**:33.

Price, R. W., and Schmitz, J., 1978, Reactivation of latent herpes simplex virus infection of the autonomic nervous system by postganglionic neurectomy, *Infect. Immun.* **19**:523.

Price, R. W., and Schmitz, J., 1979, Route of infection, systemic host resistance and integrity of ganglionic axons influence acute and latent herpes simplex virus infection of the superior cervical ganglion, *Infect. Immun.* **23**:373.

Price, R. W., Katz, B. J., and Notkins, A. L., 1975a, Latent infection of the peripheral and autonomic nervous system with herpes simplex virus, *Nature (London)* **257**:686.

Price, R. W., Walz, M. A., Wohlenberg, C., and Notkins, A. L., 1975b, Latent infection of sensory ganglia with herpes simplex virus: Efficacy of immunization, *Science* **188**:938.

Price, R. W., Rubenstein, R., and Kahn, A., 1982, Herpes simplex virus infection of isolated autonomic neurons in culture: Viral replication and spread in a neuronal network, *Arch. Virol.* **71**:127.

Puga, A., Rosenthal, J. D., Openshaw, H., and Notkins, A. L., 1978, Herpes simplex virus DNA and mRNA sequences in acutely and chronically infected trigeminal ganglia of mice, *Virology* **89**:102.

Puga, A., Cantin, E. M., Wohlenberg, C., Openshaw, H., and Notkins, A. L., 1984, Different sizes of restriction endonuclease fragments from the terminal repetitions of herpes simplex virus type 1 genome latent in trigeminal ganglia of mice, *J. Gen. Virol.* **65**:437.

Rabin, E. R., Jenson, A. B., and Melnick, J. L., 1968, Herpes simplex virus in mice: Electron microscopy of neural spread, *Science* **162**:126.

Rajčáni, J., and Conen, P. E., 1972, Observations of neural spread of herpes simplex virus in mice: An electron microscopic study, *Acta. Virol.* **16**:31.

Rajčáni, J., Sabo, A., and Blaskovic, D., 1969, Vergeichende untersuchungen zur experimentallen pathogenese des herpes simplex-virus bei der maus: Intrazerebrale, intraperitoneale und orale infektion der sauglingsmaus, *Zentralbl. Bakteriol. Parasitenkd. Infektionskr. Hyg. Abt. 1 Orig.* **211**:421.

Rajčáni, J., Ciampor, A., Sabo, A., Libikova, H., and Rosenbergova, M., 1976, Activation of latent herpesvirus hominis in explants of rabbit trigeminal ganglia: The influence of immune serum, *Arch. Virol.* **53**:55.

Rand, K. H., Rasmussen, L. E., Pollard, R. B., Arvin, A., and Merigan, T. C., 1976, Cellular immunity and herpesvirus infections in cardiac transplant patients, *New Engl. J. Med.* **296**:1372.

Ranson, S. W., 1906, Retrograde degeneration in the spinal nerves, *J. Comp. Neurol. Psychol.* **16**:265.

Rasmussen, L. E., Jordan, G. W., Stevens, D. A., and Merigan, T. C., 1974, Lymphocyte interferon production and transformation after herpes simplex infections in humans, *J. Immunol.* **112**:728.

Reeves, W. C., DiGiacomo, R. F., Alexander, E. R., and Lee, C. K., 1976, Latent herpesvirus hominis from trigeminal and sacral dorsal root ganglia of cebus monkeys, *Proc. Soc. Exp. Biol. Med.* **153**:258.

Reeves, W. C., Di Giacomo, R., and Alexander, E. R., 1981, A primate model for age and host response to genital herpetic infection: Determinants of latency, *J. Infect. Dis.* **143**:554.

Reichman, R. C., Dolin, R., Vincent, M. M., and Fauci, A. S., 1977, Cell mediated cytotoxicity in recurrent herpes simplex virus infection in man, *Proc. Soc. Exp. Biol. Med.* **155**:571.

Renis, H. E., Eidson, E. E., Mathews, J., and Gray, J. E., 1976, Pathogenesis of herpes simplex virus types 1 and 2 after various routes of inoculation, *Infect. Immun.* **14**:571.

Richter, R. B., 1944, Observations on the presence of latent herpes simplex virus in the human Gasserian ganglion, *J. Nerv. Ment. Dis.* **99**:356.

Ritchie, D. A., and Timbury, M. C., 1980, Herpes viruses and latency: Possible relevance of the structure of the viral genome, *FEMS Microbiol. Lett.* **9**:67.

Rock, D. L., and Fraser, N. W., 1983, Detection of HSV-1 genome in the central nervous system of latently infected mice, *Nature* **302**:523.

Rodda, S., Jack, I., and White, D. O., 1973, Herpes simplex virus from trigeminal ganglion, *Lancet* **1**:1395.

Rogers, T., Nowoweijski, I., and Webb, D. R., 1980, Partial characterisation of a prostaglandin-induced suppressor factor, *Cell. Immunol.* **50**:82.

Roizman, B., 1965, An inquiry into the mechanisms of recurrent herpes infections of man, in: *Perspectives in Virology*, Vol. IV (M. Pollard, ed.), p. 283, Harper &Row (Hoeber), New York.

Roizman, B., 1974, Herpesvirus, latency and cancer: A biochemical approach, *J. Reticuloendothel. Soc.* **15**:312.

Rosenberg, G. L., Snyderman, R., and Notkins, A. L., 1974, Production of chemotactic factors and lymphokines in human leukocytes stimulated with herpes simplex virus, *Infect. Immun.* **10:**111.

Rouse, B. T., and Lopez, C. (eds.), 1984, *Immunobiology of Herpes Simplex Virus Infections*, CRC Press, Boca Raton, Fla., (in press).

Russell, A. S., 1973, Cell mediated immunity to herpes simplex virus in man, *Am. J. Clin. Pathol.* **60:**826.

Russell, A. S., and Miller, C., 1978, A possible role for polymorphonuclear leucocytes in the defence against recrudescent herpes simplex virus infection in man, *Immunology* **34:**371.

Russell, A. S., and Schlaut, J., 1975, HLA transplantation antigens in subjects susceptible to recrudescent herpes labialis, *Tissue Antigens* **6:**257.

Russell, A. S., Percy, J., and Kovithavongs, T., 1975, Cell mediated immunity to herpes simplex in humans: Lymphocyte cytotoxicity measured by ^{51}Cr release from infected cell, *Infect. Immun.* **11:**355.

Russell, A. S., Kaiser, J., and Lao, V., 1976, Cell mediated immunity to herpes simplex in man. IV. The correlation of lymphocyte stimulation and inhibition of leukocyte migration, *J. Immunol. Methods* **9:**273.

Rustigian, R., Smulou, J. B., Tye, M., Gibson, W. A., and Shindell, E., 1966, Studies on latent infection of skin and oral mucosa in individuals with recurrent herpes simplex, *J. Invest. Dermatol.* **47:**218.

Sanes, J. R., and Okun, L. M., 1972, Induction of DNA synthesis in cultured neurons by ultraviolet light or methyl methane sulfonate, *J. Cell Biol.* **53:**587.

Schmidt, J. R., and Rasmussen, A. F., 1960, Activation of latent herpes simplex encephalitis by chemical means, *J. Infect. Dis.* **106:**154.

Schrier, R. D., Pizer, L. I., and Moorehead, J. W., 1983, Tolerance and suppression of immunity to herpes simplex virus: Different presentations of antigens induce different types of suppressor cells, *Infect. Immun.* **40:**514.

Schwartz, J., and Elizan, T. S., 1973, Chronic herpes simplex infection, *Arch. Neurol.* **28:**224.

Schwartz, J., Whetsell, W. O., and Elizan, T. S., 1978, Latent herpes simplex virus infection of mice: Infectious virus in homogenates of latently infected dorsal root ganglia, *J. Neuropathol. Exp. Neurol.* **37:**45.

Scriba, M., 1975, Herpes simplex infection in guinea pigs: An animal model for studying latent and recurrent herpes simplex virus infections, *Infect. Immun.* **12:**162.

Scriba, M., 1976, Recurrent genital herpes simplex virus (HSV) infection in guinea pigs, *Med. Microbiol. Immunol.* **162:**201.

Scriba, M., 1977, Extraneural localisation of herpes simplex virus in latently infected guinea pigs, *Nature (London)* **267:**529.

Scriba, M., 1981a, Persistence of herpes simplex virus (HSV) infection in ganglia and peripheral tissues of guinea pigs, *Med. Microbiol. Immunol.* **169:**91.

Scriba, M., 1981b, Vaccination against herpes simplex virus: Animal studies on the efficacy against acute, latent and recurrent infections, in: *Herpetic Ocular Diseases* (R. Sundmacher, ed.), p. 67, Springer-Verlag (Bergmann), Berlin.

Scriba, M., and Tatzber, F., 1981, Pathogenesis of herpes simplex virus infections in guinea pigs, *Infect. Immun.* **34:**655.

Sekizawa, T., Openshaw, H., Wohlenberg, C., and Notkins, A. L., 1980, Latency of herpes simplex virus in the absence of neutralizing antibody: A model for reactivation, *Science* **210:**1026.

Sequiera, L. W., Jennings, L. C., Carrasso, L. H., Lord, M. A., Curry, A., and Sutton, R. N. P., 1979, Detection of herpes simplex viral genome in brain tissue, *Lancet* **2:**609.

Severin, M. J., and White, R. J., 1968, The neural transmission of herpes simplex virus in mice: Light and electron microscopic findings, *Am. J. Pathol.* **53:**1009.

Sheridan, J. F., Donnenberg, A. D., Aurelian, L., and Elpern, D. J., 1982, Immunity to herpes simplex virus type 2. IV. Impaired lymphokine production during recrudescence correlates with an imbalance in T lymphocyte subsets, *J. Immunol.* **129:**326.

Shillitoe, E. J., Wilton, J. M. A., and Lehner, T., 1977, Sequential changes in cell-mediated immune responses to herpes simplex virus after recurrent herpetic infection in humans, *Infect. Immun.* **18:**130.

Shimeld, C., Tullo, A. B., Easty, D. L., and Thomsitt, J., 1982, Isolation of HSV from the cornea in chronic stromal keratitis, *Br. J. Ophthalmol.* **66:**643.

Smith, E. B., and McLaren, L. C., 1977, Attempts to recover herpes simplex virus from skin sites of recurrent infection, *Int. J. Dermatol.* **16:**748.

Sokawa, Y., Ando, T., and Ishihara, Y., 1980, Induction of 2' 5'-oligoadenylate synthetase and interferon in mouse trigeminal ganglia infected with herpes simplex virus, *Infect. Immun.* **28:**719.

Spear, P. G., Para, M. F., and Baucke, R. B., 1981, The Fc-binding receptor induced by herpes simplex virus, in: *The Human Herpesviruses: An Interdisciplinary Perspective* (A. J. Nahmias, W. R. Dowdle, and R. F. Schinzai, eds.), p. 237, Elsevier/North-Holland, Amsterdam.

Spencer, E. S., and Anderson, H. K., 1970, Clinically evident non terminal infections with herpes simplex viruses and the wart virus in immunosuppressed renal allograft patients, *Br. Med. J.* **3:**251.

Spruance, S. L., and Chow, F., 1980, Pathogenesis of herpes simplex labialis. I. Replication of herpes simplex virus in cultures of epidermal cells from subject with frequent recurrences, *J. Infect. Dis.* **142:**671.

Spruance, S. L., Overall, J. C., Kern, E. R., Kreuger, G. G., Pliam, V., and Miller, W., 1977, The natural history of recurrent herpes simplex labialis, *New Engl. J. Med.* **297:**69.

Spurney, R. V., and Rosenthal, M. S., 1972, Ultraviolet-induced recurrent herpes simplex virus keratitis, *Am. J. Ophthalmol.* **73:**609.

Stalder, W., and Zurukzoglu, S., 1936, Experimentelle untersuchungen über herpes: Transplantation herpes-infizierter haustellen reaktivierung von abgeheilten Künstlich infizierten haustellen, *Zentralbl. Bakteriol. Parasitenkd. Infektionskr. Hyg.* **136:**94.

Stanberry, L. R., Kern, E. R., Richards, J. T., Abbott, T. M., and Overall, J. C., 1982, Genital herpes in guinea pigs: Pathogenesis of the primary infection and description of recurrent disease, *J. Infect. Dis.* **146:**397.

Steele, R. W., Vincent, M. M., Hensen, S. A., Fuccillo, D. A., Chapa, I. A., and Canales, L., 1975, Cellular immune responses to herpes simplex virus type 1 in recurrent herpes labialis: *In vitro* blastogenesis and cytotoxicity to infected cell lines, *J. Infect. Dis.* **131:**528.

Stevens, J. G., 1975, Latent herpes simplex virus and the nervous system, *Curr. Top. Microbiol. Immunol.* **70:**31.

Stevens, J. G., 1978, Latent characteristics of selected herpesviruses, *Adv. Cancer Res.* **26:**227.

Stevens, J. G., 1980, Herpetic latency and reactivation, in: *Oncogenic Herpesviruses*, Vol. 2 (F. Rapp, ed.), p. 1, CRC Press, Boca Raton, Fla.

Stevens, J. G., and Cook, M. L., 1971, Latent herpes simplex virus in spinal ganglia of mice, *Science* **173:**843.

Stevens, J. G., and Cook, M. L., 1973a, Latent herpes simplex virus in sensory ganglia, in: *Perspectives in Virology*, Vol. VIII (M. Pollard, ed.), p. 171, Academic Press, New York.

Stevens, J. G., and Cook, M. L., 1973b, Latent herpes simplex infection, in: *Virus Research* (C. F. Fox and W. S. Robinson, eds.), p. 437, Academic Press, New York.

Stevens, J. G., and Cook, M. L., 1974, Maintenance of latent herpetic infection: An apparent role for anti-viral IgG, *J. Immunol.* **113:**1685.

Stevens, J. G., Nesburn, A. B., and Cook, M. L., 1972, Latent herpes simplex virus from trigeminal ganglia of rabbits with recurrent eye infection, *Nature New Biol.* **235:**216.

Stevens, J. G., Cook, M. L., and Jordan, M. C., 1975, Reactivation of latent herpes simplex virus after pneumococcal pneumonia in mice, *Infect. Immun.* **11:**635.

Streilein, J. W., Toews, G. B., and Bergstresser, P. R., 1979, Corneal allografts fail to express Ia antigens, *Nature (London)* **282:**326.

Sundmacher, R. (ed.), 1981, *Herpetic Eye Diseases*, Springer-Verlag (Bergmann), Berlin.

Sutter, D., and Doerfler, W., 1980, Methylation of integrated adenovirus type 12 DNA sequences in transformed cells is inversely correlated with viral gene expression, Proc. Natl. Acad. Sci. USA 77:253.

Svennerholm, B., Vahlne, A., and Lycke, E., 1981, Persistent reactivable latent herpes simplex virus infection in trigeminal ganglia of mice treated with antiviral drugs, Arch. Virol. 69:43.

Teague, O., and Goodpasture, E. W., 1923, Experimental herpes zoster, J. Med. Res. 44:185.

Tenser, R. B., and Dunstan, M. E., 1979, Herpes simplex virus thymidine kinase expression in infection of the trigeminal ganglion, Virology 99:417.

Tenser, R. B., and Hsiung, G. D., 1977, Pathogenesis of latent herpes simplex virus infection of the trigeminal ganglion in guinea pigs: Effects of age, passive immunization and hydrocortisone, Infect. Immun. 16:69.

Tenser, R. B., Miller, R. L., and Rapp, F., 1979, Trigeminal ganglion infection by thymidine kinase-negative mutants of herpes simplex virus, Science 205:915.

Tenser, R. B., Dawson, M., Ressel, S. J., and Dunstan, M. S., 1982, Detection of herpes simplex virus mRNA in latently infected trigeminal ganglion neurons by in situ hybridization, Ann. Neurol. 11:285.

ter Meulen, V., and Carter, M. J., 1982, Morbillivirus persistent infections in animals and man, in: Virus Persistence (B. W. J. Mahy, A. C. Minson, and G. K. Darby, eds.), p. 97, Cambridge University Press, London.

Thong, Y. H., Vincent, M. M., Hensen, S. A., Fuccillo, D. A., Rola-Pleszcynski, M., and Bellanti, J. A., 1975, Depressed specific cell-mediated immunity to herpes simplex virus type 1 in patients with recurrent herpes labialis, Infect. Immun. 12:76.

Tokumaru, T., 1966, A possible role of γA-immunoglobulin in herpes simplex virus infection in man, J. Immunol. 97:248.

Tooze, J. (ed.), 1973, The Molecular Biology of Tumour Viruses, Cold Spring Harbor Laboratory, New York.

Townsend, J. J., 1981a, The relationship of astrocytes and macrophages to CNS demyelination after experimental herpes simplex virus infection, J. Neuropathol. Exp. Neurol. 40:369.

Townsend, J. J., 1981b, The demyelinating effect of corneal HSV infections in normal and nude (athymic) mice, J. Neurol. Sci. 50:435.

Townsend, J. J., and Baringer, J. R., 1976, Comparative vulnerability of peripheral and central nervous tissue to herpes simplex virus, J. Neuropathol. Exp. Neurol. 35:100.

Townsend, J. J., and Baringer, J. R., 1978, Central nervous system susceptibility to herpes simplex infection. J. Neuropathol. Exp. Neurol. 37:255.

Townsend, J. J., and Baringer, J. R., 1979, Morphology of the CNS disease in immunosuppressed mice after peripheral HSV inoculation, Lab. Invest. 40:178.

Trofatter, K. F., and Daniels, C. A., 1979, Interaction of human cells with prostaglandins and cyclic AMP modulators, J. Immunol. 122:1363.

Tullo, A. B., Shimeld, C., Blyth, W. A., Hill, T. J., and Easty, D. L., 1982a, Spread of virus and distribution of latent infection following ocular herpes simplex in the non-immune and immune mouse, J. Gen. Virol. 63:95.

Tullo, A. B., Easty, D. L., Hill, T. J., and Blyth, W. A., 1982b, Ocular herpes simplex and the establishment of latent infection, Trans. Ophthalmol. Soc. U.K. 102:15.

Tullo, A. B., Shimeld, C., Blyth, W. A., Hill, T. J., and Easty, D. L., 1983a, Ocular infection with HSV in non immune and immune mice, Arch. Ophthalmol. 101:961.

Tullo, A. B., Shimeld, C. Easty, D. L., and Darville, J. M., 1983b, Distribution of latent herpes simplex virus infection in human trigeminal ganglion, Lancet 1:353.

Tullo, A. B., Keen, P., Blyth, W. A., Hill, T. J., and Easty, D. L., 1983c, Corneal sensitivity and substance P in experimental keratitis in mice, Invest. Ophthalmol. Vis. Sci. 24:596.

Underwood, G. E., and Weed, S. D., 1974, Recurrent cutaneous herpes simplex in hairless mice, Infect. Immun. 10:471.

Vahlne, A., Nyström, B., Sandberg, M., Hamberger, A., and Lycke, E., 1978, Attachment of herpes simplex virus to neurons and glial cells, J. Gen. Virol. 40:359.

Vahlne, A., Svennerholm, B., Sandberg, M., Hamberger, A., and Lycke, E., 1980, Differences in attachment between herpes simplex type 1 and type 2 viruses to neurons and glial cells, *Infect. Immun.* **28**:675.

Vahlne, A., Edstrom, S., Arstila, P., Beran, M., Ejnall, M., Nylen, O., and Lycke, E., 1981, Bell's palsy and herpes simplex virus, *Arch. Otolaryngol.* **107**:79.

Van der Merwe, C. F., and Alexander, J. J., 1982, Herpes simplex virus and duodenal ulceration, *Lancet* **2**:762.

Vestergaard, B. F., and Rune, S. J., 1980, Type specific herpes simplex virus antibodies in patients with recurrent duodenal ulcer, *Lancet* **1**:1273.

Volker-Dieben, H. J., Kok van Alphen, C. C., Schreuder, I., and D'Amaro, J., 1981, HLA antigens in recurrent corneal herpes simplex virus infection, in: *Herpetic Eye Diseases* (R. Sundmacher, ed.), p. 91, Springer-Verlag (Bergmann), Berlin.

Walz, M. A., Price, R. W., and Notkins, A. L., 1974, Latent infection with herpes simplex virus types 1 and 2: viral reactivation *in vivo* after neurectomy, *Science* **184**:1185.

Walz, M. A., Yamamoto, H., and Notkins, A. L., 1976, Immunological response restricts number of cells in sensory ganglia infected with herpes simplex, *Nature (London)* **264**:554.

Walz, M., Price, R. W., Hayashi, K., Katz, B. K., and Notkins, A. L., 1977, Effect of immunization on acute and latent infections of vaginouterine tissue with herpes simplex virus types 1 and 2, *J. Infect. Dis.* **135**:744.

Wander, A. H., Centifanto, Y. M., and Kaufman, H. E., 1980, Strain specificity of clinical isolates of herpes simplex virus, *Arch. Ophthalmol.* **98**:1458.

Warren, K. G., Gilden, D. H., Brown, S. M., Devlin, M., Wroblewska, Z., Subak-Sharpe, J., and Koprowski, H., 1977, Isolation of latent herpes simplex virus from human trigeminal ganglia including ganglia from one patient with multiple sclerosis, *Lancet* **2**:637.

Warren, K. G., Brown, S. M., Wroblewska, Z., Gilden, D., Koprowski, H., and Subak-Sharpe, J., 1978, Isolation of latent herpes simplex virus from the superior cervical and vagus ganglions of human beings, *New Engl. J. Med.* **298**:1068.

Warren, S. L., Carpenter, C. M., and Boak, R. A., 1940, Symptomatic herpes, a sequela of artificially 'induced' fever: Incidence and clinical aspects; recovery of virus from herpetic vesicles and comparison with a known strain of herpes virus, *J. Exp. Med.* **71**:155.

Watkins, J. F., 1964, Adsorption of sensitized sheep erythrocytes to HeLa cells infected with herpes simplex virus, *Nature* **202**:1364.

Watson, K., Stevens, J. G., Cook, M. L., and Subak-Sharpe, J. H., 1980, Latency competence of thirteen HSV1 temperature sensitive mutants, *J. Gen. Virol.* **49**:149.

Watson, W. E., 1974, The binding of actinomycin D to nuclei of axotomised neurons, *Brain Res.* **65**:317.

Weiss, P., Wang, H. Taylor, A. C., and Edds, M. V., 1945, Proximodistal fluid convection in endoneural spaces of peripheral nerves demonstrated by colored and radioactive (isotope) tracers, *Am. J. Physiol.* **143**:521.

Wellings, P. C., 1968, A new ocular syndrome, *Br. J. Ophthalmol.* **52**:555.

Wheeler, C. E., 1960, Further studies on the effect of neutralizing antibody upon the course of herpes simplex infections in tissue culture, *J. Immunol.* **84**:394.

Wheeler, C. E., 1975, Pathogenesis of recurrent herpes simplex infections, *J. Invest. Dermatol.* **65**:341.

Whitley, R. W., Lakeman, A. D., Nahmias, A., and Roizman, B., 1982, DNA restriction-enzyme analysis of herpes simplex virus isolates obtained from patients with encephalitis, *New Engl. J. Med.* **307**:1060.

Wigdahl, B. L., Ziegler, R. J., Sneve, M., and Rapp, F., 1983, Herpes simplex virus latency and reactivation in isolated rat sensory neurons, *Virology* **127**:159.

Wildy, P., 1967, The progression of herpes simplex virus to the central nervous system of the mouse, *J. Hyg.* **65**:173.

Wildy, P., Field, H. J., and Nash, A. A., 1982, Classical herpes latency revisited, in: *Symposium 33, Society for General Microbiology* (B. W. J. Mahy, A. C. Minson, and G. K. Darby, eds.), p. 133, Cambridge University Press, London.

Wilhelmus, K. R., Costa, D. J., Donovan, H. C., Falcon, M. G., and Jones, B. R., 1981, Prognostic indicators of herpetic keratitis, *Arch. Ophthalmol.* **99**:1578.

Wilton, J. M. A., Ivanyi, L., and Lehner, T., 1972, Cell mediated immunity in herpesvirus hominis infections, *Br. Med. J.* **1**:723.

Wohlenberg, C. R., Walz, M. A., and Notkins, A. L., 1976, Efficacy of phosphonoacetic acid on herpes simplex virus infection of sensory ganglia, *Infect. Immun.* **13**:1519.

Wohlenberg, C., Openshaw, H., and Notkins, A. L., 1979, *In vitro* system for studying efficacy of antiviral agents in preventing the reactivation of latent herpes simplex virus, *Antimicrob. Agents Chemother.* **15**:625.

Yamamoto, H., Walz, A., and Notkins, A. L., 1977, Viral-specific thymidine kinase in sensory ganglia of mice infected with herpes simplex virus, *Virology* **76**:866.

Yamamoto, T., Otani, S., and Shiraki, H., 1965, A study of the evaluation of viral infection in experimental herpes simplex encephalitis and rabies by means of fluorescent antibody, *Acta Neuropathol.* **5**:288.

Yanagi, K., Rush, M. G., and Biegeleisen, K., 1979, Integration of herpes simplex virus type 1 DNA into DNA of growth arrested BHK21 cells, *J. Gen. Virol.* **44**:657.

Youssoufian, H., Mulder, C., Hammer, S. M., and Hirsch, M. S., 1982, Methylation of the viral genome in an *in vitro* model of herpes simplex virus latency, *Proc. Natl. Acad. Sci. USA* **79**:2207.

Zenda, S. S., and Murray, B. K., 1981, cited in Frenkel (1981).

Ziegler, R. J., and Herman, R. E., 1980, Peripheral infection of rat sensory neurons in culture by herpes simplex virus, *Infect. Immun.* **28**:620.

Ziegler, R. J., and Pozos, R. S., 1981, Effects of lectins on peripheral infection by herpes simplex virus of rat sensory neurons in culture, *Infect. Immun.* **34**:588.

Zimmerman, T. J., McNeil, J. I., Richman, A., Kaufman, H., and Waltman, S. R., 1977, HLA types and recurrent corneal herpes simplex infection, *Invest. Ophthalmol. Vis. Sci.* **16**:756.

Zisman, B., Hirsch, M. S., and Allison, A. C., 1970, Selective effects of antimacrophage serum, silica and antilymphocyte serum on pathogenesis of herpes virus infection of young adult mice, *J. Immunol.* **104**:1155.

zur Hausen, H., and Schulte-Holthausen, H., 1975, Persistence of herpesvirus nucleic acid in normal and transformal human cells—A review, in: *Oncogenesis and Herpesviruses* (G. de Thé, M. A. Epstein, and H. zur Hausen, eds.), Vol. 2, p. 117, IARC, Lyon, France.

Zweerink, H. J., and Stanton, L. W., 1981, Immune response to HSV infections: Virus specific antibodies in sera from patients with recurrent facial infections, *Infect. Immun.* **31**:624.

CHAPTER 5

Herpes Simplex Viruses and Their Role in Human Cancer

I. INTRODUCTION

Herpes simplex viruses have been associated with human cancers of several sites. The great majority of studies concerned with this topic have dealt with the role of herpes simplex virus type 2 (HSV-2) in the genesis of squamous cell carcinoma of the cervix. Thus, examination of the data relating HSV-2 to cervical cancer provides the best example of the issues arising when evaluating the role of HSV in human cancers, in general.

Despite the large number of studies conducted, the role of HSV-2 in the genesis of cervical neoplasia has not been clearly defined. Much of the data from earlier studies have been reviewed (Adam et al., 1974; Aurelian, 1976; Rawls et al., 1977; Melnick and Adam, 1978; Nahmias and Sawanabori, 1978). A number of observations favor an etiologic role of the virus. Risk factors associated with developing cervical cancer are similar to those for becoming infected with HSV-2. A greater frequency of serologic evidence of past infection with HSV-2 has been repeatedly found among women with cervical cancer when compared to control women. In a prospective study, intraepithelial neoplasia of the cervix was found to occur more frequently in women infected with HSV-2 than control women. Antibodies to viral antigens expressed early in the replicative cycle have been detected more commonly among cancer patients than control women and in some instances these antibodies have correlated with tumor burden. Antibodies raised against viral antigens react with cancer cells, and viral mRNA as well as viral DNA sequences have been

WILLIAM E. RAWLS • Department of Pathology, McMaster University, Hamilton, Ontario, Canada, L8N 3Z5.

241

detected in cervical cancers. In addition, HSV-2 has been shown to transform rodent cells *in vitro* and repeated intravaginal inoculation of inactivated virus induced cervical neoplasia in mice (Wentz *et al.*, 1981). Despite these suggestive findings, a definite conclusion about the role of the virus in the genesis of the cancer is hampered by inconsistencies in the data.

Two basic questions need to be answered when attempting to evaluate the role of viruses in human cancers. First, can the risk of cancer associated with virus infection be defined? Second, can viral DNA sequences coding for proteins with putative transforming functions be found in cancer cells? Affirmative answers to these two questions strongly suggest an oncogenic role for the virus. However, for HSV-2 and cervical cancer these questions are still not answered.

II. DEFINITION OF RISK

A. Theoretical Considerations

The definition of risk requires examination of the distribution of virus infections as related to cancer cases in populations. The ease with which risk estimates can be derived from such epidemiologic studies will depend upon the possible models of causation involving the virus. This can be illustrated with the use of Table I. The risk of cervical cancer associated with HSV-2 infections would be represented by $A/(A + B)$. The virus could be necessary and sufficient, necessary but not sufficient, neither necessary nor sufficient, or not oncogenic. In the first instance, infection by the virus alone would lead to cervical cancer and no other factor would be needed. All cases of cancer and infections with HSV-2 should be represented in A of Table I. This possibility is clearly unlikely as only a small fraction of women infected with HSV-2 develop the cancer. If the virus is oncogenic, it is apparent that additional factors are needed for expression of clinical cancer.

TABLE I. Theoretical Distribution of HSV-2 Infections and Cancer Cases in a Population

Infection with HSV-2	Cervical Cancer		
	Yes	No	
Yes	A	B	$A + B$
No	C	D	$C + D$
	$A + C$	$B + D$	

The second possibility is that infection with HSV-2 is necessary but other factors must be present to promote neoplasia. In this case, A should equal $A + C$ in Table I although A will be less than $A + B$; in other words, all women with cervical cancer should have been infected with the virus but only a fraction of infected women will develop cancer. Despite the association of HSV-2 with cervical cancer by a number of techniques, evidence of viral involvement has generally been found only in a portion of the cases. It is possible that the assay techniques were not sufficiently sensitive to detect viral involvement in all cases or that a proportion of the cases were caused by genital infections with HSV-1 and thus not detected by the probes for HSV-2. However, this is an unlikely explanation, for cases of cervical cancer lacking markers for either virus have been identified. Thus, the data available suggest that HSV-2 is not a necessary factor in the causation of all cases of cervical cancer.

The third possibility is that the virus may induce some cases of cervical cancer but is not needed in all cases. The virus could represent one of several causes of cervical cancer or it could function as one of several promoting factors. In this instance, A would be less than $A + C$ in Table I and A would be less than $A + B$. The data available are compatible with this model of causation. However, a fourth possibility is that infection with HSV-2 and cervical cancer could be covariables of a yet unidentified factor(s) in which case the virus would not be directly associated with the oncogenic process.

B. Risk Estimates from Epidemiologic Studies

The genomes of HSV-1 and HSV-2 code for a number of proteins that stimulate the production of antibodies. In addition, antigenic sites on some of the proteins stimulate cell-mediated immune responses. Assays for immune reactivity to the viral antigens have been interpreted as indicating past infections with the viruses or as expression of viral antigens by transformed cells. A number of seroepidemiologic studies have been conducted in which the frequency of antibodies to HSV-2 antigens were determined among women with cervical cancer and among control women. Generally, women with cervical cancer have a greater frequency and/or higher titers of antibodies to HSV-2 than control women. Similar findings have been reported for women with severe dysplasia and carcinoma *in situ* (Thomas and Rawls, 1979; Najem *et al.*, 1982). The results of these studies suggest a greater frequency of past HSV-2 infections among cases than controls within each population studied; however, considerable variation has been found in the proportion of cases and controls with HSV-2 antibodies: 15–100% of cases were found to be seropositive while 7–71% of controls had HSV-2 antibodies. It is unlikely that differences in study design and antibody assays account for the variations in seropositivity. Rather the data suggest that the association between

HSV-2 and cervical cancer varies in different social settings. A correlation appears to exist between the proportion of cases and controls with antibodies to HSV-2 and the incidence of cervical cancer. The apparent correlation of HSV-2 antibodies with cervical cancer incidence may indicate that the virus represents one of more than one causative factors in the genesis of the cancer. A model constructed from the retrospective data and assuming multiple causation has yielded an estimated annual risk of about 50 cases of invasive cervical cancer per 100,000 infected women (Rawls and Campione-Piccardo, 1981).

The risk of cervical neoplasia associated with HSV-2 infections can be more accurately obtained by prospective studies. In such studies, the patients are usually treated when preinvasive lesions are detected. As all preinvasive lesions do not progress to invasive cancer, the results generally provide estimates of less severe neoplastic disease. A community based survey for trends in genital herpes seen by physicians revealed a secular increase in genital herpes; however, among the infected women identified, the occurrence of cervical neoplasia was no greater than that found in the general population (Chuang et al., 1983). These results contrast with those obtained in a prospective study of women with cytologic evidence of genital herpes (Naib et al., 1969; Adelusi et al., 1981). In this study, the rates of severe dysplasia, carcinoma in situ, and invasive cancer were 4-fold greater among women with evidence of HSV infections than among control women who lacked historic, cytologic, or serologic evidence of HSV-2 infections. A 12-fold greater risk was associated with primary infections occurring in women who lacked evidence of prior HSV-1 infections. Overall, about 11% of women with evidence of HSV-2 infections and followed for 1 or more years were found to develop cervical neoplasia (Adelusi et al., 1981). This rate is substantially greater than the one estimated for invasive disease from retrospective case-control studies. These observations indicate that risks of various cervical neoplastic diseases attributable to HSV-2 infections are not yet accurately defined.

Regardless of the risk of cervical neoplasia found for HSV-2 infections, data from epidemiologic investigations cannot easily distinguish between the virus being etiologic and the virus infection being a covariable of another factor involved in the genesis of the neoplasms. Repeated studies have demonstrated early age at first coitus and multiple sexual partners as risk factors for cervical cancer. A number of studies have been conducted to exclude covariability as an explanation for the association between HSV-2 and cervical cancer. These studies have included assaying for evidence of other venereally transmitted diseases (Royston and Aurelian, 1970a), selecting controls who have another venereal disease (Schneweis et al., 1975), and matching for indicators of sexual behavior such as age at first coitus and numbers of sex partners in analyzing data from case-control studies (Adam et al., 1974; Graham et al., 1982). In these studies an excess of antibodies to HSV-2 remained after controlling

TABLE II. HSV Tumor-Associated Antigens

Name	Molecular weight	Property[a]	Reference
HSV-TAA	40,000 60,000	NS; membrane-associated	Hollinshead et al. (1976)
AG-4	161,000	S; membrane-associated	Aurelian et al. (1981)
AG-e	140,000 130,000	S; cytoplasmic and nuclear	Gupta et al. (1981)
VP143	143,000	NS; cytoplasmic	Melnick et al. (1976)

[a] NS, nonstructural protein; S, structural protein of the virion.

for sexual behavior. However, the argument can be made that HSV-2 infections are a more sensitive indicator of the putative oncogenic factor than the other variables used in these studies for matching. While this appears unlikely, the recent observation that preference for owning vs. renting a home as a greater risk factor for seropositivity to HSV-2 than age at first coitus or number of sex partners (Stavraky et al., 1983) illustrates the problem of securely selecting the appropriate variables upon which to match.

III. VIRAL GENETIC INFORMATION IN CANCER CELLS

A. HSV Antigens Expressed in Cancer Cells

Several antigens have been described as possible tumor-associated viral proteins. The antigens that are produced in virus-infected cells have been defined by their preferential reactivity with sera from cervical cancer patients. Some of these antigens have been partially characterized, purified, and used to raise antisera in laboratory animals. These antisera have been found to react in immunoassays with cells obtained from cervical cancers. Table II lists the best characterized tumor-associated antigens.

One effort to identify an antigen associated with squamous carcinomas of the genital tract led to the identification of a membrane-associated antigen that reacted with sera from cervical cancer patients in a complement-fixation test (Hollinshead et al., 1972). An antigen with similar characteristics was isolated from sonicated membrane preparations of cells infected with HSV-1 or HSV-2 and the antigen has been called HSV-TAA. The antigen was partially purified by electrophoresis on polyacrylamide gels and the antigenic activity was found in the region of the gel containing two proteins with molecular weights of about 40,000 and 60,000 (Hollinshead et al., 1976). Antibodies to HSV-TAA were detected in 90% of patients with squamous cell carcinoma of various sites including the cervix. By comparison, antibodies to HSV-TAA were found in 4% of normal subjects and 11% of patients with nonsquamous cancers.

TABLE III. Occurrence of Antibodies to ICP10/AG-4 Antigen among Cervical Cancer Patients and Controls

| Location | Percentage with antibodies | | Reference |
	Cancer	Controls	
Maryland	85	12	Aurelian et al. (1973)
Pennsylvania	78	29	Notter and Docherty (1976)
North Carolina	44	0	Heise et al. (1979)
Japan	52	17	Kawana et al. (1979)
Australia	75	10–20	Arsenakis et al. (1980)
Argentina	84	0	Teyssie et al. (1980)

Among cervical cancer patients, antibodies to HSV-TAA were found with equal frequency in sera from those with active disease and those successfully treated. Antibodies to HSV-TAA prepared in rabbits stained cancer cells in two of four specimens from patients with squamous cell carcinoma (Notter et al., 1978).

AG-4 is the designation given to an antigen produced early after infection of cells with HSV-2. Studies have revealed the antigen to be a structural protein with a molecular weight of about 161,000. The antigen is expressed as an "immediate–early" protein that is made in the cytoplasm and becomes membrane associated (see Table II). Antibodies to the antigen have been detected by a quantitative complement-fixation test. In humans the antibodies are of the IgM class and have been detected transiently following primary HSV-2 infections. Antibodies to AG-4 have been found in 44 to 85% of patients with cervical cancer and in 0 to 29% of control women (Table III). Unlike HSV-TAA, antibodies have been found associated only with cervical cancer and not with squamous cell carcinomas of other sites. The antibodies have been detected in patients with tumors but are undetectable following successful therapy. Evidence has also been presented suggesting that this antigen is expressed in cervical cancer cells (Aurelian et al., 1981).

AG-e is an antigen preparation described by Bell et al. (1978). Two viral proteins of molecular weight 140,000 and 130,000 were identified in antigen preparations that were reactive in an assay of in vitro cell mediated immunity. In a leukocyte migration-inhibition assay, lymphoid cells from 75% of patients with cervical neoplasia reacted with AG-e while cells from 13% of matched control women and from 16% of patients with other neoplasms were reactive. Similar observations have been reported using antigens prepared from HSV-2-infected or uninfected ME-180 cells, a cell line established from a cervical cancer (Rivera et al., 1979). Antisera raised in rabbits to purified components of AG-e were found to stain cells exfoliated from cervical neoplastic lesions (Gupta et al., 1981).

A major nonstructural protein induced early in the replicative cycle of HSV-2 has been found to be preferentially precipitated by sera from patients with cervical cancer (Melnick *et al.*, 1976). The protein has a molecular weight of 143,000 and was shown to be a DNA-binding protein (Powell *et al.*, 1981). Antibodies to VP143 were quantitated by reacting radiolabeled antigen with human sera, precipitating the antigen–antibody complexes with goat anti-human IgG, and analyzing the precipitate by polyacrylamide gel electrophoresis. Quantitative differences were found in antibodies to this antigen between women with cervical cancer and control women (Melnick *et al.*, 1976). Antisera to purified VP143 have been shown to react with cells from cervical and vulvar neoplasms (Dreesman *et al.*, 1980; Kaufman *et al.*, 1981; Cabral *et al.*, 1983).

Additional antigens have been described but are less well characterized. Antibodies to an unstable antigen induced by HSV-1 and HSV-2 were found in patients with a variety of malignancies including cancer of the cervix (Sabin and Tarro, 1973). Evidence has been presented suggesting that this or a related antigen may be a 70,000-dalton glycoprotein (Cocchiara *et al.*, 1980). Gilman *et al.*, (1980) precipitated radiolabeled viral proteins with sera from cervical cancer patients and control women. The precipitates were analyzed by polyacrylamide gel electrophoresis and the frequency with which the individual proteins were precipitated was determined. Two proteins with molecular weight 118,000 and 38,000 were precipitated more commonly from cancer patients than from controls. The smaller protein has been of particular interest as a molecule of similar size was shown to be encoded in a fragment of HSV-2 DNA that is capable of transforming rodent cells *in vitro* (Docherty *et al.*, 1981). The protein encoded in this DNA fragment appears to be related to the virus ribonucleotide reductase, which could enhance mutagenesis (Huszar and Bacchetti, 1983; Galloway and McDougall, 1983). However, there is no direct evidence that the protein precipitated by the human sera and the protein encoded by the DNA fragment are the same, nor has evidence for the expression of this antigen in cancer cells been reported.

A number of investigators have presented evidence suggesting that HSV-2 antigens are expressed in cervical cancer cells. The studies have included examining cancer cells exfoliated from the cervical lesions as well as biopsy specimens. In the initial report by Royston and Aurelian (1970b), rabbit antisera to HSV-2 antigens prepared in HEP-2 cells reacted with more cells exfoliated from neoplastic lesions than from nonneoplastic lesions. A similar study revealed positive cells in 94% of patients with invasive cancer, in 65% of patients with dysplasia, in 41% of patients with nonneoplastic cervical lesions, and in 9% of normal women (Pasca *et al.*, 1976). Among women in Ibadan, Nigeria, cells from all 22 patients with cervical cancer reacted with antiserum to HSV-2 as compared to 0 of 24 control women (Adelusi *et al.*, 1976). Nahmias *et al.* (1975) detected antigen in 25 of 40 biopsy samples of cervical cancers

TABLE IV. Detection of Viral Antigens in Cervical Neoplastic Cells

Viral antigen	Percent of specimens with positive cells			Reference
	Dysplasia	CIS	Invasive cancer	
ICSP11–12[a]	20	40	40	Dreesman et al. (1980)
ICSP11–12	32	12	20	McDougall et al. (1982)
ICSP11–12	31	29	41	Cabral et al. (1983)
ICP12,14 (AG-e)	68	77	94	Gupta et al. (1981)

[a] Similar reaction patterns observed for antiserum to ICSP 34–35.

using anti-HSV-2 antiserum in an anticomplement immunofluorescence test.

More recently, antisera to purified proteins induced by HSV-2 have been used to probe for viral antigens in cancer cells (Table IV). Antiserum to purified HSV-TAA was found to react with cancer cells in two of four specimens examined (Notter et al., 1978). Antisera to a DNA-binding protein (VP143 or ICSP11–12) and to protein ICSP34–35 stained cells in about 40% of specimens of invasive cancer or carcinoma in situ (Dreesman et al., 1980). Similar results were obtained in another study using antiserum to ICSP11–12; staining was observed in 30, 12, and 20% of biopsies from patients with dysplasia, carcinoma in situ, and invasive carcinoma, respectively (McDougall et al., 1982). In another study, antiserum to this protein reacted with 31, 29, and 40% of biopsies from patients with severe dysplasia, carcinoma in situ, and invasive carcinoma, respectively. These specimens did not react with antiserum to a purified major capsid protein of the virus (Cabral et al., 1983). In addition, antiserum to purified AG-e was found to react with exfoliated cells from 62, 75, 77, and 94% of patients with mild, moderate, and marked atypia and invasive cancer, respectively (Gupta et al., 1981).

B. HSV DNA Sequences in Cancer Cells

The data from the above-described studies of clinical material suggest that a number of viral proteins are expressed in cervical cancer cells. Assuming the immunospecific antisera were indeed detecting virus-specific proteins, genes encoding HSV-TAA, AG-4, the two components of AG-e, and two DNA-binding proteins should be present in the tumor cells of a portion of the cervical cancer cases. An experimental approach has also been explored to define viral proteins that should be expressed in cervical cancer cells. HSV-1 and HSV-2 are capable of transforming cells in vitro (Hayward and Reyes, 1983). Two regions of the HSV-2 genome have been implicated in such transformation studies. These regions

TABLE V. HSV-2 DNA Sequences in Cervical Cancer DNA

No. tested	No. positive	Percent positive	Reference
9	3	33	Galloway and McDougall (1983)
8	1	12	Park et al. (1983)
8	2	25	Bacchetti et al. (unpublished data)
25	6	24	

map at 0.41–0.58 genome map units (MU) (Jariwalla et al., 1980) and at 0.58–0.62 MU (Reyes et al., 1979; Galloway and McDougall, 1981). For HSV-1, only a single DNA region located at 0.31–0.41 MU has been associated with transformation (Camacho and Spear, 1978; Reyes et al., 1979). Segment 0.31–0.41 MU can be considered transforming region 1 (TR1); segment 0.58–0.62 MU, transforming region 2 (TR2); and segment 0.41–0.58 MU, transforming region 3 (TR3). Theoretically, one or more of these segments of the viral genome should be present in cervical cancer cells transformed by HSV if the mechanism of transformation by these viruses is similar to that defined for other DNA viruses. Proteins encoded in these segments of the genomes should be expressed in the cancer cells.

Several groups have reported detecting HSV-specific RNA in cervical cancer cells by in situ hybridization (Eglin et al., 1981; Maitland et al., 1981; McDougall et al., 1982). From 35 to 67% of tumors examined were positive for HSV-specific RNA. The RNA transcripts expressed in the tumor cells were mapped primarily to three regions of the genome: 0.07–0.40, 0.58–0.63, and 0.82–0.85 MU. Cellular sequences homologous to the regions of the HSV genome, especially the short repeat regions, have been demonstrated (Peden et al., 1982; Puga et al., 1982); thus, the transcripts, especially those mapped at 0.82–0.85 MU, may not be virus-specific. In these studies, no set of sequences were invariably expressed in all positive specimens.

Frenkel et al. (1972) first reported finding HSV-2 DNA in a cervical cancer specimen. This specimen contained 39% of the viral genome and an estimated 3.5 copies of viral DNA were present per diploid cell. Studies of hybridization kinetics suggested the mRNA complementary to 5% of the viral genome was expressed in the cells. More recently, DNA from cervical cancer tissue has been examined for viral DNA by Southern blot hybridization using fragments of the HSV-2 genome as probes. The results of these studies are summarized in Table V. Three of nine invasive cancer tumors were found to contain sequences hybridizing with viral DNA probes (Galloway and McDougall, 1983). One sample was positive for sequences represented in the viral fragment 0.58–0.63 MU (TR2), a second for a viral probe representing 0.32–0.40 MU (TR1), and a third with sequences represented by both of these regions. In a second study, one of eight specimens was positive and the reaction was obtained with probes representing 0.58–0.61 MU (TR2) (Park *et al.*, 1983). Two of eight samples were found to contain viral DNA sequences in our laboratory (S. Bacchetti, unpublished observations). In one, the viral DNA fragments representing 0.31–0.42 MU (TR1) hybridized to the DNA obtained from a cervical cancer biopsy. However, in the second specimen, hybridization with fragments outside of the putative transforming regions was observed. Thus, 6 of 25 (24%) specimens of invasive cervical cancer were found to contain HSV-2 sequences in these three preliminary studies. Again, sequences representing any single region of the genome were not consistently observed in the positive specimens.

Appraisal of the evidence indicating that viral DNA sequences are present in neoplastic cells of the cervix is limited by the sensitivity and specificity of the assays used. The probles for HSV-2 DNA sequences are sufficiently sensitive to detect fragments equivalent to 1.5% of the viral genome at a level of 0.5 copy/cell. Thus, biopsies containing a low ratio of neoplastic to normal stromal cells may be scored as false-negatives (Park *et al.*, 1983). This problem is less likely when *in situ* hybridization is used to detect viral RNA; however, the specificity of this assay is harder to assess (Maitland *et al.*, 1981). Studies conducted with antisera to purified virus-specified proteins also permit evaluation of individual cells, but the monospecificity of the antiserum is difficult to guarantee. Host cell proteins preferentially expressed in transformed cells and in cells infected by the virus may copurify with the viral protein or may share antigenic determinants with the viral proteins. It is of interest that the percentages of specimens positive for viral RNA by *in situ* hybridization assays, the percentages of specimens with antigens detected by antiserum to ICSP11–12 (McDougall *et al.*, 1982), and the percentages of biopsies with viral DNA sequences are similar. However, the sequences of DNA found in the biopsies have not necessarily correlated with the region of the genome coding for the DNA-binding protein (Galloway and McDougall, 1983).

IV. COMMENTS

There is clearly an association between HSV-2 and cervical cancer, but the significance of this association remains unclear. Data from the epidemiologic studies suggest an increased risk of cervical cancer among women infected with HSV-2 but this risk has not been clearly defined. Evidence suggesting HSV-2 DNA sequences, RNA, and antigens in cervical cancer cells has also been presented. However, the regions of the genome supposedly detected by assays for these three types of molecules do not necessarily correspond. Studies in which viral DNA, RNA, and antigens are sought in the same specimens will be required to clarify these apparent discrepancies. These findings do suggest that it is unlikely that a single set of HSV-2 genes must be continuously expressed for cervical epithelial cells to become malignant. If the virus is involved in on-cogenesis, the transformation would have to be mediated through a "hit and run" mechanism (Galloway and McDougall, 1983) or the virus could act synergistically with other factors as an initiator or promoter of cancer development (zur Hausen, 1982). In these instances, virus-induced proteins might enhance mutation rates (Huszar and Bacchetti, 1983) during abortive infections of epithelial cells. Part or all of the HSV-2 genome could be lost upon subsequent replication cycles of the cells.

Several conclusions can be drawn from the available data, assuming that the assays for viral involvement in cervical cancer are reasonably sensitive and specific. First, oncogenic HSV-2 appears to be involved in the genesis of some but not all cases of cervical cancer, for substantial proportions of cases studied do not possess viral markers. Second, the proportion of cases related to HSV-2 appears to vary in different populations. Conservative estimates based on seroepidemiologic studies and investigations searching for viral antigens, RNA, and DNA in cancer cells suggest 20–40% of cases may be related to HSV-2. Recently, papillomavirus DNA has been found in DNA extracted from cervical cancer biopsies (Durst et al., 1983). A number of types of human papillomaviruses have been delineated and DNA sequences of types 6, 11, or 16 have been found in 72% of samples of genital tumors obtained from patients living in Germany while 44% of samples obtained from patients living in Kenya and Brazil reacted with viral probes. These observations implicate papillomaviruses as etiologic agents in at least a portion of cervical cancer cases. It has been postulated that infections with HSV-2 and papillomavirus may act synergistically in the genesis of the cancer (zur Hausen, 1982). It is also possible that the viruses may act independently with HSV-2 contributing to the genesis of some cases of cancer while papillomaviruses contribute to the genesis of other cases. Thus, if both HSV-2 and papillomaviruses are oncogenic, cervical cancers may be heterogeneous with respect to etiology.

The heterogeneity of cervical cancers has been argued from epidemiologic evidence; however, markers clearly differentiating different forms of cancer have not been defined. There are data suggesting that evidence of HSV-2 infections may have prognostic significance. In one study, the progression of cervical intraepithelial neoplasia was found to correlate with the presence of HSV-2 antibodies (Coleman et al., 1983). This suggests that intraepithelial neoplasia associated with an HSV-2 infection is more likely to progress to invasive cancer than disease in women not infected with HSV-2. It has also been observed that pretreatment antibody activity to HSV-2 antigens was greater among women with invasive cervical cancer who survived than among women who did not survive (Christenson, 1982). This could indicate that antibody activity correlated with the general competence of the host defense systems, or that the antibodies had a role in controlling tumor progression. Another possibility is that tumors associated with HSV-2 infections are less in-

252   WILLIAM E. RAWLS

vasive than tumors induced by other agents and, therefore, are associated with a better prognosis. Clearly, additional studies using defined markers of HSV-2 and papillomavirus will be needed to determine the significance of these viruses in the genesis of cervical cancer. It is likely that an understanding of the role of these viruses in cervical cancer will provide insight into the etiologic factors of squamous carcinomas of other anatomic sites.

REFERENCES

Adam, E., Rawls, W. E., and Melnick, J. L., 1974, The association of herpesvirus type 2 infection and cervical cancer, Prev. Med. 3:122.

Adelusi, B., Osunkoya, B. O., and Fabiyi, A., 1976, Herpes type 2 virus antigens in human cervical carcinoma, Obstet. Gynecol. 47:545.

Adelusi, B., Naib, Z., Muther, J., and Nahmias, A., 1981, Epidemiological studies relating genital herpes simplex virus (HSV) infection with cervical neoplasia, in: The Human Herpesviruses: An Interdisciplinary Perspective (A. J. Nahmias, W. R. Dowdle, and R. F. Schinazi, eds.), p. 627, Elsevier/North-Holland, Amsterdam.

Arsenakis, M., Georgiou, G. M., Walsh, J. K., Cauchi, M. N., and May, J. T., 1980, AG-4 complement-fixing antibodies in cervical cancer and herpes-infected patients using local herpes simplex virus type 2, Int. J. Cancer 25:67.

Aurelian, L., 1976, Sexually transmitted cancers? The case for genital herpes, J. Am. Vener. Dis. Assoc. 2:10.

Aurelian, L., Schumanna, B., Marcus, R. L., and Davis, H. J., 1973, Antibodies to HSV-2 induced tumor specific antigen in serum from patients with cervical carcinoma, Science 181:161.

Aurelian, L., Kessler, I. I., Rosenshein, N. B., and Barbour, G., 1981, Viruses and gynecologic cancers: Herpesvirus protein (ICP 10/AG-4), a cervical tumor antigen that fulfills the criteria for a marker of carcinogenicity, Cancer 48:455.

Bell, R. B., Aurelian, L., and Cohen, G. H., 1978, Proteins of herpesvirus type 2. IV. Leukocyte inhibition responses to type common antigen(s) in cervix cancer and recurrent herpetic infections, Cell. Immunol. 41:86.

Cabral, G. A., Fry, D., Marciano-Cabral, F., Lumpkin, C., Mercer, L., and Goplerud, D., 1983, A herpesvirus antigen in human premalignant and malignant cervical biopsies and explants, Am. J. Obstet. Gynecol. 145:79.

Camacho, A., and Spear, P. G., 1978, Transformation of hamster embryo fibroblasts by a specific fragment of the herpes simplex virus genome, Cell 15:993.

Christenson, B., 1982, Herpes virus-related antigens in herpes simplex virus type 2-transformed cells in the course of cervical carcinoma, Eur. J. Cancer Clin. Oncol. 18:1345.

Chuang, T. Y., Su, D., Perry, H. O., Ilstrup, D. M., and Kurland, L. T., 1983, Incidence and trend of herpes progenitalis: A 15 year population study, Mayo Clin. Proc. 58:436.

Cocchiara, R., Tarro, G., Flaminio, G., DiGioia, M., Smeraglia, R., and Geraci, D., 1980, Purification of herpes simplex virus tumor associated antigen from human kidney carcinoma, Cancer 46:1594.

Coleman, D. V., Morse, A. R., Beckwith, P., Anderson, M. C., Gardner, S. D., Knowles, W. A., and Skinner, G. R. B., 1983, Prognostic significance of herpes simplex virus antibody status in women with cervical intraepithelial neoplasia (CIN), Br. J. Obstet. Gynaecol. 90:421.

Docherty, J. J., Subak-Sharpe, J. H., and Preston, C. M., 1981, Identification of a vrius-specific polypeptide associated with a transforming fragment (Bgl II-N) of herpes simplex virus type 2 DNA, J. Virol. 40:126.

Dreesman, G. R., Burek, J., Adam, E., Kaufman, R. H., Melnick, J. L., Powell, K. L., and Purifoy, D. J. M., 1980, Expression of herpes-virus-induced antigens in human cervical cancer, *Nature* **283**:591.

Durst, M., Gissmann, L., Ikenberg, H., and zur Hausen, H., 1983, A papillomavirus DNA from a cervical carcinoma and its prevalence in cancer biopsy samples from different geographic regions, *Proc. Natl. Acad. Sci. USA* **80**:3812.

Eglin, R. P., Sharp, F., MacLean, A. B., Macnab, J. C. M., Clements, J. B., and Wilkie, N. M., 1981, Detection of RNA complementary to herpes simplex virus DNA in human cervical squamous cell neoplasms, *Cancer Res.* **41**:3597.

Frenkel, N., Roizman, B., Cassai, E., and Nahmias, A., 1972, A DNA fragment of herpes simplex 2 and its transcription in human cervical cancer tissue, *Proc. Natl. Acad. Sci. USA* **69**:3784.

Galloway, D. A., and McDougall, J. K., 1981, Transformation of rodent cells by a cloned DNA fragment of herpes simplex virus type 2, *J. Virol.* **38**:749.

Galloway, D. A., and McDougall, J. K., 1983, The oncogenic potential of herpes simplex viruses: Evidence for a "hit and run" mechanism, *Nature* **301**:21.

Gilman, S. C., Docherty, J. J., Clarke, A., and Rawls, W. E., 1980, Reaction patterns of herpes simplex virus type 1 and type 2 proteins with sera of patients with uterine cervical carcinoma and matched controls, *Cancer Res.* **40**:4640.

Graham, S., Rawls, W., Swanson, M., and McCurtis, J., 1982, Sex partners and herpes simplex virus type 2 in the epidemiology of cancer of the cervix, *Am. J. Epidemiol.* **115**:729.

Gupta, P. K., Aurelian, L., Frost, J. K., Carpenter, J. M., Klacsmann, K. T., Rosenshein, N. B., and Tyrer, H. W., 1981, Herpesvirus antigens as markers for cervical cancer, *Gynecol. Oncol.* **12**:S232.

Hayward, G. S., and Reyes, G. R., 1983, Biochemical aspects of transformation by herpes simplex viruses, *Adv. Viral Oncol.* **3**:271.

Heise, E. R., Kucera, L. S., Raben, M., and Homesley, H., 1979, Serological response patterns to herpesvirus type 2 early and late antigens in cervical carcinoma patients, *Cancer Res.* **39**:4022.

Hollinshead, A., Lee, O. B., McKelway, W., Melnick, J. L., and Rawls, W. E., 1972, Reactivity between herpesvirus type 2-related soluble cervical tumor cell membrane antigens and matched cancer and control sera, *Proc. Soc. Exp. Biol. Med.* **141**:688.

Hollinshead, A. C., Chretien, P. B., Lee, O. B., Tarpley, J. L., Kerney, S. E., Silverman, N. A., and Alexander, J. C., 1976, *In vivo* and *in vitro* measurements of the relationship of human squamous carcinomas to herpes simplex virus tumor-associated antigens, *Cancer Res.* **36**:821.

Huszar, D., and Bacchetti, S., 1983, Is ribonucleotide reductase the transforming function of herpes simplex virus 2?, *Nature (London)* **302**:76.

Jariwalla, R. J., Aurelian, L., and Ts'o, P. O. P., 1980, Tumorigenic transformation induced by a specific fragment of DNA from herpes simplex virus type 2, *Proc. Natl. Acad. Sci. USA* **77**:2279.

Kaufman, R. H., Dreesman, G. R., Burek, J., Korhonen, M. O., Matson, D. O., Melnick, J. L., Powell, K. L., Purifoy, D. J. M., Courtney, R. J., and Adam, E., 1981, Herpesvirus-induced antigens in squamous-cell carcinoma in situ of the vulva, *N. Engl. J. Med.* **305**:483.

Kawana, T., Sakamoto, S., Kasamatsu, T., and Aurelian, L., 1978, Frequency of anti-AG-4 antibody in patients with uterine cervical cancer and controls. *Gann* **69**:589.

McDougall, J. K., Crum, C. P., Fenoglio, C. M., Goldstein, L. C., and Galloway, D. A., 1982, Herpesvirus-specific RNA and protein in carcinoma of the uterine cervix, *Proc. Natl. Acad. Sci. USA* **79**:3853.

Maitland, N. J., Kinross, J. H., Busuttil, A., Ludgate, S. M., Smart, G. E., and Jones, K. W., 1981, The detection of DNA tumor virus-specific RNA sequences in abnormal human cervical biopsies by in situ hybridization, *J. Gen. Virol.* **55**:123.

Melnick, J. L., and Adam, E., 1978, Epidemiological approaches to determining whether herpesvirus is the etiological agent of cervical cancer, *Prog. Exp. Tumor Res.* **21**:49.

Melnick, J. L., Courtney, R. J., Powell, K. L., Schaffer, P. A., Benyesh-Melnick, M., Dreesman, G. R., Anzai, T., and Adam, E., 1976, Studies on herpes simplex virus and cancer, *Cancer Res.* **36**:845.

Nahmias, A. J., and Sawanabori, S., 1978, The genital herpes–cervical cancer hypothesis—10 years later, *Prog. Exp. Tumor Res.* **21**:117.

Nahmias, A. J., Del Buono, I., and Ibrahim, I., 1975, Antigenic relationship between herpes simplex viruses, human cervical cancer and HSV-associated hamster tumours, in: *Oncogenesis and Herpesviruses II* (G. de Thé, M. A. Epstein, and H. zur Hausen, eds.), Part I, pp. 309–313, International Agency for Research on Cancer, Lyon, France.

Naib, Z. M., Nahmias, A. J., Josey, W. E., and Kramer, J. H., 1969, Genital herpetic infection: Association with cervical dysplasia and carcinoma, *Cancer* **23**:940.

Najem, S. N., Barton, I. G., Al-Omar, L. S., and Potter, C. W., 1982, Antibody to *Herpesvirus hominis* in patients with carcinoma of the cervix, *Br. J. Exp. Pathol.* **63**:485.

Notter, M. F. D., and Docherty, J. J., 1976, Comparative diagnostic aspects of herpes simplex virus tumor associated antigens, *J. Natl. Cancer Inst.* **57**:483.

Notter, M. F. D., Docherty, J. J., Mortel, R., and Hollinshead, A. C., 1978, Detection of herpes simplex virus tumor-associated antigen in uterine cervical cancer tissue, *Gynecol. Oncol.* **6**:574.

Park, M., Kitchner, H. C., and Macnab, J. C. M., 1983, Detection of herpes simplex virus type 2 DNA restriction fragments in human cervical carcinoma tissue, *EMBO J.* **2**:1029.

Pasca, A. S., Kummerlander, L., Pejtsik, B., Krommer, K., and Pali, K., 1976, Herpes simplex virus-specific antigens in exfoliated cervical cells from women with and without cervical anaplasia, *Cancer Res.* **36**:2130.

Peden, K., Mounts, P., and Hayward, G. S., 1982, Homology between mammalian cell DNA sequences and human herpesvirus genomes detected by a hybridization procedure with high-complexity probe, *Cell* **31**:71.

Powell, K. L., Littler, E., and Purifoy, D. J. M., 1981, Nonstructural proteins of herpes simplex virus. II. Major virus-specific DNA-binding protein, *J. Virol.* **39**:894.

Puga, A., Cantin, E. M., and Notkins, A. L., 1982, Homology between murine and human cellular DNA sequences and the terminal repetition of S component of herpes simplex virus type 1 DNA, *Cell* **31**:81.

Rawls, W. E., and Campione-Piccardo, J., 1981, Epidemiology of herpes simplex virus type 1 and type 2 infections, in: *The Human Herpesviruses: An Interdisciplinary Perspective* (A. J. Nahmias, W. R. Dowdle, and R. E. Schinazi, eds.), pp. 137–152, Elsevier/North-Holland, Amsterdam.

Rawls, W. E., Bacchetti, S., and Graham, F. L., 1977, Relation of herpes simplex virus to human malignancies, *Curr. Top. Microbiol. Immunol.* **77**:71.

Reyes, G. R., La Femina, R., Hayward, S. D., and Hayward, G. S., 1979, Morphological transformation by DNA fragments of human herpesviruses: Evidence for two distinct transforming regions in herpes simplex virus types 1 and 2 and lack of correlation with biochemical transfer of the thymidine kinase gene, *Cold Spring Harbor Symp. Quant. Biol.* **44**:629.

Rivera, E. S., Hersh, E. M., Bowen, J. M., Barnett, J. W., Wharton, T., and Murphy, S. G., 1979, Leukocyte migration inhibition assay of tumor immunity in patients with cervical squamous cell carcinoma, *Cancer* **43**:2297.

Royston, I., and Aurelian, L., 1970a, The association of genital herpesvirus with cervical atypia and carcinoma *in situ*, *Am. J. Epidemiol.* **91**:531.

Royston, I., and Aurelian, L., 1970b, Immunofluorescent detection of herpesvirus antigens in exfoliated cells from human cervical carcinoma, *Proc. Natl. Acad. Sci. USA* **67**:204.

Sabin, A. B., and Tarro, G., 1973, Herpes simplex and herpes genitalis viruses in etiology of some human cancers, *Proc. Natl. Acad. Sci. USA* **70**:3225.

Schneweis, K. E., Haag, A., Lehmkoster, A., and Koenig, V., 1975, Sero-immunological investigations in patients with cervical cancer: Higher rate of HSV-2 antibodies than in syphilis patients and evidence of IgM antibodies to an early HSV-2 antigen, in: *On-*

cogenesis and Herpesviruses II (G. de Thé, M. A. Epstein, and H. zur Hausen, eds.), Part II, pp. 53–57, International Agency for Research on Cancer, Lyon, France.

Stavraky, K. M., Rawls, W. E., Chiavetta, J., Donner, A. P., and Wanklin, J. M., 1983, Sexual and socioeconomic factors affecting the risk of past infections with herpes simplex virus type 2, *Am. J. Epidemiol.* **118:**109.

Teyssie, A. R., De Holstein, A. B., Alonio, L., DiStefano, A., Lucero, J., Pasqualini, R. S., and De Torres, R. A., 1980, Detection of herpes simplex associated antigens and antibodies in patients with cervical carcinoma in Buenos Aires, *Cell Mol. Biol.* **26:**123.

Thomas, D. B., and Rawls, W. E., 1979, Relationship of herpes simplex virus type 2 antibodies and squamous dysplasia to cervical carcinoma *in situ*, *Cancer* **42:**2716.

Wentz, W. B., Reagan, J. W., Heggie, A. D., Fu, Y.-S., and Anthony, D. D., 1981, Induction of uterine cancer with inactivated herpes simplex virus, types 1 and 2, *Cancer* **48:**1783.

zur Hausen, H., 1982, Human genital cancer: Synergism between two virus infections or synergism between a virus infection and initiating events?, *Lancet* **2:**1370.

CHAPTER 6

Transforming Potential of Herpes Simplex Viruses and Human Cytomegalovirus

Mary J. Tevethia

I. INTRODUCTION

The herpesviruses comprise one of the most diverse groups of DNA-containing viruses. Although the details of their replication cycles significantly differ, viruses within this family share not only physical characteristics but also the same range of host cell–virus interactions. Infection of cells in culture with any of the human herpesviruses leads to one of three biological consequences: lytic growth, latency or persistence, or, less frequently, acquisition of a transformed phenotype. At the level of the infected host these interactions are manifested as pathogenesis, a reservoir for recurrent infection and potential for malignancy. At least four human herpesviruses, herpes simplex virus types 1 (HSV-1) and 2 (HSV-2), human cytomegalovirus (HCMV), and Epstein–Barr virus (EBV), have been implicated in the etiology of human neoplasia.

The relationship of the herpesviruses to human malignancies has been reviewed in depth (Rawls *et al.*, 1977; Rapp, 1980a,b; Aurelian *et al.*, 1981; Rapp and Jenkins, 1981; Rapp and Howett, 1983, 1984; Rapp and Robbins, 1984). This relationship and the repeatedly documented ability of each of the five human herpesviruses to promote growth transformation of cells in culture provide the foundation for classifying the herpesviruses among oncogenic viruses.

MARY J. TEVETHIA • Department of Microbiology and Cancer Research Center, The Pennsylvania State University College of Medicine, Hershey, Pennsylvania 17033.

Three features of the transforming interactions of herpesviruses distinguish them from other oncogenic DNA-containing viruses. First, with the exception of EBV, injection of human herpesviruses into experimental animals has not led to the appearance of malignant, invasive cancer cells. Second, no new antigenically identifiable polypeptide(s) (tumor antigens) has as yet been detected universally in cells transformed in culture or in tumors derived from transformed cells. Third, transformation of cells in culture by herpesviruses occurs at very low frequency.

The failure of the herpesviruses to fit within the conceptual framework of an oncogenic DNA virus established by the polyomaviruses, simian virus 40 (SV40) and mouse polyomavirus, and adenoviruses has retarded progress in understanding the oncogenic properties of these human pathogens for a number of years.

Recent advances in molecular technology as well as the ability to introduce exogenous DNA into cells efficiently (Graham and van der Eb, 1973) have permitted identification of specific transforming sequences in a variety of tumor viruses (reviewed in Weinberg, 1981; Tooze, 1981, 1983; Cooper, 1982). Such studies reveal the diversity of mechanisms by which viruses transform cells and in addition have both more clearly defined the functions of the well-known transforming proteins of polyomavirus and also demonstrated cooperation among virus-coded proteins in expression of transformed cell phenotypes. These technical and conceptual advances as well as the current epidemic proportions of HSV infections and association of HCMV with new or previously rare aggressive and fatal diseases (Boldogh *et al.*, 1981; Fenoglio *et al.*, 1982; Johnson *et al.*, 1982; Lawrence, 1982, Morbidity and Mortality Weekly Reports, 1982a,b) have stimulated expanded investigation into oncogenesis by these viruses.

This chapter discusses recently published evidence of the transforming properties of the three human herpesviruses HSV-1, HSV-2, and HCMV in relation to previous findings and to transformation by other extensively studied oncogenic viruses. For additional detail, the reader is referred to a number of excellent recent reviews on transformation by herpesviruses (Rapp, 1980c; Tooze, 1981; Hampar, 1981; Knipe, 1982).

II. TRANSFORMING POTENTIAL OF HSV

A. Morphological Transformation of Cells in Culture by HSV-1 and HSV-2 Virions

The ability of HSV to transform cells in culture morphologically has been documented extensively. Both the experimental procedures used and the frequencies of transformation obtained are relevant to discussions of possible mechanisms by which herpesviruses transform cells and justify a brief review.

The oncogenic potential of HSV-2 was first demonstrated by Duff and Rapp (1971a). These investigators identified foci of cells that appeared morphologically transformed after infection of hamster embryo fibroblasts (HEF) with HSV-2 strain 333 virions. The highly lytic nature of the virus required that the infectivity of the virions be reduced drastically, in this case by UV irradiation, in order to ensure survival of the cell. Morphologically distinguishable foci appeared at a frequency of approximately 0.25 focus/10^6 cells. One colony was expanded into a cell line. This line, 333-8-9, remains the most extensively studied HSV-2 transformant and has served as a positive control cell line for numerous other studies.

Subsequently, HSV-1 virions after similar treatment also were shown to transform HEF morphologically (Duff and Rapp, 1973a). In these experiments infected cells were passaged after infection so that quantitation is not precise. Nonetheless, a minimal value of 1 transformed focus per 10^6 infected cells can be estimated based on the experimental protocol. Cell lines established from transformation assays with HSV-2 and HSV-1 were examined for expression of virus-specific proteins and for oncogenicity in syngeneic animals. Predominately two assays for viral antigens were utilized. These included immunofluorescence assays on fixed transformed cells using antisera raised against HSV-1 and HSV-2, and the ability of sera from animals bearing tumors after injection of HSV-1- or HSV-2-transformed cells to neutralize the infectivity of the virus. Both in the case of HSV-2-transformed (Duff and Rapp, 1971a) and HSV-1-transformed (Duff and Rapp, 1973a) cell lines, diffuse cytoplasmic fluorescence in 1–10% of the cells was observed. Injection of newborn hamsters with HSV-2 333-8-9 cells and with HSV-1 14-012-transformed cells led to the formation of tumors in 11/30 and 18/38 animals, respectively. Sera from hamsters bearing these tumors neutralized the infectivity of HSV and reacted with HSV-1 as well as HSV-2-infected or transformed cells in immunofluorescence assays (Duff and Rapp, 1973a,b).

Since these initial reports a variety of cell types including primary embryonic cells of rat, hamster, and human origin, as well as continuous hamster and mouse cell lines and numerous methods of inactivation of infectivity have been used in transformation assays with HSV. Colonies picked from HEF infected with virions of HSV-1 and HSV-2 photodynamically inactivated with heterocyclic dyes and light (Rapp et al., 1973) or UV irradiation (Duff and Rapp, 1971a,b, 1973a; Rapp and Duff, 1973) or with HSV-1 or HSV-2 temperature-sensitive (ts) mutants at elevated temperature (Takahashi and Yamanishi, 1974; Kimura et al., 1975) have yielded transformed cell lines. Similarly, a continuous hamster cell line persistently infected with UV-irradiated HSV-2 generated transformed cell lines after treatment with anti-HSV-2 sera, incubation at elevated temperature (40°C), or both (Kutinova et al., 1973). An independent clonally derived continuous hamster cell line also yielded transformed foci after exposure to UV-irradiated HSV-2 (Hampar et al., 1980). Rat embryo cells infected with photoinactivated HSV-1 (Macnab, 1974), HSV-2 (Macnab, 1974; Kucera et al., 1977), or with untreated HSV-2 and five HSV-2 ts mutants (Macnab, 1974) at 38.5°C also produced transformed foci. In addition, UV-inactivated HSV-2 transformed continuous lines of BALB/c (Boyd and Orme, 1975) and Swiss 3T3 mouse cells (Duff and Rapp, 1975).

Human cells of embryonic origin were also shown to display transformed characteristics following infection with HSV-2 and extended maintenance at 42°C (Darai and Munk, 1973; Munk and Darai, 1973) and one transformed cell line was established after infection of human embryonic lung cells with the HSV-2 mutant ts9 at elevated temperature (Takahashi and Yamanishi, 1974). In each of the cases mentioned, evidence for the persistence of viral information in transformants was documented serologically by demonstrating the presence of cytoplasmic (Duff and Rapp, 1971a,b; Darai and Munk, 1973; Munk and Darai, 1973; Duff and Rapp, 1973a; Rapp et al., 1973; Kutinova et al., 1973; Macnab, 1974, 1979; Takahashi and Yamanishi, 1974) or membrane-associated HSV-specific antigens (Duff and Rapp, 1971a, 1973a; Kutinova et al., 1973; Macnab, 1974; Takahashi and Yamanishi, 1974; Kimura et al., 1975; Kucera et al., 1977) with hyperimmune antisera to HSV. In the majority of instances the transformed cells were oncogenic in syngeneic hosts (Duff and Rapp, 1971a,b, 1973a,b; Duff et al., 1974; Takahashi and Yamanishi, 1974; Boyd and Orme, 1975; Kimura et al., 1975; Li et al., 1975; Macnab, 1979; Hampar et al., 1980) or showed increased oncogenicity over the parental cell line (Kutinova et al., 1973). In some instances multiple passages in culture were required before the cells became tumorigenic (Kimura et al., 1975). In another instance tumorigenicity was dependent on the immune status of the host and in immunocompetent syngeneic animals was observed only after a latent period of 2.5 years (Macnab, 1979). Nonetheless, the cells derived from those tumors retained antigenically identifiable virus-specific proteins. Sera of animals bearing tumors produced by HSV-transformed cells often contained HSV-neutralizing antibodies (Duff and Rapp, 1971a,b, 1973b; Rapp and Duff, 1973; Kutinova et al., 1973; Boyd and Orme, 1975; Kimura et al., 1975) or reacted in immunofluorescence assays with HSV-infected or other HSV-transformed and tumor cells (Li et al., 1975).

At least five strains of HSV-1 and seven strains of HSV-2 have been shown to transform cells in culture to a malignant phenotype. Three observations taken together—(1) the consistent outgrowth of clones with infinite life span from a variety of primary cells and appearance of morphologically altered colonies in continuous cell lines after infection with a number of strains of HSV-1 and HSV-2, (2) the expression of virus-specific proteins in morphologically transformed cell lines, and (3) the formation of tumors in syngeneic animals following inoculation of morphologically transformed cells—strongly implicate HSV information in the initiation of transformation events that lead to malignancy.

Because of the low frequencies of transformation observed, the strength of the conclusion that HSV transform cells in culture lies in the large number of times the transforming potential has been demonstrated with numerous virus strains and a variety of cell types.

B. Frequency of Transformation

Although quantitative comparisons are made difficult by the variety of virus strains, cell types, and conditions used to limit infectivity of a virus preparation, a range of transformation frequencies from 0.25 to 20 stably transformed colonies per 10^6 infected cells has been observed. The number of morphologically distinguishable colonies that developed into stable cell lines is taken as the number of stable transformants in this estimation. The wide range of frequencies of transformation is not surprising considering the variation in virus strain, cell type, and transformation protocols employed.

The morphological transformation frequency can be examined from two vantage points. In comparison to the frequency obtained by other DNA-containing transforming viruses, such as polyomavirus or SV40, the herpesviruses appear to have very low transforming capacity. Depending upon the cell type and assay used for transformation, polyomavirus and SV40 demonstrate transformation efficiencies of approximately 1 stable transformant in 10^3 cells following infection of primary or continuous nonpermissive cells with 100 plaque-forming units (PFU)/cell (Stoker and MacPherson, 1961; MacPherson and Montagnier, 1964; Todaro and Green, 1964). Two objections to this comparison can be raised justifiably. First, the cells used for transformation assays by HSV are permissive for virus replication. For this reason virus infectivity must be inhibited by one or more of the methods described above in order for potential transformants to survive. It is not unreasonable to predict that putative virus functions that initiate or enhance transformation are rendered inactive in some proportion of the virus particles also, thereby reducing the transforming activity below its maximal level. It should be pointed out that transformation of permissive cells by SV40 is considerably more difficult to achieve than transformation of nonpermissive cells. Although few examples are available for examination, in one carefully executed and quantitated study (Gluzman et al., 1977) exposure of highly permissive monkey cells to UV-irradiated SV40 virions resulted in 10 transformed colonies/10^6 infected cells in anchorage independence assays, a frequency well below that observed in nonpermissive cells.

The structure, length, and proportion of the genome involved in transformation are likely to influence the frequency of transformation also. In the case of SV40 and polyomaviruses, only the early transcribing region, which constitutes approximately half of the 5.2-kb genome, is required for transformation by virions (reviewed in Tooze, 1981). As these viruses contain covalently closed circular DNA, a single crossover event is sufficient to integrate the entire genome. Any crossover that occurs in the late transcribing half of the genome, thereby leaving the early region intact, has the potential to transform cells. If crossover occurs randomly and with equal probability throughout the genome, then approximately half of the crossovers that take place between cellular and SV40 DNA will maintain integrity of the early region.

Integration of a linear DNA, such as that of HSV, requires two recombinational events. One of each of these events must take place on either side of the DNA segment responsible for transformation in order to accomplish its integration. If the viral genome is large compared to the transforming region, only a small proportion of the recombinational events that occur will result in integration of the relevant DNA segment(s). Therefore, the likelihood of integrating the transforming segments of HSV can be expected to be significantly lower than that observed for SV40.

Perhaps a more meaningful vantage point from which to assess the transforming potential of HSV is by comparison to biochemical transformation. Both in biochemical and in morphological transformation by HSV the recipient cells are permissive for virus growth. In both assays UV irradiation of virions or incorporation of photodynamic dyes has been used to inactivate virus infectivity. Clearly, in the case of thymidine kinase (TK) transformation, as reviewed below, a single contiguous segment of the viral genome must be integrated in order for stable transformation to occur. Routinely, frequencies of between 20 and 200 transformants/10^6 cells are observed for TK transformation of murine cells by UV-inactivated virions (Munyon et al., 1971; Rapp and Turner, 1978; Rapp et al., 1980). Given the qualifications noted above regarding quantitation, it is clear that the frequencies of morphological transformation by HSV are reasonably similar to those for TK transformation. Importantly, in one report (Hampar et al., 1981) the frequencies of TK and morphological transformation by UV-irradiated HSV-1 virions were determined simultaneously in the same TK$^-$ BALB/c mouse cell line (B2). In this experiment, morphological transformants that were tumorigenic in nude mice occurred at one-tenth the frequency of TK$^+$ transformants, in agreement with the general pattern evident from previous data.

For these reasons, although the frequency of morphological transformation achieved by HSV is orders of magnitude below that observed with polyomavirus or SV40, this frequency alone does not suggest that the herpesviruses transform cells by a mechanism(s) that is distinct from that of other oncogenic DNA viruses.

Both the low frequency of transformation and the complexity of the viral genomes have, however, made the role of specific viral sequences and gene products in the initiation and maintenance of the transformed phenotype difficult to assess.

C. Association of Viral DNA Sequences with Transformation

Primarily two approaches have associated viral DNA sequences with transformation. One approach involves determining the capacity of purified segments of the viral genome to stably alter the growth properties of cells in culture. The second relies on identifying sequences retained by transformed or tumor cells. These complementary and equally important approaches address closely related but distinct aspects of the transformation process. Transformation with defined segments of the viral genome should directly identify the minimal sequences that initiate and also may be required to maintain the transformed state. This approach assumes that all of the information required is present within a contiguous sequence of nucleotides and that no sequences in the same DNA segment hinder either expression of the essential information or cell survival. Detection of sequences within transformed cells can be expected to identify regions of the genome involved in the maintenance of transformation as well as those that augment the transformed phenotype but will not directly implicate sequences in its initiation. One advantage of this second approach is that if noncontiguous segments of the genome participate in the maintenance of transformation, the cell will select those sequences from the viral genome, retain and possibly amplify them.

The value of these combined approaches has been demonstrated amply in the mapping of the HSV-1 and HSV-2 TK genes. Cells that lack a functional TK gene (TK$^-$) can convert to a TK$^+$ phenotype by incorporating and expressing HSV-1 or HSV-2 genetic information (Munyon et al., 1971; Davis et al., 1974). Extensive studies have shown that the TK-transforming activity resides in a virally coded TK (Dubbs and Kit, 1964) that is readily distinguished from the cellular counterpart (Klemperer et al., 1967; Munyon et al., 1972). While it is clear that expression of the HSV TK gene is not directly related to morphological transformation (Camacho and Spear, 1978; Reyes et al., 1979; Hampar et al., 1981), the biochemical transformation of TK$^-$ cells has been pivotal to understanding events that occur when selective pressure is applied to cells exposed to HSV or its DNA for a phenotype determined by a single viral gene. One of the more important assessments of specific viral segments retained in cells biochemically transformed following infection with HSV was provided by Leiden et al. (1980). These investigators examined the sequences of HSV-2 DNA retained in each of four cell lines generated from infection of TK$^-$ cells with UV-irradiated virus and maintained continuously under selective pressure for TK expression for 50–100 passages. By using a modified DNA blot hybridization procedure, abundant HSV-2 sequences in transformed cell DNA were located at 0.14–0.57, 0.14–0.42, 0.21–0.32, and 0.28–0.42 map units (MU). The HSV-2 sequences identified constitute overlapping segments of the viral genome. The region of the HSV-2 genome located between 0.28 and 0.32 MU was retained in all four cell lines and, thereby, identified the physical location of the HSV-2 TK-coding sequences. By similar analysis of two HSV-1 TK transformants, the TK gene of this virus was localized to 0.28–0.32 MU also. These localizations were consistent with the map location for TK$^-$ mutants of HSV established by both marker rescue (Stow *et al.*, 1978) and analysis of intertypic recombinants (Halliburton *et al.*, 1980). All four HSV-2-transformed cell lines retained viral sequences in excess of the TK-coding region. Importantly, however, the sequences that were stably retained in each cell line apparently formed a single contiguous sequence of viral DNA adjacent to the selected marker. These findings indicate (1) that when a single viral gene is sufficient to confer a new phenotype on a cell, the viral sequences that become integrated and are retained under continuous selective pressure are contiguous with the selected gene sequence; and (2) that analysis of several transformed cell lines will identify overlapping segments of viral DNA that roughly define the limits of the transforming gene. Leiden *et al.*, (1980) noted also that sequences between 0 and 0.06 and between 0.57 and 0.82 MU are uniformly absent from the biochemical transformants examined. They suggested that at least a portion of the region beyond 0.57 MU may be obligatorily lost due to the presence of functions that are detrimental to cell survival. Consistent with this possibility is the localization of host shutoff factors between 0.52 and 0.59 MU on the HSV-2 genome (Morse *et al.*, 1978). There is no reason in the absence of data directly addressing the issue to exclude this possibility. Because the region from 0.52 to 0.59 MU overlaps segments of HSV-2 DNA that transform cells in culture morphologically (see below), it is reasonable to consider alternative explanations for the absence of particular viral sequences in stably transformed cells. The presence of DNA sequences beyond 0.57 MU, which severely limit expression of the TK gene but do not alter cell survival, also would be expected to result in their obligatory loss. It is equally possible that the exclusion of segments of the viral genome distal to 0.57 simply reflects the nature of recombinational events involved in integration of HSV DNA into cellular chromosomes.

It is likely that transformants most frequently arise as the result of the minimal number of crossover events (2) needed to integrate a linear DNA segment. Integration of the TK gene from linear viral DNA requires that one of the crossover events occur to the left of 0.28 MU. If crossovers occur with equal probability throughout the HSV genome, then the likelihood of the second crossover occurring before integrating nonessential DNA from a position on the genome distal to the TK gene increases proportionally with the distance from the selected marker. Stated simply, sequences beyond 0.57 MU may not be retained frequently in TK$^+$ transformants because of their distance from the gene responsible for the phenotype selected.

In separate experiments, restriction endonuclease-generated fragments of HSV-1 (Wigler *et al.*, 1977) and HSV-2 (Reyes *et al.*, 1979; McDougall *et al.*, 1980) DNA were used to transform TK$^-$ cells to a TK$^+$ phenotype and to determine the limits of the coding sequence. The feasibility of using viral DNA to transform TK$^-$ cells biochemically was demonstrated for HSV-2 (Bacchetti and Graham, 1977) by transfection of the cells with viral DNA that had been sheared to eliminate its infectivity. TK$^+$ transformants arose at a frequency of 3 colonies/µg per 10^6 cells for HSV-2 in human 143 TK$^-$ cells. Wigler *et al.* (1977) using mouse TK$^-$ cells as recipients demonstrated that cleavage of HSV-1 DNA with the restriction endonuclease *Eco*RI eliminated the transforming capability of the DNA, whereas cleavage with *Bam*HI resulted in the appearance of 0.5 TK$^+$ colony/µg per 10^6 cells. The transforming activity was confined to a single 3.4-kb *Bam*HI fragment. This fragment when purified from agarose gels transformed at a frequency of 10 colonies/µg equivalent per 10^6 cells. Subsequently this same DNA fragment was cloned (Colbére-Garapin *et al.*, 1979; Wilkie *et al.*, 1979) and shown to be sufficient for biochemical transformation. Sequencing of the cloned fragment (McKnight, 1980; Wagner *et al.*, 1981) indicated the presence of a single open reading frame sufficient to encode the TK and consensus control sequences 5' to the coding sequence. The *Bam*HI fragment maps between 0.29 and 0.31 MU on the HSV-1 genome. In addition, Colbére-Garapin *et al.* (1979) reported that a 2-kb subfragment of the *Bam*HI fragment generated by *Pvu*II cleavage also transformed cells. Using cloned DNA fragments containing the TK gene, frequencies of 100–2800 colonies/µg per 10^6 cells are achieved easily (Graham *et al.*, 1980; Linsley and Siminovitch, 1982).

Five observations from these and other experiments form a basis for discussion of morphological transformation by herpesviruses. First, transfection with sheared DNA appears to occur at a frequency lower than that observed after infection with inactivated virions. Although comparisons of data from different laboratories must be qualified by recognizing that neither the same recipient cell nor virus strain was used uniformly, approximately 20–200 biochemical transformants are commonly observed among 10^6 TK$^-$ cells infected with UV-inactivated virus preparations. Only 0.5–18 transformants/µg per 10^6 cells were generated by cleaved (Wigler *et al.* 1977) or sheared viral DNA (Bacchetti and Graham, 1977; Graham *et al.*, 1980). Second, by using an agarose gel-purified *Bam*HI restriction nuclease-generated fragment containing the TK gene, the number of TK$^+$ colonies was 20-fold higher per gene equivalent than that obtained with an unfractionated total *Bam*HI digest. This increase may reflect removal of the TK-coding segment from other segments that are detrimental either to cell survival (Leiden *et al.*, 1976) or to efficient expression of the TK gene and that are cotransferred during transfection of total digests. Third, by using appropriately cloned DNA and optimal transfection conditions, frequencies in excess of 2800 colonies/µg per 10^6

cells are observed. Fourth, a single copy of the TK gene integrated into cellular DNA was sufficient to generate stable TK$^+$ transformants by transfection with the 3.4-kb *Bam*HI fragment (Pellicer *et al.*, 1978). Fifth, HSV gene products other than TK are unlikely to be required for stable transformation to the TK$^+$ phenotype.

These last two observations were not entirely expected. TK belongs to the β or early class of viral proteins (Garfinkle and McAuslan, 1974; Honess and Roizman, 1974). Active transcription of the TK gene during the lytic cycle requires prior synthesis of α or immediate–early proteins (Honess and Roizman, 1974, 1975; Preston, 1979). The minimal TK transforming region, as defined by Colbére-Garapin *et al.* (1979), contains only the coding region for the TK gene and essential transcription signals in the 5' untranslated region (McKnight, 1980; Wagner *et al.*, 1981). Although the resident TK sequences still respond to HSV regulatory genes on superinfection of transformants with HSV (Lin and Munyon, 1974; Leiden *et al.*, 1976; Kit and Dubbs, 1977), these genes are not required for expression of the integrated TK gene. The level of transcription achieved when a single copy of TK gene is removed from the context of the HSV genome by incorporation into the cellular genome is sufficient to convert cells to a TK$^+$ phenotype.

Clearly these combined approaches of identifying viral sequences retained in transformed cells and determining the capacity of purified segments of the genome to transform have defined the viral sequences that were both necessary and sufficient to accomplish biochemical transformation. This same rationale has been applied to identify viral information that is essential to morphological transformation. Recently, Robinson and O'Callaghan (1983) examined the arrangement of viral sequences retained in three hamster cell lines transformed morphologically by UV-irradiated virions of equine herpesvirus type 1 (EHV-1). In this study, overlapping segments of the viral genome were detected. The sequences retained in cell lines at high passage (> passage 124) were contiguous with the consistently retained segment spanning 0.32–0.38 MU implicating this region in the maintenance of the transformed phenotype. At lower passage some cell lines retained additional viral information which was lost progressively during continued propagation.

1. HSV Proteins Expressed in Morphologically Transformed Cells

Considerable effort has been directed toward detecting HSV information retained and expressed in morphologically transformed cells. These efforts have focused on determining the extent of the genome retained, the proportion of the genome transcribed into stable message, and the identity of virus-coded proteins within and on the surface of HSV-transformed cells.

The immunological identification of virus-specific polypeptides in HSV-transformed and tumor cells has served two purposes. The first was to establish that morphological transformants which arise after exposure to HSV express viral information. The second was to identify specific viral sequences expressed within these cells. A large number of genes for virus-specific polypeptides have been mapped genetically and biochemically on the HSV-1 and HSV-2 genomes (Roizman, 1979). Identification of those proteins expressed in transformed cells provides one means of confirming the presence of specific viral sequences. Whereas polyclonal antisera and immunofluorescence assays are sufficient in the first instance, resolution of individual polypeptides requires immunoprecipitation with defined antisera or immunofluorescence assays using monospecific or monoclonal antibodies.

By using hyperimmune antisera to HSV, virus-specific antigens have been detected on the surface membranes of a large number of independently derived HSV transformed and tumor cell lines (see section IIA). The observation that sera of animals bearing tumors produced by HSV-transformed cells contain HSV-neutralizing antibodies indicated that at least one of these surface antigens represented a virus-coded glycoprotein. Recently, Lewis *et al.* (1982) examined the expression of specific glycoproteins by using monospecific antisera against gAgB (gB) and gC (gG) in immunofluorescence assay of HSV-2-transformed rat cells at high passage in cell culture or after passage in syngeneic hosts. Recently the nomenclature for the HSV glycoproteins has been modified (see P. Spear, this volume). For clarity both designations are given. The current names for the glycoproteins appear in parenthesis. The use of cells at high passage was important in this study as expression of immunologically identifiable proteins and the amount of the viral DNA retained often decreases progressively in cells transformed by herpesviruses (Minson *et al.*, 1976; Geder *et al.*, 1979; Robinson and O'Callaghan, 1983) (discussed below). It is assumed that at high cell passages polypeptides that continue to be expressed are either involved in the maintenance of transformation, or are encoded within or closely linked to a DNA segment responsible for transformed phenotype. Cells from two HSV-2-transformed lines and a tumor-derived cell line were stained with gAgB (gB) but not gC (gG) antisera. These same cell lines reacted with antisera raised against the early (B) polypeptide alternatively designated VP143 (Flannery *et al.*, 1977), ICP8 (Honess and Roizman, 1973), VP130 (Bayliss *et al.*, 1975; Wilcox *et al.*, 1980), and ICP11/10 (Purifoy and Powell, 1976). This polypeptide also has been identified in all of four tumor cell lines derived from HSV-2-transformed hamster cells (Flannery *et al.*, 1977) and on the original hamster cell transformant 333-8-9. This polypeptide VP143 corresponds to the major DNA-binding protein (DBP) of HSV-2 and has been mapped recently by intertypic recombination (Morse *et al.*, 1978; Conley *et al.*, 1981), marker rescue (Spang *et al.*, 1983), two-factor genetic crosses (Dixon *et al.*, 1983), hybrid selection and *in vitro* translation of mRNA (Conley *et al.*, 1981), and analysis of defective viruses (Frenkel *et al.*, 1980) to 0.38–0.41 MU on the prototype arrangement of the HSV-2 genome. In comparison, the gAgB(gB)-coding sequences are localized within 0.348–0.369 MU (Ruyechan *et al.*, 1979; DeLuca *et al.*, 1982). Monospecific antibody against gC(gG) [currently mapped in the short component (Roizman *et al.*, 1984)] did not react with the rat-transformed and tumor cell lines. While this finding does not necessarily indicate that sequences from the short component are absent, it does demonstrate that expression of gC (gG) is not required for maintenance of a transformed phenotype.

These results suggest that DNA sequences from 0.30 to 0.41 are expressed in HSV-2-transformed rat cells and a tumor-derived cell line at high passage number. They both illustrate the value of monospecific antibody in analyses of this type and also suggest that this region (which is consistently maintained through passage in animals) may play a role in the transformation of cells in culture by HSV-2 virions.

Additional specific antisera have identified HSV-type common antigen, associated with gD (Reed *et al.*, 1975; Cohen *et al.*, 1978), on some but not all of HSV-1- and HSV-2-transformed hamster cells. HSV-specific polypeptides have been immunoprecipitated from cells transformed by HSV also (Gupta and Rapp, 1977; Gupta *et al.*, 1980; Shu *et al.*, 1980). However, as the relationship between the immunoprecipitated polypeptides and specific virus-coded proteins of known map location has not been determined, it is difficult to relate expression of these polypeptides to specific DNA sequences. Receptor for IgG has also been identified on HSV-2-transformed cells (Westmoreland *et al.*, 1974) and may represent the virus-specified Fc receptor (gE) reported by Baucke and Spear (1979) and by Lee *et al.* (1982). This glycoprotein is encoded between 0.924 and 0.95 on the HSV genome (Para *et al.*, 1982).

Monospecific antisera serve as valuable indicators of the presence or absence of expression of specific viral proteins in transformed cells. Such information is central to uncovering the mechanisms by which HSV accomplishes transformation. However, although the identification of virus-specific proteins within transformed and tumor cells has revealed segments of the genome often retained, no polypeptide common to the majority of the transformed cells has been detected.

The absence of a consistent antigenic marker for transformation by HSV indicates either that transformation is accomplished without adding a specific protein-coding region of the virus to the cell or that antibodies against protein(s) responsible for the transformed cell phenotype are not abundant in anti-HSV-1, anti-HSV-2, or antitumor sera. It is possible that the putative transforming protein does not elicit a strong antigenic response in an infected or tumor-bearing host or that the antibodies formed in response to this protein are not reactive in immunofluorescence assays. Precedents for each possibility exist in other transforming systems. The avian leukosis virus (ALV), unlike the majority of the tumor viruses, transforms cells *in vivo* not by contributing a protein-coding sequence but by inserting its strong promoter sequence adjacent to the cellular homolog of a retrovirus transforming gene. Among viruses whose transforming capacity does depend upon continual expression of one or more virus-coded polypeptides, the transforming proteins are not uniformly detected immunologically. It is believed, for instance, that the Ela polypeptides that are expressed in adenovirus 2 (Ad2)-transformed cells are unable to elicit a strong immune response in tumor-bearing hosts (Bernards et al., 1982; Feldman and Nevins, 1983) as neither immunofluorescence nor immunoprecipitation reactivity to this polypeptide is readily detected with antitumor sera. In contrast, the small tumor (t) antigen of SV40 elicits an active immune response in tumor-bearing animals, as the protein is readily detected in immunoprecipitates of extracts from SV40-infected, tumor, or transformed cells. Yet the protein cannot be detected by immunofluorescence assays. Thus, without additional evidence, it is not possible to determine whether the inability to detect a viral polypeptide universally in cells transformed by virions more clearly reflects the paucity of specific probes for polypeptides in the form of monospecific and monoclonal antibodies or alternatively indicates that HSV transforms cells without the persistence of protein-coding sequences.

Genetic analysis has provided an alternative method of identifying portions of the viral genome retained and expressed in HSV-transformed cells. Kimura et al. (1974) demonstrated the feasibility of detecting functional HSV DNA sequences in transformed cells by complementation analyses. They showed that growth of two (tsG4 and tsD6) of eight HSV-2 ts mutants representing distinct functions during lytic growth of the virus was enhanced in both of two independently derived HSV-2-transformed cell lines at temperatures that were nonpermissive for replication of the mutants. The two mutants carry defects in late viral functions. One of these mutations, tsG4, has been localized to 0.42–0.50 MU (Knipe, 1982). Similarly, Park et al. (1980) reported complementation of four HSV-1 strain 17 ts mutants (tsG, I, A, D) and complementation of the HSV-2 strain HG52 mutant ts1 in a transformed cell line generated by treatment with sheared HSV-1 DNA (Wilkie et al., 1974). In addition, they showed intertypic complementation of four ts mutants (tsJ, I, F, A) in an HSV-2-transformed cell line (Macnab, 1974). Mutants within the HSV-1 cistrons represented have been mapped to locations between 0.35 and 0.448 MU with the exception of tsD. This mutant maps within the inverted repeat sequences of the short arm of the chromosome (Weller et al., 1983b). The HSV-2 mutant ts1 maps at approximately 0.71–0.72 MU (Chartrand et al., 1981). These results indicate that widely separated portions of the HSV-2 genome were present and produced specific functional proteins in at least one of the cell lines tested. Because cell lines used in that study expressed HSV TK activity (Macnab et al., 1980) the genomic region from 0.28 to 0.32 also had been retained. As complementation analysis is based on a single cycle of virus growth, it is likely that the transacting polypeptides were expressed in a major portion of the cells infected.

An analysis of the restriction nuclease sites in DNA from ts^+ virus obtained from plaques of HSV-2 strain HG52ts1 or HSV-1 strain 17tsG that had been grown in the HSV-1-transformed cell line revealed the presence of sequences extending from 0.54 to 1.0 MU. In addition, one ts^+ recombinant had a restriction endonuclease profile indistinguishable from that of the original transforming virus. These observations indicate that superinfecting virus replicating within the uncloned cell population acquired sequences representative of the majority if not all of the transforming viral DNA. Recombinational analysis of this type identifies viral sequences present within a population of transformed cells. These sequences need not be expressed. They also need not be present in the majority of the transformed cells. The presence of ts^+ virus in a lysate indicates that during replication in a transformed cell culture, $ts1$ located and exchanged the appropriate information to correct its genetic lesion in at least one cell. The identification of one ts^+ recombinant that regained the restriction endonuclease profile of the transforming virus is more suggestive of heterogeneity in the viral sequences retained within the cell population than of the maintenance of the entire genome in transformed cells.

2. HSV DNA Sequences Retained by Morphologically Transformed Cells

The direct search for HSV DNA sequences in cells transformed by virions also has been performed and has resulted in several important observations. Frenkel et al. (1976) examined the viral DNA sequence complexity in five HSV-2-transformed hamster cell lines, including highly, weakly, and nononcogenic lines, and one tumor line. Among these were cell lines transformed by three different HSV-2 isolates. Included in this study also were parallel passages of one cell line (333-2-29). Each cell line had been shown to express viral antigens in a small percentage of the cells. Results of hybridization kinetic analyses using the entire HSV-2 genome as probe indicated that all of the cell lines contained viral DNA sequences, although the amount of information retained varied considerably. From 8 to 37% of the viral genome was detected in 0.7–3.1 copies/cell in cell lines tested. This variation in viral DNA retained among the cell lines indicated that transformed cells contain HSV sequences in excess of those needed to maintain the transformed phenotype. Not only did independently transformed lines contain variable fractions of the viral genome but also the amount of DNA retained by parallel passages of the same cell line differed. In one case, the line contained three copies of 37% of the HSV-2 genome, whereas cells in a parallel passage contained 1.5 copies of 10% of the viral DNA.

Variability was also seen in a comparison of 333-8-9 cells at passage 80 with cells of a tumor induced by 333-8-9 cells. The tumor cells contained an increased number of copies of a smaller fraction of viral DNA than did the parental cell line. A similar variation in sequence complexity was reported by Minson et al. (1976). In that study HSV-2 sequences representing 42% of the genome were detected at early passage of 333-8-9 cells. However, the sequences decreased in complexity with passage of the line and were not present within their limits of detection (4–8% of the viral genome per diploid cell) using whole viral DNA as probe in clones established from that line. The influence of the complexity of DNA used as probe on the ability to detect viral sequences in cellular DNA by hybridization kinetics (Frenkel et al., 1976) and by DNA blot hybridization (Robinson and O'Callaghan, 1983) has been reviewed previously and may have prevented detection of small regions of the HSV genome.

As a means of identifying specific viral sequences present in HSV-2-transformed cells, Galloway et al. (1980) examined the kinetics of renaturation of each of eight restriction nuclease-generated and gel-purified HSV-2 DNA fragments in the presence of DNA from 333-8-9 cells, as well as from clonal derivatives and tumor lines of these cells. Although not all regions of the genome were represented by the fragments used as probes, the results indicate that some segments were retained in all the cell lines, whereas the parental population was heterogeneous with respect to other sequences. Sequences homologous to each of the probes representing segments of the long component of the viral chromosome were present in the 333-8-9 cells. Sequences from the short component of the genome, however, were absent from the parental cell population, all of the five clonal and two tumor-derived cell lines examined. In contrast, sequences homologous to BglII-N fragment (0.58–0.62 MU) and BglII-G fragment (0.21–0.33 MU) were consistently present. Sequences from the BglII-O and -E fragments were present in only three of the clonal or tumor lines. Because the additional probes for sequences within the long component of the viral chromosome overlap either the BglII-N or G fragment, it was not possible to determine whether additional HSV-2 sequences were retained in these cell lines.

The variability in complexity of viral DNA sequences indicates that cells within HSV-transformed lines rapidly become heterogeneous with respect to the amount of viral genome they retain. The data are consistent with the hypothesis that sequences that are not essential to maintaining a transformed phenotype are gradually lost with continued propagation. The lower sequence complexity in a tumor cell line than in the cell line from which it was derived suggests two possibilities. It is reasonable to expect, as has been suggested previously (Frenkel et al., 1976), that some sequences retained within transformed cells passaged in culture might be detrimental to tumor development. Injection of cells into an animal host would provide strong selective pressure against growth of cells containing such sequences. In contrast, growth of cells in the heterogeneous population that had spontaneously lost those sequences would be selected for strongly. Considering the variability in retention of viral information, however, it is not necessary to hypothesize the existence of sequences that limit tumor growth. Alternatively, cells that have accumulated genetic alterations within cellular sequences that favor tumorigenesis also would be selected for strongly. In this last case, the decreased complexity of viral information could be coincidental.

The heterogeneity in viral sequences retained in transformed cells indicates that not only viral sequences that may be necessary to maintain the morphologically transformed state but also extraneous functions may be retained and expressed in cells transformed by inactivated virions. This factor makes it difficult to assess the role of viral sequences in the maintenance of transformation.

One method of decreasing the complexity of viral sequences in transformed cells and thereby simplifying the interpretation of functions and sequences required for transformation is to transform cells in culture with defined segments of the genome. This approach makes the fundamental assumption that, as in the case of the TK gene, a single contiguous set of nucleotides is sufficient to initiate and maintain stable transformation of cells in culture.

D. Transformation of Cells in Culture by Genomic Fragments of HSV-1

A single region of the HSV-1 genome extending from 0.30 to 0.45 MU in the prototype arrangement has been implicated in morphological transformation by fragments of the viral DNA. Camacho and Spear (1978) compared the ability of gel-purified XbaI restriction endonuclease fragments of HSV-1 DNA to promote outgrowth of transformed colonies in low serum from secondary HEF. Their results indicated that of the seven fragments generated by XbaI cleavage, only one, XbaI-F (0.30–0.45 MU), consistently yielded transformants efficiently. Colonies appeared at an approximate frequency of 10 colonies/μg equivalent per 10^6 cells. This frequency represents a substantial increase over the 1.5 colonies/10^6 cells per μg of total XbaI digest observed. Possible explanations for such an increase have been described above. Among the several parameters of a transformed cell phenotype, the transformed cell lines displayed the ability to grow in low serum to high cell density in 10% serum, and in a semisolid matrix. At least a portion of the transfected HSV information was expressed in these cell lines as the glycoprotein gAgB (gB) was detected by immunofluorescence, immunoprecipitation, and adsorption of neutralizing antibodies from sera monospecific for gAgB (gB). The percentage of cells showing immunofluorescence varied from experiment to experiment, suggesting instability in retention of the DNA sequences or, alternatively, accumulation of alterations in the cellular genome that limited expression of the viral gene. The observation that at later passage cells transformed by the XbaI–F fragment did not contain detectable viral DNA (cited by Knipe, 1982) suggests that either only a small portion of the fragment need be retained to maintain a transformed phenotype or, alternatively, as suggested by the immunofluorescence data, only a small proportion of the cells retain the XbaI–F segment. The occurrence of unstable cell lines is frequently observed among biochemical transformants generated by transfection and is commented on below. Interestingly, the TK gene, which also lies within the XbaI–F fragment, was not expressed by morphological transformants.

Independent experiments have localized morphological transformation activity to the same segment of the HSV-1 genome using the same assay for transformation. Reyes et al. (1979) detected five foci in one dish of HEF transfected with a BglII–I fragment extending from 0.311 to 0.415 MU. The same segment of the genome also generated 11 dense foci growing on a monolayer of BALB/3T3 mouse cells in a single dish.

Cells of one line established from a dense focus of BALB/3T3 cells were serum independent, cloned efficiently in semisolid media, and grew to higher saturation densities than the parental cell line. That the cell line was stably transformed was indicated by the observation that its anchorage-independent growth was maintained through 48 in vitro passages. Viral DNA sequences could not be detected in this cell line, although the blot hybridization used should have permitted detection of 0.1 copy of the transforming fragment (Reyes et al., 1979). Among these experiments the same single transforming region was identified in three strains of HSV-1, strain F (Camacho and Spear, 1978), MP, and the St. Thomas Hospital substrain of HFEM (Reyes et al., 1979). This region has been called the morphological transforming region I (mtr-I). Its identification indicates that at least one single contiguous stretch of nucleotides is sufficient to establish a transformed phenotype following DNA transfection.

It is important that the ability of this region to transform cells be confirmed with cloned DNA segments in order to exclude the possibility that additional fragments cooperate in transforming events. In the case of low-frequency events for which strong selective pressure can be applied, this possibility cannot be excluded. It has been shown repeatedly that DNA fragments purified by one or two successive electrophoretic separations retain contaminating fragments. Although the contaminants often cannot be visualized by conventional staining of the gels, their presence is apparent in sensitive physical and biological assays. Additional sequences are readily observed when gel-purified DNA fragments are used as probes in blot hybridization assays (Galloway et al., 1980). More relevant to the experiments under consideration here are the observations of Parris et al. (1980) and Lai and Nathans (1974) that multiple gel-purified fragments will rescue ts mutants of HSV and SV40, respectively. As in the case of selection for wild-type virus in marker rescue, assays for transformation, especially transformation of primary cells, probably involve sufficiently strong selective pressure to permit low-probability events such as cooperation of a contaminating fragment to be observed.

The availability of large quantities of cloned DNA fragments, in addition, should permit estimation of the maximal frequency of transformants in individual assays of transformation. Such estimates are valuable in suggesting mechanisms by which transformation occurs and are discussed later.

1. Functions Encoded within *mtr-I*

A number of virus-specific polypeptides (Morse *et al.*, 1978; Ruyechan *et al.*, 1979) and viral transcripts (see Wagner, this volume) have been localized to the region between 0.30 and 0.45 MU. The functions of several of these proteins have been identified by analysis of ts and drug-resistant mutants. This region contains coding sequences for gAgB (gB), a late gene of unknown function, the major DNA-binding protein ICP8, and the viral DNA polymerase in that order from left to right (Weller *et al.*, 1983a; Pancake *et al.*, 1983) as well as being a putative origin of viral DNA replication (Kaerner *et al.*, 1979; Frenkel *et al.*, 1980; Locker *et al.*, 1982). The presence of these functions makes this segment of the HSV-1 genome particularly attractive as a transforming region. Cells transformed by EBV, SV40, or polyomavirus all express a DNA-binding protein that is essential to or closely associated with transformation (Tooze, 1981; Levin, 1982). In addition, the HSV-1 DNA polymerase appears to have mutator activity (Hall and Almy, 1982). If continuously or transiently expressed in transformed cells, this activity could promote alterations in the cellular genome that augment or stabilize the transformed phenotype. Finally, as expression of gAgB (gB) is common among cells transformed by virions (Lewis *et al.*, 1982) and thereby is unlikely to be detrimental to the initiation of transformation, this glycoprotein might serve as useful marker for stable incorporation of the transforming segment.

At least two additional polypeptides, ICP35 and a protein involved in DNA packaging, ICP37 (Preston *et al.*, 1983), are encoded within *mtr-I* (Morse *et al.*, 1978; Conley *et al.*, 1981). Whether any of these polypeptides or additional as yet unidentified proteins contribute to the transformed cell phenotype remains to be determined.

E. Transformation of Cells in Culture by Genomic Fragments of HSV-2

Transfection with restriction endonuclease-generated fragments (Reyes *et al.*, 1979; Jariwalla *et al.*, 1980) or cloned viral DNA fragments (Galloway and McDougall, 1981) have implicated two segments of the HSV-2 genome as transforming regions. These segments are contiguous on the physical map and may act independently in two types of transformation assays.

Among the variety of assays for the transformed phenotype, escape from senescence is generally considered the minimal modification of cellular growth that can be associated with transformation. This assay has retained importance by virtue of the likelihood that it represents a very early measurable response in the progression of events resulting in conversion of normal to malignant cells. The major limitation of the immortalization assay is that the results obtained cannot be quantitated precisely. Jariwalla *et al.* (1980) investigated the ability of fragments of the HSV-2 (strains S1 and 333) genome to immortalize Syrian hamster embryo cells. They reported that sequences within the *Bgl*II–C fragment (0.43–0.58 MU) are sufficient to transform hamster embryo cells.

Using an alternative assay of transformation, Reyes *et al.* (1979) have reported that sequences within the *Bgl*II–N fragment (0.58–0.62 MU) recovered from agarose gels are sufficient to produce foci of transformed cells on monolayers of continuous BALB/3T3 mouse cell lines. In these experiments, transformants appeared after transfection with each of four overlapping DNA fragments spanning 0.419–0.628 MU. The minimal transforming region among the fragments tested was contained within the *Bgl*II–N fragment.

In these assays transformation occurred at low frequency. A maximal efficiency of 0.3 focus/μg was reported. It is important for comparative purposes to recall that definition of this transforming segment is based on 27 foci arising in eleven 25-cm dishes of confluent cells. Blot hybridization analysis of cellular DNA from two clonally derived lines transformed by either a 0.582 to 0.628-MU or a 0.575 to 0.702-MU fragment indicated that sequences homologous to the *Bgl*II–N fragment were retained in low copy number in both lines. The appearance of transformed foci in monolayers of BALB/3T3 cells after transfection with the *Bgl*II–N fragment and retention of at least some sequences homologous to this fragment in populations of cells derived from transformed clones implicated the region of the HSV-2 genome encompassed by 0.582–0.628 MU as *mtr-II*.

Using as an assay for transformation the ability of HEF to form foci in medium containing 1% fetal calf serum, these same authors reported the occurrence of six foci in one dish of cells that had received an *Eco*RI/*Hind*III gel-purified genomic fragment representing 0.520–0.634 MU. At first inspection, this finding appears to conflict with data of Jariwalla *et al.* (1980) who demonstrated immortalization of primary HEF cells with the *Bgl*II–C (0.43–0.58 MU) but not with a pool of gel-purified fragments that contained the *Bgl*II–N fragment. Three possible resolutions of this apparent conflict can be proposed. Escape of primary hamster cells from senescence may occur by a mechanism distinct from focus formation in low serum. Alternatively, the foci that arose after treatment of primary hamster cells with an *Eco*RI/*Hind*III fragment from 0.520 to 0.634 MU may have resulted from a contaminating fragment in the DNA preparation. Cleavage of HSV-2 (333) with *Eco*RI and *Hind*III generates a fragment that comigrates with the 0.52 to 0.634-MU fragment. These two comigrating fragments together contained nearly all of the *Bgl*II–C plus *Bgl*II–N sequences and as stated by Reyes *et al.* (1979) were not resolved in the gel system and conditions employed to purify the fragments used in transformation assays.

The incorporation of multiple physically separated DNA fragments by cells following transfection using the calcium phosphate precipitation technique has been documented repeatedly (reviewed in Scangos and Ruddle, 1981) and forms the basis for cotransfection of selectable and non-selectable markers. Because another cell line in the same study that also received comigrating fragments retained sequences homologous to both fragments, it is not possible to conclude whether the foci of HEF cells that arose after exposure to this mixture of fragments originated as a result of sequences contained within the *Bgl*II–C or –N fragment or both. Last, it is possible that the sequences encompassing 0.520–0.58 MU, which are present in both the *Bgl*II–C and the *Eco*RI/*Hind*III fragment representing 0.520–0.634 MU, constitute the transforming sequences.

Both the low frequency of transformation by DNA fragments and the possibility that contaminating fragments cooperate in transformation events point to the necessity of using cloned DNA segments as donors of transforming sequences.

Two reports of transformation by cloned fragments of HSV-2 DNA have appeared recently. Using a cloned fragment of HSV-2 (333) that contained all of the sequences of the *Bgl*II–N fragment except 300 nucleotides at the far left end between the *Bgl*II site at 0.582 and the *Bam*HI site at 0.584 MU, Galloway and McDougall (1981) reported transformation of rat embryo fibroblasts and NIH 3T3 cells. Foci able to grow on a monolayer of cells in low-serum medium as well as anchorage-independent foci were obtained. The cloned *BGl*II–C fragment was not investigated in this study. Two foci derived from rat embryo cells and the two from NIH 3T3 cells exhibited multiple parameters of a transformed cell phenotype, including tumorigenicity in nude mice. The observation that the *Bgl*II–N fragment stably transformed both rat embryo and NIH 3T3 cells indicates that sequences within this segment of the genome are sufficient to transform both primary and continuous cell lines. When two rat embryo fibroblast transformants, two NIH 3T3 transformants, and two tumor cell lines derived in nude mice after inoculation of an NIH 3T3 *Bgl*II–N transformant were analyzed by DNA blot hybridization for DNA sequences homologous to the *Bgl*II–N clone, all six contained approximately 0.1 copy/diploid genome. Additional analysis revealed that only the right half of the *Bgl*II–N fragment could be detected and that this segment was attached to pBR322 sequences. These data confirmed the assignment of sequences within 0.58 and 0.62 as a transforming region of HSV-2 in the context of the assays employed.

The retention of sequences from only the right end of the *Bgl*II–N fragment in six transformed cell lines suggested that the left end sequences were rapidly or obligatorily lost. However, in a recent review article, Galloway and McDougall (1983) reported that transformation could be obtained by transfecting NIH 3T3 cells with a 2.1-kb fragment from the left-hand one-third of the *Bgl*II–N fragment.

The finding that sequences confined to the left end of the *Bgl*II–N fragment transform is in apparent conflict with previous data from the same laboratory, which indicated that only portions of the sequences from the right end of the *Bgl*II–N fragment are retained in morphologically transformed cells. This apparent conflict may simply reflect the combined effects of complexity of the probe used to detect viral sequences, the structure of the plasmid used for transformation, and the propensity of cells to incorporate and integrate multiple DNA molecules during transfection.

In those experiments involving the *Bgl*II–N fragment cloned into the *Bam*HI site of pBR322, the plasmid was converted to linear form by cleavage with *Hin*dIII prior to transfection. As the *Hin*dIII site lies within 346 base pairs from the left end of the viral *Bgl*II–N fragment within the recombinant plasmid, the pBR sequences were located almost entirely adjacent to the right (0.62 MU) of the *Bgl*II segment. This positioning of left end sequences near the end of the transfected fragment would be expected to lower the probability that they would be included among the double crossover events leading to integration. If these sequences are essential to transformation, the transformed clones that develop will contain them, nonetheless. Because cells incorporate multiple DNA molecules during transfection, it is likely that these same clones will contain, in either unstable or stable form, sequences from the right end as well, as these sequences, by virtue of their linkage to a long stretch of plasmid DNA, have a higher probability of being integrated in double crossover events than do sequences close to the molecular end. If sequences from the right end of *mtr-II* are not required for transformation, then they may be lost without compromising the transformed phenotype. The *Bgl*II–N probe may have detected the frequent, although possibly nonessential integration of right end sequences. Whether only a portion of the sequences were present in all cells or the majority of the sequences were present in a few cells could not be determined from the experiments performed.

However, the observation that three cell lines transformed by the 2.1-kb left end segment also appeared to be free of sequences from this fragment led the authors to suggest the alternative explanation that the sequence of HSV-2 that initiates transformation need not be retained in the transformed cell.

The suggestion that HSV transforms cells by a hit-and-run mechanism has been made (Hampar, 1981; Schlehofer and zur Hausen, 1982). Available data are insufficient to either support or contradict unequivocally such a process in the transformations reported thus far. If future studies confirm efficient transforming activity in the 2.1-kb fragment representing the left-hand end of the *Bgl*II–N fragment, and show that these same sequences are lost from transformed cell lines when maintained under selective pressure for a transformed phenotype, then the hit-and-run hypothesis will gain increased credence.

It is informative to recall the four minimal requirements that must be met in order to confirm a hit-and-run mechanism. First, loss of the transforming sequences must be demonstrated in cultures kept under strong selective pressure for expression of a transformed phenotype from the time of the initial assay until they are examined for viral information. Alternatively, cell lines that apparently have lost the donor segment must be shown to retain all characteristics of a transformed phenotype. This characterization is of increased importance when continuous cell lines (in contrast to primary cells) are used as DNA recipients. It is clear that DNA introduced by transfection may result in both transient expression (Linsley and Siminovitch, 1982) and unstable transforming (Scangos *et al.*, 1981) events. As established cell lines already express the genetic capacity to continue to proliferate indefinitely, loss of sequences required to maintain a transformed state is not easily recognized during routine subculturing.

Second, all sequences of viral origin, in particular transcriptional promoter sequences if they occur within the transforming segment, must be absent from or clearly shown to be nonfunctional in transformed cells in order to exclude the promoter insertion mode of transformation, as by avian leukosis virus (Hayward *et al.*, 1981). Third, the hit must be identified. The minimal transforming fragment can be expected to contain one or more protein-coding regions or unique structural features that could alter the genetic information of the cell either by altering the nucleotide sequence or by stimulating rearrangement of existing sequences. Last, definitive identification of the initiating sequences will require demonstration that they can be mutationally altered with a corresponding loss of transforming capacity. Clearly, sequencing of the minimal transforming fragment is an essential prerequisite to documenting a hit-and-run mechanism of transformation as well as to suggesting alternative mechanisms.

One unusual observation that comes from transformation assays of HSV-2 *Bgl*II DNA fragments is that two adjacent fragments, *Bgl*II–C and *Bgl*II–N, transform cells in culture. It has been suggested (Lewis *et al.*, 1982; Galloway and McDougall, 1983) that the transformation assay dictates which region of the genome displays transforming capacity. Although this suggestion remains to be confirmed by quantitative assays of transformation with cloned *Bgl*II–C and *Bgl*II–N fragments in each selection system, such a suggestion is not without precedent. Even in the simplest oncogenic DNA viruses, the polyomaviruses, alternative regions of the genome appear to be involved in converting cells to one of several transformed phenotypes. Introduction of polyomavirus or its DNA into primary cells in culture results in the outgrowth of cells with a fully transformed phenotype. Full transformants possess indefinite life span in culture, ability to grow in reduced concentrations of serum, growth as anchorage-independent colonies, and oncogenicity in syngeneic hosts. The region of the polyomavirus genome that mediates transformation encodes three proteins, the large, middle, and small tumor (T) antigens, in alternate translational reading frames. The overlapping organization of these coding sequences has complicated genetic analysis of oncogenic transformation by classical methods. Recently, however, individual coding regions of each of these proteins have been cloned into bacterial plasmids (Triesman *et al.*, 1981; Rassoulzadegan *et al.*, 1982), permitting analysis of the individual roles of each of these polypeptides in transformation. The results emphasize two important features of morphological transformation that may be relevant to other viral systems as well.

First, various assays for a transformed phenotype identify different transforming proteins, and second, individual early proteins have complementary roles in expression of a fully transformed phenotype. When plasmids containing coding sequences for middle T antigen were transferred to a continuous rat cell line (FR3T3), dense foci of transformed cells appeared at a frequency nearly identical to that observed when using the entire polyoma genome cloned into a plasmid vector. Cell lines derived from these foci grew to high density on plastic and in agar. Yet, in contrast to cells transformed by the entire polyoma genome, they were unable to grow in medium containing reduced levels of serum. Correspondingly, transformation of FR3T3 cells by the middle T plasmid alone could not be accomplished with medium containing 0.5% serum. Neither the small nor the large T antigen clone was able to transform the continuous rat cell line in a dense focus assay. However, the serum dependence of middle T-transformed cells was alleviated by introduction of a plasmid encoding large T antigen. In further experiments using a plasmid containing sequences coding for a ts large T antigen, the authors showed that continued expression of both middle and large T antigens is required to maintain a fully transformed phenotype.

The cooperative role of small t antigen of both polyomavirus and SV40 in transforming cells to grow as foci on dense monolayers or independent of anchorage has been demonstrated (reviewed in Tooze, 1981). In contrast to transformation of a continuous cell line, none of the individual genes alone was capable of transforming primary cells in culture, although the entire cloned polyoma genome accomplished this transformation readily. These alternative and complementary roles of middle and large T antigens have escaped detection previously because of the overlap in their coding sequences. In adenoviruses, identification of at least one transforming function appears to depend upon the type of cell used for its assay. The adenovirus transforming functions reside in two contiguous, nonoverlapping segments of the Ad5 genome, E1a (1.5–4.5 MU) and E1b (4.5–11.0 MU). Polypeptides coded within the E1a transcriptional unit are necessary and sufficient to convert primary cells to continuous cell lines (Jones and Shenk, 1979; Land *et al.*, 1983; Ruley, 1983; van der Elser *et al.*, 1982) while products of the E1b transcriptional unit are responsible for complete transformation of cells immortalized by E1A sequences (Ruley, 1983; van der Elser *et al.*, 1982). Thus, both transcriptional units (E1a and E1b) are required to transform primary embryonic cells fully. The possibility that HSV also employs more than one, perhaps complementary mechanisms in altering the growth properties of cells awaits the results of rigorous testing of the hypothesis.

1. Polypeptides Encoded by *mtr-II*

The major virus-specific polypeptide associated with the *Bgl*II–N fragment of HSV-2 DNA is a 38K protein (Docherty *et al.*, 1981; Galloway *et al.*, 1982) with HSV-2 immunological specificity (Docherty *et al.*, 1981). This polypeptide is encoded entirely within the *Bgl*II–N sequences and is translated from the most abundant stable mRNA species that hybridizes to this region (Docherty *et al.*, 1981). Based on studies with metabolic inhibitors, the 38K polypeptide appears to be among the β class of proteins (Galloway *et al.*, 1982). Two additional polypeptides characterized by molecular weights of 140K and 61K appear to be partially encoded within the *Bgl*II–N sequence also (Galloway *et al.*, 1982; Galloway and McDougall, 1983). Whether the 38K polypeptide is involved in the transformation process remains to be determined. The body of the message for this polypeptide lies within the viral sequences of the transforming 2.1-kb left end subclone of the *Bgl*II–N region. However, both consensus transcriptional control signals and the first two nucleotides of the translational start codon are apparently excluded from the subclone (personal communication cited in Galloway and McDougall, 1983) which contains a 300 base pair deletion of sequences from the left end of the *Bgl*II–N fragment (Galloway and McDougall, 1981; 1983). It is likely, then, that this polypeptide does not play a central role in transformation of NIH-3T3 clones to anchorage independent growth, although its role in other transformation assays cannot be excluded. It remains possible that additional less abundant or less stable messages will be mapped to the transforming region of the *Bgl*II–N region of the HSV-2 genome. The possibility that polypeptides encoded only partially within the transforming segment are responsible for altering the phenotype of the cells cannot be excluded from consideration. Precedence for activity of truncated transforming proteins exists. Recent evidence clearly shows that only the NH_2-terminal 40% of the polyoma large T antigen is required for transformation to serum independence following DNA transfection (Rassoulzadegan *et al.*, 1982). Additionally, DNA segments coding for truncated SV40 large T antigens transform cells in culture fully, although at greatly reduced frequencies relative to segments that synthesize full length T antigen (Clayton *et al.*, 1982; Colby and Shenk, 1982; Sompayrac and Danna, 1983). More recently, however, Galloway and coworkers (1984) have indicated

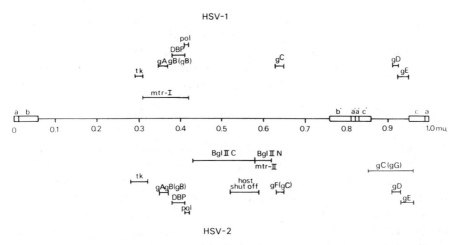

FIGURE 1. Location of transforming and selected lytic functions on the HSV-1 and HSV-2 physical maps. Recently, a new nomenclature for the glycoproteins (see P. Spear, this volume) has been adopted. For clarity both designations are given.

that a 227 base pair segment within the BglII–N region that does not appear to encode a viral polypeptide transforms primary rat embryo fibroblasts and NIH-3T3 cells. The nucleotide sequence of this segment reveals structural similarity to an insertion sequence (IS)-like element. These observations raise the possibility that HSV-2 may be capable of transforming cells in culture by an as yet unidentified and atypical mechanism for DNA containing viruses.

F. Homologies among *mtr-I* and *mtr-II* of the HSV-1 and HSV-2 Genomes

The identification of individual widely separated transforming regions in HSV-1 and HSV-2 DNA was largely unexpected. The location of *mtr-I*, *mtr-II*, and selected lytic functions is shown in Fig. 1. Analysis of intertypic recombinants has shown extensive colinearity of polypeptide-coding sequences (Halliburton *et al.*, 1977; Morse *et al.*, 1977, 1978; Marsden *et al.*, 1978) between the two serotypes. The striking similarity in organization of the HSV-1 and HSV-2 genome raises the possibility that *mtr-II* represents a translocation of *mtr-I* sequences. This possibility was initially eliminated by the results of DNA blot hybridizations (Reyes *et al.*, 1979). In these experiments gel-purified *Bgl*II–I fragment of HSV-1 DNA (*mtr-I* probe) hybridized only to HSV-2 *Bgl*II–F and –K fragments (0.314–0.416 MU). Similarly, *Bgl*II–N fragment of HSV-2 (*mtr-II*) hybridized only to *Bam*HI–N and –O fragments (0.578–0.634 MU) in HSV-1 DNA. These data indicate that a colinear counterpart for each transforming region exists in the DNA of both serotypes.

Not only is there sufficient base sequence homology between *mtr-I* and the corresponding segment of HSV-2 to permit cross-hybridization with immobilized DNA, but the segments are also functionally colinear. Comparisons of the functional defects in mutants of HSV-1 and HSV-2 by intertypic complementation coupled with fine-structure physical and recombinational mapping using cloned DNA fragments show that the order of genes within *mtr-I* sequences on both genomes is glycoprotein gAgB(gB), a late gene of unknown function, the major DNA-binding protein (alternatively named ICP8, VP143, or ICSP11/12), and DNA polymerase (Dixon *et al.*, 1983; Spang *et al.*, 1983; Weller *et al.*, 1983a; Pancake *et al.*, 1983). Thus, the genes identified within this region are functionally homologous and colinear on the HSV-1 and HSV-2 genome.

The analysis of the morphological transforming regions and their counterparts has recently been extended to distribution of base sequence homology by heteroduplex analysis of cloned viral DNA fragments. Kudler *et al.* (1983) have examined heteroduplexes of cloned fragments of HSV-1 and HSV-2 for regions of homology by electron microscopy. The regions investigated extended from 0.2 to 0.3, 0.3 to 0.4, 0.58 to 0.62, and 0.6 to 0.7 MU, and the short unique sequence region. Among these segments, the most extensive homology occurred between DNA occupying 0.2–0.3, 0.3–0.4, and 0.58–0.62 MU. A large number of regions of nonhomology (> 100 bases) were asymmetrically distributed throughout the 0.6- to 0.7-MU region except for the 5 kb adjacent to 0.6 MU and 0.7 kb adjacent to 0.7 MU. Similarly, partial nonhomology was observed throughout the short unique region. In contrast, the region 0.2–0.3 MU was completely homologous. Only three small (0.6–0.7 kb) loops were located within the 15.9-kb segment from 0.3–0.4 MU.

The region from 0.58–0.62 MU showed good or complete homology throughout with the exception of 0.37 kb of partial homology. Clearly, both the *mtr-I* region (0.3–0.4 MU) and corresponding HSV-2 segment and the *mtr-II* (0.58–0.62) and corresponding HSV-1 segment show closely parallel distribution of base sequence homology. Considering the functional and sequence colinearities observed, the inability to demonstrate transforming activity in the region of HSV-2 corresponding to *mtr-I* or in the region of HSV-1 corresponding to *mtr-II* remains puzzling.

This failure may reflect differences in transformation assays and protocols used to define *mtr-I* and *mtr-II* or functional differences that will not become obvious until the nucleotide sequences of each region in both genomes become available. Alternatively, the discrepancy may simply result from the fact that no HSV-2 DNA fragment used for transformation contained *mtr-I* in its entirety. Similarly, none of the HSV-1 fragments used encompassed *mtr-II* completely. Quantitative comparisons of cloned uninterrupted *mtr-I* and *mtr-II* from both genomes in a variety of transformation assays are essential in distinguishing among these alternatives.

G. Mechanisms of Transformation by HSV

The transformation of cells from a normal to a malignant phenotype is a spontaneously occurring multistep process (Nordling, 1953; Foulds, 1954; Armitage and Doll, 1954; Whittemore, 1978). A variety of physical, chemical, and biological agents interface with this pathway to escalate the appearance of transformed cells. Members of the retrovirus family as well as some members of almost all of the families of DNA-containing viruses are capable of stably altering the growth properties of cells in culture. Recent evidence has emphasized the variety of mechanisms employed by DNA and RNA tumor viruses to initiate, maintain, and amplify the transformed phenotype.

The majority of both DNA and RNA tumor viruses transform cells by stably adding segments of their genomes to the genetic information of the cell. By and large, the sequences added code for one or more proteins that directly or indirectly orchestrate the transition from a normal to a malignant phenotype. The relevant sequences, in the case of many RNA tumor viruses, appear to have been confiscated from the genome of normal cells in the evolutionary past as normal cells contain highly conserved homologs of these sequences (Cooper, 1982). Fifteen distinct retrovirus transforming genes have been identified (Coffin et al., 1981, Bishop, 1982; Cooper, 1982). Their cellular homologs constitute the minimal group of normal cellular genes with potential oncogenic activity. Central to the transforming activity of these genes is their juxtaposition to strong viral transcriptional promoters in retrovirus genomes (Blair et al., 1981). Apparently the increased expression of these sequences is responsible for the transformed phenotype, although qualitative changes within the expropriated sequences cannot be ruled out as a contributing factor.

Among the oncogenic DNA viruses, the adenoviruses and the papovaviruses also transform cells in culture by inserting protein-coding sequences into cellular DNA. Continued expression of the transforming sequences is required for maintenance of a transformed phenotype by each of these viruses (Tooze, 1981; Babiss et al., 1983). In contrast to retroviruses, most of the transforming proteins of adenoviruses and papovaviruses play essential and central roles early in the virus lytic cycles. Although the precise mechanisms by which these proteins act during transformation remain elusive, current evidence indicates that multiple, independently acting transforming proteins cooperate in establishing the fully transformed phenotype. The identification of each activity depends upon the transformation assay employed. In the case of polyomavirus (Rassoulzadegan et al., 1982), transformation of continuous cell lines to dense foci in high serum concentration by polyomavirus identifies the transforming activity of middle T antigen. Transformation of continuous cell lines in low-serum medium requires the cooperative activity of middle and large T antigens. Transformation of cells to anchorage independence requires an active small t antigen as well as middle T antigen. In the case of polyomavirus, these cooperating proteins are encoded in overlapping genes within a single defined segment of the viral genome. In adenoviruses, contiguous but independent regions of the genome contribute proteins that cooperate to produce a fully transformed phenotype (Graham et al., 1974; van der Eb et al., 1977; Shiroki et al., 1979; Houweling et al., 1980).

This cooperative action of virus-coded proteins in transformation is not limited to the DNA oncogenic viruses. The transforming capacity of the avian erythroblastosis virus (AEV) resides in two oncogenes one of which encodes the major transforming activity of AEV and a second whose product enhances transformation (Frykberg et al., 1983).

These results indicate that the combined effect of cell types and assays for transformation exert strong selective pressure for subsets of transforming events. In contrast to the acute-transforming retroviruses, the weakly oncogenic retroviruses do not contain a specific viral transforming gene. Significantly, these viruses do not transform cells in culture. After long latent periods, neoplasms develop in vivo following injection of virus. Within tumors induced by one such virus, avian lymphoid leukosis virus (LLV), viral promoter sequences are frequently integrated in the vicinity of the cellular gene homologous to the transforming gene of the acute-transforming virus MC-29 (myc) (Cooper and Neiman, 1981; Hayward et al., 1981), resulting in its increased expression.

Potential transforming genes of normal cells can apparently be activated without intervention of a virus by (1) linkage of noncontiguous cellular DNA segments during transfection assays in vitro (Cooper et al., 1980); (2) specific chromosomal translocations causing rearrangements of cellular oncogenes and flanking sequences in vivo (Klein, 1983); or (3) point mutation (Reddy et al., 1982; Tabin et al., 1982; Capon et al., 1983). Investigations of retroviral and cellular oncogenes also suggest that the recipient cell type imposes limitations on our ability to detect transforming genes.

The majority of the investigations have employed NIH-3T3 mouse fibroblasts as recipients of gene transfer. The choice of this cell line rests on its ability to undergo DNA-mediated transformation at a high rate, compared with other cell lines (Weinberg, 1981). It is important to realize that continuous cell lines often undergo spontaneous and chemically induced transformation at rates indicative of a single-step process (Kakunaga, 1973). The NIH-3T3 cell line may, therefore, have already accumulated all but the final alteration necessary to become fully transformed and tumorigenic. The importance of this cell line in detecting cellular oncogenes expropriated by retroviruses (Coffin et al., 1981) and in transforming sequences from a variety of animal and human tumors and leukemias (Weinberg, 1982) cannot be overstated. However, potential limitations of this assay system also must be recognized. The DNA of most human and animal tumors does not transform NIH-3T3 cells upon transfection (Krontiris and Cooper, 1981; Shih *et al.*, 1981). It is possible that the NIH-3T3 assays identify a limited and distinct subset of events involved in transformation of cells from a normal to a malignant phenotype. This likelihood is supported by the transformation of NIH-3T3 cells by DNA from LLV tumors. As mentioned above, most LLV-induced lymphomas contain viral sequences integrated adjacent to the cellular *myc* gene. Yet, when tumor cell DNA was transfected into NIH-3T3 cells, the resulting transformants did not contain *myc* sequences from the donor DNA (Cooper and Neiman, 1981). Apparently a cellular gene unrelated to *myc* and not linked to viral DNA efficiently induced transformation of the NIH-3T3 cells.

A minimum of four independent mechanisms by which viruses transform cells has been identified. These include: (1) transduction of cellular genes with oncogenic potential by incorporation into viral genomes and their resulting liberation from normal control mechanisms; (2) insertion of strong promoter sequences adjacent to cellular oncogenes; (3) introduction of virus genes coding for proteins with essential functions in virus replication and concomitant transforming capacity; and (4) genetic alterations of cellular sequences to activate cellular oncogenes.

The observation that cells exposed to inactivated HSV virions consistently give rise to morphologically transformed colonies at a low frequency, while unexposed cells do not, strongly implicates viral DNA sequences and/or gene products in the initiation of transformation. Insufficient evidence exists to suggest a single mechanism by which HSV transform cells in culture. The data available do not rule out the possibility that transformation of primary cells is accomplished by the introduction and continued expression of virus-coded sequences in a manner analogous to transformation by adenoviruses and papovaviruses. Consistent with this possibility is the observation (reviewed above) that cells transformed in culture by HSV virions both retain and express viral information. Although a general reduction in complexity is observed during prolonged propagation, the majority of cell lines investigated (even at high passage number) contained detectable viral sequences, viral RNA, or virus-specific polypeptides, where appropriate probes were used for their detection. In contrast, Galloway and McDougall (1983) have indicated that HSV-2 DNA sequences that are sufficient to permit NIH-3T3 cells to develop as anchorage-independent clones are not detectable when those clones are expanded into cell lines. The conflict that primary cells transformed by HSV virions retain specific regions of the HSV genome during prolonged propagation whereas NIH-3T3 cells transformed by the left-hand end of the *Bgl*II–N fragment lose these sequences rapidly remains to be resolved. A determination of whether these findings reflect different independent mechanisms of transformation, coincidental but stable retention of viral sequences in primary cells, or a separation of transformation initiation and maintenance events awaits additional data.

It is possible that HSV virions and DNA accomplish transformation by independent mechanisms. In the case of both polyomavirus and adenovirus, the initiation of transformation of cells in culture by virions requires viral gene products that are not necessary when transformation is assayed with purified DNA. As discussed above, a polyoma DNA segment encoding only middle T antigen is sufficient for transformation of a continuous cell line. However, efficient transformation of a continuous cell line by virions requires a functional large T antigen as well (Della Valle et al., 1981). Similarly, the 58K T antigen of adenovirus is required for transformation of primary cells by virions (Graham et al., 1978). This requirement disappears when DNA purified from virions is used in transformation assays (Rowe and Graham, 1983). Although it is not required for the maintenance of transformation (Rowe and Graham, 1983), the 58K T antigen continues to be expressed in cells transformed by virions and is the predominant antigen to which animals bearing adenovirus-induced tumors respond immunologically (Levine, 1982). The possibility that transformation by HSV DNA bypasses some initiation functions and detects only a subset of the events that participate in transformation by virions cannot be excluded. The difficult task of identifying and characterizing nontransforming mutants of HSV is a prerequisite to confirming this hypothesis.

It is equally possible that cells transformed by HSV can retain and express any viral sequences that are not detrimental to growth of cells in culture. A distinction between DNA that is simply a passenger in transformed cells and sequences that are essential to maintaining a transformed phenotype is essential to understanding the transformation process. At least one means of accomplishing this separation is available. Serial passaging of transformed cell DNA in transformation assays has been used to identify viral and cellular DNA sequences responsible for transformation by viruses or chemical carcinogens and in cells derived from human tumors (for review see Cooper, 1982; Weinberg, 1981, 1982). By using DNA from HSV-transformed cells as donor DNA in subsequent cycles of transformation, it should be possible to determine which viral DNA sequences, if any, that are retained in virion-transformed cells are essential to maintaining a transformed phenotype. The establishment of secondary and tertiary transformants should eliminate DNA that is not essential for transformation and should ultimately permit identification of transforming sequences regardless of their origin.

The rapid loss of viral DNA sequences by NIH-3T3 cells after exposure to a 2.1-kb fragment containing sequences from the left end of the BglII–N fragment of HSV-2 DNA is unexpected in light of the ability of cells transformed by virions to retain viral DNA and remains to be explained. A direct test of whether retention of sequences from the left end of the BglII–N fragment is detrimental to cell survival or limits the development of transformed clones is important to understanding this observation.

Features of the structural and functional organization of the HSV genome are consistent with several mechanisms that may initiate or enhance the transformed phenotype as well. Specifically, virus-coded proteins that enhance mutation, and virus-coded factors and sequences that mediate rearrangements of nucleotides could contribute to the transformation process.

Consistent with a hit-and-run mechanism is the observation that the HSV-1 DNA polymerase possesses mutator activity. Hall and Almy (1982) have demonstrated that the frequency with which TK mutants of HSV-1 accumulate is controlled by the viral DNA polymerase. The proposed mutator activity is consistent with the genetic variability of HSV, as compared to other herpesviruses (Buchman *et al.*, 1980; Lonsdale *et al.*, 1980). If the viral polymerase can contribute to synthesis of cellular DNA, either during normal semiconservative replication or during repair of virus-induced DNA damage (Lorentz *et al.*, 1977), the possibility that this protein induces cellular mutations that lead to transformation must be considered. Recently, Schlehofer and zur Hausen (1982) have demonstrated a 2- to 10-fold increase in the frequency of mutations in the cellular hypoxanthine–guanine phosphonoribosyltransferase (HGPRT) gene following infection with inactivated HSV-1 virions. These results suggest that one or more HSV gene products enhance the mutation rate for cellular genes. Whether this activity on cellular genes resides in the viral DNA polymerase or not remains to be determined.

Clearly, chemically transformed cells contain altered DNA sequences that are directly responsible for the transformed properties of the cell (reviewed in Weinberg, 1982). Additionally, a single point mutation in a human bladder carcinoma oncogene (Reddy *et al.*, 1982; Tabin *et al.*, 1982; Capon *et al.*, 1983) appears to confer its transforming properties to NIH-3T3 cells. There is no reason at present to reject the hypothesis that the same potentially oncogenic cellular sequences are targets for the mutagenic activity of HSV.

It is of interest to recall in this context that SV40 also induces mutations in the cells it invades. Zannis-Hadjopoulos and Martin (1983) have shown that infection of nonpermissive cells with SV40 leads to a 2- to 8-fold increased frequency of 8-azaguanine-resistant cells over the spontaneous frequency. In the case of SV40, this mutagenesis probably does not represent the primary mechanism by which the virus accomplishes transformation, as the frequency of transformants in parallel experiments increased 170-fold over the level of spontaneous transformants. Whether or not the mutator activity of HSV is primarily responsible for transformation, expression of this activity during propagation of cells could contribute to genetic alterations in the cellular genome that enhance survival or growth rate of transformed cells under restrictive growth conditions. Therefore, this activity may represent, if not the perpetuator, then a significant adjunct to the transformation process.

The observation that certain cellular genes, when removed from their normal context of nucleotides by mechanical shearing (Cooper *et al.*, 1980) or by incorporation into a retroviral genome, display transforming capacity suggests that equivalent oncogene activation could be achieved by rearrangement of blocks of normal cellular genes within cells. The activation of a cellular oncogene by DNA rearrangement has been demonstrated recently (Rechavi *et al.*, 1982). Chromosomal rearrangements have been observed frequently following infection of cells with HSV (Hampar and Ellison, 1961, 1963; Stich *et al.*, 1964; Rapp and Hsu, 1965; Waubke *et al.*, 1968; O'Neill and Rapp, 1971) and provide a potential for such activation. A second mechanism of cellular DNA rearrangement is suggested by the recent identification of *trans*-acting viral gene functions that promote inversion of DNA flanked by specific viral sequences. Mocarski and Roizman (1982) have shown that HSV gene products mediate inversion of cellular DNA sequences bounded by inverted copies of the *a* sequences of the terminal and inverted repeat sequences from the viral genome. A logical prediction, based on these observations, is that *a* sequences integrated in inverted orientation with cellular DNA between them will permit inversion of the bounded cellular DNA, thereby moving the sequences to a new genetic context. In this connection, Mushinski *et al.* (1983) have recently provided evidence that is consistent with retention of a transformed phenotype after loss of the Abelson murine leukemia virus (A-MuLV) genome from plasmacytoid lymphosarcomas. In their study, most lymphoid tumors arising after inoculation of mice with A-MuLV contained the viral genome in integrated form and abundantly produced v-*abl* message. In addition, most of the tumors produced elevated amounts of RNA from the cellular *myb* homolog and exhibited DNA rearrangements in the *myb* gene region of the chromosome. The authors showed that plasmacytoid lymphosarcomas lack the proviral gene and yet have undergone rearrangements in the c-*myb* locus resulting in its increased transcriptional activity. They conclude that A-MuLV occasionally may act during *in vivo* tumorigenesis through a hit-and-run mechanism involving transient expression of the viral genome followed by rearrangement of cellular sequences and loss of viral sequences.

The examples cited above suggest that HSV possess the potential to transform cells by a variety of mechanisms operational in other carefully investigated tumor virus systems. The prolonged retention and expression of HSV information in cells transformed by virions provide ample opportunity for multiple mechanisms to contribute to the transformed phenotype. Which, if any, of these mechanisms contribute to transformation of cells in culture and to the tumorigenesis by HSV and which one constitutes the predominant mechanism remain to be determined. It does not necessarily follow that mechanisms of transformation identified in cell culture will accurately reflect events that may occur *in vivo*. Of necessity, transformation of cells in culture by HSV has involved limiting expression of the genome by photodynamic or UV inactivation or incubation at nonphysiological temperature. The highly lytic nature of HSV suggests that transformation, if it occurs *in vivo*, may be accomplished by virions with a reduced capacity to kill the cells they invade. The class II defective virions (Locker *et al.*, 1982) are attractive candidates for two reasons. (1) These defective genomes interfere with the growth of standard virus, thereby enhancing the likelihood that the infected cell will survive. (2) They contain repeated units to DNA sequences from 0.356 to 0.423 MU in head-to-tail tandem array linked to sequences from the S component (Locker *et al.*, 1982). These sequences from the long component lie within *mtr-I* and contain at least the genes for the DNA-binding protein and viral DNA polymerase. Although virions containing class II defective genomes cannot be separated readily from standard virus, the DNA's are separable (Locker *et al.*, 1982). An assessment of the transforming capacity of class II defective virus DNA will be of considerable importance in relating transformation as it occurs *in vitro* to the potential for oncogenesis *in vivo*.

It is likely (based on our current knowledge of the complexity of the transformation process) that a clear understanding of the roles and interrelationships of viral gene products and sequences during transformation will require a coordinated investigation in at least three areas: (1) determination of genome segments that, acting together or individually, transform primary, continuous, and ultimately differentiated cells in a variety of assays for parameters of the transformed phenotype, (2) the genomic location and arrangement of viral sequences retained and expressed by transformed cells, and (3) the nature of sequences in transformed cells that transmit the transformed phenotype to cells in culture. Ultimately, a causal relationship between viral sequences or gene products and the transformed phenotype will rest on their genetic alteration and coordinate loss of transforming potential.

III. TRANSFORMING POTENTIAL OF HCMV

HCMV contains the largest genome of the herpesviruses. The structural organization of this 155×10^6-dalton linear duplex is similar to that of HSV DNA. The genome of HCMV consists of a large and a short unique segment each bounded by inverted repeats. Consequently, as in the case of HSV, four isomeric forms of the DNA accumulate during replication (DeMarchi *et al.*, 1978; Geelan *et al.*, 1978; Lakeman and Osborn, 1979; LaFemina and Hayward, 1980; Stinski and Thomsen, 1981; Spector *et al.*, 1982). Both the biology (Rapp, 1983) and the molecular biology (Stinski, 1983) of HCMV have been discussed in detail recently. Therefore, only aspects of the replication of this virus that may be relevant to its transforming capacity are reviewed here.

Three biological properties of HCMV contrast sharply with HSV host–virus interactions. Rather than inhibiting most macromolecular synthesis, HCMV stimulates cellular DNA (St. Jeor *et al.*, 1974; Furukawa *et al.*, 1975), RNA (Tanaka *et al.*, 1975), and protein synthesis (Stinski, 1977). In addition, the lytic cycle of HCMV is prolonged, requiring 5–6 days for completion versus 18 h for HSV replication (Smith and De-Harven, 1973). In contrast to the broad host range of HSV, HCMV lytically infects only human diploid fibroblasts in culture efficiently. Infection of human fibroblasts results in an ordered and sequential expression of the viral genome (Stinski, 1977, 1978; Michelson *et al.*, 1979; Tanaka *et al.*, 1979; DeMarchi *et al.*, 1980; Wathen *et al.*, 1981; Gibson, 1981; Blanton and Tevethia, 1981; Wathen and Stinski, 1982). Infected cell-specific polypeptides (ICSP) synthesized early during virus infection exert a central role in controlling expression of the HCMV genome in lytically infected cells.

Infection of nonpermissive rodent cells with HCMV results in a more limited expression of viral information. Immediate–early (IE) and most early (E) polypeptides are produced; however, virus-specific DNA polymerase (Hirai *et al.*, 1976), viral DNA, and late (L) polypeptides are not synthesized (Stinski, 1978; Michelson *et al.*, 1979).

A. Morphological Transformation of Cells in Culture by HCMV

The capacity of HCMV to transform normal cells in culture to cells capable of producing malignant tumors was first demonstrated by Albrecht and Rapp (1973). After exposure of HEF to UV-inactivated HCMV (strain C-87), dense foci of cells appeared. One of sixteen foci developed into a stable cell line, CX-90-3B. Although hamster cells are nonpermissive for replication of HCMV, infection with the virus at high multiplicity is cytotoxic to these cells. For this reason, inactivation of virion infectivity was required to generate transformed foci. The CX-90-3B cells produced tumors in both newborn and weanling hamsters.

Both the parental and the tumor-derived cell lines retained and expressed HCMV DNA, as evidence by the detection of virus-specific membrane antigens with human covalescent serum. This serum also produced diffuse cytoplasmic fluorescence in a small percentage of the parental cells but not in the tumor-derived cell line. Sera from hamsters bearing CX-90-3B tumors did, however, produce cytoplasmic fluorescence in HCMV-infected cells at late times postinfection (Albrecht and Rapp, 1973). This last result suggests that the tumor cells contained HCMV-specific antigens in addition to those detected by human convalescent sera.

Subsequently, a second independent hamster cell line transformed by UV-irradiated HCMV (C-87) was shown to be oncogenic (Boldogh *et al.*, 1978) and to express HCMV cytoplasmic and membrane antigens. The transformation of rodent cells implicates IE or E gene products in this process.

Malignant transformation by HCMV *in vitro* has not been confined to nonpermissive hamster cells, to the C-87 strain, or to UV-inactivated virus. Geder *et al.* (1976) reported that human embryonic lung cells persistently infected with HCMV also give rise to transformed foci.

A strain (Mj) of HCMV isolated from human prostate tissue (Rapp *et al.*, 1975) when inoculated into human embryonic lung (HEL) cells at a low multiplicity (< 0.001 PFU/cell) resulted in persistent infection. After a crisis period, during which the majority of the cell population died, foci of transformed cells appeared and dominated the culture. In each of two resulting cell lines, CMV-specific membrane, paranuclear, and perinuclear antigens were detected by immunofluorescence assays using human convalescent sera, and adsorption of the CMV-immune sera with CMV-infected cells removed reactivity of the sera for the transformed cell membrane. The same pattern of immunofluorescence was observed on cells derived from tumors established in athymic nude mice after injection of the HCMV-transformed cell lines. The retention of virus-specific antigens in the immortalized HEL cells following persistent infection with HCMV and in tumors derived from these cells implicated HCMV in malignant transformation of cells from the natural host by active virus. The percentage of HCMV-transformed (CMV-Mj-HEL-2) and tumor cells (CMV-Mj-HEL-2,T-1) expressing HCMV-specific antigens decreased with increasing passage level (Geder *et al.*, 1979). Although the kinetics of decrease differed between the transformed and the tumor-derived cell lines, at passages of 60 and 146, respectively, HCMV antigens detectable by immunofluorescence assays with human convalescent sera were no longer present. After observing the antigenic expression and tumorigenicity of the parental CMV-Mj-HEL-2 cell line and CMV-Mj-HEL-2,T-1 tumor cell line over 228 and 180 *in vitro* passages, respectively, Geder *et al.* (1979) concluded that a decrease in virus-specific antigens correlated with high passage number and increased tumorigenicity in nude mice.

Little information regarding viral DNA sequences retained by HCMV-transformed cells is available. The loss of antigens from cells at high passage number indicates an alteration in expression or retention of sequences in HCMV transformants. Without additional data it is not possible to distinguish between two explanations for the apparent absence of viral products. It is possible that HCMV immortalizes primary cells and provides time for spontaneous events required for progression to tumorigenicity to occur after which HCMV information is expendable. As in the case of HSV transformants, cells transformed by HCMV may become heterogeneous with respect to viral sequences that are not necessary for maintenance of transformation. It is equally possible that sequences that are required or that neighbor essential sequences may encode polypeptides that are not sufficiently abundant or stable during infection in the natural host to stimulate a strong immune response. Alternatively, one or more polypeptides expressed in HCMV-transformed cells may not be detectable by immunofluorescence assays as discussed previously.

B. Transformation by HCMV DNA and DNA Fragments

Both HCMV DNA cleaved with restriction endonucleases and cloned DNA segments of the viral genome transform NIH-3T3 cells to an anchorage-independent phenotype. Nelson *et al.* (1982) determined that HCMV (AD169) cleaved with either *Hin*dIII or *Xba*I retained its ability to transform NIH-3T3 cells. In both instances, using 20 μg of DNA, they observed clones growing in methylcellulose at a frequency of 0.8 colony/μg per 10^6 cells. In contrast, DNA cleaved with *Eco*RI generated no transformants. These results indicated that at least one *Eco*RI cleavage site resides within the sequence of nucleotides required to establish transformants. By using overlapping *Hin*dIII fragments of HCMV DNA cloned into cosmid vectors (Fleckenstein *et al.*, 1982) and liberated prior to transfection by cleavage with *Hin*dIII or *Xba*I, the transforming segment was localized to the left end of the unique sequences of the long component between 0.055 and 0.2 MU. The position of the transforming sequences was mapped more precisely with *Eco*RI subclones of this region to 0.123–0.14 MU within the *Hin*dIII–E fragment on the prototype molecule of AD169 as defined by Fleckenstein *et al.* (1982). Importantly, a 2.9-kb segment of DNA containing precisely these sequences cloned into plasmid vector transformed cells with equal efficiency. Eleven of twelve clonally derived independent cell lines transformed with HCMV DNA or recombinant cosmid-containing sequences from 0.055 to 0.2 MU were highly tumorigenic in athymic nude mice. In preliminary experiments, Nelson *et al.* (1982) were unable to detect viral DNA in transformants or tumor cells using a recombinant plasmid housing a 15.2-kb insert containing the transforming sequences. This probe is in excess of the transforming region and its complexity may have prevented detection of viral sequences within the transformed cell DNA (Frenkel *et al.*, 1976; Robinson and O'Callaghan, 1983). These results indicate that a single contiguous set of nucleotides among the unique sequences of the long component of HCMV DNA extending from 0.123 to 0.14 MU is sufficient to transform NIH-3T3 cells to anchorage independence.

One observation of Nelson *et al.* (1982) bears mentioning. The frequency of transformed clones observed when using an equal quantity of restriction nuclease digests of HCMV DNA, a cosmid containing a 34.3-kb insert, or a plasmid containing either a 6.3-kb or a 2.9-kb insert of HCMV DNA (each containing the essential transforming sequences) did not vary by more than a factor of 2, even though these DNA sources represent a 33-fold difference in size. This result contrasts with the dramatic increase in the frequency of TK^+ transformation achieved by transfecting TK^- cells with cloned TK gene sequences rather than total restriction nuclease digests of HSV DNA and suggests strongly that only a limited number of cells in the NIH-3T3 population can be transformed by the mechanism represented within 0.123–0.14 MU on the HCMV genome. The role of these transforming sequences, as identified by transfection of NIH 3T3 cells, in the possibly more complex process of conversion of normal primary to malignant cells remains to be determined. More recently, Nelson *et al.* (1984) have shown that a 490 bp subclone of this segment permits primary rat embryo fibroblasts to grow as foci in low serum and as anchorage-independent colonies. The longest open reading frame in this segment could encode a polypeptide of only 41 amino acids. It is important to point out that stable cell lines could not be established from these colonies.

Clearly, definition of transforming sequences and genes is provisional. The nature of the selective pressure applied to the transformed phenotype, the stage of cell differentiation, the cellular physiology, and the extent to which alterations required for transformation of normal cells have already accumulated dictate which viral gene products are required for stable transformation. It is of considerable importance then to develop methods for transfecting not only continuous cell lines but also primary and ultimately differentiated cells in culture.

It is reasonable to believe that outgrowth of foci from primary cells in culture and development of anchorage-independent clones from continuous cell lines after exposure to virus reflect different points of entry into a transformation pathway. In establishing an alternative transformation assay for HCMV, we investigated the ability of primary C57BL/6 mouse embryo fibroblasts (MEF) to form foci in medium containing 10% fetal bovine serum after transfection with viral DNA. Primary cultures were chosen as these cells contain the intracellular environment most likely to be encountered by viruses *in vivo*.

In this assay, C57BL/6 mouse cells were used as recipients of viral information for three reasons. First, the spontaneous transformation frequency is extremely low in these cells. Second, the immune reactivity of the C57BL/6 mouse has been characterized extensively and the immune response to virus-transformed cells has been and can be investigated effectively in this system (Tevethia *et al.*, 1980b). Third, virus-transformed mouse cells, unlike hamster transformants, are not always tumorigenic immediately after acquisition of a transformed cell phenotype (Tevethia and McMillan, 1974). Thus, dissection of events that confer transformed cell characteristics from those that enhance or limit tumorigenicity may be possible.

We have previously demonstrated that C57BL/6 MEF are readily transformed by SV40 in a focus assay (Tevethia *et al.*, 1980a). Conditions for the transformation of C57BL/6 MEF with DNA were standardized by comparing the number of transformants obtained when cells were transfected with SV40 virions, SV40 DNA, or a plasmid containing SV40 transforming sequences (0.72–0.16 MU). The results (Tevethia, 1984) indicate that MEF cells can be reproducibly transformed with equal efficiency (263 colonies/10^6 cells) by 100 PFU of virus per cell or 1 μg of either SV40 DNA or plasmid containing early region sequences. This frequency corresponds to 0.025% of the infected or transfected cells. Cell lines expanded from individual clones universally expressed the three antigenic markers of SV40 transformation—the large T, small t, and cellular p53 polypeptide [nonviral T (NVT)] antigens.

HCMV DNA extracted from virions and purified according to the method of Geelen et al. (1978) was also tested for its ability to transform C57BL/6 cells. For this purpose, a DNA preparation that had been stored for a prolonged period at 4°C was used. This type of preparation was selected for the initial experiment because evidence was already available (M. Tevethia, unpublished observation) that virion infection of nonpermissive C57BL/6 mouse cells resulted in loss of cell viability. To limit possible cytotoxicity, a DNA preparation that had accumulated random damage was used.

Transformed foci developed in three out of four flasks of C57BL/6 MEF transfected with DNA purified from HCMV virions. As expected based on the size of the genome, the number of transformed foci in cultures transfected with HCMV DNA was lower than that obtained with SV40 DNA. Transformation with the DNA preparation used occurred with a frequency of 0.5 clone/μg per 10^6 cells transfected. Although this frequency is relatively low, it should be pointed out that no spontaneous transformants have appeared in any of over 30 control flasks of C57BL/6 cells that received carrier DNA alone.

The two independent cell lines transformed by HCMV DNA (B6/HCMV) were examined for oncogenicity in nude mice. For this purpose, five mice in each group were each injected with cells at early passage number (< 10) and observed for the development of tumors. In addition, an SV40-transformed C57BL/6 cell line and a rare spontaneously transformed C57BL/6 cell line were injected into nude mice as positive and negative controls. All of the mice in each group receiving B6/HCMV cells as well as four of the mice receiving SV40-transformed C57BL/6 cells developed tumors within 4 weeks. Cells derived from tumors in nude mice receiving B6/HCMV-transformed cells were highly oncogenic in adult C57BL/6 mice. All of the mice in groups of four to six receiving cells from either of two tumor lines derived from each transformed cell line developed tumors within 2 weeks after inoculation. Nude mice inoculated with a similar number of cells of a spontaneously transformed C57BL/6 cell line did not develop tumors even after 9 weeks. The ease with which a transplantable tumor cell line could be established in adult C57BL/6 mice is significant because this strain of mouse is refractory to development of tumors by SV40- and polyomavirus-transformed syngeneic cells. Recently, an SV40-transplantable tumor has been established in C57BL/6 mice, but only after multiple selections through highly immunocompromised animals (Flyer et al., 1983). The oncogenicity of primary C57BL/6 cells at early passage following transformation by HCMV DNA strongly supports the suggestion that transformation of cells by this virus is not accompanied by acquisition of antigenicities at the cell surface that participate in tumor rejection. The absence of transplantation rejection antigens on HSV-transformed cells has been reported by Rapp and Duff (1973) and by Hay and Lausch (1979).

The low background of spontaneous transformation in C57BL/6 MEF cultures under the transfection conditions employed makes these cells attractive recipients for transformation by herpesviruses. Recently, M. K. Howett and co-workers (personal communication) accomplished onco-genic transformation of C57BL/6 cells with UV-irradiated HSV-2 virions as well as with a cloned HSV-2 *Bgl*II–N DNA fragment. The observations that HCMV DNA can transform primary mouse cells in culture under conditions that do not promote the outgrowth of spontaneous transform-ants and that the resulting cell lines often are highly tumorigenic at low passage number indicate that this virus can transform primary cells to tumorigenic cell lines without prolonged propagation.

The generation of stable oncogenic cell lines following infection of primary hamster, mouse, and human embryo fibroblasts and the contin-uous NIH-3T3 cell line with HCMV or its DNA strongly implicates viral sequences or gene products in the initiation of malignant transformation. The minimal HCMV sequences essential to this process in primary cells and those required for maintaining a transformed phenotype remain to be determined. Transfection of C57BL/6 primary cells with cloned seg-ments of HCMV DNA is in progress and should permit identification of those regions of the genome that transform primary cells in a focus assay.

This same assay system should permit the separation of sequences essential to maintaining a transformed phenotype from coincidentally retained viral DNA sequences. This separation is often accomplished by transforming cells in culture sequentially with DNA from transformed or tumor cells (reviewed in Weinberg, 1981). We have demonstrated the ability of primary C57BL/6 cells to be transformed by DNA extracted from three SV40-transformed or tumor cell lines and from the human cell line, 293 (Graham *et al.*, 1977), which contains the transforming region of adenovirus type 5. This secondary transformation occurs at a frequency of 0.2–0.4 focus/μg DNA per 10^6 cells and closely parallels the frequency observed in other systems (Cooper *et al.*, 1980; Payne *et al.*, 1982; Smith *et al.*, 1982). In all cases the secondary transformants ex-pressed the polypeptide markers of SV40 transformation (S. Tevethia and M. Tevethia, unpublished observation) or the same species of adenovirus E1a and E1b RNA and polypeptide products expressed in 293 cells (D. Spector *et al.*, unpublished data).

Resolution of the sequences and viral functions required to maintain a transformed phenotype rests on the ability to detect viral DNA, RNA, and polypeptides in transformed cells. The complexity of the viral genome and our scant knowledge of its functional organization make this detection complicated. The magnitude of the task is easily realized by recalling that the HCMV genome is half again as large as HSV DNA.

Functional analysis of the HCMV genome has been initiated recently by mapping virus-specific abundant RNA species extracted from infected cells under conditions permitting accumulation of IE, E, or L RNA. In three strains of HCMV the major abundant IE species of RNA mapped primarily to three segments of the genome, whereas E and L transcripts were detected from more extensive regions of the genome (DeMarchi *et al.*, 1980; Wathen *et al.*, 1981; McDonough and Spector, 1983). The most abundant IE RNA of HCMV maps at 0.061–0.110 MU in AD169 and 0.739–0.751 MU in Towne strain. It is important to realize that the long unique segments of the Davis and Towne strains are inverted in the prototype arrangement relative to the map of AD169 so that these regions are analogous (McDonough and Spector, 1983). This 1.95-kb species when translated *in vitro* (Stinski *et al.*, 1983) produced a 75K protein that was immunoprecipitated by monoclonal antibody E3 (Goldstein *et al.*, 1982) against the major IE polypeptide in HCMV-infected cells. This polypeptide represents the sole gene product unequivocally mapped on the HCMV genome. At least one of the IE transcription sites at 0.117–0.142 MU on the prototype arrangement of AD169 (McDonough and Spector, 1983) or the equivalent region (0.709–0.728 MU) on the prototype arrangement of the Towne strain genome (Wathen and Stinski, 1982) overlaps the transforming region as defined by loss of anchorage dependence in NIH-3T3 cells. When translated *in vitro*, hybrid-selected RNA from this region produced a 68K polypeptide abundantly (Stinski *et al.*, 1983). The transforming segment is transcribed also at early and late times during lytic growth. The locations of a transforming region, abundant IE transcripts, and the 68K and 72K polypeptide-coding sequences are shown in Fig. 2. Additional IE as well as E and L transcripts have been localized (DeMarchi, 1981; Wathen and Stinski, 1982; McDonough and Spector, 1983; Stinski *et al.*, 1983); however, their relation to specific ICSP detected by human convalescent sera is difficult to assess. This difficulty reflects a significant difference between HSV and HCMV in the detection of ICSPs.

Unlike HSV, HCMV stimulates rather than inhibits protein synthesis in host cells. Even during late stages of the infectious cycle, host cell protein synthesis remains at approximately 40% of the total synthesis in spite of the fact that large quantities of virus are produced (Stinski, 1977). The continued production of cellular proteins compromises the identification of HCMV-specific polypeptides in infected cell extracts. Primarily three methods have been employed to enhance detection of ICSPs. This enhancement has been accomplished by treating cells with cycloheximide before and during infection to permit accumulation of viral mRNA and then radiolabeling in hypertonic medium after removal of the inhibitor to stimulate preferential translation of viral mRNAs (Stinski, 1978). Alternatively, HCMV-specific polypeptides have been im-

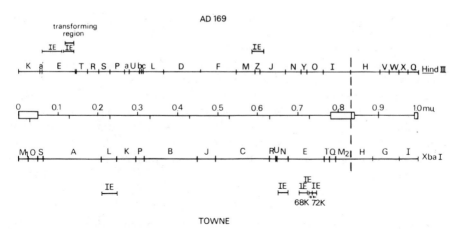

FIGURE 2. Location of a transforming region of HCMV, abundant IE transcripts, and 68K and 72K polypeptides. The *XbaI* map of the Towne strain (Thomsen and Stinski, 1981) and *Hind*III map of AD169 (Fleckenstein *et al.*, 1982) are drawn in their prototype arrangements. The L component of the AD169 map in this arrangement is inverted relative to the Towne strain map (McDonough and Spector, 1983).

munoprecipitated from infected cell extracts with human convalescent sera (Michelson *et al.*, 1979; Blanton and Tevethia, 1981). In addition, polypeptides have been synthesized *in vitro* with mRNA isolated from infected cells and selected by hybridization to HCMV DNA (Wathen *et al.*, 1981). Each of these methods has unique limitations and potentials for generating artifacts. As both different strains of HCMV and variations in SDS-polyacrylamide gel electrophoresis systems have been employed by several groups, comparisons of studies of ICSPs are difficult. Nonetheless, at least 4 IE and 18E ICSPs have been detected in extracts of permissive cells (Stinski, 1978; Michelson *et al.*, 1979; Blanton and Tevethia, 1981) during the first 24 hr of infection. In comparison, *in vitro* translation of IE mRNA yielded 9 proteins. An additional 10 polypeptides were translated from mRNA isolated from infected cells after 6 or 14 hr of protein synthesis (Wathen *et al.*, 1981). Late in the infectious cycle (36 hr after infection) virion structural proteins and glycoproteins accumulate and E proteins are no longer detectable (Stinski, 1977, 1978). Many of these L proteins have the electrophoretic mobilities of proteins associated with purified virions (Stinski, 1977, 1978). None are detected in infected human fibroblasts treated with the viral DNA synthesis inhibitor phosphonoacetic acid or in nonpermissive guinea pig cells (Stinski, 1978).

Although notable differences exist in results of studies of ICSPs from various laboratories, two consistent observations emerge. First, IE proteins of HCMV play a central, although often transient role in productive infection, as the further sequential expression of the viral genome requires their synthesis (Stinski, 1978; Michelson *et al.*, 1979; Blanton and

Tevethia, 1981). Second, virus-specified E proteins, in contrast to IE proteins, are expressed for prolonged periods of time. Little change in the profile of ICSPs is detected between 6 and 24 hr postinfection (Stinski, 1978; Wathen et al., 1981; M. Tevethia and L. Brogan, unpublished data). We have examined the hamster cell line CX-90-3B, a tumor derived from these cells (CX-90-3B-T1), the human cell line CMV-Mj-HEL, and three B6/HCMV cell lines for the presence of IE and E polypeptides by immunoprecipitation of radiolabeled cell extracts with human convalescent serum. The serum used was capable of distinguishing 22 IE and E polypeptides in HCMV-infected cultures (Blanton and Tevethia, 1981); however, no virus-specific polypeptides were observed in extracts of any of the transformed cells. The apparent absence of ICSP is not surprising. It is likely that the ICSPs identified are a minimal representation of the virus-coded polypeptides. ICSPs detected by immunoprecipitation during early stages of infection must be considered a minimal estimate of the actual number produced for several reasons. First, sera differ both qualitatively and quantitatively in their ability to immunoprecipitate polypeptides from infected cells. It is likely that some polypeptides that are present in low abundance or expressed transiently, may not elicit a strong antibody response in a naturally infected host. Second, ICSPs bound tightly to DNA (Tanaka et al., 1979) or to nuclear structures may not be extracted efficiently from infected cells under the conditions used.

The variability in antibody level for specific ICSPs in human convalescent sera from different individuals (Michelson et al., 1979; Blanton and Tevethia, 1981) and from a single individual at different times as well as the indications that some ICSP bands in polyacrylamide gels represent more than one polypeptide emphasize the need for monoclonal antibodies against IE and E polypeptides to detect viral gene products in infected and transformed cells.

The factors that have assisted in probing functions within the morphological transforming regions of HSV, i.e., the large number of mutants identifying specific viral functions, the capability of localizing protein-coding sequences by intertypic recombinational analysis, and the ability to map mutations on the viral chromosome by marker rescue are not available at this point for HCMV. Several HCMV-transformed cell lines derived from primary cells infected with virions of HCMV DNA and tumor-derived lines are available. Analysis of the viral sequences retained in these cells can be expected to indicate regions of the genome potentially involved in maintaining a transformed state. The very nature of the methods used to generate these cell lines suggests that the arrangement of viral information may be more complex than has been observed in cells transformed by UV-irradiated HSV and EHV. Three of these cell lines arose after transfecting C57BL/6 MEF with randomly damaged viral DNA.

Perucho et al. (1980) have shown clearly that during transfection by the calcium phosphate coprecipitation method, viral DNA molecules become linked to carrier DNA as well as to each other prior to integration (reviewed in Scangos and Ruddle, 1981). In a population of DNA molecules representing more than one segment of the genome, atypical arrangements of fragments can be anticipated. Similarly, the CMV-Hel-Mj cell lines may retain viral DNA segments that, although separated on the normal viral genome, are adjacent in the genome that actually accomplished the transformation. This expectation arises from the observations of Mocarski and Stinski (1979) that persistently infected human cells contain as many as 120 genome equivalents of viral DNA per cell and that persistently infected cultures produce defective viral genomes. The presence of multiple viral genomes in individual cells during long periods of subculturing should permit multiple independent exchanges between viral and cellular DNAs. In addition, structural DNA rearrangements occur among the defective genomes (Mocarski and Stinski, 1979). Any rearrangement that juxtaposes sequences that can initiate and sequences that augment or amplify expression of the transformed phenotype are likely to be selected for during transformation of primary cells. The stable retention of viral DNA sequences from distant regions of the genome in transformants generated by transfection of sheared DNA and by defective interfering particles has been demonstrated previously (Leiden et al., 1980; O'Callaghan et al., 1983).

The combined data of Leiden et al. (1980) and O'Callaghan et al. (1983) suggest that cell lines transformed by UV-irradiated virus in contrast to lines generated by transfection with sheared DNA or defective particles present the most readily interpretable arrangement of viral DNA sequences. For this reason we have also established HCMV-transformed lines following infection of C57BL/6 MEF with UV-irradiated HCMV virions.

The assumption that the C57BL/6 cell lines developed after infection with DNA or UV-irradiated virions are virally transformed rests heavily on the observation that spontaneous transformation of these primary cells occurs extremely rarely. To accumulate evidence for the role of HCMV in the outgrowth of these clonal lines, we examined them for the expression of the Fc receptor (Keller et al., 1976; Rahman et al., 1976; Westmoreland et al., 1976; Sakuma et al., 1977). Preliminary results indicate that both of two C57BL/6 HCMV-transformed lines at early passage as well as the human cell line of Geder et al. (1976) transformed by the Major strain bind human IgG to the cell surface. Normal mouse sera did not bind in immunofluorescence assays. This attachment to the Fc receptor was blocked by normal rabbit but not by normal mouse sera as expected (Keller et al., 1976). Cells of a C57BL/6 SV40-transformed cell line, in contrast, did not bind human IgG. That HCMV Fc receptors appear on the cell lines tested indicates that these cells incorporated HCMV DNA and have expressed at least a portion of that DNA.

While these preliminary studies do not at present extend our understanding of the process of transformation by HCMV, they do present an additional assay that is responsive to transformation by transfected viral, recombinant plasmid, and transformed cellular DNAs. This system may aid in elucidating the mechanisms by which HCMV transforms primary cells in culture.

It is premature to suggest a mechanism(s) by which HCMV transforms cells in culture. As in the case of HSV, infection of cells with HCMV results in chromosomal aberrations (Lüleci et al., 1980). The close structural similarity between HCMV and HSV DNA suggests that HCMV DNA will contain sequences and express virus-coded products that enhance DNA rearrangements also.

HCMV, like HSV, encodes a unique polymerase (Huang, 1975; Hirai and Watanabe, 1976; Nishiyama et al., 1983). It is possible that this enzyme has mutator activity, although direct evidence favoring this assumption is not available. An understanding of the mechanisms by which these HCMV sequences transform cells in culture will require expansion of both functional and genetic analysis of the viral genome as well as additional analyses of the transforming potential of this virus.

IV. CONCLUSION

The introduction of genetic material from HSV or HCMV into cells in culture either by means of virion infection or DNA transfection consistently leads to the outgrowth of oncogenically transformed cells. While our understanding of this process is fragmentary, five observations can be made.

Cells transformed by herpesviruses retain and often express information from widely separated regions of the viral genome. During continuous propagation, transformed cells become heterogeneous with respect to the viral sequences they retain, although the population as a whole may still contain genetic information from multiple segments of the viral DNA. Analysis of the sequences retained by virion-mediated, biochemical, or EHV morphological transformants at high passage number using DNA enriched for viral sequences (Leiden et al., 1980) or carefully selected probes (Robinson and O'Callaghan, 1983) reveals an overlapping set of viral sequences from which a minimal retained segment can be deduced. Additionally, at least some transforming functions of HSV and HCMV are contained within single segments of the genome, as transformation can be accomplished with individual cloned fragments. Finally, transformation of cells by inactivated virions or genomic DNA occurs at a low frequency. This frequency is not unexpected considering the size of the viral genome nor does it lessen the potential significance of transformation by these viruses. The ability of herpesviruses to establish latency and thereby to maintain a reservoir for recurrent infection provides multiple opportunities for low-frequency events to occur in vivo. Clearly, the efficiency of transformation in cell culture does not uniformly predict oncogenicity *in vivo*. For instance, avian leukosis virus, which will not transform cells in culture effectively, produces tumors in its permissive host. The consistent observation that cells transformed by HSV or HCMV are oncogenic even in immunocompetent animals and that the tumor cells metastasize readily indicates that, if it occurs *in vivo*, the consequence of transformation (whatever the frequency) would be significant to the host.

The mechanism(s) by which herpesviruses accomplish transformation of normal cells to malignant cells remains elusive. The basis for speculation on this mechanism is a patchwork of evidence in which no consistent pattern can yet be discerned. Based on knowledge accumulated with other tumor viruses, herpesviruses appear to possess the potential to transform cells by several processes. The virus codes for proteins with mutagenic activity, as well as proteins that stimulate rearrangements and amplification of DNA sequences bounded by and containing appropriate viral sequences.

In addition, both HSV and HCMV viral genomes contain sequences that are homologous to cellular sequences (Peden *et al.*, 1982; Puga *et al.*, 1982). The extent to which the sequences represent good base pair or functional homology with proto-oncogenes or alternatively regions of preferred interaction with the cellular chromosome remains to be established. The processes of mutation, gene amplification, and translocation have each been correlated with development of neoplasia and may act during herpesvirus infection to either initiate or augment the transformed phenotype. Preliminary evidence is accumulating to suggest that HSV may transform NIH-3T3 cells to an anchorage-independent phenotype by a hit-and-run mechanism. Documentation of such a process would provide a significant addition to the list of mechanisms by which DNA tumor viruses transform cells in culture.

There is no reason to believe at present, however, that herpesviruses use a single mechanism to transform cells in culture. Nor is evidence available to exclude the possibility that herpesviruses can alter the growth properties of cells by adding a protein-coding sequence whose presence is essential to maintenance of the transformed phenotype, as suggested by the retention and expression of specific segments by herpesvirus genomes during prolonged propagation and passage in animals.

Specific segments of HSV and HCMV DNA transform cells in culture. The reports that different regions of the HSV-2 genome transform cells in assays of two parameters of the transformed phenotype and indications from the papovavirus and adenovirus systems that the selective pressure applied by various transformation assays identifies different transforming genes suggest that quantitative comparisons of transformation by herpesviruses or viral DNA fragments for a number of transformation parameters in more than one cell type are essential to elucidating the mechanisms of the transformation processes. It is now apparent that transformation in NIH-3T3 cells assays a subset of transforming events. While it may be true that the same sequences or gene products that transform these preneoplastic cells are sufficient to transform primary cells, this correlation can no longer be assumed, but must be demonstrated. Clearly, coordinated efforts to determine those DNA sequences that initiate, maintain, and enhance tumorigenesis in a variety of cell types should increase our understanding of the complex biological phenomenon of cell transformation and malignancy by these significant human pathogens.

ACKNOWLEDGMENTS. The secretarial services of Carol Buck and Elaine Neidigh and editorial assistance of Melissa Reese are gratefully acknowledged. Support for these investigations was provided by Research Grants CA-24694, CA-27503, and CA-18450 from the National Institutes of Health.

REFERENCES

Albrecht, T., and Rapp, F., 1973, Malignant transformation of hamster embryo fibroblasts following exposure to ultraviolet-irradiated human cytomegalovirus, *Virology* **55**:53.

Armitage, P., and Doll, R., 1954, The age distribution of cancer and a multi-stage theory of carcinogenesis, *Br. J. Cancer* **8**:1.

Aurelian, L., Manak, M. M., McKinlay, M., Smith, C. C., Klacsman, K. T., and Gupta, P. K., 1981, The herpesvirus hypothesis: Are Koch's postulates satisfied?, *Gynecol. Oncol.* **12**:556.

Babiss, L. E., Ginsberg, H. S., and Fisher, P. B., 1983, Cold-sensitive expression of transformation by a host range mutant of type 5 adenovirus, *Proc. Natl. Acad. Sci. USA* **80**:1352.

Bacchetti, S., and Graham, F. L., 1977, Transfer of the gene for thymidine kinase to thymidine kinase-deficient human cells by purified herpes simplex viral DNA, *Proc. Natl. Acad. Sci. USA* **74**:1590.

Baucke, R. B., and Spear, P. G., 1979, Membrane proteins specified by herpes simplex viruses. V. Identification of an Fc-binding glycoprotein, *J. Virol.* **32**:779.

Bayliss, G. J., Marsden, H. S., and Hay, J., 1975, Herpes simplex virus proteins: DNA-binding proteins in infected cells and in the virus structure, *Virology* **68**:124.

Bernards, R., Houweling, A., Schrier, P. I., Bos, J. L., and van der Eb, A. J., 1982, Characterization of cells transformed by Ad5/Ad12 hybrid early region 1 plasmids, *Virology* **120**:422.

Bishop, J. M., 1982, Retroviruses and cancer genes, *Adv. Cancer Res.* **37**:2.

Blair, D. G., Oskarsson, M., Wood, G., McClements, M. L., Fischinger, P. J., and Van de Woude, G., 1981, Activation of the transforming potential of a normal cell sequence: A molecular model for oncogenes, *Science* **212**:941.

Blanton, R. A., and Tevethia, M. J., 1981, Immunoprecipitation of virus-specific immediate-early and early polypeptides from cells lytically infected with human cytomegalovirus strain AD169, *Virology* **112**:262.

Boldogh, I., Gönczöl, É., and Váci, L., 1978, Transformation of hamster embryonic fibroblast cells by UV-irradiated human cytomegalovirus, *Acta Microbiol. Acad. Sci. Hung.* **25**:269.

Boldogh, I., Beth, E., Huang, E.-S., Kyalwazi, S. K., and Giraldo, G., 1981, Kaposi's sarcoma. IV. Detection of CMV DNA, CMV RNA and CMNA in tumor biopsies, *Int. J. Cancer* **28**:469.

Boyd, A. L., and Orme, T. W., 1975, Transformation of mouse cells after infection with ultraviolet-irradiation-inactivated herpes simplex virus type 2 *Int. J. Cancer* **16**:526.

Buchman, T. G., Simpson, T. Nosal, C., Roizman, B., and Nahmias, A. J., 1980, The structure of herpes simplex virus DNA and its application to molecular epidemiology, *Ann. N.Y. Acad. Sci.* **354:**279.

Camacho, A., and Spear, P. G., 1978, Transformation of hamster embryo fibroblasts by a specific fragment of the herpes simplex virus genome, *Cell* **15:**993.

Capon, D. J., Chen, E. Y., Levinson, A. D., Seeburg, P. H., and Goeddel, D. V., 1983, Complete nucleotide sequence of the T24 human bladder carcinoma oncogene and its normal homologue, *Nature (London)* **302:**33.

Chartrand, P., Wilkie, N. M., and Timbury, M. C., 1981, Physical mapping of temperature-sensitive mutations of herpes simplex virus type 2 by marker rescue, *J. Gen. Virol.* **52:**121.

Clayton, D., Murphy, D., Lovett, M., and Rigby, P., 1982, A fragment of the SV40 T-antigen transforms, *Nature (London)* **299:**59.

Coffin, J. M., Varmus, H. E., Bishop, J. M., Essex, M., Hardy, W. D., Jr., Martin, G. S., Rosenberg, N. E., Scolnick, E. M., Weinberg, R. A., and Vogt, P. K., 1981, Proposal for naming host cell-derived inserts in retrovirus genomes, *J. Virol.* **40:**953.

Cohen, G., Katze, M., Hydrean-Stern, C., and Eisenberg, R. J., 1978, Type-common CP-1 antigen of herpes simplex virus is associated with a 59,000 molecular weight envelope glycoprotein, *J. Virol.* **27:**172.

Colbère-Garapin, F., Chousterman, S., Horodniceanu, F., Kourilsky, P., and Garapin, A., 1979, Cloning of the active thymidine kinase gene of herpes simplex virus type 1 in *Escherichia coli* K-12, *Proc. Natl. Acad. Sci. USA* **76:**3755.

Colby, W. W., and Shenk, T., 1982, Fragments of the simian virus 40 transforming gene facilitate transformation of rat embryo cells, *Proc. Natl. Acad. Sci. USA* **79:**5189.

Conley, A. J., Knipe, D. M., Jones, P. C., and Roizman, B., 1981, Molecular genetics of herpes simplex virus. VII. Characterization of a temperature-sensitive mutant produced by *in vitro* mutagenesis and defective in DNA synthesis and accumulation of γ polypeptides, *J. Virol.* **37:**191.

Cooper, G. M., 1982, Cellular transforming genes, *Science* **217:**801.

Cooper, G. M., and Neiman, P. E., 1981, Two distinct candidate transforming genes of lymphoid leukosis virus-induced neoplasms, *Nature (London)* **292:**857.

Cooper, G. M., Okenquist, S., and Silverman, L., 1980, Transforming activity of DNA of chemically transformed and normal cells, *Nature (London)* **284:**418.

Darai, G., and Munk, K., 1973, Human embryonic lung cells abortively infected with herpes virus hominis type 2 show some properties of cell transformation, *Nature New Biol.* **241:**268.

Davis, B. D., Munyon, W., Buchsbaum, R., and Chawda, R., 1974, Virus type-specific thymidine kinase in cells biochemically transformed by herpes simplex virus types 1 and 2, *J. Virol.* **13:**1400.

Della Valle, G., Fenton, R. G., and Basilico, C., 1981, Polyoma large T antigen regulates the integration of viral DNA sequences into genomes of transformed cells, *Cell* **23:**347.

DeLuca, N., Bzik, D. J., Bond, V. C., Person, S., and Snipes, W., 1982, Nucleotide sequences of herpes simplex virus type 1 (HSV-1) affecting virus entry, cell fusion, and production of glycoprotein gB (VP7), *Virology* **122:**411.

DeMarchi, J. M., 1981, Human cytomegalovirus DNA: Restriction enzyme cleavage maps and map location for immediate-early, early, and late RNAs, *Virology* **114:**23.

DeMarchi, J. M., Blankenship, M. L., Brown, G. D., and Kaplan, A. S., 1978, Size and complexity of human cytomegalovirus DNA, *Virology* **89:**643.

DeMarchi, J. M., Schmidt, C. A., and Kaplan, A. S., 1980, Patterns of transcription of human cytomegalovirus in permissively infected cells, *J. Virol.* **35:**277.

Dixon, R. A. F., Sabourin, D. J., and Schaffer, P. A., 1983, Genetic analysis of temperature-sensitive mutants which define genes for the major herpes simplex virus type 2 DNA-binding protein and a new late function, *J. Virol.* **45:**343.

Docherty, J. J., Subak-Sharpe, J. H., and Preston, C. M., 1981, Identification of a virus-specific polypeptide associated with a transforming fragment (BglII–N) of herpes simplex virus type 2 DNA, *J. Virol.* **40:**126.

Dubbs, D. R., and Kit, S., 1964, Mutant strains of herpes simplex deficient in thymidine kinase inducing activity, *Virology* **22**:493.

Duff, R., and Rapp, F., 1971a, Oncogenic transformation of hamster cells after exposure to herpes simplex virus type 2, *Nature New Biol.* **233**:48.

Duff, R., and Rapp, F., 1971b, Properties of hamster embryo fibroblasts transformed *in vitro* after exposure to ultraviolet-irradiated herpes simplex virus type 2, *J. Virol.* **8**:469.

Duff, R., and Rapp, F., 1973a, Oncogenic transformation of hamster embryo cells after exposure to inactivated herpes simplex virus type 1, *J. Virol.* **12**:209.

Duff, R., and Rapp, F., 1973b, The induction of oncogenic potential by herpes simplex viruses, in: *Perspectives in Virology VIII* (M. Pollard, ed.), pp. 189–210, Academic Press, New York.

Duff, R., and Rapp, F., 1975, Quantitative assay for transformation of 3T3 cells by herpes simplex virus 2, *J. Virol.* **15**:490.

Duff, R., Kreider, J. W., Levy, B. M., Katz, M., and Rapp, F., 1974, Comparative pathology of cells transformed by herpes simplex virus type 1 and type 2, *J. Natl. Cancer Inst.* **53**:1159.

Feldman, L. T., and Nevins, J. R., 1983, Localization of the adenovirus E1A$_a$ protein, a positive-acting transcriptional factor, in infected cells, *Mol. Cell. Biol.* **3**:829.

Fenoglio, C. M., Oster, M. W., Gerfo, P. L., Reynolds, T., Edelson, R., Patterson, J. A. K. Madeiros, E., and McDougall, J. K., 1982, Kaposi's sarcoma following chemotherapy for testicular cancer in a homosexual man: Demonstration of cytomegalovirus RNA in sarcoma cells, *Hum. Pathol.* **13**:955.

Flannery, V. L., Courtney, R. J., and Schaffer, P. A., 1977, Expression of an early, nonstructural antigen of herpes simplex virus in cells transformed *in vitro* by herpes simplex virus, *J. Virol.* **21**:284.

Fleckenstein, B., Müller, I., and Collins, J., 1982, Cloning of the complete human cytomegalovirus genome in cosmids, *Gene* **18**:39.

Flyer, D., Pretell, J., Campbell, A. E., Liao, W. S. L., Tevethia, M. J., Taylor, J. M., and Tevethia, S. S., 1983, Biology of simian virus 40 (SV40) transplantation antigen (TrAg). X. Tumorigenic potential of mouse cells transformed by SV40 in high responder C57BL/6 mice and correlation with the persistence of SV40 TrAg, early proteins and viral sequences, *Virology* **131**:207.

Foulds, I., 1954, The experimental study of tumor progression: A review, *Cancer Res.* **14**:327.

Frenkel, N., Locker, H., Cox, B., Roizman, B., and Rapp, F., 1976, Herpes simplex virus DNA in transformed cells: Sequence complexity in five hamster cell lines and one derived hamster tumor, *J. Virol.* **18**:885.

Frenkel, N., Locker, H., and Vlazny, D. A., 1980, Studies of defective herpes simplex viruses, *Ann. N.Y. Acad. Sci.* **354**:347.

Frink, R. J., Eisenberg, R., Cohen, G., and Wagner, E. K., 1983, Detailed analysis of the portion of the herpes simplex virus type 1 genome encoding glycoprotein C, *J. Virol.* **45**:634.

Frykberg, L., Palmieri, S., Beug, H., Graf, T., Hayman, M. J., and Vennström, B., 1983, Transforming capacities of avian erythroblastosis virus mutants deleted in erbA and erbB oncogenes, *Cell* **32**:227.

Furukawa, T., Tanaka, S., and Plotkin, S. A., 1975, Stimulation of macromolecular synthesis in guinea pig cells by human CMV, *Proc. Soc. Exp. Biol. Med.* **148**:211.

Galloway, D. A., and McDougall, J. K., 1981, Transformation of rodent cells by a cloned DNA fragment of herpes simplex virus type 2, *J. Virol.* **38**:749.

Galloway, D. A., and McDougall, J. K., 1983, The oncogenic potential of herpes simplex virus: Evidence for a 'hit-and-run' mechanism, *Nature (London)* **302**:21.

Galloway, D. A., Copple, C. D., and McDougall, J. K., 1980, Analysis of viral DNA sequences in hamster cells transformed by herpes simplex virus type 2, *Proc. Natl. Acad. Sci. USA* **77**:880.

Galloway, D. A., Goldstein, L. C., and Lewis, J. B., 1982, Identification of proteins encoded by a fragment of herpes simplex virus type 2 DNA that has transforming activity, *J. Virol.* **42**:530.

Galloway, D. A., Nelson, J. A., and McDougall, 1984, Small fragments of herpesvirus DNA with transforming activity contain IS-like structures. *Proc. Natl. Acad. Sci., USA* (in press).

Garfinkle, B., and McAuslan, B. R., 1974, Regulation of herpes simplex virus-induced thymidine kinase, *Biochem. Biophys. Res. Commun.* **58:**822.

Geder, L., Lausch, R., O'Neill, F., and Rapp, F., 1976, Oncogenic transformation of human embryo lung cells by human cytomegalovirus, *Science* **192:**1134.

Geder, L., Laychock, A. M., Gorodecki, J., and Rapp, F., 1979, Alterations in biological properties of different lines of cytomegalovirus-transformed human embryo lung cells following *in vitro* cultivation, in: *Oncogenesis and Herpesviruses* (G. de Thé, W. Henle, and F. Rapp, eds.), pp. 591–601, IARC, Lyon, France.

Geelan, J. L. M. C., Walig, C., Wertheim, P., and van der Noordaa, J., 1978, Human cytomegalovirus DNA. I. Molecular weight and infectivity, *J. Virol.* **26:**813.

Gibson, W., 1981, Immediate-early proteins of human cytomegalovirus strains AD169, Davis and Towne differ in electrophoretic mobility, *Virology* **112:**350.

Gluzman, Y., Davison, J., Oren, M., and Winocour, E., 1977, Properties of permissive monkey cells transformed by UV-irradiated simian virus 40, *J. Virol.* **22:**256.

Goldstein, L. C., McDougall, J., Hackman, R., Meyers, J. D., Thomas, E. D., and Nowinski, R. C., 1982, Monoclonal antibodies to cytomegalovirus: Rapid identification of clinical isolates and preliminary use in diagnosis of CMV pneumonia. *Infect. Immun.* **38:**273.

Graham, F. L., and van der Eb, A. J., 1973, A new technique for the assay of infectivity of human adenovirus 5 DNA, *Virology* **52:**456.

Graham, F. L., Abrahams, P. J., Mulder, C., Heijneker, H. L., Warnaar, S. O., DeVries, F. A. J., Fiers, W., and van der Eb, A. J., 1974, Studies on *in vitro* transformation by DNA and DNA fragments of human adenoviruses and simian virus 40, *Cold Spring Harbor Symp. Quant. Biol.* **39:**637.

Graham, F. L., Smiley, J., Russell, W. C., and Nairn, R., 1977, Characterization of a human cell line transformed by DNA from human adenovirus type 5, *J. Gen. Virol.* **36:**59.

Graham, F. L., Harrison, T., and Williams, J., 1978, Defective transforming capacity of adenovirus 5 host-range mutants, *Virology* **86:**10.

Graham, F. L., Bacchetti, S., McKinnon, R., Stanners, C., Cordell, B., and Goodman, H. M., 1980, Transformation of mammalian cells with DNA using the calcium technique, in: *Introduction of Macromolecules into Viable Mammalian Cells*, pp. 3–25, Liss, New York.

Gupta, P., and Rapp, F., 1977, Identification of virion polypeptides in hamster cells transformed by HSV type 1, *Proc. Natl. Acad. Sci. USA* **74:**372.

Gupta, P., Lausch, R. N., Hay, K. A., and Rapp, F., 1980, Expression of type-common envelope antigens by herpes simplex virus type 2-transformed hamster cells, *Intervirology* **14:**50.

Hall, J., and Almy, R. E., 1982, Evidence for control of herpes simplex virus mutagenesis by the viral DNA polymerase, *Virology* **116:**535.

Halliburton, I., 1980, Intertypic recombinants of herpes simplex viruses, *J. Gen. Virol.* **48:**1.

Halliburton, I. W., Randall, R. E., Killington, R. A., and Watson, D. H., 1977, Some properties of recombinants between type 1 and type 2 herpes simplex virus, *J. Gen. Virol.* **36:**471.

Halliburton, I. W., Morse, L. S., Roizman, B., and Quinn, K. E., 1980, Mapping of the thymidine kinase genes of type 1 and type 2 herpes viruses using intertypic recombinants, *J. Gen. Virol.* **49:**235.

Hampar, B., 1981, Transformation induced by herpes simplex virus: A potentially novel type of virus–cell interaction, *Adv. Cancer Res.* **35:**27.

Hampar, B., and Ellison, S. A., 1961, Chromosomal aberrations induced by an animal virus, *Nature (London)* **192:**145.

Hampar, B., and Ellison, S. A., 1963, Cellular alterations in the MCH line of Chinese hamster cells following infection with herpes simplex virus, *Proc. Natl. Acad. Sci. USA* **49:**474.

Hampar, B., Boyd, A. L., Derge, J. G., Zweig, M., Eader, L., and Showalter, S. D., 1980, Comparison of properties of mouse cells transformed spontaneously by ultraviolet-irradiated herpes simplex virus or simian virus 40, *Cancer Res.* **40:**2213.

Hampar, B., Derge, J. G., Boyd, A. L., Tainsky, M. A., and Showalter, S. D., 1981, Herpes simplex virus (type 1) thymidine kinase gene does not transform cells morphologically, *Proc. Natl. Acad. Sci. USA* **78:**2616.

Hay, K. A., and Lausch, R. N., 1979, Uninhibited growth and metastases of herpes simplex virus transformed cells in virus-sensitized hosts, *Int. J. Cancer* **23:**337.

Hayward, W. J., Need, B. G., and Astrin, S. M., 1981, Activation of a cellular *onc* gene by promoter insertion in ALV-induced lymphoid leukosis, *Nature (London)* **290:**475.

Hirai, K., and Watanabe, Y., 1976, Induction of α-type DNA polymerases in human cytomegalovirus-infected WI-38 cells, *Biochim. Biophys. Acta* **447:**238.

Hirai, K., Furukawa, T., and Plotkin, S. A., 1976, Induction of DNA polymerase in WI-38 and guinea pig cells infected with human cytomegalovirus (HCMV), *Virology* **70:**251.

Honess, R. W., and Roizman, B., 1973, Proteins specified by herpes simplex virus. XI. Identification and relative molar rates of synthesis of structural and nonstructural herpes virus polypeptides in the infected cell, *J. Virol.* **12:**1347.

Honess, R. W., and Roizman, B., 1974, Regulation of herpes virus macromolecular synthesis. I. Cascade regulation of the synthesis of three groups of viral proteins, *J. Virol.* **14:**8.

Honess, R. W., and Roizman, B., 1975, Regulation of herpesvirus macromolecular synthesis: Sequential transition of polypeptide synthesis requires functional viral polypeptides, *Proc. Natl. Acad. Sci. USA* **72:**1276.

Houweling, A., van den Elsen, P. J., and van der Eb, A. J., 1980, Partial transformation of primary rat cells by the leftmost 4.5% fragment of adenovirus 5 DNA, *Virology* **105:**537.

Huang, E. S., 1975, Human cytomegalovirus, III. Virus-induced DNA polymerase, *J. Virol.* **16:**298.

Jariwalla, R. J., Aurelian, L., and Ts'o, P. O. P., 1980, Tumorigenic transformation induced by a specific fragment of DNA from herpes simplex virus type 2, *Proc. Natl. Acad. Sci. USA* **77:**2279.

Johnson, R., Horwitz, S. N., and Frost, P., 1982, Disseminated Kaposi's sarcoma in a homosexual male, *J. Am. Med. Assoc.* **247:**1739.

Jones, N., and Shenk, T., 1979, Isolation of adenovirus type 5 host range deletion mutants defective for transformation of rat embryo cells, *Cells* **17:**683.

Kaerner, H. C., Maichle, I. B., Oh, A., and Schroder, C. H., 1979, Origin of two different classes of defective HSV-1 Angelotti DNA, *Nucleic Acids Res.* **6:**1467.

Kakunaga, T., 1973, A quantitative system for assay of malignant transformation by chemical carcinogens using a clone derived from Balb/3T3, *Int. J. Cancer* **12:**463.

Keller, R., Peitchel, R., Goldman, J. N., and Goldman, M., 1976, An IgG-Fc receptor induced in cytomegalovirus-infected human fibroblasts, *J. Immunol.* **116:**772.

Kimura, S., Esparza, J., Benyesh-Melnick, M., and Schaffer, P. A., 1974, Enhanced replication of temperature-sensitive mutants of herpes simplex virus type 2 (HSV-2) at the nonpermissive temperature in cells transformed by HSV-2, *Intervirology* **3:**162.

Kimura, S., Flannery, V. L., Levy, B., and Schaffer, P. A., 1975, Oncogenic transformation of primary hamster cells by herpes simplex virus type 2 (HSV-2) and an HSV-2 temperature-sensitive mutant, *Int. J. Cancer* **15:**786.

Kits, S., and Dubbs, D. R., 1977, Regulation of herpesvirus thymidine kinase activity in LM(tk⁻) cells transformed by ultraviolet light-irradiated herpes simplex virus, *Virology* **76:**331.

Klein, G., 1983, Specific chromosomal translocations and the genesis of B-cell-derived tumors of mice and man, *Cell* **32:**311.

Klemperer, H. G., Haynes, G. R., Shedden, W. I. H., and Watson, D. H., 1967, A virus specific thymidine kinase in BHK21 cells infected with herpes simplex virus, *Virology* **31:**120.

Knipe, D. M., 1982, Cell growth transformation by herpes simplex virus, *Prog. Med. Virol.* **28:**114.

Krontiris, T., and Cooper, G. M., 1981, Transforming activity of human tumor DNAs, *Proc. Natl. Acad. Sci. USA* **78:**1181.

Kucera, L. S., Gusdon, J. P., Edwards, I., and Herbst, G., 1977, Oncogenic transformation of rat embryo fibroblasts with photoinactivated herpes simplex virus: Rapid *in vitro* cloning of transformed cells, *J. Gen. Virol.* **35:**473.

Kudler, L., Jones, T. R., Russell, R. J., and Hyman, R. W., 1983, Heteroduplex analysis of cloned fragments of herpes simplex virus DNAs, *Virology* **124**:86.

Kutinova, L., Vonka, V., and Broucek, J., 1973, Increased oncogenicity and synthesis of herpes virus antigens in hamster cells exposed to herpes simplex type-2 virus, *J. Natl. Cancer Inst.* **50**:759.

LaFemina, R. L., and Hayward, G. S., 1980, Structural organization of the DNA molecules from human cytomegalovirus, in: *Animal Virus Genetics, ICN–UCLA Symposia on Molecular and Cellular Biology,* Vol. 18 (B. N. Fields, R. Jaenisch, and C. F. Fox, eds.), pp. 39–55.

Lai, C.-J., and Nathans, D., 1974, Mapping of temperature-sensitive mutants of simian virus 40: Rescue of mutants by fragments of viral DNA, *Virology* **60**:466.

Lakeman, A. D., and Osborn, J. E., 1979, Size of infectious DNA from human and murine cytomegaloviruses, *J. Virol.* **30**:414.

Land, H., Parada, L. F., and Weinberg, R. A., 1983, Tumorigenic conversion of primary embryo fibroblasts requires at least two cooperating oncogenes, *Nature,* **304**:596.

Lawrence, S., 1982, AIDS—No relief in sight, *Science* **122**:202.

Lee, G. T.-Y., Para, M. F., and Spear, P. G., 1982, Location of the structural genes for glycoproteins gD and gE and for other polypeptides in the S component of herpes simplex virus type 1 DNA, *J. Virol.* **43**:41.

Leiden, J. M., Buttyan, R., and Spear, P. G., 1976, Herpes simplex virus gene expression in transformed cells. I. Regulation of the thymidine kinase gene in transformed L cells by products of superinfecting virus, *J. Virol.* **20**:413.

Leiden, J. M., Frenkel, N., and Rapp, F., 1980, Identification of the herpes simplex virus DNA sequences present in six herpes simplex virus thymidine kinase-transformed mouse cell lines, *J. Virol.* **33**:272.

Levine, A. J., 1982, Transformation associated tumor antigens, *Adv. Cancer Res.* **37**:75.

Lewis, J. G., Kucera, L. S., Eberle, R., and Courtney, R. J., 1982, Detection of herpes simplex virus type 2 glycoproteins expressed in virus transformed rat cells, *J. Virol.* **42**:275.

Li, J.-L. H., Jerkofsky, M. A., and Rapp, F., 1975, Demonstration of oncogenic potential of mammalian cells transformed by DNA-containing viruses following photodynamic inactivation, *Int. J. Cancer* **15**:190.

Lin, S. S., and Munyon, W., 1974, Expression of the viral thymidine kinase gene in herpes simplex virus-transformed L cells, *J. Virol.* **14**:1199.

Linsley, P. S., and Siminovitch, L., 1982, Comparison of phenotypic expression with genotypic transformation by using cloned, selectable markers, *Mol. Cell Biol.* **2**:593.

Locker, H., Frenkel, N., and Halliburton, I., 1982, Structure and expression of class II defective herpes simplex genomes encoding infected cell polypeptide 8, *J. Virol.* **43**:574.

Lonsdale, D. M., Brown, S. M., Lang, J., Subak-Sharpe, J. H., Koprowski, H., and Warren, K. G., 1980, Variations in herpes simplex virus isolated from human ganglia and a study of clonal variation in HSV-1, *Ann. N.Y. Acad. Sci.* **354**:291.

Lorentz, A. K., Munk, K., and Darai, G., 1977, DNA repair replication in human embryonic lung cells infected with herpes simplex virus, *Virology* **82**:401.

Lüleci, G., Sakízlí, M., and Günlap, A., 1980, Selective chromosomal damage caused by human cytomegalovirus, *Acta Virol.* (*Engl. Ed.*) **24**:341.

McDonough, S. H., and Spector, D. H., 1983, Transcription in human fibroblasts permissively infected by human cytomegalovirus strain AD169, *Virology* **125**:31.

McDougall, J. K., Masse, T. H., and Galloway, D. A., 1980, Location and cloning of the herpes simplex virus type 2 thymidine kinase gene, *J. Virol.* **33**:1221.

McKnight, S. L., 1980, The nucleotide sequence and transcript map of the herpes simplex virus thymidine kinase gene, *Nucleic Acids Res.* **8**:5949.

Macnab, J. C. M., 1974, Transformation of rat embryo cells by temperature-sensitive mutants of HSV, *J. Gen. Virol.* **24**:143.

Macnab, J. C. M., 1979, Tumor production by HSV-2 transformed lines in rats and the varying response to immunosuppression, *J. Gen. Virol.* **43**:39.

Macnab, J. C. M., Visser, L., Jamison, A. T., and Hay, J., 1980, Specific viral antigens in rat cells transformed by herpes simplex virus type 2 and in rat tumors induced by inoculation of transformed cells, *Cancer Res.* **40:**2074.

MacPherson, I., and Montagnier, L., 1964, Agar suspension culture for the selective assay of cells transformed by polyoma virus, *Virology* **23:**291.

Marsden, H. S., Stow, N. D., Preston, V. G., Timbury, M. C., and Wilkie, N. M., 1978, Physical mapping of herpes simplex virus-induced polypeptides, *J. Virol.* **28:**624.

Michelson, S., Horodniceanu, F., Kress, M., and Tardy-Panit, M., 1979, Human cytomegalovirus induced immediate early antigens: Analysis in sodium dodecyl sulfate-polyacrylamide gel electrophoresis after immunoprecipitation, *J. Virol.* **32:**259.

Minson, A. C., Thouless, M. E., Eglin, R. P., and Darby, G., 1976, The detection of virus DNA sequences in a herpes type 2 transformed hamster cell line (333-8-9), *Int. J. Cancer* **17:**493.

Mocarski, E. S., and Roizman, B., 1982, Herpesvirus-dependent amplification and inversion of cell associated viral thymidine kinase gene flanked by viral α sequences and linked to an origin of viral DNA replication, *Proc. Natl. Acad. Sci. USA* **79:**5626.

Mocarski, E. S., and Stinski, M., 1979, Persistence of the cytomegalovirus genome in human cells, *J. Virol.* **31:**761.

Morbidity and Mortality Weekly Reports, 1982a, Update on acquired immune deficiency syndrome (AIDS)—United States, **31:**507.

Morbidity and Mortality Weekly Reports, 1982b, Unexplained immunodeficiency and opportunistic infections in infants—New York, New Jersey, California, **31:**665.

Morse, L., Buchman, T., Roizman, B., and Schaffer, P., 1977, Anatomy of herpes simplex virus DNA. IX. Apparent exclusion of some parental DNA arrangements in the generation of intertypic (HSV-1 × HSV-2) recombinants, *J. Virol.* **24:**231.

Morse, L. S., Pereira, L., Roizman, B., and Schaffer, P. A., 1978, Anatomy of HSV DNA. X. Mapping of viral genes by analysis of polypeptides and functions specified by HSV-1 × HSV-2 recombinants, *J. Virol.* **26:**389.

Munk, W., and Darai, G., 1973, Human embryonic lung cells transformed by HSV, *Cancer Res.* **33:**1535.

Munyon, W., Kraiselburd, E., Davis, D., and Mann, J., 1971, Transfer of thymidine kinase to thymidine kinaseless L cells by infection with ultraviolet-irradiated herpes simplex virus, *J. Virol.* **7:**813.

Munyon, W., Buchsbaum, R., Paoletti, J., Mann, J., Kraiselburd, E., and Davis, D., 1972, Electrophoresis of thymidine kinase activity synthesized by cells transformed by herpes simplex virus, *Virology* **49:**683.

Mushinski, J. F., Potter, M., Bauer, S. R., and Reddy, E. P., 1983, DNA rearrangement and altered RNA expression of the c-*myb* oncogene in mouse plasmacytoid lymphosarcomas, *Science* **220:**795.

Nelson, J. A., Fleckenstein, B., Galloway, D. A., and McDougall, J. K., 1982, Transformation of NIH3T3 cells with cloned fragments of human cytomegalovirus strain AD169, *J. Virol.* **43:**83.

Nelson, J. A., Fleckenstein, B., Jahn, G., Galloway, D. A., and McDougall, J. K., 1984, Structure of the transforming region of human cytomegalovirus AD169, *J. Virol.* **49:**109.

Nishiyama, Y., Maeno, K., and Yoshida, S., 1983, Characterization of human cytomegalovirus-induced DNA polymerase and the associated 3′-to-5′ exonuclease, *Virology* **124:**221.

Nordling, C. O., 1953, A new theory on the cancer-inducing mechanism, *Br. J. Cancer* **7:**68.

O'Callaghan, D. J., Gentry, G. A., and Randall, C. C., 1983, The equine herpesviruses, in: *The Herpesviruses* (B. Roizman, ed.), Vol. 2, pp. 215–305, Plenum Press, New York.

O'Neill, F. J., and Rapp, F., 1971, Early event required for induction of chromosomal abnormalities in human cells by herpes simplex virus, *Virology* **44:**544.

Pancake, B. A., Aschman, D. P., and Schaffer, P. A., 1983, Genetic and phenotypic analysis of HSV-1 mutants conditionally resistant to immune lysis, *J. Virol.* **47:**568.

Para, M. F., Goldstein, L., and Spear, P. G., 1982, Similarities and differences in the Fc-binding glycoprotein g(E) of herpes simplex virus types 1 and 2 and tentative mapping of the viral gene for this glycoprotein, *J. Virol.* **41:**137.

Park, M., Lonsdale, D. M., Timbury, M. C., Subak-Sharpe, J. H., and Macnab, J. C. M., 1980, Genetic retrieval of viral genome sequences from herpes simplex transformed cells, *Nature,* **285:**412.

Parris, D. S., Dixon, R. A. F., and Schaffer, P. A., 1980, Physical mapping of herpes simplex virus type 1 ts mutants by marker rescue: Correlation of physical and genetic maps, *Virology* **100:**275.

Payne, G. S., Bishop, J. M., and Varmus, H. E., 1982, Multiple arrangements of viral DNA and an activated host oncogene in bursal lymphomas, *Nature (London)* **295:**209.

Peden, K., Mounts, P., and Hayward, G. S., 1982, Homology between mammalian cell DNA sequences and human herpesvirus genomes detected by a hybridization procedure with high-complexity probe, *Cell* **31:**71.

Pellicer, A., Wigler, M., Axel, R., and Silverstein, S., 1978, The transfer and stable integration of the HSV thymidine kinase gene into mouse cells, *Cell* **14:**133.

Perucho, M., Hanahan, D., and Wigler, M., 1980, Genetic and physical linkage of exogenous sequences in transformed cells, *Cell* **22:**309.

Preston, C. M., 1979, Control of herpes simplex virus type 1 mRNA synthesis in cells infected with wild-type virus or a temperature-sensitive mutant tsK, *J. Virol.* **29:**275.

Preston, V. G., Coates, J. V., and Rixon, F. J., 1983, Identification and characterization of a herpes simplex virus gene product required for encapsidation of virus DNA, *J. Virol.* **45:**1056.

Puga, A., Canlin, E. M., and Notkins, A., 1982, Homology between murine and human cellular DNA sequences and the terminal repetition of the S component of herpes simplex virus type 1 DNA, *Cell* **31:**81.

Purifoy, D. J. M., and Powell, K. L., 1976, DNA-binding proteins induced by herpes simplex virus type 2 in HEp-2 cells, *J. Virol.* **19:**717.

Rahman, A. A., Teschner, M., Sethi, K., and Brandis, H., 1976, Appearance of IgG (Fc) receptor(s) on cultured human fibroblasts infected with human cytomegalovirus, *J. Immunol.* **117:**253.

Rapp, F. (ed.), 1980a, *Oncogenic Herpesviruses,* Vols. I and II, CRC Press, Boca Raton, Fla.

Rapp, F., 1980b, Persistence and transmission of cytomegalovirus, in: *Comprehensive Virology,* Vol. 3 (H. Fraenkel-Conrat and R. R. Wagner, eds.), pp. 193–232, Plenum Press, New York.

Rapp, F., 1980c, Transformation by herpes simplex viruses, in: *Viruses in Naturally Occurring Cancers* (M. Essex, G. Todaro, and H. zur Hausen, eds.), pp. 63–80, Cold Spring Harbor Laboratory, New York.

Rapp, F., 1983, The biology of cytomegaloviruses, in: *The Herpesviruses* (B. Roizman, ed.), Vol. 2, pp. 1–66, Plenum Press, New York.

Rapp, F., and Duff, R., 1973, Transformation of hamster embryo fibroblasts by herpes simplex virus type 1 and type 2, *Cancer Res.* **33:**1527.

Rapp, F., and Howett, M. K., 1984, Herpesviruses and cancer, in: *Concepts in Viral Pathogenesis* (A. L. Notkins and M. A. B. Oldstone, eds.), Springer-Verlag, New York (in press).

Rapp, F., and Howett, M. K., 1983, Involvement of herpes simplex virus in cervical carcinoma, in: *Biochemical and Biological Markers of Neoplastic Transformation* (P. Chandra, ed.), Plenum Press, New York, **57:**555.

Rapp, F., and Hsu, T. C., 1965, Viruses and mammalian chromosomes. IV. Replication of herpes simplex virus in diploid Chinese hamster cells, *Virology* **25:**401.

Rapp, F., and Jenkins, F. J., 1981, Genital cancer and viruses, *Gynecol. Oncol.* **12:**525.

Rapp, F., and Robbins, D., 1983, Cytomegalovirus and human cancer, in: *CMV Pathogenesis and Prevention of Human Infection* (S. A. Plotkin, S. Michelson, J. S. Pagano, and F. Rapp, eds.) Birth Defects: Original Article Series, March of Dimes Defects Foundation, Alan R. Liss, Inc., New York, pp. 175–192.

Rapp, F., and Turner, N., 1978, Biochemical transformation of mouse cells by herpes simplex virus types 1 and 2: Comparison of different methods for inactivation of viruses, *Arch. Virol.* **56**:77.

Rapp, F., Li, J.-L. H., and Jerkofsky, M., 1973, Transformation of mammalian cells by herpes simplex viruses following photodynamic inactivation, *Virology* **55**:339.

Rapp, F., Geder, L., Murasko, D., Lausch, R., Ladda, R., Huang, E. S., and Webber, M. M., 1975, Long-term persistence of cytomegalovirus genome in cultured human cells of prostatic origin, *J. Virol.* **16**:982.

Rapp, F., Turner, N., and Schaffer, P. A., 1980, Biochemical transformation by temperature-sensitive mutants of herpes simplex virus type 1, *J. Virol.* **34**:704.

Rassoulzadegan, M., Cowie, A., Carr, A., Glaichenhaus, N., Kamen, R., and Cuzin, F., 1982, The roles of individual polyoma virus early proteins in oncogenic transformation, *Nature (London)* **300**:713.

Rawls, W. E., Bacchetti, S., and Graham, F. L., 1977, Relation of herpes simplex viruses to human malignancies, *Curr. Top. Microbiol. Immunol.* **77**:71.

Rechavi, G., Givol, D., and Canaani, E., 1982, Activation of a cellular oncogene by DNA rearrangement: Possible involvement of an IS-like element, *Nature (London)* **300**:607.

Reddy, E. P., Reynolds, R. K., Santos, E., and Barbacid, M., 1982, A point mutation is responsible for the acquisition of transforming properties by the T24 human bladder carcinoma oncogene, *Nature (London)* **300**:149.

Reed, C., Cohen, G., and Rapp, F., 1975, Detection of a virus specific antigen on the surface of herpes simplex virus transformed cells, *J. Virol.* **15**:668.

Reyes, G. R., LaFemina, R., Hayward, S. D., and Hayward, G. S., 1979, Morphological transformation by DNA fragments of human herpesviruses: Evidence for two distinct transforming regions in herpes simplex virus types 1 and 2 and lack of correlation with biochemical transfer of the thymidine kinase gene, *Cold Spring Harbor Symp. Quant. Biol.* **44**:629.

Robinson, R. A., and O'Callaghan, D. J., 1983, A specific viral DNA sequence is stably integrated in herpesvirus oncogenically transformed cells, *Cell* **32**:569.

Roizman, B., 1979, The organization of the herpes simplex virus genomes, *Annu. Rev. Genet.* **13**:25.

Roizman, B., Norrild, B., Chan, C. and Pereira, L., 1984, Identification and preliminary mapping with monoclonal antibodies of a herpes simplex virus 2 glycoprotein lacking a known type 1 counterpart, *Virology* **133**:242.

Rowe, D. T., and Graham, F. C., 1983, Transformation of rodent cells by DNA extracted from transformation-defective adenovirus mutants, *J. Virol.* **46**:1039.

Ruley, H. E., 1983, Adenovirus early region 1A enables viral and cellular transforming genes to transform primary cells in culture *Nature* **304**:602.

Ruyechan, W. T., Morse, L. S., Knipe, D. M., and Roizman, B., 1979, Molecular genetics of herpes simplex virus. II. Mapping of the major viral glycoprotein precursors and their products in type 1-infected cells, *J. Virol.* **29**:677.

Sakuma, S., Furukawa, T., and Plotkin, S., 1977, The characterization of IgG receptor induced by human cytomegalovirus (39769), *Proc. Soc. Exp. Biol. Med.* **155**:168.

Scangos, G., and Ruddle, F. H., 1981, Mechanisms and applications of DNA-mediated gene transfer in mammalian cells—A review, *Gene* **14**:1.

Scangos, G. A., Huttner, K. M., Juricek, D. K., and Ruddle, F. H., 1981, Deoxyribonucleic acid-mediated gene transfer in mammalian cells: Molecular analysis of unstable transformants and their progression to stability, *Mol. Cell Biol.* **1**:111.

Schlehofer, J. R., and zur Hausen, H., 1982, Induction of mutations within the host cell genome by partially inactivated herpes simplex virus type 1, *Virology* **122**:471.

Shih, C., Padhy, L. C., Murray, M., and Weinberg, R. A., 1981, Transforming genes of carcinomas and neuroblastomas introduced into mouse fibroblasts, *Nature (London)* **290**:261.

Shiroki, K., Shimojo, H., Sawada, Y., Uemizu, Y., and Fujinaga, K., 1979, Incomplete transformation of rat cells by a small fragment of adenovirus 12 DNA, *Virology* **95**:127.

Shu, M., Kessous, A., Poiver, N., and Simard, R., 1980, Immunoprecipitation of polypeptides from hamster embryo cells transformed by herpes simplex virus type 2, *Virology* **104**:303.

Smith, B. L., Anisowicz, A., Chodosh, L. A., and Sager, R., 1982, DNA transfer of focus- and tumor-forming ability into nontumorigenic CHEF cells, *Proc. Natl. Acad. Sci. USA* **79**:1964.

Smith, J. D., and DeHarven, E., 1973, Herpes simplex virus and human cytomegalovirus replication in WI-38 cells. I. Sequence of viral replication, *J. Virol.* **12**:919.

Sompayrac, L., and Danna, K. J., 1983, Simian virus 40 deletion mutants that transform with reduced efficiency, *Mol. Cell. Biol.* **3**:484.

Spang, A. E., Godowski, P. J., and Knipe, D. M., 1983, Characterization of herpes simplex virus 2 temperature-sensitive mutants whose lesions map in or near the coding sequences for the major DNA-binding protein, *J. Virol.* **45**:332.

Spector, D. H., Hock, L., and Tamashiro, J. C., 1982, Cleavage maps for human cytomegalovirus DNA strain AD169 for restriction endonucleases EcoRI, BglII, and HindIII, *J. Virol.* **42**:558.

Stich, H. F., Hsu, T. C., and Rapp, F., 1964, Viruses and mammalian chromosomes. I. Localization of chromosomal aberrations after infection with herpes simplex virus, *Virology* **22**:439.

Stinski, M. F., 1977, Synthesis of proteins and glycoproteins in cells infected with human cytomegalovirus, *J. Virol.* **23**:751.

Stinski, M. F., 1978, Sequences of protein synthesis in cells infected by human cytomegalovirus: Early and late virus-induced polypeptides, *J. Virol.* **26**:686.

Stinski, M. F., 1983, Molecular biology of cytomegaloviruses, in: *The Herpesviruses* (B. Roizman, ed.), Vol. 2, pp. 67–113, Plenum Press, New York.

Stinski, M., and Thomsen, D. R., 1981, Cloning of the human cytomegalovirus genome as endonuclease Xbal fragments, *Gene* **16**:207.

Stinski, M. F., Thomsen, D. R., Stenberg, R. M., and Goldstein, L. C., 1983, Organization and expression of immediate early genes of human cytomegalovirus, *J. Virol.* **46**:1.

St. Jeor, S., Albrecht, T. B., Funk, F. D., and Rapp, F., 1974, Stimulation of cellular DNA synthesis by human cytomegalovirus, *J. Virol.* **13**:353.

Stoker, M., and MacPherson, I., 1961, Studies on transformation of hamster cells by polyoma virus *in vitro*, *Virology* **14**:357.

Stow, N. D., Subak-Sharpe, J. H., and Wilkie, N. M., 1978, Physical mapping of herpes simplex virus type 1 mutations by marker rescue, *J. Virol.* **28**:182.

Tabin, C. J., Bradley, S. M., Bargmann, C. I., Weinberg, R. A., Papageorge, A. G., Scolnick, E. M., Dhar, R., Lowy, D. R., and Chang, E. H., 1982, Mechanism of activation of a human oncogene, *Nature (London)* **300**:143.

Takahashi, M., and Yamanishi, K., 1974, Transformation of hamster embryo and human embryo cells by temperature-sensitive mutants of herpes simplex virus type 2, *Virology* **61**:306.

Tanaka, S., Furakawa, T., and Plotkin, S. A., 1975, Human cytomegalovirus stimulates host cell RNA synthesis, *J. Virol.* **15**:297.

Tanaka, S., Otsuka, M., Ihara, S., Maeda, F., and Watanabe, Y., 1979, Induction of pre-early nuclear antigens in HEL cells infected with human cytomegalovirus, *Microbiol. Immunol.* **23**:263.

Tevethia, M. J., 1984, Immortalization of primary mouse embryo fibroblasts with SV40 virions, viral DNA and a subgenomic DNA fragment in a quantitative assay, *Virology* (in press).

Tevethia, S. S., and McMillan, V. L., 1974, Acquisition of malignant properties by SV40-transformed mouse cells: Relationship to type-C viral antigen expression, *Intervirology* **3**:269.

Tevethia, S. S., Flyer, D. C., Tevethia, M. J., and Topp, W. C., 1980a, Biology of simian virus 40 (SV40) transplantation antigen (TrAg). VII. Induction of SV40 TrAg in nonpermissive mouse cells by early viable SV40 deletion (0.54/0.59) mutants, *Virology* **107**:488.

Tevethia, S. S., Greenfield, R. S. Flyer, D. C., and Tevethia, M. J., 1980b, Simian virus 40 transplantation rejection antigen: Relationship to SV40 specific proteins, *Cold Spring Harbor Symp. Quant. Biol.* **44:**235.

Thomsen, D. R., and Stinski, M. F., 1981, Cloning of the human cytomegalovirus genome as endonuclease XbaI fragments, *Gene* **16:**207.

Todaro, G. J., and Green, H., 1964, An assay for cellular transformation by SV40, *Virology* **23:**117.

Tooze, J. (ed.), 1981, *Molecular Biology of Tumor Viruses*, Part 2, *DNA Tumor Viruses* (rev.), Cold Spring Harbor Laboratory, New York.

Tooze, J. (ed.), 1983, *Molecular Biology of Tumor Viruses*, Part 1, *RNA Tumor Viruses*, Cold Spring Harbor Laboratory, New York.

Triesman, R., Novak, U., Favaloro, J., and Kamen, R., 1981, Transformation of rat cells by an altered polyoma virus genome expressing only the middle-T protein, *Nature (London)* **292:**595.

van der Eb, A. J., Mulder, C., Graham, F. L., and Houweling, A., 1977, Transformation with specific fragments of adenovirus DNAs. I. Isolation of specific fragments with trans-forming activity of adenovirus 2 and 5 DNA, *Gene* **2:**115.

van der Elsen, P. J., de Pater, S, Houweling, A., van der Veer, J., and van der Eb, A., 1982, The relationship between region E1a and E1b of human adenoviruses in cell transfor-mation, *Gene* **18:**175.

Wagner, M. J., Sharp, J. A., and Summers, W. C., 1981, Nucleotide sequence of the thymidine kinase gene of herpes simplex virus type 1, *Proc. Natl. Acad. Sci. USA* **78:**1441.

Wathen, M. W., and Stinski, M. F., 1982, Temporal patterns of human cytomegalovirus transcription: Mapping of viral RNAs synthesized at immediate early, early, and late times after infection, *J. Virol.* **41:**462.

Wathen, M. W., Thomsen, D. R., and Stinski, M. F., 1981, Temporal regulation of human cytomegalovirus transcription at immediate early and early times after infection, *J. Virol.* **38:**446.

Waubke, R., zur Hausen, H., and Henle, W., 1968, Chromosomal and autoradiographic stud-ies of cells infected with herpes simplex virus, *J. Virol.* **2:**1047.

Weinberg, R. A., 1981, Use of transfection to analyze genetic information and malignant transformation, *Biochim. Biphys. Acta* **651:**25.

Weinberg, R. A., 1982, Oncogenesis of spontaneous and chemically induced tumors, *Adv. Cancer Res.* **36:**149.

Weller, S. K., Lee, K., Sabourin, D. J., and Schaffer, P. A., 1983a, Genetic analysis of tem-perature-sensitive mutants which define the gene for the major herpes simplex virus type 1 DNA binding protein, *J. Virol.* **45:**354.

Weller, S. K., Pancake, B. A., Sacks, W. R., Aschman, D., Coen, D., and Schaffer, P. A., 1983b, Genetic analysis of temperature-sensitive mutants of HSV-1: the combined use of com-plementation and physical mapping for cistron assignment, *Virology* **130:**290.

Westmoreland, D., Watkins, J. F., and Rapp, F., 1974, Demonstration of a receptor for IgG in Syrian hamster cells transformed with herpes simplex virus, *J. Gen. Virol.* **25:**167.

Westmoreland, D., St. Jeor, S., and Rapp, F., 1976, The development by cytomegalovirus-infected cells of binding affinity for normal human immunoglobulin, *J. Immunol.* **116:**1566.

Whittemore, A. S., 1978, Quantitative theories of oncogenesis, *Adv. Cancer Res.* **27:**55.

Wigler, M., Silverstein, S., Lee, L.-S., Pellicer, A., Cheng, Y.-C., and Axel, R., 1977, Transfer of purified herpes virus thymidine kinase gene to cultured mouse cells, *Cell* **11:**223.

Wilcox, K. W., Kohn, A., Sklyanskaya, E., and Roizman, B., 1980, Herpes simplex virus phosphoproteins. I. Phosphate cycles on and off some viral polypeptides and can alter their affinity for DNA, *J. Virol.* **33:**167.

Wilkie, N. M., Clements, J. B., Macnab, J. C. M., and Subak-Sharpe, J. H., 1974, The structure and biological properties of herpes simplex virus DNA, *Cold Spring Harbor Symp. Quant. Biol.* **39:**657.

Wilkie, N. M., Clements, J. B., Boll, W., Mantei, N., Longsdale, D., and Weissman, C., 1979, Hybrid plasmids containing an active thymidine kinase gene of herpes simplex virus 1, *Nucleic Acids Res.* **7:**859.

Zannis-Hadjopoulos, M., and Martin, R. G., 1983, Relationship of simian virus 40 tumor antigens to virus-induced mutagenesis, *Mol. Cell. Biol.* **3:**421.

CHAPTER 7

Glycoproteins Specified by Herpes Simplex Viruses

Patricia G. Spear

I. INTRODUCTION

Membrane glycoproteins specified by enveloped viruses are important determinants of viral pathogenicity. They are exposed on the surfaces of virions and on the surfaces of infected cells. They mediate entry of the virus into cells and cell-to-cell spread of infection and also influence tissue tropism and host range. As a consequence of the foregoing, viral membrane glycoproteins are also probably the most important elicitors of protective immune responses.

The purpose of this review is to summarize available information (as of year end, 1983) about the number, structures, synthesis, and functions of glycoproteins specified by herpes simplex viruses types 1 and 2 (HSV-1 and HSV-2). Other topics not covered here, including immunogenicity and antigenic structure of the HSV glycoproteins, have been reviewed elsewhere (Watson and Honess, 1977; Spear, 1980, 1984; Norrild, 1980; Glorioso and Levine, 1984).

II. THE HSV GLYCOPROTEINS AND THEIR STRUCTURAL GENES

A. Genetic Loci Encoding Glycoproteins

Numerous virus-induced glycosylated polypeptides can be detected in cells infected with HSV-1 or HSV-2 and resolved by electrophoresis

PATRICIA G. SPEAR • Department of Molecular Genetics and Cell Biology, The University of Chicago, Chicago, Illinois 60637.

(Heine *et al.*, 1972; Honess and Roizman, 1975) or by isoelectric focusing (Cohen *et al.*, 1980; Haarr and Marsden, 1981, Palfreyman *et al.*, 1983). Because the primary translation product of a single HSV glycoprotein gene is processed to yield multiple electrophoretically differentiable glycosylated forms (see below), enumeration of the genetically distinct glycoprotein products requires mapping and enumeration of the genes encoding them. Although there has been considerable progress along these lines, all of the HSV-1 and HSV-2 glycoprotein genes have probably not yet been identified and mapped. Consequently, the final count is not yet in.

At least five different genetic loci encoding membrane glycoproteins have been mapped to the HSV-2 genome and four to the HSV-1 genome, as summarized in Fig. 1 (see figure legend and Appendix for a discussion of HSV glycoprotein nomenclature and of recent changes in terminology). Approximate map locations for most of the glycoprotein genes shown in Fig. 1 were inferred originally from analyses of intertypic recombinant viruses (Marsden *et al.*, 1978; Ruyechan *et al.*, 1979; Para *et al.*, 1982b, 1983; Hope *et al.*, 1982). More precise information about locations resulted from use of cloned viral DNA fragments for marker transfer or rescue of mutations affecting antigenic determinants of gB-1 (Holland *et al.*, 1983b; Kousoulas *et al.*, 1984); for marker rescue of temperature-sensitive (ts) mutations affecting gB-1 structure and processing (Ruyechan *et al.*, 1979; Little *et al.*, 1981; DeLuca *et al.*, 1982; Holland *et al.*, 1983b; Kousoulas *et al.*, 1984); for selection of gC-1, gD-1, and gE-1 mRNAs identified by antigenic analyses of their *in vitro* translation products (Lee *et al.*, 1982a,b; Frink *et al.*, 1983); and for insertion into the thymidine kinase (TK) gene of appropriate HSV strains to monitor expression of gC-1, gC-2, or gD-1 from the inserts (Lee *et al.*, 1982b; Gibson and Spear, 1983; Zezulak and Spear, 1984b). Finally, the most precise map locations come from nucleotide sequence analyses, which have been reported for the genes encoding gB-1 (Bzik *et al.*, 1984), gC-1 (Frink *et al.*, 1983), and gD-1 (Watson *et al.*, 1982), as will be discussed more fully below.

The gene for gC-2 encodes a glycoprotein whose immature and mature forms are all significantly smaller in size than those of gC-1. When first described, the product of the gC-2 gene appeared, on the basis of size and antigenic structure, to be distinct from all known HSV-1 and HSV-2 glycoproteins and was originally designated gF (Balachandran *et al.*, 1981, 1982). It has been shown, however, that gC-2 (or gF) is antigenically related to gC-1 (Zweig *et al.*, 1983; Zezulak and Spear, 1983). Moreover, the genes for gC-2 (Para *et al.*, 1983; Zezulak and Spear, 1984b) and gC-1 (Lee *et al.*, 1982b; Frink *et al.*, 1983) have been independently mapped to colinear partially homologous regions of the HSV-2 and HSV-1 genomes, respectively, as shown in Fig. 1. Although gC-2 and gC-1 are clearly related products of the two serotypes, it appears that their genes have diverged more than the genes encoding other related glycoproteins (see below).

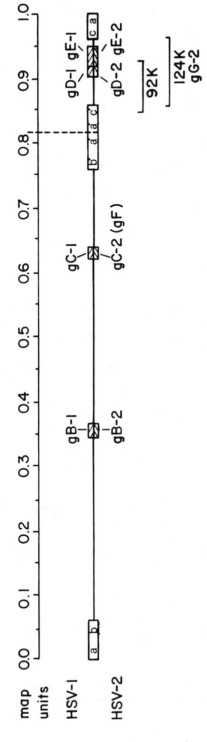

FIGURE 1. Map of the HSV-1 and HSV-2 genomes showing the locations of genes encoding viral glycoproteins. HSV-1 genes are shown above the line and HSV-2 genes below the line. Nomenclature of the HSV glycoproteins and recent changes in terminology are discussed in the Appendix. Note that the HSV-2 glycoprotein here designated gC-2 (Zweig et al., 1983; Para et al., 1983; Zezulak and Spear, 1983, 1984b) was originally named gF (Balachandran et al., 1981, 1982). Also, the HSV-2 glycoprotein here designated gG-2 (Roizman et al., 1984) was originally called gC (Ruyechan et al., 1979). The glycoprotein originally designated gA (Spear, 1976) is not shown because it proves to be a form of gB (Eberle and Courtney, 1980b; Pereira et al., 1981; Balachandran et al., 1982). Citations for the mapping data are given in the text.

The genes for gB-2, gD-2, and gE-2 have not yet been independently mapped as precisely as those encoding their HSV-1 counterparts. These HSV-2 genes are assumed to be located at positions on the HSV-2 genome colinear with the positions of the gB-1, gD-1, and gE-1 genes on the HSV-1 genome. This assumption is based on the demonstrated antigenic and structural relatedness of HSV-1 and HSV-2 glycoproteins assigned the same alphabetic designation (see below) and on evidence obtained from analyses of HSV-1 × HSV-2 recombinant viruses that the genomes are for the most part, if not entirely, colinear (Morse et al., 1977; Preston et al., 1978).

There is at least one, and possibly two, HSV-2 glycoproteins for which no HSV-1 counterpart has yet been identified. Marsden et al. (1978, 1984) mapped an HSV-2 glycoprotein with a molecular weight of about 92,000 (92K) to the S component of the genome, as depicted in Fig. 1, and showed by several criteria that this glycoprotein is distinct from the various forms of gD-2 and gE-2. Ruyechan et al. (1979) originally mapped a 124K HSV-2 glycoprotein (called gC at that time) to the region of the genome from about 0.65 to 0.70 map units. New results from the same laboratory indicate that this previously reported location is not the correct one; the 124K HSV-2 glycoprotein has now been remapped to the region indicated in Fig. 1 and was assigned the new name gG-2 (Roizman et al., 1984). Mapping results and other observations raise the possibility that the 92K and 124K glycoproteins are actually the same species, despite the large difference in their reported molecular weights. Marsden et al. (1984) state that the glycoprotein they mapped has an apparent molecular weight of 92K on gradient polyacrylamide gels cross-linked with bis-acrylamide whereas its apparent molecular weight is higher (about 120K) on polyacrylamide gels cross-linked with N,N'-diallyltartardiamide. The molecular weight of 124K reported by Roizman et al. (1984) for the glycoprotein they mapped was based on use of gels cross-linked with N,N'-diallyltartardiamide.

The possibility remains that glycoprotein genes other than those shown on the map in Fig. 1 will ultimately be identified. First, it seems unlikely that an HSV-1 counterpart for gG-2 does not exist. Second, Showalter et al. (1981) described HSV-1 and HSV-2 glycoproteins of about 110K molecular weight that are antigenically related to each other but appear to be distinct from all other known HSV glycoproteins, based on immunoprecipitation experiments done with monoclonal antibodies. Third, Palfreyman et al. (1983) described an HSV-1 glycoprotein (designated gY) that has an electrophoretic mobility similar to gC-1, is resolvable from gC-1 by isoelectric focusing, and appears to be antigenically unrelated to gB, gC, gD, and gE. The genes for these glycoproteins, should they be distinct from previously identified genes, have not yet been mapped. Most of the glycosylated polypeptides detected by electrophoretic analyses of extracts from infected cells can be identified as products of known glycoprotein genes. Consequently, any glycosylated products encoded by other genes must be relatively minor species (in terms of radiolabel incorporation) that perhaps comigrate with other species in the kinds of electrophoretic analyses usually performed.

B. Structures of the Glycoproteins

1. Polypeptide Moieties

Partial amino acid sequences have been reported for gD-1 and gD-2 (Eisenberg et al., 1984). The only other information available about amino acid sequence has been inferred from nucleotide sequences published for gB-1 (Bzik et al., 1984), gC-1 (Frink et al., 1983), and gD-1 (Watson et al., 1982). In each case, the open reading frame identified as encoding the glycoprotein is without introns. Evidence that the correct reading frame has been identified is based on construction of a hybrid gene and expression of a fusion protein antigenically related to the HSV glycoprotein, for gD-1 (Watson et al., 1982), or on inability to find any other open reading frame of appropriate characteristics consistent with mapping results summarized above, for the other glycoproteins. Figure 2 summarizes in diagrammatic form some of the pertinent characteristics of the polypeptides believed to be translated from the gB-1, gC-1, and gD-1 genes. What follows is a description of some of the more interesting features pointed out originally by the authors cited in the legend to Fig. 2 and below.

a. Signal Sequences

Signal sequences are found at the NH_2-termini of translation products for many membrane proteins and secreted proteins and are believed to mediate attachment of the nascent peptide on polysomes to membranes of the rough endoplasmic reticulum (RER) and vectorial transport of the growing polypeptide chain across the membrane; often the signal sequence is cleaved off during processing of the translation product (reviewed by Blobel et al., 1979; Kreil, 1981). As summarized by Kreil (1981), NH_2-terminal domains shown to serve as signal sequences vary in length from about 15 to 30 residues; have a core of hydrophobic residues no less than 9 in length; usually have a charged residue such as Arg near the NH_2-terminus, in addition to the α-amino group; and usually increase in hydrophilicity toward the COOH-terminal end.

Based on this analogy with well characterized secreted and membrane proteins, it seems likely that the hydrophobic NH_2-terminal domains of the translation products for gB-1, gC-1, and gD-1 (stippled domains in Fig. 2) constitute part of a signal sequence for each product. The features characteristic of signal sequences are found in each case (except that the NH_2-terminal hydrophobic domain for gB-1 seems longer than usual). Also, for gD-1 at least, the first 25 amino acids of the translation

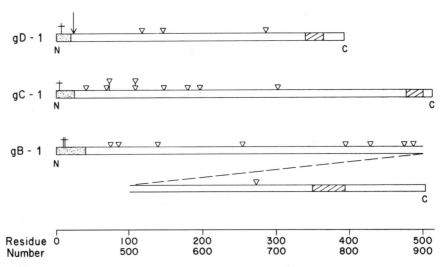

FIGURE 2. Diagram illustrating some of the important features of translation products encoded by the genes for gD-1, gC-1, and gB-1, based on nucleotide sequences published by Watson *et al.* (1982), Frink *et al.* (1983), and Bzik *et al.* (1984), respectively. The stippled bars represent hydrophobic domains that could be part of signal sequences and the daggers point to the positions of Arg residues (characteristically found near the NH$_2$-termini of signal sequences). The arrow shows where proteolytic cleavage occurs during the cotranslational or posttranslational processing of gD-1. The triangles indicate the positions of Asn residues that are part of the sequence Asn-X-Thr or Asn-X-Ser and could be modified by the addition of *N*-linked oligosaccharides. The hatched bars represent the positions of hydrophobic domains that could span the lipid bilayer.

product (including the entire NH$_2$-terminal hydrophobic domain) are cleaved off during posttranslational processing steps that yield the mature glycoprotein (Eisenberg *et al.*, 1984). Posttranslational cleavage also occurs at the equivalent site in gD-2 (Eisenberg *et al.*, 1984). It has not yet been determined whether similar cleavages occur during processing of the gB-1 and gC-1 translation products. Experiments to test the hypothesis that the putative signal sequences are actually necessary for attachment of the translation products to membranes and vectorial transport have not yet been reported.

b. Potential Sites for the Addition of N-Linked Oligosaccharides

N-linked oligosaccharides are added to Asn residues that are part of the sequence Asn-X-Ser or Asn-X-Thr (Eylar, 1965; Marshall, 1974). There are three such potential glycosylation sites in the translation product for gD-1 and nine each for gB-1 and gC-1 (Fig. 2). It must be empirically determined which potential sites are actually glycosylated. In the case of gD-1 (and gD-2), evidence has been presented for the existence of three *N*-linked oligosaccharide chains, indicating that all three potential sites can be used (Cohen *et al.*, 1983; Matthews *et al.*, 1983). The sites that are recognized for the addition of *N*-linked oligosaccharides in gB-1 and gC-1 have not yet been identified. Most or all of the HSV glycoproteins also contain *O*-linked oligosaccharides (see Section II.B.2.b.). General rules have not been established to permit predictions as to which Ser or Thr residues might be used for the addition of *O*-linked carbohydrates.

c. Membrane-Spanning Domains and Orientation in the Membrane

Certain integral membrane glycoproteins, such as the G protein of vesicular stomatitis virus (VSV), have been shown to have a hydrophobic domain near the COOH-terminus and to be oriented in the membrane so that a small hydrophilic sequence at the COOH terminus is in contact with the cytoplasm. The hydrophobic domain spans the lipid bilayer and the larger NH_2-terminal domain, containing all oligosaccharide chains, is exposed to the exterior of the cell (Rose *et al.*, 1980; Rose and Gallione, 1981).

Based on locations of hydrophobic domains and positions of potential glycosylation sites in the HSV-1 glycoproteins depicted in Fig. 2, it seems likely that gB-1, gC-1, and gD-1 will each be found to assume an orientation within membranes similar to that of the VSV G protein. Evidence that this is the case has been presented for gD-1 and gD-2. Matthews *et al.* (1983) found that synthesis of either the gD-1 or the gD-2 translation product *in vitro*, in the presence of dog pancreas microsomes, resulted in *N*-glycosylation at all three potential sites and association of the glycosylated product with the microsomes such that the domain containing carbohydrate was protected from proteolysis by added trypsin whereas a nonglycosylated fragment of 3K molecular weight was cleaved off by trypsin. This is the result to be expected if the gD-1 or gD-2 product spans the microsome membrane at the hydrophobic region indicated in Fig. 2, with the NH_2-terminus located inside the microsome (equivalent topologically to cisternae of the RER) and the COOH-terminus exposed to the outside of the microsome (equivalent to the cytosol). Matthews *et al.* (1983) also showed that a type-specific antigenic determinant destroyed by cleavage of the 3K fragment from gD-1 is inaccessible to antibody on intact infected cells expressing gD-1 on their surfaces, whereas another antigenic determinant unaffected by proteolysis of the fragment is accessible to antibody on intact infected cells. Taken together, these findings provide strong evidence for the transmembrane orientation of gD-1 suggested by the information summarized in Fig. 2.

Exploitation of the available nucleotide sequences and amino acid sequences, to obtain information about three-dimensional structure and the locations of various functional domains on the glycoproteins, has only just begun. For example, conditional lethal ts mutations and mutations affecting epitopes recognized by neutralizing antibodies have been mapped to the NH_2-terminal portion of the gB-1 gene and a Syn mutation to very near the COOH-terminus (DeLuca *et al.*, 1982; Holland *et al.*, 1983b; Kousoulas *et al.*, 1984; Bzik *et al.*, 1984). Between these two regions, a domain affecting rate of entry of virus has also been mapped (DeLuca *et al.*, 1982). In addition, epitopes recognized by monoclonal antibodies have been mapped to proteolytic fragments of gD-1 and gD-2 (Eisenberg *et al.*, 1982a; Matthews *et al.*, 1983). It seems safe to predict an explosion of new findings along these lines in the near future.

2. Prosthetic Groups

There has been progress in identifying the kinds of prosthetic groups that are attached to the polypeptide chains of HSV glycoproteins and also in defining, so far by indirect tests rather than by direct sequence analysis, some aspects of their structures. The kinds of prosthetic groups that have been detected on one or another of the HSV glycoproteins include *N*-linked oligosaccharides, *O*-linked oligosaccharides, and fatty acid. [Campadelli-Fiume and Serafini-Cessi (Chapter 8, this volume) provide additional information on this topic.]

a. N-Linked Oligosaccharides

For all HSV glycoproteins examined to date (including gB, gC, gD, gE, and gG-2), evidence of *N*-linked glycosylation has been obtained. Although most of the studies reported have been done with HSV-1, it seems likely that similar results would be obtained with HSV-2. It has been shown that tunicamycin, which specifically blocks production of the dolichol phosphate intermediate essential for transfer of oligosaccharide to Asn residues of polypeptides (Takatsuki *et al.*, 1975; Tkacz and Lampen, 1975; Struck and Lennarz, 1980), blocks partially or completely glycosylation of the HSV glycoproteins mentioned above (Pizer *et al.*, 1980; Olofsson and Lycke, 1980; Peake *et al.*, 1982; Norrild and Pedersen, 1982; Kousoulas *et al.*, 1983; Hope and Marsden, 1983; Roizman *et al.*, 1984). In addition, it has been shown that endo-β-*N*-acetylglucosaminidase H (Endo H) releases oligosaccharides from immature forms of the HSV glycoproteins, with a concomitant increase in electrophoretic mobility of the polypeptides indicative of decreased molecular weight, whereas all mature forms of the glycoproteins (except gB) are resistant to the action of Endo H (Serafini-Cessi and Campadelli-Fiume, 1981; Wenske *et al.*, 1982; Cohen *et al.*, 1983; Johnson and Spear, 1983; Zezulak and Spear, 1983; Matthews *et al.*, 1983). Because Endo H cleaves selectively *N*-linked oligosaccharides of the high-mannose type, but not of the complex type (Tarentino and Maley, 1974), these results suggest that the high-mannose-type oligosaccharides detected on immature forms of HSV glycoproteins are processed to yield complex-type oligosaccharides in the mature forms, as has been shown to occur for many glycoproteins (Kornfeld and Kornfeld, 1980). In the case of gB, however, at least some of the high-mannose-type *N*-linked oligosaccharides remain in a form that can be cleaved by Endo H (Wenske *et al.*, 1982; Johnson and Spear, 1983).

As mentioned above, the number of *N*-linked oligosaccharide chains added is three for gD-1 and gD-2 (Cohen *et al.*, 1983; Matthews *et al.*, 1983) and not known for the other glycoproteins. It is assumed for most of the HSV glycoproteins that the *same* number of *N*-linked oligosaccharide chains is added cotranslationally to each and every translation product of a single gene. Although this assumption will probably prove to be true for most of the glycoproteins, evidence to the contrary has been reported for gC-2. It was found that approximately equal amounts of two electrophoretically differentiable forms of gC-2 were detectable immediately after a short pulse-label and that these two discrete forms appeared to differ only in the number of *N*-linked oligosaccharide chains (Zezulak and Spear, 1983). Although the significance of this finding is not known, further exploration of this phenomenon in relation to multiple pathways of processing of HSV glycoproteins (see below) is warranted.

b. *O-Linked Oligosaccharides*

Evidence has been presented that HSV glycoproteins also contain *O*-linked oligosaccharides although, for at least one glycoprotein (gC-1), contradictory conclusions have been published. Results reported from three laboratories (Olofsson *et al.*, 1981b, 1983a; Johnson and Spear, 1983; Wenske and Courtney, 1983) indicate that oligosaccharides can be released from mature forms of gC-1 by β-elimination [treatment with alkaline borohydride under conditions selective for release of *O*-linked oligosaccharides (Spiro, 1966; Marshall and Neuberger, 1977)] whereas negative results were obtained for such experiments done in a fourth laboratory (Kumarasamy and Blough, 1983). There is no obvious explanation for the contradictory findings and more work must be done to resolve the inconsistency. Other observations support the presence of *O*-linked oligosaccharides on gC-1, however, including (1) the affinity of a fraction of gC-1 for *Helix pomatia* lectin or soybean lectin (Olofsson *et al.*, 1981a, 1983a), both of which bind to GalNAc residues (Hammarström *et al.*, 1977), moieties found in *O*-linked but not in *N*-linked oligosaccharides (Kornfeld and Kornfeld, 1976); (2) the reduced but detectable incorporation of glucosamine into oligosaccharides of gC-1 in the presence of amounts of tunicamycin sufficient to block *N*-glycosylation (Wenske and Courtney, 1983; Olofsson *et al.*, 1983b); and (3) release of oligosaccharides from gC-1 by GalNAc oligosaccharidase (Johnson and Spear, 1983), an enzyme that cleaves *O*-linked oligosaccharides (Huang and Aminoff, 1972; Pomato and Aminoff, 1978; Johnson and Spear, 1983).

Tunicamycin-resistant oligosaccharides obtained by protease digestion of gC-1 could be differentiated on the basis of their affinities for *H. pomatia* lectin and wheat germ agglutinin (specific for GlcNAc or sialic acid), indicating heterogeneity in structures of the oligosaccharides

(Olofsson *et al.*, 1983b). Moreover, among stable forms of gC-1 that accumulate in infected cells in the absence of tunicamycin, two populations containing O-linked oligosaccharides could be discerned—one that bound to *H. pomatia* lectin and one that did not (Olofsson *et al.*, 1983a). As these authors suggested, the two populations may be products of different pathways in the processing of gC-1 although the possibility that one is a precursor to the other cannot yet be ruled out.

Presence of O-linked oligosaccharides on mature forms of other HSV glycoproteins is indicated by the release of oligosaccharides from gD-1 due to treatment with alkaline borohydride (Johnson and Spear, 1983); by shifts (increases) in the electrophoretic mobilities of mature forms of gB-1, gC-1, gC-2, gD-1, and gE-1 after treatment with GalNAc oligosaccharidase (Johnson and Spear 1983; Zezulak and Spear, 1983); and by acceptor activity of immature forms, but not mature forms, of gB-1, gC-1, and gD-1 for the enzymatic transfer *in vitro* of GalNAc from UDP-GalNAc (Serafini-Cessi *et al.*, 1983a). It should be noted that, although O-glycosylation of gC-1 can apparently occur in the presence of tunicamycin, no one has yet reported glycosylation of other HSV glycoproteins in the presence of this drug. O-glycosylation of other HSV glycoproteins (in the absence of drug) is not ruled out by these negative results, however, on the basis of the following considerations. (1) The effects of inhibiting N-glycosylation are different for different membrane proteins; in some cases the translation product remains trapped in the RER, whereas in others the translation product can be transported to other compartments of the cells and to the cell surface (Gibson *et al.*, 1980). (2) The addition or elongation of O-linked oligosaccharides, at least for HSV glycoproteins, appears to occur in the Golgi apparatus (Johnson and Spear, 1983). (3) As a consequence of (1) and (2), an HSV glycoprotein could be O-glycosylated in the presence of tunicamycin only if it is transported from the RER. Consistent with the notion that selective O-glycosylation of gC-1 in the presence of tunicamycin might be due simply to selective transport of gC-1 under these conditions is the finding that antigenic determinants of gC-1, but not of other HSV-1 glycoproteins, could be detected on the surfaces of infected cells incubated in the presence of tunicamycin (Norrild and Pedersen, 1982).

Notwithstanding the above, the evidence for O-linked glycosylation of HSV glycoproteins other than gC-1 (and gD-1) is indirect and should be confirmed by other tests. Investigations of composition and structure are also needed, both for the N-linked and O-linked oligosaccharides, in part because the information obtained may define markers for different pathways of glycoprotein processing.

c. Sulfation

Inorganic sulfate is incorporated into all the major HSV-1 and HSV-2 glycoproteins, particularly into gE-1 and gE-2 (Hope *et al.*, 1982; Hope and Marsden, 1983). The nature of the linkage between sulfate and the glycoproteins is not known and, as pointed out by these authors, several possibilities exist. Sulfations of N-linked oligosaccharides (Prehm *et al.*, 1979), of O-linked oligosaccharides (reviewed by Kornfeld and Kornfeld, 1980), and of Tyr residues in proteins (Huttner, 1982) have all been described. Tunicamycin was shown to inhibit the incorporation of radioactive sulfate into HSV-1 glycoproteins, although reduced amounts of sulfate could be detected in an abnormal form of gE-1 (Hope and Marsden, 1983). As noted by these authors, this result indicates that at least some of the sulfate on gE-1 is incorporated into O-linked oligosaccharides or into the polypeptide backbone. They favor the idea that, for the other glycoproteins, sulfate is incorporated into the N-linked oligosaccharides, but recognized the possibility that attachment of O-linked oligosaccharides (and sulfation of these oligosaccharides if it occurs) may be dependent on N-glycosylation.

d. Fatty Acylation

It has been shown that ^{3}H from palmitate becomes incorporated into gE-1 (Johnson and Spear, 1983), indicating that fatty acid may become covalently attached to this protein as has been documented for other membrane glycoproteins (Schmidt and Schlesinger, 1979; Schmidt *et al.*, 1979) and for other proteins such as membrane-bound forms of SV40 T antigen (Henning and Lange-Mutschler, 1983). The physiological significance of fatty acylation of membrane proteins is not yet understood. Schmidt (1983) has recently reviewed this subject.

C. Intermolecular Interactions

It has been shown that gB-1 and gB-2 can be isolated from virions and infected cells in the form of SDS-stable heat-dissociable homodimers and higher oligomers (Sarmiento and Spear, 1979; Haffey and Spear, 1980; Eberle and Courtney, 1982). At least two electrophoretically differentiable forms of the gB polypeptide (one previously called gA) accumulate in infected cells and both appear to form dimers and other oligomers. The mutant HSV-1(HFEM)tsB5 fails to produce gB dimers characteristic of wild-type viruses at permissive temperature and instead produces gB oligomers of slower electrophoretic mobility (Sarmiento *et al.*, 1979; Haffey and Spear, 1980). It is not known whether this mutant phenotype results from the Syn mutation in the gB gene, from the ts mutation in this gene, or from some other mutation. It is also not known how the oligomeric conformation of gB relates to gB function.

Oligomers, probably dimers, of a high-molecular-weight HSV-2 glycoprotein have also been described (Eberle and Courtney, 1982). Dissociation of the oligomers required heat and a reducing agent. The glycoprotein in question was designated gC at the time but is probably the species recently renamed gG-2 by Roizman *et al.* (1984).

Under certain conditions a small fraction of extracted gD-1 (but not gD-2) has an apparent molecular weight of 120–130K, instead of the usual 60K, and is not dissociated by boiling in SDS and β-mercaptoethanol. Eisenberg *et al.* (1982b) detected this high-molecular-weight form after immunoaffinity chromatography of gD-1 and showed that its tryptic peptides were indistinguishable from those of 60K gD-1. Gibson and Spear (1983) detected a similar high-molecular-weight form after radioiodination of infected cell surfaces and immunoprecipitation of cell extracts by an anti-gD-1 monoclonal antibody. It is not yet clear whether this form of gD-1 actually exists in infected cells or whether its appearance in cell extracts results from some artifact of preparative procedures used.

Much remains to be learned about the intermolecular interactions of the HSV glycoproteins and the functional significance of these interactions.

III. SYNTHESIS AND PROCESSING OF THE GLYCOPROTEINS

A. Kinetics of Polypeptide Synthesis

Very limited information has been published about the kinetics of HSV glycoprotein synthesis, despite the availability of methods and materials to permit quantitation of rates of synthesis. Balachandran *et al.* (1982) performed immunoprecipitation experiments with monoclonal antibodies to determine the kinetics of synthesis of several HSV-2 glycoproteins. They found that (1) synthesis of gB-2, gD-2, and gE-2 could be detected earliest, at 1–3 hr after infection; (2) rates of synthesis of gD-2 and gE-2 sharply declined after 5–7 hours of infection whereas the rate of synthesis of gB-2 steadily increased until at least 11 hr after infection; and (3) gF-2 and gC-2 (designated here gC-2 and gG-2, respectively) were barely detectable at 1–3 hr of infection, after which their rates of synthesis steadily increased until at least 11 hr after infection. Therefore, the HSV-2 glycoproteins can be divided into at least three groups based on kinetics of synthesis.

For HSV-1 glycoproteins, it has been shown that gD-1 can be detected earlier in the replicative cycle than gC-1 (Cohen *et al.*, 1980) and that the rate of gD-1 synthesis sharply declines after 4–6 hr of infection, whereas the rate of gC-1 synthesis increases steadily (Johnson and Spear, 1984), analogous to the results described above for the homologous glycoproteins of HSV-2. The rate of gB-1 synthesis may also decline late in the infectious cycle (Spear, 1976).

B. Regulation of Polypeptide Synthesis

Regulation of the several HSV glycoproteins can be differentiated on the basis of whether any synthesis at all can be detected in the absence of viral DNA replication. Use of drugs or mutations to block viral DNA replication permits synthesis of gB-1 and gD-1, but at reduced levels compared to control values, whereas gC-1 synthesis is not detectable (Peake et al., 1982; Gibson and Spear, 1983). According to the classification scheme published by Honess and Roizman (1974), gC-1 belongs to the γ regulatory class of HSV polypeptides whereas gB-1 and gD-1 do not fit neatly into the original classification. Wagner and co-workers (see review in this volume) have assigned polypeptides resembling gB-1 and gD-1 to a class designated βγ because, like β polypeptides, they are synthesized at maximal rates during intermediate stages of the replicative cycle and can be expressed in the absence of viral DNA replication but, like γ polypeptides, maximal rates of synthesis depend on viral DNA replication.

More detailed information about regulation of expression of the glycoprotein genes should begin to accumulate at a rapid pace now that structures of the genes and of the mRNAs transcribed from the genes are being defined. Frink et al. (1983) have described the 2.7-kb unspliced mRNA from which gC-1 is translated and have also identified several different spliced mRNAs that are transcribed from the same region and appear to have the same 5' and 3' termini as the gC-1 mRNA. There are several open reading frames (at least three overlapping the gC-1 reading frame) that could be translated from these spliced mRNAs. It seems likely that some of the putative gene products encoded in these open reading frames are coordinately expressed with gC-1 and the possibility exists that their functions may relate in some way to the function of gC-1.

Different sizes have been reported for the gD-1 mRNA, ranging from 2.3 kb (Ikura et al., 1983) to 3.0 kb (Watson et al., 1983). The mRNA analyzed in the former study was shown to be spliced, downstream of the coding region for gD-1, whereas the mRNA analyzed in the latter study was reported to be unspliced. It was suggested that the splicing pattern of the gD-1 mRNA may change with time after infection (Ikura et al., 1983), which could in part explain the discrepancy in conclusions presented in the two studies. In addition, different cell types were used by the two groups of researchers.

Only limited information is available on the relationship between mRNA synthesis and polypeptide synthesis. It has been shown that inhibition of gC-1 and gD-1 synthesis resulting from a block in viral DNA replication is paralleled by a significant decrease in accumulation of gC-1 and gD-1 mRNAs (Peake et al., 1982; Ikura et al., 1983; Johnson and Spear, 1984). On the other hand, the decline in rate of gD-1 synthesis during a normal replicative cycle occurs in spite of continued accumulation of gD-1 mRNA that is functional for translation in vitro (Johnson and Spear, 1984). This suggests the existence of a mechanism for translational control of gD expression in the infected cell.

C. Sequence of Events and Intracellular Locations for Polypeptide Synthesis and Processing

It is assumed, in part from analogies with other membrane proteins (Blobel et al., 1979, Kreil, 1981), that the polypeptide chains of the HSV glycoproteins are made on membrane-bound polysomes in the RER, that the nascent chains are vectorially transported during translation into the lumen of the RER (until transport ceases with anchoring of the polypeptide in the membrane), and that addition of N-linked high-mannose-type oligosaccharides occurs cotranslationally.

Consistent with this assumption are the following observations. For gD-1 and gD-2, at least, the translation products made in vitro associate with dog pancreas microsomes, but only if the microsomes are present during translation; moreover, the products made in the presence of microsomes contain N-linked high-mannose-type oligosaccharides and are partially sequestered inside the microsomes (Matthews et al., 1983). In general, newly synthesized forms of the HSV glycoproteins made in vivo cannot be extracted from infected cells without the use of detergents. In addition, under most conditions, newly synthesized polypeptides already contain N-linked oligosaccharides as soon as they can be detected in infected cells (Pizer et al., 1980; Wenske et al., 1982; Johnson and Spear, 1983; Zezulak and Spear, 1983). An exception to this statement has been noted, however. Late after infection of cells maintained at 34°C, but not at 37 or 39°C, polypeptides characteristic of nonglycosylated precursors of gB-1 and gC-1 could be detected in nuclear fractions, suggesting that N-glycosylation had been uncoupled from translation under these conditions (Compton and Courtney, 1983). Additional studies are required to understand the significance of this observation.

By following the fate of pulse-labeled polypeptides, it can be shown that posttranslational processing of the glycoprotein precursors is accompanied by a large shift (decrease) in electrophoretic mobility and that this shift occurs within 15–120 min of synthesis of the polypeptide (Spear, 1976; Baucke and Spear, 1979; Eisenberg et al., 1979; Cohen et al., 1980; Eberle and Courtney, 1980b; Haarr and Marsden, 1981; Wenske et al., 1982; Zezulak and Spear, 1983). The decrease in electrophoretic mobility is probably due in part to an increase in molecular weight and change in charge resulting from addition and extension of oligosaccharide chains, as discussed below. The actual increase in molecular weight cannot be estimated from the change in mobility, however, because of anomalies in the behavior of glycoproteins in SDS-polyacrylamide gels.

Although signal sequences may be cleaved from the nascent polypeptides, as discussed above, such cleavages would probably occur prior to completion of translation and consequently would not contribute to changes in electrophoretic mobilities observed during maturation of the glycoproteins. Processing of the HSV glycoproteins does not appear to involve other kinds of proteolytic cleavage. Differences in the sizes of HSV glycoproteins produced by different cell types have been noted (Balachandran et al., 1982; Pereira et al., 1981, 1982a; Showalter et al., 1981; Zweig et al., 1983) and it has been shown that proteases active in some cell types but not others can account for some of the differences (Pereira et al., 1982a). Use of appropriate protease inhibitors in extraction buffers, however, can eliminate most of the differences in apparent size observed; (Zezulak and Spear, 1984a).

Several posttranslational processing events have been correlated with the shift in electrophoretic mobility observed during maturation of the HSV glycoproteins. Because [^{3}H]palmitate can be incorporated into the more rapidly migrating immature form of gE-1 during a short labeling period, but is found predominantly in the more slowly migrating mature forms, it appears that fatty acylation occurs just prior to the shift in electrophoretic mobility (Johnson and Spear, 1983). It has also been found that sensitivity to Endo H is lost, while sensitivity to GalNAc oligosaccharidase is gained concomitant with processing events that cause the shift in mobility (Wenske et al., 1982; Johnson and Spear, 1983; Zezulak and Spear, 1983). Apparently, therefore, these events include conversion of N-linked oligosaccharides from the high-mannose type to the complex type and acquisition of O-linked oligosaccharides. Because treatment with GalNAc oligosaccharidase increases the electrophoretic mobility of the more slowly migrating mature forms to that of immature forms, it appears that the large posttranslational shift in electrophoretic mobility is due largely to acquisition of O-linked oligosaccharides. It should be noted that removal of sialic acid residues by neuraminidase also increases the electrophoretic mobility of mature forms, to a lesser extent than does treatment with GalNAc oligosaccharidase (Johnson and Spear, 1983), and reduces the charge heterogeneity of mature forms resolvable by isoelectric focusing (Cohen et al., 1980; Haarr and Marsden, 1981).

Because fatty acylation of glycoproteins (Schmidt and Schlesinger, 1980; Dunphy et al., 1981) and conversion of high-mannose-type to complex-type N-linked oligosaccharides (Schachter and Roseman, 1980) have been shown to occur in the Golgi apparatus, it could be concluded that O-glycosylation of the HSV glycoproteins, or most if not all of the enzymatic reactions involved in chain extension, also occurs in the Golgi apparatus (Johnson and Spear, 1983). Studies done with monensin, which interferes with Golgi function and blocks transport of material from the Golgi apparatus to the cell surface (Tartakoff and Vassalli, 1977; Uchida et al., 1979), support this conclusion; monensin was shown to inhibit O-glycosylation of HSV glycoproteins (Johnson and Spear, 1982, 1983) as well as processing of the N-linked oligosaccharides (Wenske et al., 1982). Moreover, in the presence of monensin, HSV glycoproteins were not

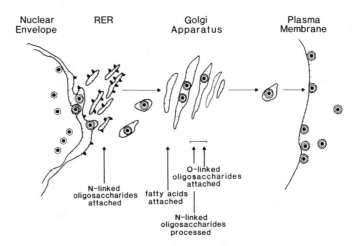

FIGURE 3. Diagram illustrating the two pathways by which HSV glycoproteins are transported to the cell surface and indicating the intracellular sites at which specific events in posttranslational processing of the glycoproteins occur (Johnson and Spear, 1983). Adapted from a drawing made by D. Johnson.

transported to the cell surface (Johnson and Spear, 1982), indicating that, as for other glycoproteins, transport to the cell surface is obligatorily via the Golgi apparatus.

There is reason to believe that a fraction of the HSV glycoproteins being transported through the Golgi apparatus is present in virions whereas the remainder are in cellular organelle membranes and will ultimately become incorporated into the cell surface (Fig. 3). Studies with monensin suggest that egress of virions from infected cells is via the Golgi apparatus. In the presence of this inhibitor, transport of virus to the cell surface was blocked; although infectious virus was produced, this virus contained immature forms of the envelope glycoproteins and accumulated in large abnormal intracytoplasmic vacuoles probably derived from Golgi membranes (Johnson and Spear, 1982).

At least two lines of argument suggest that glycoproteins incorporated into virions at the time of envelopment may be immature forms that are processed to the mature forms *in situ* as the virions are transported from the perinuclear space to the cell surface (Fig. 3). First, envelopment occurs at the inner nuclear membrane (see reviews by Roizman and Furlong, 1974; Spear, 1980), and nuclear fractions of infected cells are enriched for immature forms of the glycoproteins (Compton and Courtney, 1984). Second, infectious virions containing immature forms of glycoproteins can accumulate under conditions designed to block maturation of the glycoproteins, including treatment with monensin (Johnson and Spear, 1982), treatment with ammonium ions (Kousoulas *et al.*, 1983), and replication of virus in a mutant cell line deficient in a glycosyl transferase (Campadelli-Fiume *et al.*, 1982).

Because there are apparently two modes of transport of HSV glyco-proteins to the surfaces of infected cells (in virions and in organelle mem-branes), there could very well be subtle, but as yet undetected, differences in the nature of posttranslational modifications occurring in each mode. It is difficult to assess how much of each glycoprotein is transported via each pathway because virions tend to remain attached to the surfaces of infected cells following egress (Morgan *et al.*, 1968; Katsumoto *et al.*, 1981; Johnson and Spear, 1982). Most procedures designed to quantitate viral glycoproteins on the cell surface would not distinguish between molecules present in adherent virions and those present in the plasma membrane. That at least some of the HSV glycoproteins are actually pres-ent in the infected cell plasma membrane is evident from analyses of isolated membranes (Heine *et al.*, 1972) or from the binding to cell sur-faces of specific antibodies (Nii *et al.*, 1968) and probes for Fc receptors (Nakamura *et al.*, 1978; Para *et al.*, 1980) as detected by electron mi-croscopy.

Given the genetic complexity of HSV, it is possible that some of the enzymes required for cotranslational and posttranslational modification of the glycoproteins are viral in origin. It is presumed, however, in the absence of evidence to the contrary, that most or all of the enzymes required are specified by the host cell. Studies described above (Johnson and Spear, 1983) have helped to establish the intracellular location of enzymes involved in the addition and/or extension of O-linked oligosac-charides.

IV. ACTIVITIES AND ROLES OF THE GLYCOPROTEINS

A list can be drawn up of various activities and functions that will probably prove to be mediated by one or another of the HSV glycoproteins, based on analogy with other enveloped viruses, direct observations made in the HSV system, and various *a priori* considerations. This list includes adsorption and penetration of virus; expression of Fc receptors, comple-ment receptors, and perhaps other receptors on infected cell surfaces; envelopment and egress of progeny virus; cell fusion and other means of direct cell-to-cell transmission of virus infection (Fig. 4). There are un-doubtedly also other activities and functions of the viral glycoproteins that have yet to be recognized. The remainder of this section will focus on progress made in assigning functions to individual HSV-1 and HSV-2 glycoproteins and in assessing the role of carbohydrate in glycoprotein function.

A. Virion Infectivity

1. Neutralizing Antibodies

There is the expectation that information about virion components necessary for infectivity may emerge from investigation of the specific-

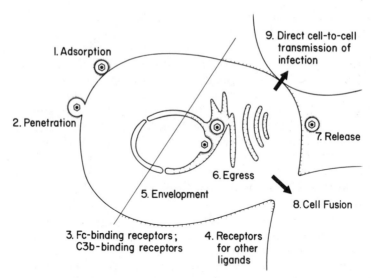

FIGURE 4. Diagram illustrating the various activities and processes associated with HSV glycoproteins.

ities and activities of neutralizing antibodies. This expectation has not yet been satisfactorily fulfilled, in the case of HSV, in part because efficient means of producing purified immunogens have just recently become available and also because neutralization is still a poorly understood process particularly for enveloped viruses.

Preparations of gB-1, gC-1, gC-2, and gE-1, purified after denaturation by SDS, have all been shown to induce the production of neutralizing antibodies (Powell et al., 1974; Eberle and Courtney, 1980a; Para et al., 1982a; Zezulak and Spear, 1983). In the absence of complement, this neutralizing activity was poor, however, for all except anti-gB-1 antibodies. Isolated gD-1 and gD-2, prepared by a variety of methods, also induce the production of neutralizing antibodies (Watson and Wildy, 1969; Honess and Watson, 1974; Cohen et al., 1978; Eisenberg et al., 1982). Comparison of three different HSV-1 glycoproteins as immunogens (gB-1, gC-1 and gD-1) revealed that gD-1 induced the highest titer of neutralizing antibodies (Norrild and Vestergaard, 1977; Vestergaard and Norrild, 1978). Because in this study the glycoproteins were isolated as immune precipitates in agar and amounts of immunogen used were difficult to quantitate, it is important that such studies be repeated with controlled doses of HSV glycoproteins purified in other ways.

Monoclonal antibodies specific for gB, gC, and gD have been reported to have neutralizing activity (Pereira et al., 1980; Showalter et al., 1981; Rector et al., 1982; Balachandran et al., 1982; Holland et al., 1983a). Where quantitative comparisons of neutralizing activity have been presented, the highest activities in the absence of complement have generally been associated with the anti-gD monoclonal antibodies.

For HSV, it can be questioned whether identification of the target of a neutralizing antibody also identifies a component necessary for infectivity. Although anti-gC-1 and anti-gC-2 antibodies can neutralize wild-type virus, mutant virions apparently devoid of gC-1 and gC-2 can be infectious, in cell culture (Heine *et al.*, 1974; Cassai *et al.*, 1975/76; Holland *et al.*, 1983a, 1984; Zezulak and Spear, 1984b) and in animals (Centifanto-Fitzgerald *et al.*, 1982).

2. Role of Oligosaccharides

Use of appropriate inhibitors or other means to block glycosylation of the viral glycoproteins has permitted some assessment of the role of the oligosaccharides in infectivity.

Both tunicamycin and deoxyglucose inhibit the production of *infectious* virus to a much greater extent than they inhibit the production of virions, i.e., virions with greatly reduced specific infectivity are made in the presence of the drugs (Courtney *et al.*, 1973; Pizer *et al.*, 1980; Katz *et al.*, 1980; Peake *et al.*, 1982; Svennerholm *et al.*, 1982; Spivack *et al.*, 1982; Kousoulas *et al.*, 1983). Studies done with tunicamycin-modified and deoxyglucose-modified virions indicate that their capacity to adsorb to cells is much less affected than is their capacity to penetrate into cells (Svennerholm *et al.*, 1982; Spivack *et al.*, 1982). All these results indicate that addition of normal high-mannose N-linked oligosaccharides is a minimal requirement for infectivity. Tunicamycin blocks the addition of the N-linked oligosaccharides whereas deoxyglucose not only blocks the addition of, but also alters the structure of, the core N-linked chains.

In contrast, conditions that block later stages in processing of the HSV glycoproteins have very little adverse effect on the infectivity of the virions produced. Specifically, monensin inhibits the processing of high-mannose N-linked oligosaccharides to the complex type (Wenske *et al.*, 1982) and the addition of O-linked oligosaccharides (Johnson and Spear, 1983) without significantly reducing the specific infectivity of the virions produced (Johnson and Spear, 1982). Similarly, in mutant cell lines deficient for specific glycosyltransferases, infectious virus containing immature forms of the glycoproteins is produced (Campadelli-Fiume *et al.*, 1982; Serafini-Cessi *et al.*, 1983b). Finally, infectious virus produced in the presence of ammonium ions contains glycoproteins that, on the basis of electrophoretic mobilities, appear to be immature (Kousoulas *et al.*, 1983).

In the studies done with monensin, it was shown that the block in maturation of virion glycoproteins was associated with a block in transport of virus to the cell surface (Johnson and Spear, 1982). Egress of virus appeared to be impaired also in the mutant cell lines mentioned above (Serafini-Cessi *et al.*, 1983b). It would be interesting to know whether the same applies to virus produced in the presence of ammonium ions.

3. Adsorption

The viral structure (antireceptor) responsible for the adsorption of virions to receptors on cell surfaces has not yet been identified or characterized with respect to composition. Findings that gC-1⁻ (Heine *et al.*, 1974; Manservigi *et al.*, 1977) and gC-2⁻ (Zezulak and Spear, 1984b) mutants are viable suggest, however, that this glycoprotein is not an essential component of the antireceptor.

Results of recent studies suggest an approach to identifying viral gene products that influence the specificity of the antireceptor. It seems that HSV-1 and HSV-2 bind to different receptors on cell surfaces, based on findings that the ratio of receptors for HSV-1 and HSV-2 differs for different cell types (Vahlne *et al.*, 1978, 1979, 1980) and that virions or viral products of one serotype can block the binding of homologous, but not of heterologous, virus (Vahlne *et al.*, 1979, 1980). This offers the possibility of mapping viral genes responsible for the intertypic difference in binding properties.

Additional approaches that could and should be exploited for characterization of the HSV-1 and HSV-2 antireceptors for cell surfaces include production and characterization of monoclonal antibodies that can block viral adsorption, isolation and characterization of viral mutants defective in adsorption, and purification and characterization of isolated viral products with the appropriate binding characteristics.

4. Penetration

a. Mode and Site of Penetration

It is generally accepted that penetration of enveloped viruses (or, more precisely, their nucleocapsids) into eukaryotic cells results from fusion between the virion envelope and a cell membrane. It seems that enveloped viruses fall into two classes depending upon whether this fusion can occur at the cell surface or only after endocytosis of the virion (reviewed by Lenard and Miller, 1983). Evidence has been presented for several viruses (including vesicular stomatitis virus, Semliki Forest virus, and influenza virus) that fusion occurs *after* endocytosis; activation of fusogenic activity to enable penetration apparently requires the low pH characteristic of endosomes and lysosomes and this activation can be blocked by agents (such as chloroquine and ammonium ions) that raise lysosomal pH (Helenius *et al.*, 1980; White *et al.*, 1980; Matlin *et al.*, 1981, 1982; Yoshimura *et al.*, 1982). Specific proteolytic cleavage of particular viral envelope glycoproteins may also be a requirement for activation of fusogenic activity (Homma and Ohuchi, 1973; Scheid and Choppin, 1974; Lazarowitz and Choppin, 1975; Klenk *et al.*, 1975).

For HSV, although there are no indications that activation of fusogenic activity requires cleavage of any viral glycoprotein, this possibility has not been rigorously ruled out. The requirement of endocytosis and low pH for penetration appears *not* to apply to HSV, just as it apparently does not apply to paramyxoviruses (Nagai *et al.*, 1983), because ammonium ions do not block the infectivity of HSV (Holland and Person, 1977).

Although electron microscopy (Morgan *et al.*, 1968) and biochemical experiments (Sarmiento *et al.*, 1979; DeLuca *et al.*, 1981) suggest that infectious penetration of HSV may occur by fusion at the cell surface, this has yet to be definitively established. Evidence that HSV virion envelope components (Fc receptors) can be detected in the cell surface membrane immediately after viral penetration, and in the absence of viral gene expression, has been presented (Para *et al.*, 1980). The possibility exists, however, that these viral envelope components came to be in the cell surface because of recycling of membrane from endocytotic vesicles (Goldstein *et al.*, 1979). Additional experiments of the kind described previously (Para *et al.*, 1980), using inhibitors of membrane recycling (Basu *et al.*, 1981), may help to resolve the issue.

b. Envelope Components Required for Penetration

Evidence that gB plays a role in penetration has come from the study of ts HSV-1 mutants (tsB5 and tsJ12) that are blocked in the processing of gB-1 precursors to mature forms at nonpermissive temperature (Sarmiento *et al.*, 1979; Little *et al.*, 1981). These two mutants belong to the same complementation group (Schaffer *et al.*, 1978) and early mapping studies suggested that their ts lesions might be in the structural gene for gB-1 (Ruyechan *et al.*, 1979; Little *et al.*, 1981), a suspicion recently confirmed by precise mapping of the mutations and the gB-1 gene (DeLuca *et al.*, 1982; Holland *et al.*, 1983b; Kousoulas *et al.*, 1984) and by direct nucleotide sequence comparisons of the wild-type and mutant genes (Bzik *et al.*, 1984).

The block in replication of these mutants is late in the infectious cycle because viral DNA, most or all viral proteins except mature gB-1, and even virions are produced at nonpermissive temperature. The virions produced are noninfectious, however. Because these noninfectious virions can bind to cells and the block to their infectivity can be overcome by treating the virus–cell complexes with polyethylene glycol (a membrane-fusing agent), it has been suggested that the defect in these mutants is in membrane-fusing activity required for viral penetration (Sarmiento *et al.*, 1979; Little *et al.*, 1981). Hence, gB-1 is implicated in this membrane-fusing activity and in penetration. An additional observation in support of this proposed role for gB-1 is association of the tsB5 ts lesion with temperature-sensitivity of HSV-induced cell fusion (see below). Whether penetration also requires the presence and activity of other envelope proteins remains to be determined.

5. A Cell Line Resistant to Penetration by Some HSV Strains

The Rous sarcoma virus-transformed cell line XC is one of the few cell lines known to be nonpermissive for HSV replication. Contradictory reports have appeared in the literature as to the nature of the block to replication. In some studies, evidence was presented that penetration did not occur (Docherty et al., 1973; Campbell et al., 1974), whereas in others it was shown that viral genes were transcribed and expressed (Garfinkle and McAuslan, 1973, 1974; Padgett et al., 1978). Becker et al. (1974) related the expression of HSV genes (e.g., penetration) to the cell line used to produce the infecting virus.

Recently, Epstein and Jacquemont (1983) have presented evidence that the stage at which replication is blocked in XC cells depends on the strain of HSV used. Virions of some HSV-1 strains can initiate infection and induce the expression of viral proteins whereas, for other HSV-1 strains, entry of virus appears to be blocked at the stage of penetration and the block can be bypassed by use of polyethylene glycol (Epstein et al., 1983). This finding could account for the discrepancies in results reported earlier and, in addition, offers the possibility of identifying the HSV gene or genes responsible for determining whether penetration of XC cells can occur. These authors have concluded, on the basis of preliminary mapping data, that a locus affecting penetration may map between coordinates 0.70 and 0.83 on the HSV-1 genome and that reduced levels of gC-1, relative to gB-1, may be associated with enhanced ability to penetrate XC cells (Epstein et al., 1984).

B. Envelopment of Virions and Their Egress from Infected Cells

It seems reasonable to propose that at least some domains of some of the HSV glycoproteins are essential to the envelopment process, assuming that budding and envelopment are mediated at least in part by interactions between domains of the glycoproteins that extend into the nucleoplasm and domains of tegument or nucleocapsid proteins. Although no evidence has been obtained that any particular glycoprotein is essential to the envelopment process, the kinds of mutants required to test this possibility are not yet available for most of the glycoproteins. For example, although tsB5 virions produced at nonpermissive temperature are deficient in mature forms of gB, they appear to contain immature or aberrant forms of gB-1 (Sarmiento et al., 1979). Therefore, a role for at least some form of gB-1 in envelopment cannot be ruled out.

There are viral mutants that may perhaps rule out a requirement for gC-1 in the envelopment process. Syncytial strains such as HSV-1(MP) are viable despite the apparent inability to produce gC-1 (Manservigi et al., 1977) and absence of gC-1 from virions (Heine et al., 1974); it remains to be determined, however, precisely what genetic defect accounts for the gC$^-$ phenotype and whether any heretofore unrecognizable form of gC-1 is actually produced. Other viable mutants altered in gC-1 expression have recently been isolated by selection with neutralizing anti-gC monoclonal antibodies (Holland *et al.*, 1983a, 1984). Some of these mutants secrete truncated forms of gC-1 suggesting that the mutant proteins may lack membrane anchor sequences (and thus would not be stably incorporated into membranes or virions) while retaining a signal sequence to permit secretion; alternatively, the mutant proteins could have a membrane anchoring domain but be abnormally sensitive to proteolytic cleavage that results in secretion of a gC-1 fragment (Holland *et al.*, 1984). Additional characterization of the mutants is necessary to clarify the issue and to determine whether they can rule out a requirement for gC-1 in envelopment.

Once envelopment occurs at the inner nuclear membrane, the virion must be transported from the perinuclear space to the outside of the cell. It has been proposed on the basis of electron microscopic studies that egress of virus is via a process of "reverse phagocytosis" (Morgan *et al.*, 1959) or via a network of tubules thought to be formed in infected cells, but not in uninfected cells, to connect the perinuclear region with the extracellular space (Schwartz and Roizman, 1969). Alternatively, on the basis of electron microscopic studies with another herpesvirus, it has been proposed that egress is a process involving fusion of perinuclear virus with the outer lamella of the nuclear envelope and reenvelopment of the resulting cytoplasmic nucleocapsid at a cytoplasmic membrane (Stackpole, 1969). Electron micrographs have been presented as evidence that envelopment of HSV nucleocapsids can occur at cytoplasmic membranes (Epstein, 1962; Siminoff and Menefee, 1966).

I favor a specific version of the "reverse phagocytosis" hypothesis mentioned above, namely that envelopment occurs once at the inner nuclear membrane and egress of virions is via cisternae of the RER, the Golgi apparatus, and transport vesicles operating between RER and Golgi and Golgi and cell surface. In the first place, the sequential double envelopment scheme seems clumsy and requires that cytoplasmic nucleocapsids of progeny virus be treated and processed differently from cytoplasmic nucleocapsids of parental virus (which must be disassembled to enable gene expression). Moreover, the electron micrographs that were interpreted as showing envelopment at cytoplasmic membranes could instead be showing deenvelopment concomitant with superinfection. Second, experiments have shown that monensin, which is known to block the transport of material from the Golgi apparatus to the cell surface (Tartakoff and Vassalli, 1977, 1978; Uchida *et al.*, 1979), also blocks the transport of HSV virions to the cell surface and causes the accumulation of infectious virions in abnormally large cytoplasmic vacuoles (Johnson and Spear, 1982). Unless there are transport pathways that bypass the Golgi apparatus and are also blocked by monensin (which are conceivable, but not known to exist), these results suggest that egress of HSV is obligatorily via the Golgi apparatus.

Whatever the actual details of the transport mechanism governing egress of virions, it seems possible that specific interactions of virion envelope constituents with other structures are important to this process. For example, directionality of egress may be influenced by interactions between molecules in the virion surface and in membranes (cisternal face) of the RER, Golgi apparatus, or transport vesicles. Perhaps unidirectionality of transport is in part ensured by changes, occurring during transport, in the structures of the viral glycoproteins resulting from addition, trimming, and extension of oligosaccharides as described above.

Finally, it seems likely that some sort of mechanism exists to reduce the probability that progeny virions will fuse with membranes of the infected cell during or after egress. Such a mechanism is postulated to account for the fact that infectious HSV does accumulate in infected cells and on the surfaces of infected cells, despite intimate contact with cell membranes, instead of being eclipsed by penetration into the cytoplasm. One could postulate that infectivity of the virion is latent or blocked until the virion is released from the cell or, alternatively, that membranes of the infected cell are modified to reduce the probability of penetration by infectious virus. It is also possible that, even in the uninfected cell, membranes bounding different compartments of the cell may differ in their capacity to fuse with virus.

One HSV-1 mutant has been isolated that may exhibit a defect either in egress from infected cells or in the mechanism postulated to block eclipse of progeny virus. This mutant, designated HSV-1(mP)50B, was recognized by its decreased rate of plaque development at low temperature (31°C) when compared with the parental strain; accumulation of intracellular infectious virus seems normal although abnormally high numbers of empty coreless capsids were detected at nuclear pores late in infection (Tognon et al., 1981). Mapping studies indicate that the mutation is in the S component of the genome (Tognon et al., 1981), possibly in the gene for gE (Para et al., 1982b). More work is required to define the nature of this mutation and its consequences.

C. Expression of Receptors for the Fc Region of IgG and for the C3b Component of Complement

Cells infected with HSV express new receptors, one kind of which has affinity for the Fc region of IgG and another for the activated third component of complement. Evidence summarized below indicates that these receptors are viral glycoproteins or, more precisely, that specific viral glycoproteins form at least part of each receptor. The physiological significance of these receptors for HSV replication and for HSV pathogenesis is not understood nor is it known whether the ligands used to detect the receptors (IgG, C3b) are actually the physiologically important ligands.

It is particularly intriguing, of course, that receptor activities often found together on certain cells of the lymphoreticular system can be detected on the surfaces of many cell types infected by HSV [Fc receptors are also induced by infections with cytomegalovirus (Keller *et al.*, 1976; Rahman *et al.*, 1976; Westmoreland *et al.*, 1976) and with varicella–zoster virus (Ogata and Shigeta, 1979)]. It has been speculated that such receptors may interfere with or modulate immune responses to HSV-infected cells (Costa and Rabson, 1976; Lehner *et al.*, 1975; Adler *et al.*, 1978) or that binding of appropriate ligands to the receptors might influence what happens inside the infected cell (Westmoreland and Watkins 1974; Lehner *et al.*, 1975; Costa *et al.*, 1977). Although experiments have shown that aggregated nonimmune IgG can interfere with immune cytolysis *in vitro* (Adler *et al.*, 1978), it remains to be seen whether such phenomena are demonstrable or important *in vivo*. And although it was reported that high concentrations of nonimmune IgG can reduce the yields of infectious HSV produced in cell culture (Costa *et al.*, 1977), the effect was not large and it remains to be determined that this effect was actually due to binding of IgG to the Fc receptors.

1. Fc Receptor

The first indication of HSV-induced expression of Fc receptors came from findings that antibody-sensitized erythrocytes, but not uncoated erythrocytes, could bind to HSV-infected cells (Watkins, 1964). Subsequently, it was shown that many cell types infected with either HSV-1 or HSV-2 bound to IgG from a variety of species and that the binding was to Fc regions of IgG (Yasuda and Milgrom, 1968; Westmoreland and Watkins, 1974; Feorino *et al.*, 1977; Costa *et al.*, 1978; McTaggart *et al.*, 1978; Nakamura *et al.*, 1978; Bourkas and Menezes, 1979; Cines *et al.*, 1982). It can be inferred that the receptor recognizes some feature shared in common by IgGs from different mammalian species but little else can be said about the nature of the chemical group bound, nor is it known whether the receptor discriminates among subclasses of IgG. Fc receptors are expressed on the surfaces of virions (Para *et al.*, 1980) as well as on the surfaces of infected cells.

Evidence implicating gE as the Fc receptor came from the demonstration by affinity chromatography that solubilized gE-1 and gE-2 have Fc-binding activity (Baucke and Spear, 1979; Para *et al.*, 1982b) and from finding that F(ab')$_2$ fragments of an anti-gE-1 serum could at least partially block Fc-binding activity (Para *et al.*, 1982a). As anticipated, gE has been shown to be present and exposed on the surfaces both of virions (Para *et al.*, 1982a) and of infected cells (Baucke and Spear, 1979). It is not yet known whether the Fc-binding activity or any other activity of gE is essential for HSV replication.

2. C3b Receptor

It was recently shown that endothelial cells infected with HSV-1 develop receptors for the C3b component of complement as well as Fc receptors (Cines *et al.*, 1982; Friedman *et al.*, 1984). The presence of the C3b receptors was demonstrated by the binding of complement-coated bacteria or by quantitating the binding of IgM-sensitized erythrocytes to which purified components of complement had been added. Evidence that gC-1 forms part of, or is, the C3b receptor comes from findings that a gC⁻ viral mutant is also C3b receptor-negative and that monoclonal antibodies specific for gC-1, but not for other viral glycoproteins, block C3b-binding activity (Friedman *et al.*, 1984). Viability of the gC⁻, C3b receptor-negative mutant indicates that this receptor activity is not essential for HSV-1 replication in cell culture.

In light of differences observed between gC-1 and gC-2, as described above, it is of interest that C3b receptors could *not* be detected on endothelial cells infected with HSV-2 (Friedman *et al.*, 1984). Although the study of HSV-1 C3b receptors has so far been restricted to infected endothelial cells, it seems likely that these receptors will also be detected on other infected cell types.

An observation made several years ago should be mentioned in connection with the results described above. It was found that Raji lymphoblastoid cells coated with purified human C3 could bind to HSV-1-infected cells, but not to uninfected cells, and that this binding was inhibited by the protease inhibitor *TLCK* (Dierich *et al.*, 1979). The authors proposed that a protease activity on the surfaces of HSV-1-infected cells converted the Raji cell-bound C3 to a form that in turn could bind also to the infected cell. The question arises as to whether C3 activation (cleavage) as well as the binding of activated C3 can be attributed specifically to HSV-1-infected cells.

D. Cell Fusion

Most, if not all, enveloped viruses can induce the fusion of infected cells under appropriate conditions. Probably this is a consequence of the following circumstances—namely that (1) capacity of an enveloped virus to induce membrane fusion is essential for infectivity; (2) fusogenic activity is associated with the presence of specific glycoproteins in the virion envelope; and (3) these glycoproteins may also be found in the infected cell surface.

Presence of the appropriate viral glycoproteins in the cell surface may not be sufficient to induce cell fusion, however, for a variety of reasons. For example, fusogenic activity of some viruses is apparently activated only at low pH so that fusion of infected cells requires exposure to low pH (White *et al.*, 1981). In addition, density, conformation, and activities of viral glycoproteins in the cell surface and in virions are unlikely to be identical; the possibility exists, as has been suggested previously (Manservigi et al., 1977; Ruyechan et al., 1979), that interactions of viral glycoproteins with other virion constituents or virion products may govern the induction of fusogenic activity.

Clinical isolates of HSV usually do not exhibit the capacity to fuse cultured cells although isolation of fusion-inducing strains from patients has occasionally been reported (Terni and Roizman, 1970). Fusion-inducing variants of HSV can readily be identified and isolated during propagation of virus in vitro (reviewed by Roizman, 1962; Spear, 1980). Studies to be described below document the occurrence of nonlethal mutations in HSV that result in capacity of the mutant to induce cell fusion. The phenotype resulting from such mutations is designated Syn whereas Syn$^+$ denotes the wild-type nonfusing phenotype. Expression of the mutant phenotype can be cell-dependent, i.e., some mutants are Syn on one cell type and Syn$^+$ on another. Syn mutations are termed nonlethal because they permit replication and propagation of virus in cell culture and in experimental animals; such mutations could very well be "lethal" for survival of virus in the human population, however.

Evidence that mutations in one or more of several different HSV genes can result in expression of the Syn phenotype emerged from mapping studies to be described below and from the results of complementation tests (Little and Schaffer, 1981). Syn$^+$ revertants or pseudorevertants can also be isolated from Syn mutants. It has been shown that 16 Syn$^+$ variants isolated from one Syn mutant could be classified into four complementation groups (Yamamoto and Kabuta, 1977).

For HSV, definition of the factors that govern fusogenic activity, both in the virion and in the infected cell surface, requires study of the products altered by the mutations described above as well as other kinds of experiments designed to assess activities of the viral glycoproteins.

1. Syn Mutations

Table I summarizes available information about the genomic locations of selected HSV-1 Syn mutations that have been mapped. In only one instance has the mutated gene product been identified. The Syn mutation and ts mutation of HSV-1(HFEM)tsB5 are distinct and segregable from each other (Honess et al., 1980; DeLuca et al., 1982; Kousoulas et al., 1984) but both are in the gene for gB. Specifically, the Syn mutation appears to be very near the COOH-terminus, possibly in the domain of the polypeptide that may extend into the cytoplasm (compare the mapping data of 'DeLuca et al. (1982) and Kousoulas et al. (1984) with the nucleotide sequence analysis presented by Bzik et al. (1984), Fig. 2). Consequently, the mutation could be expected to affect the interactions of gB with components on the inner aspect of the plasma membrane (and

TABLE I. Map Coordinates of Selected Syn Mutations in HSV-1

Mutant name	Locus name[a]	Map coordinates of Syn mutation	Reference
HFEM-tsB5	syn3	0.30–0.40	Ruyechan et al. (1979)
		0.345–0.355	DeLuca et al. (1982)
		0.345–0.350	Kousoulas et al. (1984)
MP	syn1	0.68–0.82	Ruyechan et al. (1979)
		0.735–0.740	Pogue-Geile et al. (1984), Bond and Person (1984)
	syn2	0.68–0.82	Ruyechan et al. (1979)
KOS-78R		0.724–0.747	Little and Schaffer (1981)
KOS-804		0.040–0.064	Little and Schaffer (1981)
17-TK1301 17-TK1302 17-TK1303		Deletions at ≈0.30 in the TK gene	Sanders et al. (1982)

[a] See Ruyechan et al. (1979).

virion envelope). As discussed above and below, gB appears to play a role both in viral penetration and in cell fusion.

Mutations in at least three other regions of the genome can result in the Syn phenotype (Table I). Although the relevant mutated gene products have not yet been identified, a comparison of Table I and Fig. 1 reveals that none of these other Syn mutations are in genes encoding known glycoproteins.

Ruyechan et al. (1979) showed that the Syn phenotype of strain HSV-1(MP) results from at least two segregable mutations. Recombinants obtained after the transfer of DNA fragments from HSV-1(MP) to a Syn$^+$ strain fell into two classes: members of one class could fuse Vero cells, but not HEp-2 cells, whereas members of the other class resembled HSV-1(MP) in ability to fuse many cell types. The mutated locus responsible for the Syn phenotype on Vero cells was designated syn1 and the second mutated locus that, alone or in combination with mutated syn1, was responsible for the Syn phenotype on HEp-2 cells was designated syn2. Both the syn1 and syn2 loci were originally mapped on the HSV-1 genome to the region from 0.68 to 0.82 MU (Ruyechan et al., 1979). Recently, the mutation defining the syn1 locus was more precisely mapped to the region from 0.735 to 0.740 MU (Bond and Person, 1984; Pogue-Geile et al., 1984). Evidence was presented that a second mutation, which lies outside the region between 0.702 and 0.752 MU, does not, by itself, cause the Syn phenotype, but that when present with the syn1 mutation, causes the Syn2 phenotype (Pogue-Geile et al., 1984).

Genetic evidence that the Syn phenotype of another mutant also results from the presence of at least two segregable mutations has been published (Yamamoto *et al.*, 1972, 1975; Yamamoto and Kabuta, 1976). These mutations have apparently not been mapped.

Strain HSV-1 (MP) is gC⁻ (Heine *et al.*, 1974; Manservigi *et al.*, 1977) as well as Syn (Hoggan and Roizman, 1959). Other Syn mutants are also gC⁻ (Cassai *et al.*, 1975/76). These observations led to the suggestion that absence of gC-1 expression might be at least in part responsible for the Syn phenotype (Manservigi *et al.*, 1977). This turns out not to be correct for HSV-1(MP). Insertion of a functional gC-1 gene into the TK gene of HSV-1(MP) resulted in expression of gC-1 without change in the Syn phenotype (Lee *et al.*, 1982b). There does seem to be some as yet unexplained relationship between expression of the gC⁻ and Syn phenotypes, however. In addition to the gC⁻ Syn mutants of HSV-1, a Syn mutant of HSV-2 was recently shown to be gC⁻ (Zezulak and Spear, 1984b). Moreover, the *syn2* locus of HSV-1(MP) and a mutation responsible for the gC⁻ phenotype segregated together in recombinants obtained by marker transfer (Ruyechan *et al.*, 1979) and by mixed infection (Manservigi *et al.*, 1977). A mutation responsible for the gC⁻ phenotype of HSV-1(MP) has been mapped to the vicinity of the gC-1 structural gene (Pogue-Geile *et al.*, 1984) whereas the *syn2* locus has not yet been precisely mapped.

Puzzling results have been obtained in characterization of the mutant HSV-1(KOS)78R (Little and Schaffer, 1981). The Syn mutation in this strain maps to the same region of the genome as does the *syn1* locus mutated in HSV-1(MP) (Table I). Moreover, HSV-1(KOS)78R fuses Vero cells, but not HEp-2 cells, as do recombinant viruses carrying only the mutated *syn1* locus of HSV-1(MP). The results of complementation studies, however, suggest that HSV-1(KOS) 78R and HSV-1(MP) are mutated in different Syn loci. It will be of interest to define the precise relationship among the Syn mutations in these two strains. Another interesting feature of HSV-1(KOS)78R is the temperature-sensitivity of its Syn phenotype (Syn at 39°C, Syn⁺ at 34°C). No other Syn mutant described is of this kind.

Another mutant, also isolated and characterized by Little and Schaffer (1981), fuses Vero cells (and other cell types), but not HEp-2 cells. The Syn mutation in this strain has been mapped to the region of the genome from 0.040 to 0.064 MU.

Finally, it has been shown that deletions in the TK gene can result in expression of the Syn phenotype. Some of these mutants fail to produce polypeptides of 43K and 19K but it is not known whether one of these deficiencies or some other undetected defect is responsible for the Syn phenotype (Sanders *et al.*, 1982).

2. Roles of the Viral Glycoproteins

The first indications that glycoproteins play some role in HSV-induced cell fusion emerged from studies done with inhibitors of glycosylation (Gallaher *et al.*, 1973; Keller, 1976; Knowles and Person, 1976;

Holland and Person, 1977; Kousoulas et al., 1978; Campadelli-Fiume et al., 1980). Any inhibitor that blocks the synthesis or processing of glycoproteins after infection also blocks fusion induced by the appropriate Syn mutants of HSV. Fusion is also not observed when Syn mutants are used to infect mutant cell lines defective in glycosyltransferases (Campadelli-Fiume et al., 1982; Serafini-Cessi et al., 1983b) or when the infected cells (mutant or wild-type cells) are repeatedly exposed to neuraminidase after infection (Serafini-Cessi et al., 1983b). All these observations indicate that normal glycosylation, and perhaps sialylation, of either cellular or viral products are required after infection in order for virus-induced cell fusion to occur.

On the assumption that viral glycoproteins in the cell surface are at least in part responsible for the induction of cell fusion, the absence of fusion under the conditions mentioned above could result from failure of the appropriate glycoproteins to reach the cell surface, from alterations in the structure and function of glycoproteins present in the cell surface, or from direct inhibitory effects of the drugs used on the process of cell fusion itself. In most instances, experiments that could differentiate among these alternatives have apparently not been done. In the case of ammonium chloride-mediated inhibition of cell fusion (Holland and Person, 1977), however, it was reported that immature forms of the HSV glycoproteins were made and transported to the cell surface and that these forms of the glycoproteins could induce cell fusion once the ammonium chloride was removed (Kousoulas et al., 1982).

Two specific HSV-1 glycoproteins, gB-1 and gD-1, have been implicated in the process of cell fusion, on the basis of studies done with mutants and with monoclonal antibodies. It has been shown for the mutant HSV-1(HFEM)tsB5, and for recombinants carrying the ts lesion of this mutant, that temperature-sensitivity of gB-1 processing and accumulation correlates with temperature-sensitivity of cell fusion (Manservigi et al., 1977; Haffey and Spear, 1980). This correlation also holds for partial phenotypic revertants of HSV-1(HFEM)tsB5 (Haffey and Spear, 1980). For all these strains of virus that are ts for cell fusion, the gB-1 gene probably has a Syn mutation as well as the ts lesion, as discussed above. Although it is important to determine whether temperature-sensitivity of cell fusion would be observed when the ts lesion is in gB-1 and the Syn mutation is in a gene other than gB-1, the results obtained to date suggest that gB-1 is essential for cell fusion as well as for viral penetration.

Findings implicating gD-1 in cell fusion have come from the use of monoclonal antibodies (Noble et al., 1983). All 7 anti-gD antibodies tested inhibited the fusion of Vero cells induced by a Syn mutant of HSV-1 (HFEM) whereas negative results (inability to block cell fusion) were obtained with 12 anti-gB, 8 anti-gC, and 2 anti-gE antibodies. The negative results obtained with anti-gB antibodies are of interest in light of the studies mentioned above. It was demonstrated that failure of the anti-gB antibodies to block cell fusion was not due to their failure to bind to the surfaces of infected cells. Any hypothesis as to the role of gB in cell fusion must take into account the fact that at least some anti-gB monoclonal antibodies do not interfere with fusion-inducing activity at the cell surface whereas all anti-gD antibodies so far tested do. The precise roles of gB-1 and gD-1 is HSV-induced cell fusion cannot be deduced from the results of studies done to date.

E. Functional Relatedness of the HSV-1 and HSV-2 Glycoproteins

The HSV-1 and HSV-2 glycoproteins assigned the same alphabetic designations are antigenically related (reviewed by Spear, 1984) and are generally assumed to be functionally related, also. This assumption is supported by evidence in only a few instances.

For example, gE-1 and gE-2 are both highly sulfated (Hope et al., 1982; Hope and Marsden, 1983) and have independently been shown to exhibit Fc-binding activity (Baucke and Spear, 1979; Para et al., 1982b). As discussed above, gC-1 and gC-2 appear to differ from each other in size and antigenic structure more than do most of the other related HSV-1 and HSV-2 glycoproteins. There may also be some functional differences between gC-1 and gC-2 although this has not yet been established. C3b receptor activity has been associated with gC-1 (Friedman et al., 1984) whereas no such receptor activity has yet been detected on cells infected with HSV-2. A point of similarity between gC-1 and gC-2 is nonlethality in cell culture of gC-1$^-$ and gC-2$^-$ mutations (Heine et al., 1974; Manservigi et al., 1977; Cassai et al., 1975/76; Zezulak and Spear, 1984b).

Although evidence has been obtained that gB-1 plays some role both in viral penetration and in HSV-1-induced cell fusion (Manservigi et al., 1977; Sarmiento et al., 1979; Haffey and Spear, 1980; Little et al., 1981) and that gD-1 is involved in cell fusion (Noble et al., 1983), no such evidence has been obtained independently for gB-2 and gD-2, respectively. No activities or functions have yet been assigned to gG-2 nor has an HSV-1 counterpart yet been identified.

It is evident that much remains to be learned about the functions of the HSV glycoproteins and perhaps equally evident that the methods and materials are available to permit significant progress in the near future.

V. APPENDIX

Over the past several years a consensus has emerged among workers in the field about development and use of a nomenclature for HSV-1 and HSV-2 glycoproteins. The principles governing this nomenclature were previously summarized in a published report of workshop discussions

TABLE II. Nomenclature for Glycoproteins of Herpes Simplex Viruses

---

1. All forms of a single translation product should be assigned the same alphabetic designation, preceded by the small letter "g," and will be collectively known as gB, or gC, etc.
2. It may be necessary to identify and name each of the multiple forms of a single translation product, as follows:
   a. Unstable precursor forms should be assigned the prefix "p" (e.g., pgC), and multiple unstable forms can be differentiated by adding the apparent molecular weight ($\times 10^{-3}$) in parentheses [e.g., pgC(120)].
   b. Multiple *stable* forms of a glycoprotein can also be differentiated by adding the apparent molecular weight ($\times 10^{-3}$) in parentheses [e.g., gC(125)].
   The apparent molecular weight designations are not intended to be permanent and invariant features of the name because the values assigned will vary from system to system and from lab to lab.
3. Alphabetic designations assigned to HSV-1 glycoproteins prior to March 1980 (gB, gC, gD, gE) should be retained (except gA, which is a form of gB).
4. Any newly discovered HSV-1 or HSV-2 glycoproteins should be assigned the next alphabetic designation in sequence (excluding gA and gF), *provided* the new glycoprotein is shown to be encoded by a genetic locus different from those encoding all previously named HSV-1 and HSV-2 glycoproteins.
5. The suffix 1 or 2 can be used to distinguish between the HSV-1 and HSV-2 glycoproteins, as follows: gC-1 and gC-2.

(Cohen *et al.*, 1981). Recent findings have made it clear, however, that, in two instances, names in current usage do not conform to two of the principles of the adopted nomenclature—namely that all forms of a single glycoprotein translation product have the same alphabetic designation and that the same alphabetic designation be used for functionally or antigenically related glycoproteins specified by HSV-1 and HSV-2. Specifically, the HSV-1 glycoprotein originally designated gA (Spear, 1976) is now known to be one of the multiple electrophoretically differentiable forms of gB (Eberle and Courtney, 1980b; Pereira *et al.*, 1981; Balachandran *et al.*, 1982). Also, the HSV-2 glycoprotein designated gF (Balachandran *et al.*, 1981) is now known to be the antigenically and genetically related counterpart of HSV-1 gC (Para *et al.*, 1983; Zweig *et al.*, 1983; Zezulak and Spear, 1983, 1984b).

At a recent International Herpesvirus Workshop (held from July 31 to August 5, 1983, in Oxford, England), a group of interested participants discussed these matters relating to nomenclature of HSV glycoproteins. There was general agreement that the principles previously agreed upon for nomenclature continue to have utility and value and that it would be worthwhile to modify current usage of names in order to achieve conformance to the original scheme.

Table II presents a reiteration of the agreed-upon scheme for nomenclature of HSV glycoproteins, modified slightly in form, but not in intent, from the original version (Cohen *et al.*, 1981) in accordance with recent developments. What follows are the specific recommendations made for changes in current usage of names:

- Use of the designation gA for any form of gB should be discontinued.
- The HSV-2 glycoprotein designated gF should be renamed gC (gC-2).
- The HSV-2 glycoprotein previously designated gC (Ruyechan *et al.*, 1979), but now thought to be unrelated to HSV-1 gC, should be renamed.
- The designations gA and gF should not be used again for naming HSV glycoproteins, in order to minimize confusion with earlier usage.

ACKNOWLEDGMENTS. I thank all those individuals who generously provided preprints of their work and greatly facilitated the writing of this review. I am grateful to Chan Stroman and Janice Hoshizaki for competent assistance in the preparation of the manuscript. Work done in my laboratory and discussed in this review was supported by grants from the American Cancer Society and the National Cancer Institute (CA-21776 and CA-19264).

REFERENCES

Adler, R., Glorioso, J. C., Cossman, J., and Levine, M., 1978, Possible role of Fc receptors on cells infected and transformed by herpesvirus: Escape from immune cytolysis, *Infect. Immun.* **21**:442.

Balachandran, N., Harnish, D., Killington, R. A., Bacchetti, S., and Rawls, W. E., 1981, Monoclonal antibodies to two glycoproteins of herpes simplex virus type 2, *J. Virol.* **39**:438.

Balachandran, N., Harnish, D., Rawls, W. E., and Bacchetti, S., 1982, Glycoproteins of herpes simplex virus type 2 as defined by monoclonal antibodies, *J. Virol.* **44**:344.

Basu, S. K., Goldstein, J. L., Anderson, R. G. W., and Brown, M. S., 1981, Monensin interrupts the recycling of low density lipoprotein receptors in human fibroblasts, *Cell* 24:493.

Baucke, R. B., and Spear, P. G., 1979, Membrane proteins specified by herpes simplex viruses. V. Identification of an Fc-binding glycoprotein, *J. Virol.* **32**:779.

Becker, Y., Shlomai, J., Asher, Y., Weinberg, E., Cohen, Y., Olshevsky, U., and Kotler, M., 1974, Interaction of herpes simplex virus type 1 with Rous sarcoma virus-transformed rat cells [XC and R(B77) cell lines], *Intervirology* 4:325.

Blobel, G., Walter, P., Chang, C. N., Goldman, B. M., Erikson, A. H., and Lingappa, V. R., 1979, Translocation of proteins across membranes: The signal hypothesis and beyond, *Symp. Soc. Exp. Biol.* **33**:9.

Bond, V. C., and Person, S., 1984, Fine structure physical map locations of alterations that affect cell fusion in herpes simplex virus type I, *Virology* **132**:368.

Bourkas, A. E., and Menezes, J., 1979, Studies on the induction of IgG-Fc receptors and synthesis of IgM in primary and chronically infected lymphoid (Raji) cells by herpes simplex virus, *J. Gen. Virol.* **44**:361.

Bzik, D. J., Fox, B. A., DeLuca, N. A., and Person, S., 1984, Nucleotide sequence specifying the glycoprotein gene, gB, of herpes simplex virus type 1, *Virology* **133**:301.

Campadelli-Fiume, G., Sinibaldi-Vallebona, P., Cavrini, V., and Mannini-Palenzona, A., 1980, Selective inhibition of herpes simplex virus glycoprotein synthesis by a benz-amidino-hydrazone derivative, *Arch. Virol.* **66**:179.

Campadelli-Fiume, G., Poletti, L., Dall'Olio, F., and Serafini-Cessi, F., 1982, Infectivity and glycoprotein processing of herpes simplex virus type 1 grown in a ricin-resistant cell line deficient in *N*-acetylglucosaminyl transferase I, *J. Virol.* **43**:1061.

Campbell, W. F., Murray, B. K., Biswal, N., and Benyesh-Melnick, M., 1974, Restriction of herpes simplex virus type I replication in oncornavirus transformed cells, *J. Natl. Cancer Inst.* **52**:757.

Cassai, E., Manservigi, R., Corallini, A., and Terni, M., 1975/76, Plaque dissociation of herpes simplex virus: Biochemical and biological characters of the viral mutants, *Intervirology* **6**:212.

Centifanto-Fitzgerald, Y. M., Yamaguchi, T., Kaufman, H. E., Tognon, M., and Roizman, B., 1982, Ocular disease pattern induced by herpes simplex virus is genetically determined by a specific region of viral DNA, *J. Exp. Med.* **155**:475.

Cines, D. B., Lyss, A. P., Bina, M., Corkey, R., Klialidis, N. A., and Friedman, H. M., 1982, Fc and C3 receptors induced by herpes simplex virus on cultured human endothelial cells, *J. Clin. Invest.* **69**:123.

Cohen, G., Halliburton, I., and Eisenberg, R., 1981, Glycoproteins of herpesviruses, in: *The Human Herpesviruses: An Interdisciplinary Perspective* (A. J. Nahmias, W. R. Dowdle, and R. F. Shinazi, eds.), pp. 549–554, Elsevier/North-Holland, Amsterdam.

Cohen, G. H., Katze, M., Hydrean-Stern, C., and Eisenberg, R. J., 1978, Type-Common CP-1 antigen of herpes simplex virus is associated with a 59,000-molecular-weight envelope glycoprotein, *J. Virol.* **27**:172.

Cohen, G. H., Long, D., and Eisenberg, R. J., 1980, Synthesis and processing of glycoproteins gD and gC of herpes simplex virus type 1, *J. Virol.* **36**:429.

Cohen, G. H., Long, D., Matthews, J. T., May, M., and Eisenberg, R., 1983, Glycopeptides of the type-common glycoprotein gD of herpes simplex virus types 1 and 2, *J. Virol.* **46**:679.

Compton, T., and Courtney, R. J., 1983, Synthesis and localization of the nonglycosylated precursor glycoproteins in herpes simplex virus type 1 (HSV-1)-infected cells, Abstract of paper presented at the International Herpesvirus Workshop, Oxford, England.

Compton, T., and Courtney, R. J., 1984, Virus-specific glycoproteins associated with the nuclear fraction of herpes simplex virus type 1-infected cells, *J. Virol.* **49**:594.

Costa, J. C., and Rabson, A. S., 1975, Role of Fc receptors in herpes simplex virus infection, *Lancet* **1**:77.

Costa, J., Rabson, A. S., Yee, C., and Tralka, T. S., 1977, Immunoglobulin binding to herpes virus-induced Fc receptors inhibits virus growth, *Nature* **269**:251.

Costa, J., Yee, C., Nakamura, Y., and Rabson, A., 1978, Characteristics of the Fc receptor induced by herpes simplex virus, *Intervirology* **10**:32.

Courtney, R. J., Steiner, S. M., and Benyesh-Melnick, M., 1973, Effects of 2-deoxy-D-glucose on herpes simplex virus replication, *Virology* **52**:447.

DeLuca, N., Bzik, D., Person, S., and Snipes, W., 1981, Early events in herpes simplex virus type 1 infection; Photosensitivity of fluorescein isothiocyanate-treated virions, *Proc. Natl. Acad. Sci. USA* **78**:912.

DeLuca, N., Bzik, D. J., Bond, V. C., Person, S., and Snipes, W., 1982, Nucleotide sequences of herpes simplex type 1 (HSV-1) affecting virus entry, cell fusion, and production of glycoprotein gB (VP7), *Virology* **122**:411.

Dierich, M. P., Landen, B., Schulz, T., and Falke, D., 1979, Protease activity on the surface of HSV-infected cells, *J. Gen. Virol.* **45**:241.

Docherty, J. J., Mitchel, W. R., and Thompson, C. J., 1973, Abortive herpes simplex virus replication in Rous sarcoma virus transformed cells (37665), *Proc. Soc. Exp. Biol. Med.* **144**:697.

Dunphy, W. G., Fries, E., Urbani, L. J., and Rothman, J. E., 1981, Early and late functions associated with the Golgi apparatus reside in distinct compartments, *Proc. Natl. Acad. Sci. USA* **78**:7453.

Eberle, R., and Courtney, R. J., 1980a, Preparation and characterization of specific antisera to individual glycoprotein antigens comprising the major glycoprotein region of herpes simplex virus type 1, *J. Virol.* **35**:902.

Eberle, R., and Courtney, R. J., 1980b, gA and gB glycoproteins of herpes simplex virus type I: Two forms of a single polypeptide, *J. Virol.* **36**:665.

Eberle, R., and Courtney, R. J., 1982, Multimeric forms of herpes simplex virus type 2 glycoproteins, *J. Virol.* **41**:348.

Eisenberg, R. J., Hydrean-Stern, C., and Cohen, G. H., 1979, Structural analysis of precursor and product forms of type-common envelope glycoprotein D (CP-1 antigen) of herpes simplex virus type 1, *J. Virol.* **31**:608.

Eisenberg, R. J., Long, D., Pereira, L., Hampar, B., Zweig, M., and Cohen, G. H., 1982a, Effect of monoclonal antibodies on limited proteolysis of native glycoprotein gD of herpes simplex virus type 1, *J. Virol.* **41**:478.

Eisenberg, R. J., Ponce de Leon, M., Pereira, L., Long, D., and Cohen, G. H., 1982b, Purification of glycoprotein gD of herpes simplex virus types 1 and 2 by use of monoclonal antibody, *J. Virol.* **41**:1099.

Eisenberg, R. J., Long, D., Hogue-Angeletti, R., and Cohen, G. H., 1984, Amino terminal sequence of glycoprotein D of herpes simplex virus types 1 and 2, *J. Virol.* **49**:265.

Epstein, A. L., and Jacquemont, B., 1983, Virus polypeptide synthesis induced by herpes simplex virus in non-permissive rat XC cells, *J. Gen. Virol.* **64**:1499.

Epstein, A., Jacquemont, B., and Machuca, I., 1983, Differences in penetration into non-permissive XC cells between different strains of herpes simplex virus type 1. *Ann. Virol.* (Inst. Pasteur) **134E**: 439.

Epstein, A., Jacquemont, B. and Machuca, I., 1984, Infection of a restrictive cell line (XC cells) by intratypic recombinants of HSV-1: Relationship between penetration of the virus and relative amounts of glycoprotein C, *Virology* **132**:315.

Epstein, M. A., 1962, Observations on the mode of release of herpes virus from infected HeLa cells, *J. Cell Biol.* **12**:589.

Eylar, E. H., 1965, On the biological role of glycoproteins, *J. Theor. Biol.* **10**:89.

Feorino, P. M., Shore, S. L., and Reimer, C. B., 1977, Detection by indirect immunofluorescence of Fc receptors in cells acutely infected with herpes simplex virus, *Int. Arch. Allergy Appl. Immunol.* **53**:222.

Friedman, H. M., Cohen, G. H., Eisenberg, R. J., Seidel, C. A., and Cines, D. B., 1984, Glycoprotein C of HSV-1 functions as a C3b receptor on infected endothelial cells, *Nature* **309**:633.

Frink, R. J., Eisenberg, R., Cohen, G., and Wagner, E. K., 1983, Detailed analysis of the portion of the herpes simplex virus type 1 genome encoding glycoprotein C, *J. Virol.* **45**:634.

Gallaher, W. R., Levitan, D. B., and Blough, H. A., 1973, Effect of 2-deoxy-D-glucose on cell fusion induced by Newcastle disease and herpes simplex viruses, *Virology* **55**:193.

Garfinkle, B., and McAuslan, B. R., 1973, Non-cytopathic, nonproductive infection by herpes simplex viruses types 1 and 2, *Intervirology* **1**:362.

Garfinkle, B., and McAuslan, B. R., 1974, Regulation of herpes simplex virus-induced thymidine kinase, *Biochem. Biophys. Res. Commun.* **58**:822.

Gibson, M. G., and Spear, P. G., 1983, Insertion mutants of herpes simplex virus have a duplication of the gD gene and express two different forms of gD, *J. Virol.* **48**:396.

Gibson, R., Kornfeld, S., and Schlesinger, S., 1980, A role of oligosaccharides in glycoprotein synthesis, *Trends Biochem. Sci.* **5**:290.

Glorioso, J., and Levine, M., 1984, Monoclonal antibodies in studies of herpes simplex virus, in: *Monoclonal Antibodies in Biology and Clinical Medicine* (S. Ferone, ed.), Noyes, Park Ridge, New Jersey, in press.

Goldstein, J. L., Anderson, R. G. W., and Brown, M. S., 1979, Coated pits, coated vesicles, and receptor-mediated endocytosis, *Nature* **279**:679.

Haarr, L., and Marsden, H. S., 1981, Two-dimensional gel analysis of HSV type 1-induced polypeptides and glycoprotein processing, *J. Gen. Virol.* **52**:77.

Haffey, M. L., and Spear, P. G., 1980, Alterations in glycoprotein gB specified by mutants and their partial revertants in herpes simplex virus type 1 and relationship to other mutant phenotypes, *J. Virol.* **35**:114.

Hammarström, S., Murphy, L. A., Goldstein, I. J., and Etzler, M. E., 1977, Carbohydrate binding specificity of four N-acetyl-D-galactosamine-" specific" lectins: *Helix pomatia* A hemagglutinin, soybean agglutinin, lima bean agglutinin and *Dolichos biflorus* lectin, *Biochemistry* 16:2750.

Heine, J. W., Spear, P. G., and Roizman, B., 1972, Proteins specified by herpes simplex virus. VI. Viral proteins in the plasma membrane, *J. Virol.* **9**:431.

Heine, J. W., Honess, R. W., Cassai, E., and Roizman, B., 1974, Proteins specified by herpes simplex virus. XII. The virion polypeptides of type 1 strains, *J. Virol.* **14**:640.

Helenius, A., Kartenbeck, J., Simons, K., and Fries, E., 1980, On the entry of Semliki Forest virus into BHK-21 cells, *J. Cell Biol.* **84**:404.

Henning, R., and Lange-Mutschler, J., 1983, Tightly associated lipids may anchor SV40 large T antigen in plasma membrane, *Nature* **305**:736.

Hoggan, M. D., and Roizman, B., 1959, The isolation and properties of a variant of herpes simplex producing multinucleated giant cells in monolayer cultures in the presence of antibody, *Am. J. Hyg.* **70**:208.

Holland, T. C., and Person, S., 1977, Ammonium chloride inhibits cell fusion induced by *syn* mutants of herpes simplex virus type 1, *J. Virol.* **23**:213.

Holland, T. C., Marlin, S. D., Levine, M., and Glorioso, J., 1983a, Antigenic variants of herpes simplex virus selected with glycoprotein-specific monoclonal antibodies, *J. Virol.* **45**:672.

Holland, T. C., Sandri-Goldin, R. M., Holland, L. E., Marlin, S. D., Levine, M., and Glorioso, J. C., 1983b, Physical mapping of the mutation in an antigenic variant of herpes simplex virus type 1 by use of an immunoreactive plaque assay, *J. Virol.* **46**:649.

Holland, T. C., Homa, F., Marlin, S. D., Levine, M., and Glorioso, J., 1984, Herpes simplex virus type 1 gC⁻ mutants exhibit multiple phenotypes, including secretion of truncated glycoproteins, *J. Virol.*, in press.

Homma, M., and Ohuchi, M., 1973, Trypsin action on the growth of Sendai virus in tissue culture cells. III. Structural difference of Sendai viruses grown in eggs and tissue culture cells, *J. Virol.* **12**:1457.

Honess, R. W., and Roizman, B., 1974, Regulation of herpesvirus macromolecular synthesis. I. Cascade regulation of the synthesis of three groups of viral proteins, *J. Virol.* **14**:8.

Honess, R. W., and Roizman, B., 1975, Proteins specified by herpex simplex virus. XIII. Glycosylation of viral polypeptides, *J. Virol.* **16**:1308.

Honess, R. W. and Watson, D. H., 1974, Herpes simplex virus-specific polypeptides studied by polyacrylamide gel electrophoresis of immune precipitates, *J. Gen. Virol.* **22**:171.

Honess, R. W., Buchan, A., Halliburton, I. W., and Watson, D. H., 1980, Recombination and linkage between structural and regulatory genes of herpes simplex virus type I: Study of the functional organization of the genome, *J. Virol.* **34**:716.

Hope, R. G., and Marsden, H. S., 1983, Processing of glycoproteins induced by herpes simplex virus type 1: Sulphation and nature of the oligosaccharide chains, *J. Gen. Virol.* **64**:1943.

Hope, R. G., Palfreyman, J., Suh, M., and Marsden, H. S., 1982, Sulphated glycoproteins induced by herpes simplex virus, *J. Gen. Virol.* **58**:399.

Huang, C. C., and Aminoff, D., 1972, Enzymes that destroy blood group specificity. V. The oligosaccharidase of *Clostridium perfringens*, *J. Biol. Chem.* **247**:6737.

Huttner, W. B., 1982, Sulphation of tyrosine residues—A widespread modification of proteins, *Nature* **299**:273.

Ikura, K., Betz, J. L., Sadler, J. R., and Pizer, L. I., 1983, RNAs transcribed from a 3.6-kilobase *Sma*I fragment of the short unique region of the herpes simplex virus type 1 genome, *J. Virol.* **48**:460.

Johnson, D. C., and Spear, P. G., 1982, Monensin inhibits the processing of herpes simplex virus glycoproteins, their transport to the cell surface and the egress of virions from infected cells, *J. Virol.* **43**:1102.

Johnson, D. C., and Spear, P. G., 1983, O-linked oligosaccharides are acquired by herpes simplex virus glycoproteins in the Golgi apparatus, Cell 32:987.

Johnson, D. C., and Spear, P. G., 1984, Evidence for translational regulation of herpes simplex virus type 1 gD expression, J. Virol. 51:389.

Katsumoto, T., Hirano, A., Kurimura, T., and Takagi, A., 1981, In situ electron microscopical observation of cells infected with herpes simplex virus, J. Gen. Virol. 52:267.

Katz, E., Margalith, E., and Duksin, D., 1980, Antiviral activity of tunicamycin on herpes simplex virus, Antimicrob. Agents Chemother. 17:1014.

Keller, J. M., 1976, The expression of the syn⁻ gene of herpes simplex virus type 1. II. Requirements for macromolecular synthesis, Virology 72:402.

Keller, R., Peitchel, R., Goldman, J. N., and Goldman, M., 1976, An IgG-Fc receptor induced in cytomegalovirus-infected human fibroblasts, J. Immunol. 116:772.

Klenk, H.-D., Rott, R., Orlich, M., and Blödorn, J., 1975, Activation of influenza virus by trypsin treatment, Virology 68:426.

Knowles, R. W., and Person, S., 1976, Effects of 2-deoxyglucose, glucosamine, and mannose on cell fusion and the glycoproteins of herpes simplex virus, J. Virol. 18:644.

Kornfeld, R., and Kornfeld, S., 1976, Comparative aspects of glycoprotein structure, Annu. Rev. Biochem. 45:217.

Kornfeld, R., and Kornfeld, S., 1980, Structure of glycoproteins and their oligosaccharide units, in: The Biochemistry of Glycoproteins and Proteoglycans (W. J. Lennarz, ed.), pp. 1–34, Plenum Press, New York.

Kousoulas, K. G., Person, S., and Holland, T. C., 1978, Timing of some of the molecular events required for cell fusion induced by herpes simplex virus type 1, J. Virol. 27:505.

Kousoulas, K. G., Person, S., and Holland, T. C., 1982, Herpes simplex virus type 1 cell fusion occurs in the presence of ammonium chloride-inhibited glycoproteins, Virology 123:257.

Kousoulas, K. G., Bzik, D. J., DeLuca, N., and Person, S., 1983, The effect of ammonium chloride and tunicamycin on the glycoprotein content and infectivity of herpes simplex virus type 1, Virology 125:468.

Kousoulas, K. G., Pellett, P. E., Pereira, L., and Roizman, B., 1984, Mutations affecting conformation or sequence of neutralizing epitopes identified by reactivity of viable plaques segregate from syn and ts domains of HSV-1(F) gB gene, Virology 135:379.

Kreil, G., 1981, Transfer of proteins across membranes, Annu. Rev. Biochem. 50:317.

Kumarasamy, R., and Blough, H. A., 1982, Characterization of oligosaccharides of highly purified glycoprotein gC of herpes simplex virus type 1 (HSV-1), Biochem. Biophys. Res. Commun. 109:1108.

Lazarowitz, S. G., and Choppin, P., 1975, Enhancement of the infectivity of influenza A and B viruses by proteolytic cleavage of the hemagglutinin polypeptide, Virology 68:440.

Lee, G. T.-Y., Para, M. F., and Spear, P. G., 1982a, Location of the structural genes for glycoproteins gD and gE and for other polypeptides in the S component of herpes simplex virus type 1 DNA, J. Virol. 43:41.

Lee, G. T.-Y., Pogue-Geile, K. L., Pereira, L., and Spear, P. G., 1982b, Expression of herpes simplex virus glycoprotein C from a DNA fragment inserted into the thymidine kinase gene of this virus, Proc. Natl. Acad. Sci. USA 79:6612.

Lehner, T., Wilton, J. M. A., and Shillitoe, E. J., 1975, Immunological basis for latency, recurrences, and putative oncogenicity of herpes simplex virus, Lancet ii:60.

Lenard, J., and Miller, D. K., 1983, Entry of enveloped viruses into cells, in: Receptor-mediated Endocytosis and Processing (P. Cuatrecasas and T. Roth, eds.), pp. 121–138, Chapman & Hall, London.

Little, S. P., and Schaffer, P. A., 1981, Expression of the syncytial (syn) phenotype in HSV-1, strain KOS: Genetic and phenotypic studies of mutants in two syn loci, Virology 112:686.

Little, S. P., Jofre, J. T., Courtney, R. J., and Schaffer, P. A., 1981, A virion-associated glycoprotein essential for infectivity of herpes simplex virus type 1, Virology 115:149.

McTaggart, S. P., Burns, W. H., White, D. O., and Jackson, D. C. 1978, Fc receptors induced by herpes simplex virus. I. Biologic and biochemical properties, *J. Immunol.* **121**:726.

Manservigi, R., Spear, P. G., and Buchan, A., 1977, Cell fusion induced by herpes simplex virus is promoted and suppressed by different viral glycoproteins, *Proc. Natl. Acad. Sci. USA* **74**:3913.

Marsden, H. S., Stow, N. D., Preston, V. G., Timbury, M. C., and Wilkie, N. M., 1978, Physical mapping of herpes simplex virus-induced polypeptides, *J. Virol.* **28**:624.

Marsden, H. S., Buckmaster, A., Palfreyman, J. W., Hope, R. G., and Minson, A. C., 1984, Characterization of the 92,000 dalton glycoprotein induced by herpes simplex virus type 2, *J. Virol.* **50**:547.

Marshall, R. D., 1974, The nature and metabolism of the carbohydrate–peptide linkage of glycoproteins, *Biochem. Soc. Symp.* **40**:17.

Marshall, R. D., and Neuberger, A., 1977, Aspects of the structure and metabolism of glycoproteins, *Adv. Carbohydr. Chem. Biochem.* **25**:407.

Matlin, K. S., Reggio, H., Helenius, A., and Simons, K., 1981, Infection entry pathway of influenza virus in a canine kidney cell line, *J. Cell Biol.* **91**:601.

Matlin, K. S., Reggio, H., Helenius, A., and Simons, K., 1982, Pathway of vesicular stomatitis virus entry leading to infection, *J. Mol. Biol.* **156**:609.

Matthews, J. T., Cohen, G. H., and Eisenberg, R. J., 1983, Synthesis and processing of glycoprotein D of herpes simplex virus types 1 and 2 in an in vitro system, *J. Virol.* **48**:521.

Morgan, C. H., Rose, M., Holden, M., and Jones, E. P., 1959, Electron microscopic observations on the development of herpes simplex virus, *J. Exp. Med.* **110**:643.

Morgan, C., Rose, H. M., and Mednis, B., 1968, Electron microscopy of herpes simplex virus. I. Entry, *J. Virol.* **2**:507.

Morse, L. S., Buchman, T. G., Roizman, B., and Schaffer, P. A., 1977, Anatomy of herpes simplex virus DNA. IX. Apparent exclusion of some parental DNA arrangements in the generation of intertypic (HSV-1 × HSV-2) recombinants, *J. Virol.* **24**:231.

Nagai, Y., Hamaguchi, M., Toyoda, T., and Yoshida, T., 1983, The uncoating of paramyxoviruses may not require a low pH mediated step, *Virology* **130**:263.

Nakamura, Y., Costa, J., Tralka, T. S., Yee, C. L., and Rabson, A. S., 1978, Properties of the cell surface Fc-receptor induced by herpes simplex virus, *J. Immunol.* **121**:1128.

Nii, S., Morgan, C., Rose, H. M., and Hsu, K. C., 1968, Electron microscopy of herpes simplex virus. IV. Studies with ferritin-conjugated antibodies, *J. Virol.* **2**:1172.

Noble, A. G., Lee, G. T.-Y., and Spear, P. G., 1983, Anti-gD monoclonal antibodies inhibit cell fusion induced by herpes simplex virus type 1, *Virology* **129**:218.

Norrild, B., 1980, Immunochemistry of herpes simplex virus glycoproteins, *Curr. Top. Microbiol. Immunol.* **90**:67.

Norrild, B., and Pedersen, B., 1982, Effect of tunicamycin on the synthesis of herpes simplex virus type 1 glycoproteins and their expression on the cell surface, *J. Virol.* **43**:395.

Norrild, B. and Vestergaard, B. F., 1977, Polyacrylamide gel electrophoretic analysis of herpes simplex virus type 1 immunoprecipitates obtained by quantitative immunoelectrophoresis in antibody-containing agarose gel, *J. Virol.* **22**:113.

Ogata, M., and Shigeta, S., 1979, Appearance of immunoglobulin G Fc receptors on cultured human cells infected with varicella–zoster virus, *Infect. Immun.* **26**:770.

Olofsson, S., and Lycke, E., 1980, Glucosamine metabolism of herpes simplex virus infected cells: Inhibition of glycosylation by tunicamycin and 2-deoxy-D-glucose, *Arch. Virol.* **65**:201.

Olofsson, S., Jeansson, S., and Lycke, E., 1981a, Unusual lectin-binding properties of a herpes simplex virus type 1-specific glycoprotein, *J. Virol.* **38**:564.

Olofsson, S., Blomberg, J., and Lycke, E., 1981b, O-glycosidic carbohydrate–peptide linkages of herpes simplex virus glycoproteins, *Arch. Virol.* **70**:321.

Olofsson, S., Norrild, B., Andersen, A. B., Pereira, L., Jeansson, S., and Lycke, E., 1983a, Populations of herpes simplex virus glycoprotein gC with and without affinity for the N-acetyl-galactosamine specific lectin of *Helix pomatia*, *Arch. Virol.* **76**:25.

Olofsson, S., Sjöblom, I., Lundström, M., Jeansson, S., and Lycke, E., 1983b, Glycoprotein C of herpes simplex virus type 1: Characterization of O-linked oligosaccharides, *J. Gen. Virol.* **64:**2735.

Padgett, R. A., Moore, D. F., and Kingsbury, D. T., 1978, Herpes simplex virus nucleic acid synthesis following infection of non-permissive XC cells, *J. Gen. Virol.* **40:**605.

Palfreyman, J. W., Haarr, L., Cross, A., Hope, R. G., and Marsden, H. S., 1983, Processing of herpes simplex virus type 1 glycoproteins: Two-dimensional gel analysis using monoclonal antibodies, *J. Gen. Virol.* **64:**873.

Para, M. F., Baucke, R. B., and Spear, P. G., 1980, Immunoglobulin G(Fc)-binding receptors on virions of herpes simplex virus type 1 and transfer of these receptors to the cell surface by infection, *J. Virol.* **34:**512.

Para, M. F., Baucke, R. B., and Spear, P. G., 1982a, Glycoprotein gE of herpes simplex virus type 1: Effects of anti-gE on virion infectivity and on virus-induced Fc-binding receptors, *J. Virol.* **41:**129.

Para, M. F., Goldstein, L., and Spear, P. G., 1982b, Similarities and differences in the Fc-binding glycoprotein (gE) of herpes simplex virus types 1 and 2 and tentative mapping of the viral gene for this glycoprotein, *J. Virol.* **41:**137.

Para, M. F., Zezulak, K. M., Conley, A. J., Weinberger, M., Snitzer, K., and Spear, P. G., 1983, Use of monoclonal antibodies against two 75,000-molecular-weight glycoproteins specified by herpes simplex virus type 2 in glycoprotein identification and gene mapping, *J. Virol.* **45:**1223.

Peake, M. L., Nystrom, P., and Pizer, L. I., 1982, Herpesvirus glycoprotein synthesis and insertion into plasma membranes, *J. Virol.* **42:**678.

Pereira, L., Klassen, T., and Baringer, J. R., 1980, Type-common and type-specific monoclonal antibody to herpes simplex virus type 1, *Infect. Immun.* **29:**724.

Pereira, L., Dondero, D., Morrild, B., and Roizman, B., 1981, Differential immunologic reactivity and processing of glycoproteins gA and gB of herpes simplex virus types 1 and 2 made in Vero and HEp-2 cells, *Proc. Natl. Acad. Sci. USA* **78:**5202.

Pereira, L., Dondero, D., and Roizman, B., 1982a, Herpes simplex virus glycoprotein gA/B: Evidence that the infected Vero cell products comap and arise by proteolysis *J. Virol.* **44:**88.

Pizer, L. I., Cohen, G. H., and Eisenberg, R. J., 1980, Effect of tunicamycin on herpes simplex virus glycoproteins and infectious virus production, *J. Virol.* **34:**142.

Pogue-Geile, K. L., Lee, G. T.-Y., Shapira, S. K., and Spear, P. G., 1984, Fine mapping of mutations in the fusion-inducing MP strain of herpes simplex virus type 1, *Virology* (in press).

Pomato, N., and Aminoff, D., 1978, α-D-N-Acetylgalactosaminyl-oligosaccharidase of *Clostridium perfringens, Fed. Proc.* **37:**1602.

Powell, K. L., Buchan, A., Sim, C., and Watson, D. H., 1974, Type-specific protein in herpes simplex virus envelope reacts with neutralizing antibody, *Nature* **249:**360.

Prehm, P., Scheid, A., and Choppin, P. W., 1979, The carbohydrate structure of the glycoproteins of the paramyxovirus SV5 grown in bovine kidney cells, *J. Biol. Chem.* **254:**9669.

Preston, V. G., Davison, A. J., Marsden, H. S., Timbury, M. C., Subak-Sharpe, J. H., and Wilkie, N. M., 1978, Recombinants between herpes simplex virus types 1 and 2: Analyses of genome structures and expression of immediate early polypeptides, *J. Virol.* **28:**499.

Rahman, A. A., Teschner, M., Sethi, K. K., and Brandis, H., 1976, Appearance of IgG (Fc) receptor(s) on cultured human fibroblasts infected with human cytomegalovirus, *J. Immunol.* **117:**253.

Rector, J. T., Lausch, R. N., and Oakes, J. E., 1982, Use of monoclonal antibodies for analysis of antibody-dependent immunity to ocular herpes simplex virus type 1 infection, *Infect. Immun.* **38:**168.

Roizman, B., 1962, Polykaryocytosis, *Cold Spring Harbor Symp. Quant. Biol.* **27:**327.

354   PATRICIA G. SPEAR

Roizman, B., and Furlong, D., 1974, The replication of herpes viruses, in: *Comprehensive Virology*, Vol. 3 (H. Fraenkel-Conrat and R. R. Wagner, eds.), pp. 229–403, Plenum Press, New York.

Roizman, B., Norrild, B., Chan, C., and Pereira, L., 1984, Identification and preliminary mapping with monoclonal antibodies of a herpes simplex virus 2 glycoprotein lacking a known type 1 counterpart, *Virology* **133**:242.

Rose, J. K., and Gallione, C. J., 1981, Nucleotide sequences of the mRNAs encoding the vesicular stomatitis virus G and M proteins determined from cDNA clones containing the complete coding regions, *J. Virol.* **39**:519.

Rose, J. K., Welch, W. J., Sefton, B. M., Esch, F. S., and Ling, N. C., 1980, Vesicular stomatitis glycoprotein is anchored in the viral membrane by a hydrophobic domain near the COOH-terminus, *Proc. Natl. Acad. Sci. USA* **77**:3884.

Ruyechan, W. T., Morse, L. S., Knipe, D. M., and Roizman, B., 1979, Molecular genetics of herpes simplex virus. II. Mapping of the major viral glycoproteins and of the genetic loci specifying the social behavior of infected cells, *J. Virol.* **29**:677.

Sanders, P. G., Wilkie, N. M., and Davison, A. J., 1982, Thymidine kinase deletion mutants of herpes simplex virus type 1, *J. Gen. Virol.* **63**:277.

Sarmiento, M., and Spear, P. G., 1979, Membrane proteins specified by herpes simplex viruses. IV. Conformation of the virion glycoprotein designated VP7(B$_2$), *J. Virol.* **29**:1159.

Sarmiento, M., Haffey, M., and Spear, P. G., 1979, Membrane proteins specified by herpes simplex viruses. III. Role of glycoprotein VP7 (B$_2$) in virion infectivity, *J. Virol.* **29**:1149.

Schachter, H., and Roseman, S., 1980, Mammalian glycosyltransferases: Their role in the synthesis and function of complex carbohydrates and glycolipids, in: *The Biochemistry of Glycoproteins and Proteoglycans* (W. J. Lennarz, ed.), pp. 86–160, Plenum Press, New York.

Schaffer, P. A., Carter, V. C., and Timbury, M. C., 1978, A collaborative complementation study of temperature-sensitive mutants of herpes simplex virus types 1 and 2, *J. Virol.* **27**:490.

Scheid, A., and Choppin, P. W., 1974, Identification of biological activities of paramyxovirus glycoproteins: Activation of cell fusion, hemolysis, and infectivity by proteolytic cleavage of an inactive precursor protein of Sendai virus, *Virology* **57**:475.

Schmidt, M. F. G., 1983, Fatty acid binding: A new kind of posttranslational modification of membrane proteins, *Curr. Top. Microbiol. Immunol.* **102**:101.

Schmidt, M., and Schlesinger, M., 1979, Fatty acid binding to vesicular stomatitis virus glycoprotein: A new type of post-translational modification of the viral glycoprotein, *Cell* **17**:813.

Schmidt, M. F. G., and Schlesinger, M. J., 1980, Relation of fatty acid attachment to the translation and maturation of vesicular stomatitis and Sindbis virus membrane glycoproteins, *J. Biol. Chem.* **255**:3334.

Schmidt, M. F. G., Bracha, M., and Schlesinger, M. J., 1979, Evidence for covalent attachment of fatty acids to Sindbis virus glycoproteins, *Proc. Natl. Acad. Sci. USA* **76**:1687.

Schwartz, J., and Roizman, B., 1969, Concerning the egress of herpes simplex virus from infected cells: Electron and light microscopic observations, *Virology* **38**:42.

Serafini-Cessi, F., and Campadelli-Fiume, G., 1981, Studies on benzhydrazone, a specific inhibitor of herpesvirus glycoprotein synthesis: Size distribution of glycopeptides and endo-β-N-acetylglucosaminidase-H treatment, *Arch. Virol.* **70**:331.

Serafini-Cessi, F., Dall'Olio, F., Scannavini, M., Costanzo, F., and Campadelli-Fiume, G., 1983a, N-acetylgalactosaminyl-transferase activity involved in O-glycosylation of herpes simplex virus type 1 glycoproteins, *J. Virol.* **48**:325.

Serafini-Cessi F., Dall'Olio, F., Scannavini, M., and Campadelli-Fiume, G., 1983b, Processing of herpes simplex virus-1 glycans in cells defective in glycosyl transferases of the Golgi system: Relationship to cell fusion and virion egress, *Virology* **131**:59.

Showalter, S. D., Zweig, M., and Hampar, B., 1981, Monoclonal antibodies to herpes simplex virus type 1 proteins, including the immediate-early proteins ICP4, *Infect. Immun.* **34**:684.

Siminoff, P., and Menefee, M. G., 1966, Normal and 5-bromodeoxyuridine-inhibited development of herpes simplex virus: An electron microscope study, *Exp. Cell Res.* **44**:241.

Spear, P. G., 1976, Membrane proteins specified by herpes simplex virus. I. Identification of four glycoprotein precursors and their products in type 1-infected cells, *J. Virol.* **17**:991.

Spear, P. G., 1980, Herpesviruses, in: *Cell Membranes and Viral Envelopes*, Vol. 2 (H. A. Blough and J. M. Tiffany, eds.), pp. 709–750, Academic Press, New York.

Spear, P. G., 1984, Antigenic structure of herpes simplex viruses, in: *Immunochemistry of Viruses: The Basis for Serodiagnosis and Vaccines* (M. H. V. Van Regenmortel and A. R. Neurath, eds.), Elsevier, Amsterdam, in press.

Spiro, R. G., 1966, Characterization of carbohydrate units of glycoproteins, *Methods Enzymol.* **8**:26.

Spivack, J. G., Prusoff, W. H., and Tritton, T. R., 1982, A study of the antiviral mechanism of action of 2-deoxy-D-glucose: Normally glycosylated proteins are not strictly required for herpes simplex virus attachment but increase viral penetration and infectivity, *Virology* **123**:123.

Stackpole, C. W., 1969, Herpes-type virus of the frog renal adenocarcinoma. I. Virus development in tumor transplants maintained at low temperature, *J. Virol.* **4**:75.

Struck, D. K., and Lennarz, W. J., 1980, The function of saccharide-lipids in synthesis of glycoproteins, in: *The Biochemistry of Glycoproteins and Proteoglycans* (W. J. Lennarz, ed.), pp. 35–83, Plenum Press, New York.

Svennerholm, B., Olofsson, S., Lunden, R., Vahlne, A., and Lycke, E., 1982, Adsorption and penetration of enveloped herpes simplex virus particles modified by tunicamycin or 2-deoxy-D-glucose, *J. Gen. Virol.* **63**:343.

Takatsuki, A., Kohno, K., and Tamura, G., 1975, Inhibition of biosynthesis of polyisoprenol sugars in chick embryo microsomes by tunicamycin, *Agric. Biol. Chem.* **39**:2089.

Tarentino, A. L., and Maley, F., 1974, Purification and properties of an endo-β-N-acetyl-glucosaminidase from *Streptomyces griseus*, *J. Biol. Chem.* **249**:811.

Tartakoff, A. M., and Vassalli, P., 1977, Plasma membrane immunoglobulin secretion: Arrest is accompanied by alterations in the Golgi complex, *J. Exp. Med.* **146**:1332.

Tartakoff, A. M., and Vassalli, P., 1978, Comparative studies of intracellular transport of secretory proteins, *J. Cell Biol.* **79**:694.

Terni, M., and Roizman, B., 1970, Variability of herpes simplex virus: Isolation of two variants from simultaneous eruptions at different sites, *J. Infect. Dis.* **121**:212.

Tkacz, J. S., and Lampen, J. O., 1975, Tunicamycin inhibition of polyisoprenol N-acetyl-glucosaminyl pyrophosphate formation in calf liver microsomes, *Biochem. Biophys. Res. Commun.* **65**:248.

Tognon, M., Furlong, D., Conley, A. J., and Roizman, B., 1981, Molecular genetics of herpes simplex virus. V. Characterization of a mutant defective in ability to form plaques at low temperatures and in a viral function which prevents accumulation of coreless capsids at nuclear pores late in infection, *J. Virol.* **40**:870.

Uchida, N., Smilowitz, M., and Tanzer, M. L., 1979, Monovalent ionophores inhibit secretion of procollagen and fibronectin from cultured human fibroblasts, *Proc. Natl. Acad. Sci. USA* **76**:1868.

Vahlne, A., Nyström, B., Sandberg, M., Hamberger, A., and Lycke, E., 1978, Attachment of herpes simplex virus to neurons and glial cells, *J. Gen. Virol.* **40**:359.

Vahlne, A., Svennerholm, B., and Lycke, E., 1979, Evidence for herpes simplex virus type-selective receptors on cellular plasma membranes, *J. Gen. Virol.* **44**:217.

Vahlne, A., Svennerholm, B., Sandberg, M., Hamberger, A., and Lycke, E., 1980, Differences in attachment between herpes simplex type 1 and type 2 viruses to neurons and glial cells, *Infect. Immun.* **28**:675.

Vestergaard, B. F. and Norrild, B., 1978, Crossed immunoelectrophoretic analysis and viral neutralizing activity of five monospecific antisera against five different herpes simplex virus glycoproteins, *IARC Sci. Publ.* **24**:225.

Watkins, J. F., 1964, Adsorption of sensitized sheep erythrocytes to HeLa cells infected with herpes simplex virus, *Nature* **202**:1364.

Watson, D. H., and Honess, R. W., 1977, Polypeptides and antigens of herpes simplex virus: Their nature and relevance in chemotherapy and epidemiology of herpes infections, *Antimicrob. Chemother.* **3**(Suppl. A):33.

Watson, D. H. and Wildy, P., 1969, The preparation of 'monoprecipitin' antisera to herpes virus specific antigens, *J. Gen. Virol.* **4**:163.

Watson, R. J., Weis, J. H., Salstrom, J. S., and Enquist, L. W., 1982, Herpes simplex virus type 1 glycoprotein D Gene: Nucleotide sequence and expression in *Escherichia coli,* *Science* **218**:381.

Watson, R. J., Colberg-Poley, A. M., Marcus-Sekura, C. J., Carter, B. J., and Enquist, L. W., 1983, Characterization of the herpes simplex virus type 1 glycoprotein D mRNA and expression of this protein in *Xenopus* oocytes, *Nucleic Acids Res.* **11**:1507.

Wenske, E. A., and Courtney, R. J., 1983, Glycosylation of herpes simplex virus type 1 gC in the presence of tunicamycin, *J. Virol.* **46**:297.

Wenske, E. A., Bratton, M. W., and Courtney, R. J., 1982, Endo-β-N-acetylglucosaminidase H sensitivity of precursors to herpes simplex virus type 1 glycoproteins gB and gC, *J. Virol.* **44**:241.

Westmoreland, D., and Watkins, J. F., 1974, The IgG receptor induced by herpes simplex virus: Studies using radioiodinated IgG, *J. Gen. Virol.* **24**:167.

Westmoreland, D., St. Jeor, S., and Rapp, F., 1976, The development by cytomegalovirus-infected cells of binding affinity for normal human immunoglobulin, *J. Immunol.* **116**:1566.

White, J., Kartenbeck, J., and Helenius, A., 1980, Fusion of Semliki Forest virus with the plasma membrane can be induced by low pH, *J. Cell Biol.* **87**:264.

White, J., Matlin, K., and Helenius, A., 1981, Cell fusion by Semliki Forest, influenza and vesicular stomatitis virus, *J. Cell Biol.* **89**:674.

Yamamoto, S., and Kabuta, H., 1976, Genetic analysis of polykaryocytosis by herpes simplex virus. II. Recombination between viruses with non-fusing ability and fusing ability, *Kurume Med. J.* **23**:209.

Yamamoto, S., and Kabuta, H., 1977, Genetic analysis of polykaryocytosis by herpes simplex virus. III. Complementation and recombination between non-fusing mutants and construction of a linkage map with regard to the fusion function, *Kurume Med. J.* **24**:163.

Yamamoto, S., Kabuta, H., and Suenaga, Y., 1972, Mutants of herpes simplex virus. II. IDU-resistant mutants and a preliminary experiment on genetic recombination, *Kurume Med. J.* **19**:237.

Yamamoto, S., Kabuta, H., Imamoto, M., and Matsumoto, H., 1975, Genetic analysis of polykaryocytosis by herpes simplex virus. I. Experiments using recombinants from intertypic cross, *Kurume Med. J.* **22**:71.

Yasuda, J., and Milgrom, F., 1968, Hemadsorption by herpes simplex virus infected cell cultures, *Int. Arch. Allergy* **33**:151.

Yoshimura, A., Kuroda, K., Kawasaki, K., Yamashina, S., Maeda, T., and Ohnishi, S.-I., 1982, Infectious cell entry of influenza virus, *J. Virol.* **43**:284.

Zezulak, K. M., and Spear, P. G., 1983, Characterization of a herpes simplex virus type 2 75,000-molecular-weight glycoprotein antigenically related to herpes simplex virus type 1 glycoprotein C, *J. Virol.* **47**:553.

Zezulak, K. M., and Spear, P. G., 1984a, Limited proteolysis of herpes simplex virus glycoproteins that occurs during their extraction from Vero cells, *J. Virol.* **50**:258.

Zezulak, K. M., and Spear, P. G., 1984b, Mapping of the structural gene for the herpes simplex virus type 2 counterpart of herpes simplex virus type 1 glycoprotein C and identification of a type 2 mutant which does not express this glycoprotein, *J. Virol.* **49**:741.

Zweig, M., Showalter, S. D., Bladen, S. V., Heilman, C. J., Jr., and Hampar, B., 1983, Herpes simplex virus type 2 glycoprotein gF and type 1 glycoprotein gC have related antigenic determinants, *J. Virol.* **47**:185.

CHAPTER 8

Processing of the Oligosaccharide Chains of Herpes Simplex Virus Type 1 Glycoproteins

GABRIELLA CAMPADELLI-FIUME
AND FRANCA SERAFINI-CESSI

I. INTRODUCTION

The presence of oligosaccharide chains (glycans) with specific structures appears to be required for the expression of several functions associated with herpes simplex virus type 1 (HSV-1) glycoproteins. Among these functions are infectious virus yield, virion infectivity, virion egress, and virus-induced cell fusion. This chapter focuses on the processing of glycans of the glycoproteins specified by HSV-1, on the correlations that have been established between glycan structure and expression of viral functions, and on the available evidence that HSV-1 proteins are glycosylated mainly by enzymes of cellular origin. A detailed description of the HSV genes specifying HSV glycoproteins and characterization of the structure and function of HSV glycoproteins are given by Spear (Chapter 7, this volume).

Unlike most viruses, which encode one or two glycoproteins (e.g., vesicular stomatitis virus, Sindbis virus, influenza virus, and retroviruses), HSV-1 codes for four major and an as yet undetermined number

GABRIELLA CAMPADELLI-FIUME • Institute of Microbiology and Virology, University of Bologna, 40126 Bologna, Italy.   FRANCA SERAFINI-CESSI • Institute of General Pathology, University of Bologna, 40126 Bologna, Italy.

of minor glycoprotein species. Most of the studies reported to date on oligosaccharide chains of herpesvirus glycoproteins have dealt mainly with infected cell lysates containing all of the species of HSV glycoproteins. Detailed characterizations of the glycan structure of each species of herpesvirus glycoprotein and, for each species, of precursors and mature forms have not been reported. As this field is relatively new, our knowledge is far from being extensive and this report is, in a sense, a review of the state of the science in this area. Availability of monoclonal antibodies to HSV glycoproteins and the ease of obtaining highly purified glycoproteins by affinity chromatography to immobilized antibodies will undoubtedly stimulate such studies.

The focus of much of the research on HSV glycoproteins centers on their use as subunit vaccines to prevent HSV-1 and HSV-2 infections. At present there is no clear indication as to the extent to which glycans contribute to antigenicity of HSV glycoproteins. It has been reported that some monoclonal antibodies raised against glycoproteins do not interact with the unglycosylated peptides obtained by *in vitro* mRNA translation or with underglycosylated proteins produced in the presence of tunicamycin. Studies on the carbohydrate moiety of glycoproteins will contribute to our understanding of the antigenic specificities of HSV glycoproteins.

II. HSV-1 GLYCOPROTEINS

HSV-1 specifies several glycoproteins; these were initially differentiated on the basis of apparent molecular weight in SDS-polyacrylamide gel electrophoresis (SDS-PAGE) (Spear *et al.*, 1970; Spear and Roizman, 1972; Heine *et al.*, 1972, 1974). The current designations of the four major species of antigenically distinct glycoproteins are gB (or gA/gB), gC, gD, and gE (Spear, 1976; Baucke and Spear, 1979). Recently the application of two-dimensional gel electrophoresis has revealed another species, which has the same apparent molecular weight as gC but does not appear to be antigenically related to other known glycoproteins. This species has been tentatively designated as gY (Hope and Marsden, 1983). HSV-1-infected cells and virions contain minor species of glycoproteins, as yet not extensively characterized (Spear and Roizman, 1972; Heine *et al.*, 1972, 1974; Hope and Marsden, 1983).

Spear (1976) and Baucke and Spear (1979) showed that each glycoprotein species exhibits at least two forms—the precursors pgC, pgD, and pgE, which appear after a short pulse, and the mature forms gC, gD, and gE, which appear after a chase and have a higher apparent molecular weight than immature forms. A case in point is gA, considered at one time to be an independent major species of HSV-1 glycoproteins. Later studies showed that gA is a precursor to gB (Eberle and Courtney, 1980) and that gA and gB could not be differentiated by a bank of independently derived monoclonal antibodies made in response to infection with HSV-1 or HSV-2 (Pereira *et al.*, 1981). Consistent with these observations, it was reported that gA and gB polypeptides comap within the same restricted region of HSV-1 DNA (Ruyechan *et al.*, 1979; Pereira *et al.*, 1982). At the International Herpesvirus Workshop in Oxford, England, August 1983, it was agreed that gA would be designated as pgB.

Table I summarizes some biochemical and biophysical properties of HSV-1 glycoproteins. It reports the apparent molecular weight in SDS-PAGE of fully mature glycoproteins and of their partially glycosylated precursors. Different values for the apparent molecular weights of precursor and mature gC were reported by various authors, which may be due to different virus–cell systems used and/or to different conditions employed in the electrophoretic analysis. The molecular weight of nonglycosylated forms is also shown (1) as obtained from precursors after treatment with endo-β-N-acetylglucosaminidase H (Endo-H) (see Section III.B) (Wenske *et al.*, 1982, (2) as products of *in vitro* mRNA translation (Inglis and Newton, 1982, Lee *et al.*, 1982a,b; Watson *et al.*, 1982; Frink *et al.*, 1983), and (3) as predicted from DNA sequence analysis (Watson *et al.*, 1982; Frink *et al.*, 1983, Bzik *et al.*, 1984). The latter analysis allowed an estimation of the number of potential N-glycosylation sites (Table I). Altogether, the data provide strong evidence that differences in apparent molecular weight among nonglycosylated forms, immature forms, and mature glycoproteins are due mainly to stepwise addition of oligosaccharide chains. Table I includes the isoelectric points of precursors and mature glycoproteins as determined by isoelectric focusing. The lower values displayed by mature forms depend on the addition of sialic acid residues (Cohen *et al.*, 1980). HSV-1 glycoproteins become labeled when the following precursors are used: glucosamine [which is incorporated as GlcNAc*, as sialic acid (SA) (Honess and Roizman, 1975), and as GalNAc (Olofsson *et al.*, 1981a)], mannose, galactose (Brennan *et al.*, 1976), and fucose (Honess and Roizman, 1975; Campadelli-Fiume *et al.*, 1980). Honess and Roizman (1975) studied the fucose/glucosamine ratio in glycopeptides of various size-classes and found it to be higher in higher-molecular-weight glycopeptides, consistent with more recent findings that fucose addition is a rather late event in glycan processing. Presence of fucose in gC was recently reported by Kumarasamy and Blough (1982).

Glycoproteins gB, gC, gD, gE, and gY are sulfated. Evidence suggesting that inorganic sulfate is attached to N-linked oligosaccharides was provided by the observation that sulfate incorporation is decreased in the presence of tunicamycin. Sulfation appears to occur late in the maturation process (Hope and Marsden, 1983).

---

* Abbreviations: Gal, galactose; GalNAc, N-acetylgalactosamine; Glc, glucose; GlcN, glucosamine; GlcNAc, N-acetylglucosamine; Fuc, fucose; Man, mannose; SA, sialic acid; Asn, aparagine; Ser, serine; Thr, threonine.

TABLE I. Biochemical and Biophysical Properties of the Major HSV-1 Glycoprotein Species

	gA/B	gC	gD	gE	References
Apparent molecular weight					
Mature form	120,000	115,000–130,000	59,000–65,000	80,000	Spear (1976), Eisenberg et al. (1979), Baucke and Spear (1979), Cohen et al. (1980), Pizer et al. (1980), Palfreyman et al. (1983)
Precursor	110,000	86,000–105,000	51,000	66,000	
Nonglycosylated[a]					
In vitro translation product	97,000	75,000	50,000	ND[b]	Wenske et al. (1982) Inglis and Newton (1982), Lee et al. (1982a,b), Watson et al. (1982), Frink et al. (1983)
	85,000	69,000–74,000	46,000–52,000	66,000	
Predicted from DNA sequence	100,000	60,000	43,000	ND	Watson et al. (1982), Frink et al. (1983), Bzik et al., 1984.
N-glycosylation sites	9	8	3	ND	Inglis and Newton (1982), Watson et al. (1982), Bzik et al. (1984), Cohen et al. (1983)
Isoelectric point					
Mature form	ND	4.9–6.4	5.9–6.6	ND	Cohen et al., 1980
Precursor	ND	7.55	7.45	ND	

[a] The apparent molecular weight of nonglycosylated protein was obtained from partially glycosylated precursors treated with Endo-H to remove high-mannose oligosaccharides [Man_n-GlcNac].
[b] ND, not determined.

gE was shown to be the only HSV-1 glycoprotein to incorporate significant quantities of label from [^{3}H]palmitate and therefore to carry fatty acid molecules, whose attachment appears to occur in the Golgi apparatus (Johnson and Spear, 1983).

As would be predicted for components of virion envelopes and of plasma membranes, HSV-1 glycoproteins have the characteristics of integral membrane glycoproteins. Nucleotide sequence analyses of gB, gC and gD genes indicate the following (Bzick et al., 1984; Watson et al., 1982; Frink et al., 1983):

1. The COOH-terminal region is hydrophobic and is followed by a strongly basic region; thus, it constitutes a transmembrane sequence that serves to anchor the glycoprotein in the membrane.
2. The NH$_2$-terminal portion contains hydrophobic or nonpolar amino acids and acts as a signal sequence that assists in the cotranslational translocation of the polypeptide across the microsomal membranes; as such, it should be removed during translation (see Sabatini et al., 1982).

III. BIOSYNTHESIS AND CHARACTERIZATION OF THE OLIGOSACCHARIDE CHAINS OF HSV-1 GLYCOPROTEINS

A. General Pathway of Viral Glycoprotein Synthesis

Pertinent to this chapter is the general pathway of oligosaccharide chain addition to proteins, as was inferred from studies on well-known viral glycoproteins, such as gG of vesicular stomatitis virus (VSV), gE1 and gE2 of Sindbis virus, and hemagglutinin and neuraminidase of influenza virus (for reviews, see Hubbard and Ivatt, 1981; Berger et al., 1982). Nonglycosylated glycoprotein precursors are synthesized by membrane-bound polyribosomes of the rough endoplasmic reticulum (RER). During synthesis they are translocated into the lumen of the RER by the signal sequence. Glycosylation includes cotranslational and posttranslational events. Asparagine-linked (N-linked) oligosaccharide chains are the most frequent glycomoieties in viral glycoproteins. N-linked glycosylation initiates by en bloc transfer of glycans with composition Glc_3-Man_9-$GlcNAc_2$ from a dolichol phosphate lipid carrier to asparagine residues of the nascent polypeptide chains (Li et al., 1978; Tabas et al., 1978). The recognized sequence in the peptide is Asn-X-Thr/Ser, where X can be almost any other amino acid (Marshall, 1972). After transfer, the oligosaccharide chains are trimmed by a number of glycosidases (both glucosidases and mannosidases) to yield polymannosyl chains referred to as high-mannose glycans (Hunt et al., 1978; Spiro and Spiro, 1982). Trimming starts in the RER and is terminated after transport of the glycopro-

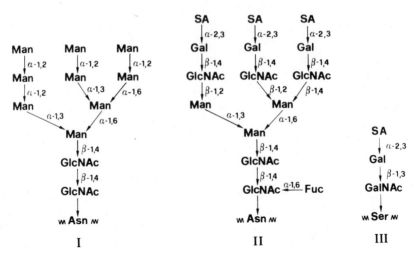

FIGURE 1. Structure of two N-linked and one O-linked oligosaccharide. (I) High-mannose chain and (II) triantennary complex-type chain from VSV gG; (III) O-linked chain from porcine submaxillary mucine.

tein to the Golgi apparatus. In some cases, assembly of N-linked glycans does not proceed beyond this stage, and glycans with varying numbers of mannose residues have been reported to occur in different viral glycoproteins (Kornfeld and Wold, 1981; Klenk et al., 1983; Hsieh et al., 1983b). More often, high-mannose glycans are processed by glycosyltransferases of the Golgi system to complex-type glycans, which consist of a pentasaccharide core (Man$_3$-GlcNAc$_2$) and of a number of side chains (antennae) whose most frequent composition is SA-Gal-GlcNAc. In VSV, glycoprotein G, triantennary complex-type glycans accumulate (Reading et al., 1978), whereas in Sindbis virus glycoproteins, diantennary species are present (Burke and Keegestra, 1979). Fucose, when present, is added after the assembly of side chains, generally to the innermost GlcNAc (see Hubbard and Ivatt, 1981). Figure 1 illustrates the structure of a high-mannose (I) and of a triantennary complex-type glycan (II).

O-glycosylation of viral proteins has been reported to occur less frequently than N-glycosylation (Olofsson et al., 1981a; Nieman and Klenk, 1981; Shida and Dales, 1981; Johnson and Spear, 1983; Edson and Thorley-Lawson, 1983). Therefore, no detailed information is available on the structure of O-linked chains in viral glycoproteins. In mammalian glycoproteins such as mucins, the first step in the assembly of O-linked glycans consists of the transfer of GalNAc to the hydroxyl group of threonine or serine by the enzyme N-acetylgalactosaminyltransferase. This step is followed by the sequential addition of sugars (e.g., Gal, GlcNAc, Fuc, and SA) by glycosyl transferases located in the Golgi apparatus (see Beyer et al., 1981). Figure 1 shows the structure of an O-linked glycan from porcine submaxillary mucine (III) (Baig and Aminoff, 1972).

B. Assembly of N-Linked Oligosaccharides in Precursors and in Mature Forms

Honess and Roizman (1975) were the first to study the oligosaccharide chains of HSV-1 glycoproteins by analyzing the glycopeptides obtained after Pronase digestion of infected cell lysates. They showed that glycoprotein precursors acquire oligosaccharides (molecular weight about 2000) that contain glucosamine but little or no fucose or sialic acid, whereas mature glycoproteins carry sialylated oligosaccharides of higher molecular weight, originated by a stepwise addition of sugars.

Eisenberg et al. (1979) subsequently showed that processing of pgD to gD does not involve any major alteration in polypeptide structure, but involves modifications in the carbohydrate portion. pgD was shown to contain an 1800-molecular-weight oligomannosyl moiety that is converted to higher molecular weight and more heterogeneous carbohydrate components carrying terminal sialic acid. This same group reported further studies on gD and gC species (Cohen et al., 1980). For both glycoprotein species, the precursors were found to be basic glycoproteins, whereas mature glycoproteins exhibited heterogeneity, detected in two-dimensional electrophoresis as a series of spots (6 for gD and 15–20 for gC) of increasing negative charge and molecular weight. Because neuraminidase treatment decreased the size, number, and acidic charge of the spots, it was suggested that processing was due in part, but not entirely, to addition of sialic acid. Essentially similar results were obtained for gC, gD as well as for gB by Haarr and Marsden (1981) and by Palfreyman et al. (1983) in studies in which the glycoproteins were identified with monoclonal antibodies.

Working on gB and gC, Eberle and Courtney (1980) examined the [^{3}H]glucosamine/[^{14}C]amino acid ratio in precursors and in mature glycoproteins. The ratio was found to be higher in the mature forms, suggesting that mature glycoproteins differ from precursors in the number and/or the complexity of the oligosaccharide chains present in a common polypeptide moiety.

On the whole, these reports constituted the first indications that biosynthesis of HSV-1 glycoproteins followed a stepwise process analogous to that reported for N-linked oligosaccharides of VSV and Sindbis virus glycoproteins. Of particular relevance was the observation by Pizer et al. (1980), subsequently confirmed by several groups (Olofsson and Lycke, 1980; Serafini-Cessi and Campadelli-Fiume, 1981; Peake et al., 1982; Norrild and Pedersen,1982; Kousoulas et al., 1983a; Wenske and Courtney, 1983), that glycosylation of HSV-1 glycoproteins was inhibited by tunicamycin, an antibiotic that inhibits the production of N-

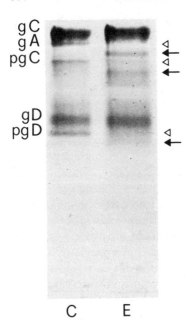

C   E

FIGURE 2. Effect of Endo-H treatment on electrophoretic mobility of precursors of HSV-1(F) glycoproteins. ^{14}C-glucosamine-labeled HSV-1(F)-infected HEp-2 cells were either (C) untreated or (E) treated for 20 hr with Endo-H. Open symbols indicate the positions of bands that disappeared. Closed symbols indicate the new bands that originated from gA, pgC, and pgD.

acetylglucosaminepyrophosphoryldolichol and consequently blocks the transfer of high-mannose oligosaccharides to proteins (Heifetz et al., 1979). This observation suggested that glycans of HSV-1 glycoproteins are of the N-linked type.

Subsequently, Serafini-Cessi and Campadelli-Fiume (1981) reported that the partially glycosylated precursors (pgC, pgB, and pgD) are susceptible to Endo-H, which cleaves N-linked oligosaccharides of the high mannose type between the two proximal GlcNAc residues (Tarentino and Maley, 1974). This was inferred from the increase in electrophoretic mobility of precursors (Fig. 2) and from a concomitant release of radiolabeled acid-soluble carbohydrates following treatment with the enzyme and indicated that precursors contain glycans rich in mannose. This result was confirmed and extended in experiments involving Endo-H digestion of immunoprecipitated glycoproteins (Wenske et al., 1982; Johnson and Spear, 1983). Unlike the immature forms, the mature forms of glycoproteins were found not to be Endo-H-sensitive, implying that they carry complex-type glycans resulting from the processing of high-mannose oligosaccharides present in the immature forms (Serafini-Cessi and Campadelli-Fiume, 1981; Wenske et al., 1982; Person et al., 1982; Johnson and Spear, 1983). gB seems to undergo a different processing in that even the fully processed form carries Endo-H-sensitive glycans, indicating that not all of its high-mannose glycans undergo conversion (Wenske et al., 1982; Johnson and Spear, 1983).

Except for analysis on glycopeptides of gC (Kumarasamy and Blough, 1982; Serafini-Cessi et al., 1984b) (see Section III.D), there have been no

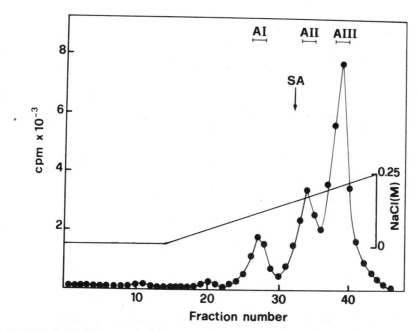

FIGURE 3. DEAE-Sephacel chromatography of complex-type Pronase-digested glycopeptides previously fractionated by Bio-Gel filtration from HSV-1(MP)-infected BHK cells labeled with [^{14}C]glucosamine from 5 to 18 hr after infection. AI, AII, and AIII correspond to the elution position of mono-, di-, and highly sialylated glycopeptides, respectively. The arrow indicates the elution position of sialic acid (SA). Reproduced from Campadelli-Fiume et al. (1982).

reports of detailed analyses of the glycans of each of the HSV-1 glycoprotein species.

Structural studies have dealt mainly with complex-type glycans of all HSV-1 glycoproteins present in infected cell lysates. Analyses of the size distribution on Bio-Gel P-10 of Pronase-digested glycopeptides obtained from cells labeled for a long time interval (6–18 hr postinfection) showed the accumulation of complex-type glycans whereas neutral high-mannose glycans predominated after a short (14–17 hr postinfection) labeling interval (Serafini-Cessi and Campadelli-Fiume, 1981). By using a combination of ion-exchange chromatography and neuraminidase treatment, complex-type glycopeptides were resolved into three species (Fig. 3) designated AI, AII, and AIII and which display increasing degrees of sialylation (Campadelli-Fiume et al., 1982). It has been shown that N-linked glycans bind to Con A-Sepharose according to the degree of substitution of the mannosyl core. Thus, high-mannose glycans are strongly bound by the lectin-gel whereas diantennary glycans are weakly retained and tri- and tetraantennary structures do not bind at all (Krusius et al., 1976). The results of analyses of the binding of glycopeptides AI, AII, and

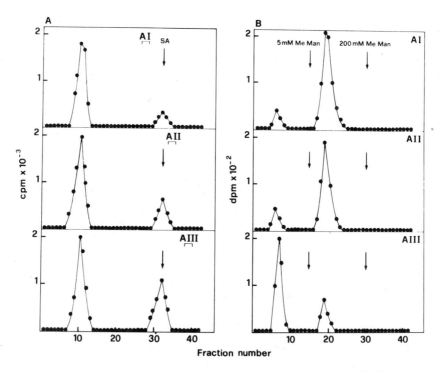

FIGURE 4. Characterization of complex-type glycopeptides of HSV-1(MP) glycoproteins. (A) DEAE-Sephacel chromatography of complex-type glycopeptides AI, AII, and AIII of Fig. 3 after neuraminidase digestion. The horizontal bars indicate the elution position of each glycopeptide prior to neuraminidase digestion; the arrow indicates the elution position of sialic acid (SA). Elution conditions were as shown in Fig. 3. Modified from Campadelli-Fiume et al. (1982). (B) Con A-Sepharose fractionation of AI, AII, and AIII. Most of AI and AII species were eluted with 5 mM α-methylmannose, indicating that they were weakly bound to the gel; the vast majority of AIII was not retained by the column.

AIII to Con A-Sepharose suggested that AI and AII have a diantennary structure whereas AIII has a triantennary or a more highly branched structure (Fig. 4B). Because the amount of sialic acid released by neuraminidase from AI is about one-half of that released from AII (Fig. 4A), we suggested that only one of the branches of the AI glycopeptides is sialylated; the other branch lacks terminal sialic acid (incomplete form) (Dall'Olio et al., 1982). Recently it has been shown (Cummings and Kornfeld, 1982; Hammarström et al., 1982) that leucoagglutinin, a lectin from Phaseolus vulgaris, interacts specifically with complex-type glycans in which one of the α-mannose of the trimannosyl core is substituted at position C-2 and C-6 with β-GlcNAc, as in tetraantennary species. We found that AIII glycopeptides were not at all retained by leucoagglutinin linked to agarose. Thus, it can be inferred that tetraantennary glycans are absent in HSV-1 glycoproteins (Serafini-Cessi et al., 1984b).

C. Relationship between *N*- and *O*-Glycosylation

N-linked glycans are the predominant chains in HSV-1 glycoproteins. It has been reported that HSV-1 glycoproteins also carry *O*-linked glycans. The main evidence in favor of the presence of glycans linked to serine or threonine was the release of low-molecular-weight radioactive compounds following mild alkaline borohydride treatment of ^{14}C-glucosamine-labeled HSV glycoproteins. This treatment releases *O*-linked chains by β-elimination. However, while this procedure has little effect on the *N*-linkage of glycans to asparagine residues, it also splits some residues of sialic acid from complex-type glycans (Pesonen *et al*, 1982). Labeled carbohydrates were also released by treatment of HSV-1 glycoproteins with GalNAc oligosaccharidase (Johnson and Spear, 1983). Although the structure of *O*-linked HSV-1 chains was not characterized in detail, it was reported that both sialylated and neutral glycans are β-eliminated from gC and gD (Johnson and Spear, 1983). Olofsson *et al*. (1981a) identified labeled galactosamine after acid hydrolysis of ^{14}C-glucosamine-labeled HSV-1 glycoproteins; GalNAc preferentially occurs in *O*-linked glycans (Montreuil, 1980). The same group also demonstrated that gC binds to *Helix pomatia* linked to Sepharose (Olofsson *et al*., 1981b). This lectin interacts with high affinity with terminal nonreducing GalNAc, although it displays a broad carbohydrate-binding specificity (Hammarström *et al*., 1977).

Serafini-Cessi *et al*. (1983a) reported on the presence in HSV-1-infected cell lysates of an *N*-acetylgalactosaminyltransferase activity able to add this sugar *in vitro* to HSV-1 glycoproteins, as identified by immunoprecipitation. GalNAc was found to be selectively incorporated into immature forms of HSV glycoproteins, pgC, pgD, and pgB. It appeared to be *O*-linked to the peptide because it was completely released as monosaccharide by mild alkaline borohydride treatment. The inability of mature glycoproteins to accept the sugar *in vitro* was related to the presence in mature glycoproteins of complex-type glycans. It was suggested that full processing of *N*-linked glycans, which takes place during maturation of glycoproteins, may hinder the accessibility of serine/threonine sites to the transferase.

In order to ascertain whether the relationship observed *in vitro* between *O*-glycosylation and processing of *N*-linked glycans also occurred during the biosynthesis of HSV glycoproteins in cultured cells, Serafini-Cessi *et al*. (1984a) investigated the occurrence of *O*-linked glycans in HSV glycoproteins produced in a system that accumulates immature glycoproteins. As reported in Section III.E, HSV-1-infected Ric^{R}14 cells accumulate viral glycoproteins carrying high-mannose glycans. Of particular interest was the finding in gC purified from HSV-infected Ric^{R}14 cells of an unusual glycopeptide that carried one high-mannose glycan susceptible to Endo-H and one sialylated oligosaccharide chain cleaved by mild alkaline borohydride treatment. A glycopeptide with this structure did not appear to be present in mature gC from parental, wild type, cells. It was suggested that because full processing of high-mannose glycans is blocked in Ric^{R}14 cells, O-glycosylation may occur in a substitutive fashion in a serine/threonine site very close to the asparagine residue that carries a high-mannose glycan.

D. Glycans of pgC and gC

Kumarasamy and Blough (1982) reported the results of a preliminary characterization of complex-type glycans of gC purified from HSV-1 virions. Affinity chromatography to Con A-Sepharose was used to distinguish diantennary glycopeptides from glycopeptides with a higher number of side chains. By labeling with [^{3}H]glucosamine, [^{3}H]mannose, and [^{3}H]fucose, it was shown that fucosylated diantennary and fucosylated tri/tetraantennary glycans are present in gC. Failure to detect glycopeptides strongly bound to Con A-Sepharose was indicative of absence of high-mannose glycans in gC extracted from herpes virions.

In order to obtain more detailed information on the biosynthetic pathway and on the structure of oligosaccharides of gC, Serafini-Cessi et al. (1984b) undertook the characterization of glycans present both in immature and in mature forms of gC. The immature form (pgC) was obtained from lysates of HSV-1(F)-infected BHK cells labeled for 20 min with [^{3}H] mannose, whereas mature gC was obtained after a long label (7–18 hr postinfection) followed by a 3-hr chase. The two labeling intervals were chosen so that only pgC was labeled in the first case, and only mature gC in the long label–chase protocol. Purification of pgC and gC was achieved in one step by affinity chromatography on immobilized monoclonal antibody HCl (Pereira et al., 1980; Eisenberg et al., 1982). The characterization of glycans was done according to the procedure schematically described in Fig. 5.

Pronase-digested glycopeptides of pgC were eluted from a Bio-Gel P-10 column as a symmetric narrow peak in the region of high-mannose glycopeptides (Fig. 6). They were all strongly bound to Con A-Sepharose. Treatment with Endo-H shifted all radioactivity to a region of lower molecular weight. These results indicated that only high-mannose glycans were present in pgC. The compostion and size of the oligomannosyl chains of these glycans were determined in some detail by thin-layer chromatography, according to Godelaine et al. (1981). The glycans exhibited a marked heterogeneity (Fig. 7) in that five high-mannose glycan species differing in the number of mannose residues (from Man$_9$ to Man$_5$) were separated. The predominant species were Man$_8$-GlcNAc and Man$_7$-GlcNAc, whereas the amount of Man$_9$-GlcNAc was negligible. This result is consistent with previous findings that one of the terminal non-reducing α(1-2)-mannose residues of polymannosyl units is cleaved by an

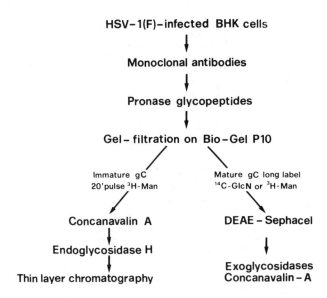

FIGURE 5. Schematic representation of the procedure for the separation and characterization of the oligosaccharide chains of immature and mature gC.

early mannosidase probably located in the RER (Spiro and Spiro, 1982). It seems likely that pgC detected after a short pulse represents a form of gC already routed from the RER to the Golgi apparatus.

Analysis of the glycan structure of mature gC (Fig. 6) showed that complex-type glycopeptides also exhibit a high degree of heterogeneity. Thus, fractionation of Pronase-digested glycopeptides by DEAE-Sephacel chromatography separated three different glycopeptide species according to their degree of sialylation. In particular, triantennary (AIII), diantennary (AII), and monosialylated diantennary (AI) glycopeptides were found to be present in gC, as was the case for Pronase digests of HSV-1-infected cell lysates. The structure of AI glycopeptides was proposed on the basis of the following results:

1. The glycopeptides were weakly bound to Con A-Sepharose, suggesting that the structure of the oligomannosyl core is the same as that of diantennary glycans.
2. The amount of sialic acid released by neuraminidase treatment was half of that released from diantennary AII glycopeptides.
3. N-acetylglucosaminidase treatment released a small amount of GlcNAc, which increased significantly when the treatment was done simultaneously with β-galactosidase.

The results are consistent with the hypothesis that AI glycopeptides carry one branch capped with sialic acid whereas the incomplete branch terminates in part with Gal (F. Dall'Olio, F. Serafini-Cessi, and G. Campadelli-Fiume, unpublished results). The presence of tetraantennary gly-

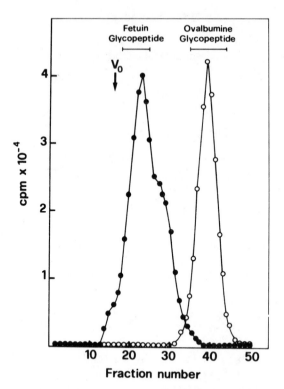

FIGURE 6. Bio-Gel P-10 gel filtration of Pronase digests from pgC and gC. (O) pgC was purified from HSV-1(F)-infected BHK cells pulse-labeled for 20 min with [³H]mannose at 5.5 hr after infection. (●) gC was purified from cells labeled with [³H]mannose from 7 to 18 hr after infection and chased for 3 hr. V_0, void volume, as determined with blue dextran. The two bars indicate the elution regions of complex-type glycopeptide (from fetuin) and of high-mannose glycopeptide (from ovalbumine). The column (1 × 80 cm) was eluted with 0.1 M NH_4HCO_3.

cans in gC was ruled out as glycopeptides were not at all retained by leucoagglutinin-agarose (see Section III.B). Only traces of glycopeptides were bound with strong affinity to Con A-Sepharose, implying that high-mannose glycans were almost completely processed to complex-type glycans during the long label–chase interval.

The heterogeneity of complex-type glycans may be related to the presence of several N-glycosylation sites in the peptide backbone. For example, eight potential N-glycosylation sites were predicted for gC on the basis of DNA sequencing data (Frink et al., 1983). It has been shown for other viral glycoproteins that different oligosaccharides are not randomly distributed and that each individual glycosylation site carries predominantly one type of glycan; this suggests that the polypeptide structure and conformation around the glycosylation site affect the processing of the glycan species and that the final structure of the glycan depends

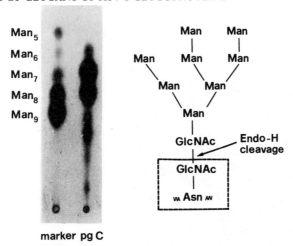

marker pg C

FIGURE 7. Thin-layer chromatography of high-mannose glycans from pgC. ^{3}H-mannose-labeled oligosaccharides were released from Pronase-digested glycopeptides by Endo-H treatment and were separated by thin-layer chromatography on silica gel plates. The oligosaccharides were revealed by fluorography. Oligosaccharides from A unit of thyroglobulin after Endo-H treatment and reduction with NaB^{3}H$_4$ were used as markers.

primarily on its accessibility to the enzymes involved in processing (Klenk *et al.*, 1983; Hsieh *et al.*, 1983a,b). At present it seems likely that the same regulation mechanism operates for HSV-1 gC as well and may account for the observed heterogeneity of complex type glycans.

E. Host-Cell Dependence of Glycosylation

A central question in the studies of HSV glycoproteins is the origin of the enzymes involved in their glycosylation. In an attempt to resolve this question, extensive analyses were done on oligosaccharide processing and glycosylation of HSV-1 glycoproteins made in mutant cells defective in specific glycosyltransferases. It was expected that if the virus were to use host-cell glycosyltransferases, the defects in cellular enzymes would be reflected in the altered structure of glycans linked to viral proteins made in these cells.

Two lines of ricin-resistant (Ric^R) mutant cells, Ric^{R}14 and Ric^{R}21, were used in these studies (Vischer and Hughes, 1981). Ric^{R}14 cells lack almost completely active N-acetylglucosaminyltransferase I, the enzyme that adds the first GlcNAc to high-mannose glycans and in this way initiates the conversion to complex-type glycans. The electrophoretic profile of glycoproteins made in Ric^{R}14 cells indicated that mature glycoproteins were absent and that only the corresponding underglycosylated forms accumulated (Fig. 8). In comparison with glycopeptides from infected parent BHK cells, glycopeptides from infected Ric^{R}14 cells

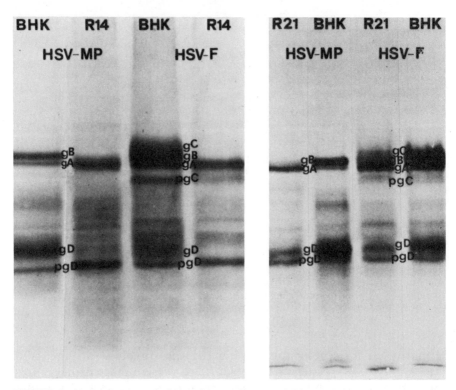

FIGURE 8. Autoradiogram of electrophoretically separated HSV-1 glycoproteins synthesized in BHK and Ric[R]14 cells (left panel) and in BHK and Ric[R]21 cells (right panel). Cells were infected with HSV-1(MP) or (F) and labeled with [14C]glucosamine. Left and right panels reproduced from Campadelli-Fiume et al. (1982) and Serafini-Cessi et al. (1983b), respectively.

showed a dramatic increase in neutral high-mannose glycans and in incompletely processed glycans, concomitant with a strong decrease in highly sialylated oligosaccharides (Campadelli-Fiume et al., 1982).

Experiments with Ric[R]21 cells yielded results consistent with these observations. Ric[R]21 cells are partially defective in enzymes of the Golgi system that perform late sugar addition, namely N-acetylglucosaminyl-transferase I and II and galactosyltransferase (Vischer and Hughes, 1981), but are able to produce small quantities of mature HSV glycoproteins (Fig. 8). Accordingly, in infected cells the accumulation of di- and triantennary complex-type glycans was reduced in amount, whereas the accumulation of imcomplete species was increased (Serafini-Cessi et al., 1983b).

Further evidence for host-dependent variations in HSV-1 glycan biosynthesis was obtained by comparing the content of the various glycopeptide species between infected and the corresponding uninfected cells. This approach was based on the observation that different cell lines vary

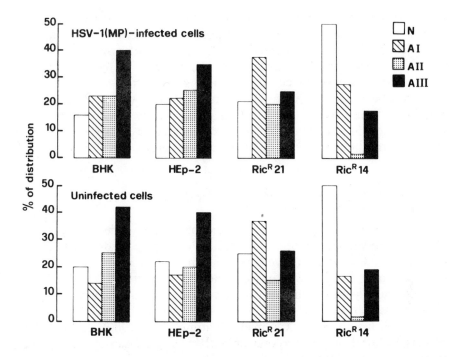

FIGURE 9. Relative distribution of radioactivity in the four glycopeptide species [neutral (N), AI, AII, AIII] obtained from HSV-1(MP)-infected and uninfected Ric^{R}14, Ric^{R}21, HEp-2, and BHK cells. Infected or mock-infected cells were labeled with [^{14}C]glucosamine from 7 to 18 hr after infection. The various glycopeptide species were obtained by Bio-Gel filtration and subsequently fractionated by DEAE-Sephacel chromatography (see Fig. 3).

in their capacity to process oligosaccharides of viral glycoproteins and consequently one glycan species prevails in each cell line. Figure 9 shows that the relative distribution in the four glycan species [neutral (N), AI, AII, AIII] from BHK, HEp-2, Ric^{R}21, and Ric^{R}14 cells was strikingly similar before and after infection.

In sum, these findings suggest that HSV utilizes the glycosylation machinery of the host. However, in comparison with the small viruses, such as VSV, Sindbis virus, and influenza virus, the coding capacity of HSV DNA is huge. Furthermore, the functions of many of the proteins specified by the HSV genome are largely unknown. In consequence, it is not possible to rule out the possibility that one or more steps in the complex process of glycosylation and of glycoprotein transport are regulated by viral gene products.

Recently, Berman *et al.* (1983) constructed an expression vector containing the gene for gD and established stable cell lines that constitutively synthesize the glycoprotein. The authors concluded from pulse–chase experiments with labeled glucosamine and from estimates of the apparent molecular weight of the gene products in SDS-PAGE that the gD made in these cultures is glycosylated in a fashion similar to that of the authentic gD made in lytically infected cells. Of particular interest was the report that when the DNA sequence coding for the COOH-terminal anchoring region was removed, glycosylated gD was secreted in the cell medium. The results of these studies are consistent with the hypothesis that host cells are able to carry out the glycosylation of HSV-1 proteins, although they do not prove that in lytically infected cells viral glycosylation is carried out entirely under the control of host proteins.

F. Benzhydrazone, an Inhibitor of HSV-1 Glycosylation

Tunicamycin and 2-deoxy-D-glucose have been widely used as inhibitors of HSV-1 glycoprotein synthesis. Both compounds block glycosylation of mammalian as well as viral proteins. Cavrini et al. (1979) observed that a bis-amidinohydrazone derivative (1H-benz[F]indene 1.3(2H)dione-bis-amidinohydrazone) designated benzhydrazone (BH) reduces replication of HSV-1 and hinders the appearance of HSV-1(MP)-induced cell fusion. Later, Campadelli-Fiume et al. (1980) reported that BH acts by inhibiting incorporation of labeled glucosamine, mannose, and fucose into HSV-1 glycoproteins. The compound appeared to be a selective inhibitor of herpesvirus glycosylation in that it does not reduce HSV protein synthesis or the glycosylation of uninfected cells, of Sindbis virus, and of paramyxovirus proteins. In order to ascertain what step of the glycosylation process was affected by BH, Serafini-Cessi and Campadelli-Fiume (1981) compared Endo-H sensitivity and size-class distribution of Pronase glycopeptides of HSV-1 glycoproteins made in the presence or absence of BH. The results suggested that the drug prevents high-mannose oligosaccharide addition to proteins and thus causes an early block in N-glycosylation process similar to that induced by tunicamycin. Subsequently, Tognon et al. (1984) isolated an HSV mutant, designated HSV1(13)S11, which appeared to be resistant to BH inasmuch as incorporation of [^{14}C]glucosamine was not inhibited in infected HEp-2 cells exposed to the drug. BH resistance was shown to be encoded in the DNA of the mutant virus and could be transferred into the genome of wild-type BH-sensitive HSV-2(G), giving rise to intertypic recombinants. The available data do not discriminate among the mutually exclusive hypotheses that (1) BH specifically inhibits a viral gene product required for glycosylation of HSV proteins in HEp-2 cells, (2) BH is an active inhibitor of glycosylation only when activated by an HSV gene product(s) or that (3) BH does not readily penetrate into uninfected cells or cells infected with other viruses but does penetrate into HSV-infected cells as a consequence of modifications introduced into plasma membranes by virus-specific membrane proteins.

IV. RELATIONSHIP BETWEEN GLYCAN STRUCTURE AND EXPRESSION OF HSV-1 FUNCTIONS

HSV glycoproteins have long been known to be essential for expression of some HSV functions (Spear and Roizman, 1980). The function of HSV glycoproteins is dealt with extensively in Chapter 7, and will not be considered further here. The focus of this section centers on whether glycans with specific structures are required for glycoprotein-associated activities.

A. Yield of Infectious Virus

Studies on the effect of tunicamycin on yield of infectious viruses provided evidence that glycosylation is a requirement for infectivity of several viruses (see Gibson et al., 1978). In contrast, VSV produced in the presence of the antibiotic was found to have a specific infectivity comparable to that of virus containing fully processed gG (Gibson et al., 1978). Pizer et al. (1980) observed that yield of infectious HSV-1 is abolished in cells treated with tunicamycin, where only nonglycosylated analogs of the viral glycoproteins were synthesized. Subsequent studies pointed out that the step defective in tunicamycin-treated cells is virion envelopment. By electron microscopy, nucleocapsids were observed in both the nucleus and the cytoplasm, whereas only a small number of enveloped particles were detected on the cell surface (Peake et al., 1982). Svennerholm et al. (1982) and Kousoulas et al. (1983a) succeeded in purifying noninfectious viral particles from tunicamycin-treated cells. De novo synthesis of glycosylated proteins was blocked and glycoproteins were not detected on cell surfaces either by radioiodination (Peake et al., 1982) or as targets in antibody-dependent cell-mediated cytoxicity tests (Norrild and Pedersen, 1982).

It was of interest to investigate whether infectious HSV-1 was produced under conditions in which glycan processing was arrested at the high-mannose level. Courtney et al. (1973) had reported that 2-deoxy-D-glucose hinders HSV-1 replication. We found, however, that yields of infectious HSV-1 as well as specific infectivity of herpes virions produced in infected Ric^R14 and Ric^R21 cells were not significantly affected by the dramatic changes in glycan composition of the glycoproteins. This led to the conclusion that fully processed oligosaccharides are not an absolute requirement for these functions (Campadelli-Fiume et al., 1982; Serafini-Cessi et al., 1983b). Studies on the effects of monensin (Johnson and Spear, 1982; Kousoulas et al., 1983b) and ammonium chloride (Kousoulas et al., 1983a) led to similar conclusions. Monensin is an ionophore that disrupts the Na^+ gradient across the membranes. Its effect on glycoprotein processing varies to some extent depending on the glycoprotein type. Johnson and Spear (1982) showed that it blocks processing of immature forms of

HSV-1 glycoproteins, as well as transport of the viral glycoproteins through the Golgi apparatus. Infectivity of herpes virions and production of infectious progeny were not significantly altered even by treatment of cells with ammonium chloride, a compound that hinders full processing of HSV-1 glycoproteins (Kousoulas et al., 1983a).

It appears that (1) absence of N-linked glycans precludes the envelopment and hence the yield of infectious HSV-1; (2) presence of complex-type glycans is not an obligatory requirement for herpes virion infectivity; and (3) high-mannose oligosaccharides suffice for this function.

B. Virion Egress

Considerable evidence has accumulated in several viral systems that the processing of glycans and the intracellular transport of glycoproteins to the plasma membrane through the Golgi are interrelated and that curtailment of processing affects transport. That this applies to HSV-1 glycoproteins may be inferred from experiments with infected cells treated with tunicamycin (Peake et al., 1982; Norrild and Pedersen, 1982) or with monensin (Johnson and Spear, 1982). Electron microscopic studies have shown that herpes virions first acquire their envelope by budding through the inner membrane, which is an extension of the RER. As mentioned above, intracytoplasmic HSV particles accumulate in tunicamycin-treated cells (Peake et al., 1982) and herpes virions accumulate in large intracytoplasmic vacuoles, probably derived from the Golgi cisternae, in cells treated with monensin (Johnson and Spear, 1982). Therefore, herpes viron egress appears to be hampered under conditions of defective glycoprotein transport. Serafini-Cessi et al. (1983b) found that egress of herpes virions was reduced in both Ric^R21 and Ric^R14 cells (Table II). This reduction correlated well with defective glycan processing (see Section III, E). Subsequently, Compton and Courtney (1984) reported that in purified nuclei from HSV-1-infected cells, the major HSV-1 glycoproteins are present predominantly as precursors carrying high mannose glycans. The data suggest that proteins with fully processed oligosaccharides are required for the egress of herpes virions (Serafini-Cessi et al., 1983b) and that during the transit through the cytoplasm, glycans of the HSV envelope are processed from the high mannose to the complex type by interaction with the Golgi system (Johnson and Spear, 1982).

C. Virus-Induced Cell Fusion

Expression of HSV-1-induced cell-fusion activity is inhibited by exposing infected cells to various compounds, most of which act by inhibiting glycosylation. This is the case, for example, for 2-deoxy-D-glucose, whose inhibitory effect on fusion was reported first (Gallaher et al., 1973;

TABLE II. Yield of Cell-Associated and Extracellular Infectious HSV-1(F) from BHK, Ric^{R}14, and Ric^{R}21 Cells[a,b]

Cell type	Time after infection (hr)	Yield (PFU/cell)		Extracellular medium/cell associated
		Cell associated	Extracellular medium	
BHK[c]	16	90.0	150.0	1.67
	24	94.0	600.0	6.38
Ric^{R}14[c]	16	30.0	4.4	0.15
	24	40.0	32.0	0.80
BHK[d]	16	57.0	63.0	1.11
	24	59.0	211.0	3.58
Ric^{R}21[d]	16	13.0	0.5	0.04
	24	50.0	6.0	0.12

[a] Reproduced in part from Serafini-Cessi et al. (1983b).
[b] Infected cells were incubated at 37°C for the indicated times. At the end of the incubation the medium over the cells was collected and extracellular virus titers were determined. Cell-associated virus was assayed after freezing and thawing three times.
[c] Monolayers were infected at a multiplicity of 10 PFU/cell.
[d] Monolayers were infected at a multiplicity of 3 PFU/cell.

Knowles and Person, 1976), tunicamycin (Pizer et al., 1980), BH (Campadelli-Fiume et al., 1980), ammonium chloride (Holland and Person, 1977), and, as reported more recently, monensin (Kousoulas et al., 1983b). Whereas tunicamycin and BH block the earlier steps of N-glycosylation (Heifetz et al., 1979; Serafini-Cessi and Campadelli-Fiume, 1981), the other compounds primarily prevent processing of precursors to mature HSV-1 glycoproteins. Studies by Manservigi et al. (1977) and by Haffey and Spear (1980) with the temperature-sensitive mutant HSV-1 tsB5 and with some of its recombinants and revertants showed that the presence of mature gB at permissive temperature correlated with expression of cell-fusion activity, whereas the presence of immature pgB correlated with the absence of cell fusion. Our experiments (Campadelli-Fiume et al., 1982) with ricin-resistant cells showed the following. Fusion was not induced in Ric^{R}14 cells and this correlated with the absence of mature glycoproteins (Fig. 8). Ric^{R}21 cells, however, displayed a low but significant ability to undergo HSV-1(MP)-induced cell fusion (Serafini-Cessi et al., 1983b); in these cells there was a reduction but not a total absence of fully processed glycoproteins (see Fig. 8) and of polybranched glycans. In sum, the studies on the infected cells treated with glycosylation inhibitors and on ricin-resistant cells indicate that interference with processing of glycans to a fully glycosylated state is associated with inability of the cells to express herpesvirus-induced cell-fusion activity. This conclusion is further supported by the observation that exposure of HSV-1 (MP)-infected BHK and Ric^{R}21 cells to neuraminidase reduced the appearance of cell fusion. Addition of the enzyme to cell culture media released sialic acid from cell surface membranes, suggesting that this sugar is crucial for the type of cell–cell interaction that permits the cells to fuse (Serafini-Cessi *et al.*, 1983b). Herpesvirus-induced fusion results from the interaction of infected recruiter cells with neighboring cells (either infected or uninfected), which become recruited in the polykaryocyte (Roizman, 1962). In all experimental conditions described above (i.e., with glycosylation inhibitors, in ricin-resistant cells and following neuraminidase treatment) glycoproteins were present in underglycosylated form on both the recruiter-infected cell and on the cells to be recruited. This raises the possibility that the presence of sialylated glycans is crucial not only on HSV glycoproteins but also on the membrane receptors of the cells that are to be recruited.

ACKNOWLEDGMENTS. The authors' work was supported by grants from the Ministry of Public Education and the National Research Council, Grant 83.00992.51 Progetto Finalizzato Ingegneria Genetica, and Grants 83.00373.04 and 82.02828.04 and Progetto Finalizzato Oncologia.

REFERENCES

Baig, M. M., and Aminoff, D., 1972, Glycoproteins and blood group activity. I. Oligosaccharides of serologically inactive hog submaxillary glycoproteins, *J. Biol. Chem.* **247**:6111–6118.

Baucke, R. B., and Spear, P. G., 1979, Membrane proteins specified by herpes simplex viruses. V. Identification of an Fc-binding glycoprotein, *J. Virol.* **32**:779–789.

Berger, E. C., Buddecke, E., Kamerling, J. P., Kobata, A., Paulson, J. C., and Vliegenthart, J. F. C., 1982, Structure, biosynthesis and functions of glycoprotein glycans, *Experientia* **38**:1129–1162.

Berman, P. W., Dowbenko, D., Simonsen, C. C., and Lasky, L. A., 1983, Detection of antibodies to herpes simplex virus using a continuous cell line expressing cloned glycoprotein D, *Science* **222**:524–527.

Beyer, T. A., Sadler, J. E., Rearick, J. I., Paulson, J. C., and Hill, R. L., 1981, Glycosyltransferases and their use in assessing oligosaccharide structure and structure–function relationships, *Adv. Enzymol.* **52**:23–175.

Brennan, P. J., Steiner, S. M., Courtney, R. J., and Skelly, J., 1976, Metabolism of galactose in herpes simplex virus-infected cells, *Virology* **69**:216–228.

Burke, D. and Keegestra, K., 1979, Carbohydrate structure of Sindbis virus glycoprotein E2 from virus grown in hamster and chicken cells, *J. Virol.* **29**:546–554.

Bzik, D., Fox, B., DeLuca, N., and Person, S., 1984, Nucleotide sequence of the glycoprotein gene, gB, of herpes simplex virus type 1, *Virology* **133**:301–314.

Campadelli-Fiume, G., Sinibaldi-Vallebona, P., Cavrini, V., and Mannini-Palenzona, A., 1980, Selective inhibition of herpes simplex virus glycoprotein synthesis by a benzamidinohydrazone derivative, *Arch. Virol.* **66**:179–191.

Campadelli-Fiume, G., Poletti, L., Dall'Olio, F., and Serafini-Cessi, F., 1982, Infectivity and glycoprotein processing of herpes simplex virus type 1 grown in a ricin-resistant cell line deficient in N-acetylglucosaminyl transferase I, *J. Virol.* **43**:1061–1071.

Cavrini, V., Gatti, R., Roveri, P., Giovanninetti, G., Mannini-Palenzona, A., and Baserga, M., 1979, Antiviral compounds. XII. Synthesis and *in vitro* antiherpetic activity of some cyclic α- and β-diketone bis-amidinohydrazones, *Eur. J. Med. Chem.* **14**:343–346.

Cohen, G. H., Long, D., and Eisenberg, R. J., 1980, Synthesis and processing of glycoprotein gD and gC of herpes simplex virus type 1. *J. Virol.* **36**:429–439.

Cohen, G. H., Long, D., Matthews, J. T., May, M., and Eisenberg, R., 1983, Glycopeptides of type common glycoprotein gD of herpes simplex virus type 1 and 2, *J. Virol.* **46**:679–689.

Compton, T., and Courtney, R. J., 1984, Virus specific glycoproteins associated with the nuclear fraction of herpes simplex virus type 1-infected cells, *J. Virol.* **49**:594–597.

Courtney, R. J., Steiner, S. M., and Benyesh-Melnich, M., 1973, Effects of 2-deoxy-D-glucose on herpes simplex virus replication, *Virology* **52**:447–455.

Cummings, R. D., and Kornfeld, S., 1982, Characterization of the Structural determinants required for the high affinity interaction of asparagine-linked oligosaccharides with immobilized *Phaseolus vulgaris* leucoagglutinin and erythroagglutinin lectins, *J. Biol. Chem.* **257**:11230–11234.

Dall'Olio, F., Serafini-Cessi, F., Campadelli-Fiume, G., and Scannavini, M., 1982, Herpes simplex virus glycoproteins from wild-type and ricin-resistant BHK cells: Characterization of glycopeptides by Concanavalin A-Sepharose, in: *Lectins-Biology, Biochemistry and Clinical Biochemistry*, Vol. 3 (T. C. Bøg-Hansen and G. A. Spengler, eds), pp. 351–359, de Gruyter, Berlin.

Eberle, R., and Courtney, R. J., 1980, gA and gB glycoproteins of herpes simplex virus type 1: Two forms of a single polypeptide, *J. Virol.* **36**:665–675.

Edson, C. M., and Thorley-Lawson, D. A., 1983, Synthesis and processing of the three major envelope glycoproteins of Epstein–Barr virus, *J. Virol.* **46**:547–556.

Eisenberg, R. J., Hydrean-Stern, C., and Cohen, G. H., 1979, Structural analysis of precursors and products forms of type-common envelope-glycoprotein D (CP-1 antigen) of herpes simplex virus type 1, *J. Virol.* **31**:608–620.

Eisenberg, R. J., Ponce de Leon, M., Pereira, L., Long, D., and Cohen, G. H., 1982, Purification of glycoprotein gD of herpes simplex virus types 1 and 2 by use of monoclonal antibody, *J. Virol.* **41**:1099–1104.

Frink, R. J., Eisenberg, R., Cohen, G., and Wagner, E. K., 1983, Detailed analysis of the portion of the herpes simplex virus type 1 genome encoding glycoprotein C, *J. Virol.* **45**:634–647.

Gallaher, W. R., Levitan, D. B., and Blough, H. A., 1973, Effect of 2-deoxy-D-glucose on cell fusion induced by Newcastle disease and herpes simplex viruses, *Virology* **55**:193–201.

Gibson, R., Leavitt, R., Kornfeld, S., and Schlesinger, S., 1979, Synthesis and infectivity of vesicular stomatitis virus containing nonglycosylated G protein, *Cell* **13**:671–679.

Godelaine, D., Spiro, M. J., and Spiro, R. G., 1981, Processing of the carbohydrate unit of thyroglobulin, *J. Biol. Chem.* **256**:10161–10168.

Haarr, L., and Marsden, H. S., 1981, Two-dimensional gel analysis of HSV type 1-induced polypeptides and glycoprotein processing, *J. Gen. Virol.* **52**:77–92.

Haffey, M., and Spear, P. G., 1980, Alteration in glycoprotein gB specified by mutants and their partial revertants in herpes simplex virus type 1 and relationship to other mutant phenotypes, *J. Virol.* **35**:114–128.

Hammarström, S., Murphy, L. A., Goldstein, I. J., and Etzler, M. J., 1977, Carbohydrate binding specificity of *N*-acetyl-D-galactosamine-"specific" lectins: *Helix pomatia* A hemagglutinin, soy bean agglutinin, lima bean lectin, and *Dolichos biflorus* lectin, *Biochemistry* **16**:2750–2755.

Hammarström, S., Hammarström (née Dillner), M. L., Sunblad, G., Arnarp, J., and Lönngren, J., 1982, Mitogenic leucoagglutinin from *Phaseolus vulgaris* binds to a pentasaccharide unit in *N*-acetyllactosamine type glycoprotein glycans, *Proc. Natl. Acad. Sci. U.S.A.* **79**:1611-1615.

Heifetz, A., Keenan, R. W., and Elbein, A. D., 1979, Mechanism of action of tunicamycin on the UDP-GlcNAc: dolichylphosphate GlcNAc-1-phosphate transferase, *Biochemistry* **18**:2186–2192.

Heine, J. W., Spear, P. G., and Roizman, B., 1972, Proteins specified by herpes simplex virus. VI. Viral proteins in plasma membranes, *J. Virol.* **9**:431–439.

Heine, J. W., Honess, R. W., Cassai, E., and Roizman, B., 1974, Proteins specified by herpes simplex virus. XII. The virion polypeptides of type I strains, *J. Virol.* **14**:640–651.

Holland, T. C., and Person, S., 1977, Ammonium chloride inhibits cell fusion induced by *syn* mutants of herpes simplex virus type 1, *J. Virol.* **23**:213–215.

Honess, R. W., and Roizman, B., 1975, Proteins specified by herpes simplex virus. XIII. Glycosylation of viral polypeptides, *J. Virol.* **16**:1308–1326.

Hope, R. G., and Marsden, H. S., 1983, Processing of glycoproteins induced by herpes simplex virus type 1: Sulphation and nature of the oligosaccharide linkages, *J. Gen. Virol.* **64**:1943–1953.

Hsieh, P., Rosner, M. R., and Robbins, P. W., 1983a, Host-dependent variation of asparagine-linked oligosaccharides at individual glycosylation sites of Sindbis virus glycoproteins, *J. Biol. Chem.* **258**:2548–2554.

Hsieh, P., Rosner, M. R., and Robbins, P. W., 1983b, Selective cleavage by endo-β-N-acetylglucosaminidase H at individual glycosylation sites of Sindbis virion envelope glycoproteins, *J. Biol. Chem.* **258**:2555–2561.

Hubbard, S. C., and Ivatt, R. J., 1981, Synthesis and processing of asparagine-linked oligosaccharides, *Annu. Rev. Biochem.* **50**:555–583.

Hunt, L. A., Etchison, J. R., and Summers, D. F., 1978, Oligosaccharide chains are trimmed during synthesis of the envelope glycoprotein of vesicular stomatitis virus, *Proc. Natl. Acad. Sci. USA* **75**:754–758.

Inglis, M. M., and Newton, A. A., 1982, Identification of polypeptide precursors to HSV-1 glycoproteins by cell-free translation, *J. Gen. Virol.* **58**:217–222.

Johnson, D. C., and Spear, P. G., 1982, Monensin inhibits the processing of herpes simplex virus glycoproteins, their transport to the cell surface, and the egress of virions from infected cells, *J. Virol.* **43**:1102–1112.

Johnson, D. C., and Spear, P. G., 1983, O-linked oligosaccharides are attached to herpes simplex virus glycoproteins in the Golgi apparatus, *Cell* **32**:987–997.

Klenk, H. D., Keil, W., Niemann, H., Geyer, R., and Schwarz, R. T., 1983, The characterization of influenza A viruses by carbohydrate analysis, *Curr. Top. Microbiol. Immunol.* **104**:247–257.

Knowles, R. W., and Person, S., 1976, Effect of 2-deoxyglucose, glucosamine and mannose on cell fusion and the glycoproteins of herpes simplex virus, *J. Virol.* **18**:644–651.

Kornfeld, R., and Wold, W. S. M., 1981, Structure of the oligosaccharides of the glycoprotein coded by early region E3 of adenovirus 2, *J. Virol.* **40**:440–449.

Kousoulas, K. G., Bzik, D. J., DeLuca, N., and Person, S., 1983a, The effect of ammonium chloride and tunicamycin on the glycoprotein content and infectivity of herpes simplex virus type 1, *Virology* **125**:468–474.

Kousoulas, K. G., Bzik, D. J., and Person, S., 1983b, Effect of the ionophore monensin on herpes simplex virus type 1-induced cell fusion, glycoprotein synthesis and virion infectivity, *Intervirology* **20**:56–60.

Krusius, T., Finne, J., and Rauvala, H., 1976, The structural basis of the different affinities of two types of acidic N-glycosidic glycopeptides for Concanavalin A-Sepharose, *FEBS Lett.* **71**:117–120.

Kumarasamy, R., and Blough, H. A., 1982, Characterization of oligosaccharides of highly purified glycoprotein gC of herpes simplex virus type 1 (HSV-1), *Biochem. Biophys. Res. Commun.* **109**:1108–1115.

Lee, G. T. Y., Para, M. F., and Spear, P. G., 1982a, Location of the structural genes for glycoproteins gD and gE and for other polypeptides in the S component of herpes simplex virus type 1 DNA. *J. Virol.* **43**:41–49.

Lee, G. T. Y., Pogue Geile, K. L., Pereira, L., and Spear, P. G., 1982b, Expression of herpes simplex virus glycoprotein C from a DNA fragment inserted into the thymidine kinase gene of this virus. *Proc. Natl. Acad. Sci. U.S.A.* **79**:6612–6616.

Li, E., Tabas, I., and Kornfeld, S., 1978, The synthesis of complex type oligosaccharides. I. Structure of the lipid-linked oligosaccharide precursor of the complex type oligosaccharides of the vesicular stomatitis virus G protein, *J. Biol. Chem.* **253**:7762–7770.

Manservigi, R., Spear, P. G., and Buchan, A., 1977, Cell fusion induced by herpes simplex virus is promoted and suppressed by different viral glycoproteins, *Proc. Natl. Acad. Sci. USA* **74**:3913–3917.

Marshall, R., 1972, Glycoproteins, *Annu. Rev. Biochem.* **41**:673–702.

Montreuil, J., 1980, Primary structure of glycoprotein glycans: basis for the molecular biology of glycoproteins, *Adv. Carbohydr. Chem. Biochem.* **37**:158–213.

Nieman, H., and Klenk, H. D., 1981, Coronavirus glycoprotein E1, a new type of viral glycoprotein, *J. Mol. Biol.* **153**:993–1010.

Norrild, B., and Pedersen, B., 1982, Effect of tunicamycin on the synthesis of herpes simplex virus type 1 glycoproteins and their expression on the cell surface, *J. Virol.* **43**:395–402.

Olofsson, S., and Lycke, E., 1980, Glucosamine metabolism of herpes simplex virus infected cells: Inhibition of glycosylation by tunicamycin and 2-deoxy-D-glucose, *Arch. Virol.* **65**:201–209.

Olofsson, S., Blomberg, J., and Lycke, E., 1981a, O-glycosidic carbohydrate–peptide linkages of herpes simplex virus glycoproteins, *Arch. Virol.* **70**:321–329.

Olofsson, S., Jeansson, S., and Lycke, E., 1981b, Unusual lectin-binding properties of a herpes simplex virus type 1-specific glycoprotein, *J. Virol.* **38**:564–570.

Palfreyman, J. W., Haarr, L., Cross, A., Hope, R. G., and Marsden, H. S., 1983, Processing of herpes simplex virus type 1 glycoproteins: Two-dimensional gel analysis using monoclonal antibodies, *J. Gen. Virol.* **64**:873–886.

Peake, M. L., Nystrom, P., and Pizer, L. I., 1982, Herpesvirus glycoprotein synthesis and insertion into plasma membranes, *J. Virol.* **42**:678–690.

Pereira, L., Klassen, T., and Baringer, J. R., 1980, Type-common and type specific monoclonal antibody to herpes simplex virus type 1, *J. Virol.* **29**:724–732.

Pereira, L., Dondero, D., Norrild, B., and Roizman, B., 1981, Differential immunologic reactivity and processing of glycoproteins gA and gB of herpes simplex virus types 1 and 2 made in Vero and HEp-2 cells, *Proc. Natl. Acad. Sci. USA* **78**:5202–5206.

Pereira, L., Dondero, D., and Roizman, B., 1982, Herpes simplex virus glycoprotein gA/B: Evidence that the infected Vero cell products comap and arise by proteolysis, *J. Virol.* **44**:88–97.

Person, S., Kousoulas, K. C., Knowles, R. W., Read, G. S., Holland, T. C., Keller, P. M., and Warner, S. C., 1982, Glycoprotein processing in mutants of HSV-1 that induce cell fusion, *Virology* **117**:293–306.

Pesonen, M., Rönnholm, R., Kruismanen, E., and Pettersson, R. F., 1982, Characterization of the oligosaccharides of Inkoo virus envelope glycoproteins, *J. Gen. Virol.* **63**:425–434.

Pizer, L. I., Cohen, G. H., and Eisenberg, R. J., 1980, Effect of tunicamycin on herpes simplex virus glycoproteins and infectious virus production, *J. Virol.* **34**:142–153.

Reading, C. L., Penhoet, E. E., and Ballou, C. E., 1978, Carbohydrate structure of vesicular stomatitis virus glycoprotein, *J. Biol. Chem.* **253**:5600–5612.

Roizman, B., 1962, Polykaryocytosis, *Cold Spring Harbor Symp. Quant. Biol.* **27**:327–342.

Ruyechan, W. T., Morse, L. S., Knipe, D. M., and Roizman, B., 1979. Molecular genetics of herpes simplex virus. II. Mapping of the major virus glycoproteins and of the genetic loci specifying the social behavior of infected cells, *J. Virol.* **29**:677–697.

Sabatini, D. D., Kreibich, G., Morimoto, T., and Adenik, M., 1982, Mechanisms for the incorporation of proteins in membranes and organelles, *J. Cell Biol.* **92**:1–22.

Serafini-Cessi, F., and Campadelli-Fiume, G., 1981, Studies on benzhydrazone, a specific inhibitor of herpesvirus glycoprotein synthesis: Size distribution of glycopeptides and endo-β-N-acetylglucosaminidase H treatment, *Arch. Virol.* **70**:331–343.

Serafini-Cessi, F., Dall'Olio, F., Scannavini, M., Costanzo, F., and Campadelli-Fiume, G., 1983a, N-acetylgalactosaminyltransferase activity involved in O-glycosylation of herpes simplex virus type 1 glycoproteins, *J. Virol.* **48**:325–329.

Serafini-Cessi, F., Dall'Olio, F., Scannavini, M., and Campadelli-Fiume, G., 1983b, Processing of herpes simplex virus 1 glycans in cells defective in glycosyl transferases of the Golgi system: relationship to cell fusion and virion egress, *Virology* **131**:59–70.

Serafini-Cessi, F., Dall'Olio, F., and Campadelli-Fiume, G., 1984a, O-glycosylation of herpes simplex virus glycoproteins carrying high-mannose glycans, *Biochem. Soc. Trans.* **12**:328–329.

Serafini-Cessi, F., Dall'Olio, F., Pereira, L., and Campadelli-Fiume, G., 1984b, Processing of N-linked oligosaccharides from precursor to mature glycoprotein gC of herpes simplex virus type 1, *J. Virol.* (in press).

Shida, H., and Dales, S., 1981, Biogenesis of vaccinia: Carbohydrate of the hemagglutinin molecule, *Virology* **111**:56–76.

Spear, P. G., 1976, Membrane proteins specified by herpes simplex viruses. I. Identification of four glycoprotein precursors and their products in type-1 infected cells, *J. Virol.* **17**:991–1008.

Spear, P. G., and Roizman, B., 1972, Proteins specified by herpes simplex virus. V. Purification and structural proteins of the herpesvirion, *J. Virol.* **9**:143–159.

Spear, P. G., Keller, J. M., and Roizman, B., 1970, Proteins specified by herpes simplex virus. II. Viral glycoproteins associated with cellular membranes, *J. Virol.* **5**:123–131.

Spiro, R. G., and Spiro, M. J., 1982, Studies on the synthesis and processing of the asparagine-linked carbohydrate units of glycoproteins, *Philos. Trans. R. Soc. London Ser. B.* **300**:117–127.

Svennerholm, B., Olofsson, S., Lundén, R., Vahlne, A., and Lycke, E., 1982, Adsorption and penetration of enveloped herpes simplex virus particles modified by tunicamycin or 2-deoxy-D-glucose, *J. Gen. Virol.* **63**:343-349.

Tabas, I., Schlesinger, S., and Kornfeld, S., 1978, Processing of high mannose oligosaccharides to form complex type oligosaccharides on the newly synthesized polypeptides of the vesicular stomatitis virus G protein and IgG heavy chain, *J. Biol. Chem.* **253**:716–722.

Tarentino, A. L., and Maley, F., 1974, Purification and properties of an endo-β-N-acetylglucosaminidase from *Streptomyces griseus*, *J. Biol. Chem.* **294**:811–816.

Tognon, M., Manservigi, R., Cavrini, V., and Campadelli-Fiume, G., 1984, Characterization of a herpes simplex type mutant resistant to benzhydrazone, a selective inhibitor of herpesvirus glycosylation, *Proc. Natl. Acad. Sci. USA* **81**:2440–2443.

Vischer, P., and Hughes, R. C., 1981, Glycosyl transferases of baby-hamster-kidney (BHK) cells and ricin-resistant mutants: N-glycan biosynthesis, *Eur. J. Biochem.* **117**:275–284.

Watson, R. J., Weis, J. H., Salstrom, J. S., and Enquist, L. W., 1982, Herpes simplex virus type 1 glycoprotein D gene: Nucleotide sequence and expression in *Escherichia coli*, *Science* **218**:381–384.

Wenske, E. A., and Courtney, R. J., 1983, Glycosylation of herpes simplex virus type 1 gC in the presence of tunicamycin, *J. Virol.* **46**:297–301.

Wenske, E. A., Bratton, M. W., and Courtney, R. J., 1982, Endo-β-N-acetylglucosaminidase H sensitivity of precursors to herpes simplex virus type 1 glycoproteins gB and gC, *J. Virol.* **44**:241–248.

CHAPTER 9

Glycoproteins Specified by Human Cytomegalovirus

LENORE PEREIRA

I. INTRODUCTION

Human cytomegalovirus (CMV) belongs to the herpesviruses, a group of large DNA viruses. CMV virions contain a DNA core, an icosahedral capsid, a poorly defined tegument or matrix, and an outer membrane designated as the envelope. This review focuses on the viral glycoproteins at the surface of the envelope and the infected cell. Inasmuch as the envelope is the outermost structure of the virion and contains the viral glycoproteins, it seems reasonable to attribute to them the function of adsorption and penetration of virions into the host cell. Studies with monoclonal antibodies to CMV have shown that infected cells contain at least four antigenically distinct glycoproteins (Pereira *et al.*, 1982b; Pereira and Hoffman, 1984a). At the present time, there is little information on their structure and function, their synthesis and processing, or the location of the glycoprotein gene templates on the DNA. I will summarize the data currently available and will describe in depth recently completed studies from my laboratory using monoclonal antibodies for characterizing the immunological and electrophoretic properties of CMV glycoproteins.

LENORE PEREIRA • Viral and Rickettsial Disease Laboratory, California Department of Health Service, Berkeley, California 94704.

II. CMV GLYCOPROTEINS

A. Glycoprotein Synthesis in Infected Cells

Due to slow posttranslational processing events, CMV glycoproteins appear to be highly heterogeneous in molecular weight. Estimates of the actual number of viral glycoproteins made in infected cells vary. Stinski (1977) reported that seven viral glycoproteins were virus specific by immune precipitation. Two glycoproteins were preferentially made in the early phase. Their synthesis was not inhibited by phosphonoacetic acid, an inhibitor of viral DNA synthesis. CMV structural glycoproteins, identified by immune precipitation with antisera to purified virions and dense bodies, were made at late times after infection. Pereira *et al.* (1982a) reported that eight electrophoretically distinct glycoproteins are made in CMV-infected cells. Those with apparent molecular weights of 130,000, 110,000, 96,000, 66,000, 50,000 and 25,000 were immune precipitated by sera from patients with CMV infections (Pereira *et al.*, 1982a, 1983).

Electrophoretic profiles illustrating the time course of CMV glycoprotein synthesis in infected cells are shown in Fig. 1. Glycoproteins made late in infection were judged to be the products of viral genes based on the following observations. (1) The electrophoretic properties of late glycoproteins differ from those made early in infection. (2) The peak rate of glycoprotein synthesis, 72–96 hr postinfection, coincides with the decline of host cell protein synthesis. (3) Precipitates formed by convalescent sera and by monoclonal antibodies to CMV in immune reactions with extracts of CMV-infected cells contain viral glycoproteins (Pereira *et al.*, 1982a, 1983).

B. Glycoproteins in the Plasma Membrane, Dense Bodies, and Virions

CMV glycoproteins carry the immunological determinants of the virion envelope and these are shared with the membranes of infected cells. Antisera prepared against purified virions and dense bodies neutralize virus and react by immunofluorescence with antigens in infected cells (Forghani *et al.*, 1976; Stinski, 1976). Iodination of the surfaces of virions and dense bodies showed that the viral glycoproteins are in the surface membrane. Based on studies with antisera produced against the membranes of CMV-infected cells, Stinski *et al.* (1979) showed that the plasma membrane contains virus-specific antigen as early as 24–48 hr postinfection. At later times, these antigens accumulate in the plasma membrane, endoplasmic reticulum, and nuclear membrane. Ultrastructural studies showed also that extracellular virions and dense bodies stain with antisera to CMV-infected cell membranes.

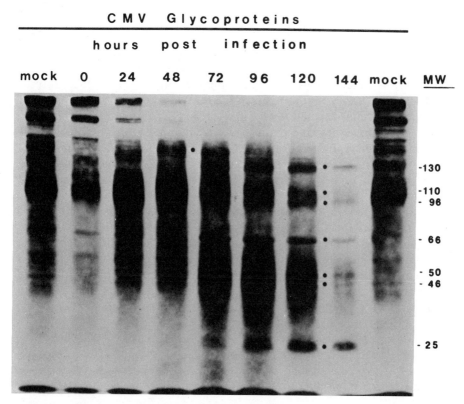

FIGURE 1. Autoradiographic profiles of ^{14}C-glucosamine-labeled polypeptides synthesized at different times after infection in cells infected with CMV strain AD169 and electrophoretically separated in SDS-polyacrylamide gels.

Studies with monoclonal antibodies showed that the targets for neutralizing antibodies are antigenic determinants present at the surface of intact CMV-infected cells and infectious virions (Pereira *et al.*, 1982b). This conclusion is based on several lines of evidence. (1) Monoclonal antibodies react by immunofluorescence with the intact surface membrane of CMV-infected cells. (2) Monoclonal antibodies capable of abolishing viral infectivity immune precipitate two glycoproteins from extracts of CMV-infected cells. (3) Viral glycoproteins have been localized to the surface membrane of infected cells by precipitation with monoclonal antibodies.

Analysis of purified CMV virions and dense bodies showed that the glycoproteins have similar electrophoretic properties. Fiala *et al.* (1976) identified four glycoproteins (210,000, 100,000, 62,000 and 57,000 apparent molecular weight) shared between virions and dense bodies of CMV strain AD169. Kim *et al.* (1976) found that virions and dense bodies of CMV strain C87 contained glycoproteins with apparent molecular weight of 165,000, 135,000, 100,000, 66,000, 50,000, and 22,000. Stinski (1977) reported that virions and dense bodies of CMV strain Towne contained nine glycoproteins of apparent molecular weight 145,000, 132,000, 120,000, 115,000, 90,000, 70,000, 64,000, 55,000, and 16,000. Recent studies by Gibson (1983) comparing protein counterparts of human and simian CMV virions, identified three electrophoretically distinct glycoproteins with apparent molecular weight of 145,000, 62,000, and 57,000 that are produced in cells infected with CMV strain 751. An interesting observation emerging from this study was that incorporation of glucosamine into the virion proteins of human CMV strains was consistently lower than for simian strain Colburn. In addition, the glycoproteins of human CMV exhibited more apparent heterogeneity in size and charge than their simian counterparts.

Various laboratories have reported that CMV dense bodies contain a glycoprotein with an apparent molecular weight of 62,000–68,000. Clark et al. (1984) reported the isolation of a glycoprotein of molecular weight 64,000 by high-pressure liquid chromatography. Amino acid analysis showed that the glycoprotein contained glycine and proline in relatively high amounts whereas galactosamine represented about 2.3% (w/w) of the protein. Very low levels of glucosamine were detected. The tryptic peptide map contained about 40 distinct peaks.

Like other herpesviruses, CMV-infected cells have been shown to bind the Fc receptor of immunoglobulin (Keller et al., 1976; Westmoreland et al., 1976). Although it is likely that the receptor activity can be attributed to a viral glycoprotein, none has been identified to date.

III. CHARACTERIZATION OF CMV GLYCOPROTEINS WITH MONOCLONAL ANTIBODIES

A. Monoclonal Antibodies to CMV

Studies on the properties of CMV glycoproteins have been hindered by the complexity of protein synthesis in infected cells. CMV fails to shut off host cell protein synthesis until late in infection, and this continued synthesis of host proteins makes it difficult to identify virus-specific glycoproteins. CMV glycoproteins are not made in large quantities and the electrophoretic properties of glycoprotein gene products are modified by both rapid and slow posttranslational processing.

In order to characterize CMV glycoproteins, it has been both necessary and desirable to use monoclonal antibodies as immunological markers for viral glycoproteins. Monoclonal antibodies to CMV glycoproteins have been produced by a number of workers in the field. The observation that three glycoproteins were immune precipitated by monoclonal antibodies to CMV was first reported by this laboratory (Pereira et al., 1982b). Furthermore, select monoclonal antibodies from the panel had neutralizing activity and reacted with the surface membrane of intact

CMV-infected cells. Recently, monoclonal antibodies to a 66,000-dalton glycoprotein (Kim et al., 1983) and those with neutralizing activity were also reported (Amadei et al., 1983).

Work done in this laboratory exploited the site-specific nature of monoclonal antibodies to CMV for the purpose of (1) grouping antigenically related glycoproteins with different electrophoretic properties (Pereira et al., 1982b), (2) characterizing the synthesis and processing of CMV glycoproteins (Pereira et al., 1985a,b), (3) purifying the viral glycoproteins from extracts of CMV-infected cell by immunoaffinity chromatography (Pereira et al., 1985c), and (4) diagnosis of CMV infections (Volpi et al., 1983; Pereira, 1984). The following sections will briefly describe studies on the properties of CMV glycoproteins.

B. Characterization of Antigenically Distinct Glycoproteins

Monoclonal antibodies to CMV were produced by fusing mouse myeloma cells with the spleen cells of mice immunized with CMV strain AD169 (Pereira et al., 1982b). Hybridomas synthesizing monoclonal antibodies to CMV were selected by immunofluorescence tests on infected cells. Initially, seven antigenically and electrophoretically distinct groups of polypeptides were precipitated by reacting hybridoma culture fluids with extracts of infected cells. Subsequent fusions yielded hybridomas reactive with five additional polypeptides (Pereira and Hoffman, 1984b).

At this point, it is convenient to introduce the nomenclature for CMV glycoproteins to be used in describing studies on their properties. Antigenically distinct glycoproteins were given alphabetical letter designations according to guidelines established for naming the glycoproteins of herpes simplex viruses. Using monoclonal antibodies as immunological markers, we identified four CMV glycoproteins, which we designated as gA, gB, gC, and gD (Pereira and Hoffman, 1984a). Antigenically related, electrophoretically distinct forms were assigned numbers in order of decreasing apparent molecular weight, e.g., gA-1, gA-2, etc. Figure 2 shows the autoradiographic profile of CMV glycoproteins in lysates of infected cells denatured in SDS and electrophoresed in a 9% polyacrylamide slab gel cross-linked with diallyltartardiamide. It is interesting to note that the number of CMV glycoproteins and their electrophoretic properties are similar to those of herpes simplex viruses. CMV glycoproteins range in apparent molecular weight from approximately 160,000 to 25,000.

1. Properties of Glycoproteins Immunoprecipitated by Monoclonal Antibodies

To characterize the electrophoretic properties of antigenically distinct CMV glycoproteins, immune precipitates of radiolabeled infected

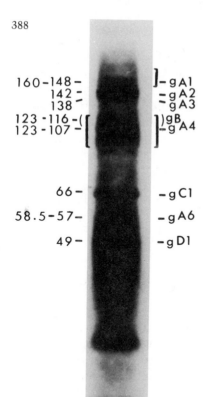

FIGURE 2. Autoradiographic image of electro-
phoretically separated glycoproteins in lysates of
CMV-infected cells. (Right) Glycoproteins A, B,
C, and D; (left) molecular weights $\times\ 10^{-3}$.

cells were analyzed in SDS-polyacrylamide gels. Figure 3 shows the precipitates obtained with representative monoclonal antibodies to gA, gC,
and gD. We found that multiple forms with shared antigenic determinants were precipitated for each of these glycoproteins. Although gB labeled poorly with methionine, it incorporated glucosamine. The glucosamine-labeled form of gB migrated as a broad band (Fig. 4). To show that
the polypeptides were glycosylated, infected cell lysates radiolabeled with
mannose or glucosamine were precipitated with monoclonal antibodies
and electrophoretically separated (Fig. 4). The profiles revealed differences
in the degree of glycosylation. (1) Various forms of each glycoprotein
contained glucosamine gA was more heavily labeled than gC or gD. Analysis of the gA family showed that gA-1, gA-2, gA-3, and gA-4 incorporated
more glucosamine than gA-6 whereas gA-5 failed to label with glucosamine. (2) One or more forms of each group of related glycoproteins contained mannose.

   Table I summarizes the properties of CMV glycoproteins as far as we
know them.

2. Reactivity of Monoclonal Antibodies with Denatured, Electrophoretically Separated, Immobilized Glycoproteins

   To show that the multiple forms of CMV glycoproteins share antigenic determinants and were not merely coprecipitating mixtures of un-

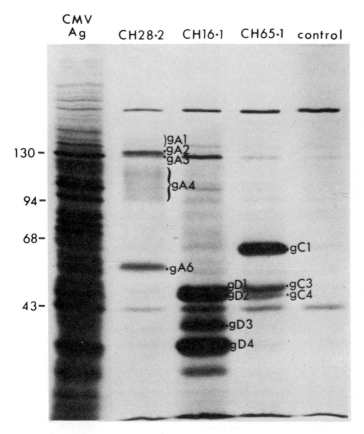

FIGURE 3. Autoradiographic images of antigenically distinct glycoproteins immune precipitated by monoclonal antibodies from extracts of ^{35}S-methionine-labeled, CMV-infected cells. Lanes are marked with hybridoma clone numbers.

related proteins, we reacted monoclonal antibodies with electrophoretically separated, immobilized polypeptides. For these experiments, CMV-infected cell lysates were subjected to electrophoresis in SDS-polyacrylamide gels, transferred to nitrocellulose paper, and reacted with panels of monoclonal antibodies to gA, gB, gC, and gD. Immune reactions with representative monoclonal antibodies to each glycoprotein are shown in (Fig. 5). The features of the results deserve comment.

1. Monoclonal antibody to each glycoprotein reacted with multiple, electrophoretically distinct forms of immobilized polypeptides. The data supported results obtained by immune precipitation.

2. Monoclonal antibody CH28-2 reacted strongly with gA-2/gA-3, gA-6, and weakly with gA-1 and gA-7. Additional members in the panel of antibodies to gA were differentiated on the basis of reactivity with the SDS-denatured glycoproteins. The antibodies to gA were divided into groups representing three antigenic domains as summarized in Table II.

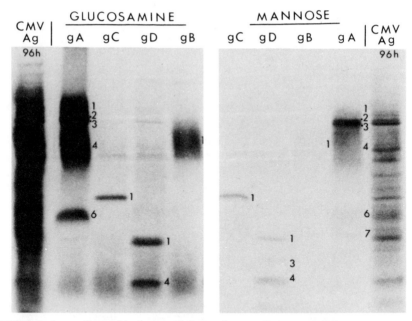

FIGURE 4. Autoradiographic images of electrophoretically separated glycoproteins in immune precipitates formed by monoclonal antibodies and extracts of ^{14}C-mannose- or glucosamine-labeled, CMV-infected cells. Lanes are marked with glycoproteins or antigen (Ag). Antigenically related, electrophoretically distinct forms are shown by numbers to the right of the lanes.

3. We detected strong immune reactions of monoclonal antibody CH65-1 with gC-1, gC-3, and gC-4 and of monoclonal antibody CH16-1 with gD-1, gD-2, gD-3, and gD-4. CH33-1, the sole monoclonal antibody thus far precipitating gB, failed to react with the SDS-denatured glycoprotein, suggesting that it is directed to a conformational site on the glycoprotein.

C. Purification of Glycoproteins on Monoclonal Antibody Affinity Columns

Studies on herpes simplex viruses have demonstrated the feasibility of purifying the viral glycoproteins using immunoaffinity columns constructed from monoclonal antibodies to gD (Eisenberg et al., 1982) and gC (Arvin et al., 1983; Coleman et al., 1983; Olofsson et al., 1983). The purified glycoproteins were suitable for biochemical analysis of the polypeptide and carbohydrate moiety, elicited neutralizing antibody in immunized animals, and reacted with convalescent sera in diagnostic tests. We recently showed that CMV glycoproteins were efficiently purified from CMV-infected cell extracts on immunoaffinity columns constructed

TABLE I. Characteristics of CMV Glycoproteins

Form	Apparent M_r ($\times 10^{-3}$)	Comments
gA1	160–148	Contains mannose and glucosamine, appears after a chase
gA2	142	Contains mannose and glucosamine, made during a pulse
gA3	138	Contains mannose and glucosamine, made during a pulse, denatured in the presence of β-mercaptoethanol
gA4	123–107	Appears after a chase, accumulates in the presence of deoxyglucose, contains glucosamine and mannose
gA5	95	Unglycosylated precursor, accumulates in the presence of tunicamycin
gA6	58.5	Present after chase, contains mannose and glucosamine, accumulates in the presence of deoxyglucose
gA7	56.5	Accumulates in the presence of deoxyglucose
gB1	123–116	Contains glucosamine and trace amounts of mannose
gB2	80	Made during a pulse
gC1	66	Contains glucosamine and mannose
gC2	55	
gC3	50	
gC4	46	
gD1	49	Contains mannose and glucosamine
gD2	48	Contains mannose
gD3	34	Contains mannose
gD4	25	Contains mannose and glucosamine

from monoclonal antibodies (Pereira *et al.*, 1985a). Purified CMV glycoproteins retained their immunological and electrophoretic properties and were excellent antigens in serological tests measuring antibody in patient sera and with monoclonal antibodies (Pereira, *et al.*, 1985c). Electrophoretic profiles of radiolabeled CMV glycoproteins purified using immunoaffinity columns are shown in Fig. 6. Immune reactions of monoclonal antibodies with electrophoretically separated, immobilized glycoproteins showed that multiple forms with shared antigenic determinants were efficiently recovered from immunoaffinity columns and retained their immunological and electrophoretic properties. We also found that sera from patients with CMV infections reacted with the purified glycoproteins and that the glycoproteins were immunogenic in mice insofar as they elicited immunoprecipitating antibody. Biochemical analysis of purified CMV glycoproteins should yield valuable information on their structural and functional domains.

IV. MAPPING CMV GENES

At this time, only a few CMV genes have been localized on the viral DNA (Nowak *et al.*, 1984; Stinski, *et al.*, 1983). Major obstacles to map-

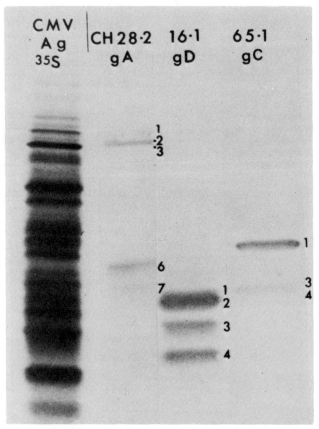

FIGURE 5. Autoradiographic images of electrophoretically separated, CMV-infected cell polypeptides and images of monoclonal antibodies reactive with gA, gC, and gD. Procedures were according to Braun *et al.* (1983). Infected cells were labeled 72 hr postinfection with [^{35}S]methionine, solubilized in disruption buffer containing SDS and β-mercaptoethanol, subjected to electrophoresis, and transferred to nitrocellulose. Each monoclonal antibody was reacted with the immobilized polypeptides, followed by binding of horseradish peroxidase-coupled rabbit anti-mouse immunoglobulin. The filters were then exposed to 4-chloro-1-naphthol and hydrogen peroxide. The left lane shows the autoradiographic images of labeled transferred polypeptides. Lanes marked with hybridoma clone numbers show images of immune-reactive monoclonal antibodies.

ping CMV genes have been the large size of the viral genome, slow growth of the virus in cell culture, continued synthesis of host proteins during infection and a lack of mutants. A novel system for mapping CMV genes has recently been developed by Mocarski *et al.* (1985) based on the prokaryotic expression vector lambda phage, λgt11 and monoclonal antibodies to CMV proteins. In this study, monoclonal antibodies to glycoprotein D were used to map the group of antigenically related bands, designated as the ICP36 family, identified by immune reactions with electrophorectically separated polypeptides immobilized on nitrocellu-

TABLE II. Panel of Monoclonal Antibodies to gA

Group	Monoclonal antibodies	RIP[a]	Transfer[b]		Neutralization		Cell[c] surface reactive	Ig
			ga2/ga3	ga6	$-C'$	$+C'$		
I	CH45–1	+	+	–	+	+	+	G1
	CH86–3	+	+	–	+	+	+	G1
II	CH87–1	+	–	–	+	+	–	G2a
	CH92–1	+	–	–	+	+	–	G2a
	CH105–7	+	–	–	–	+	–	G2a
	CH130–9	+	–	–	+	+	–	G2a
	CH51–4	+	–	–	–	+	–	G2b
	CH177–3	+	–	–	–	+	–	G2a
	CH112–1	+	–	–	–	+	–	G1
	CH114–5	+	–	–	–	+	–	G2b
	CH143–13	+	–	–	–	+	–	G1
	CH253–1	+	–	–	–	+	–	G2a
III	CH28–2	+	+	+	–	–	–	G1
	CH158–5	+	+	+	–	–	–	M
	CH216–2	+	+	+	–	–	–	G1
	CH244–4	+	–	–	–	+	–	G1

[a] Radioimmune precipitation.
[b] Reactive with denatured, immobilized glycoproteins.
[c] Reactive by immunofluorescence.
[d] Complement, $-C'$ (without), $+C'$ (with).

lose filters (Pereira *et al.*, 1958b). The system utilizes monoclonal antibodies to identify λgt11(CMV) clones carrying 400–500 base pair CMV DNA fragments that code for small portions of viral genes. In phase, open reading frames that are found in the CMV DNA inserts are expressed as β-galatosidase fusion proteins in phage-infected *E. coli*. Immunoreactive phage clones carrying inserts are used in hybridization analyses of genomic DNA digests to precisely localize the viral genes on the genome.

The λgt11(CMV) library was screened with monoclonal antibodies CH13-2 and CH16-1 and reactive phage plaques were detected by an immune assay using biotinylated anti-mouse antibody and an avidin–biotin–horseradish peroxidase conjugate. DNA extracted from plaque purified phage was used to probe Southern blots of CMV (Towne and AD169) DNA fragments (Fig. 7). All phage clones hybridized to *BAM*HI F, *Hind*III J, *Xba*I L and *Eco*RI a of CMV strain Towne and to an identical region of strain AD169 (map coordinates 0.228 to 0.340). The phage inserts and the 2800 base pair *Eco*RI a fragment were subclones into plasmid vectors. The authenticity of the viral genes mapped using this technique was demonstrated by showing that the inserts hybridize to a single 5000 base polyadenylated transcript detected during the early and late phases of CMV growth. *In vitro* translation of this transcript directed the synthesis of a polypeptide with the molecular weight and antigenic specificity of

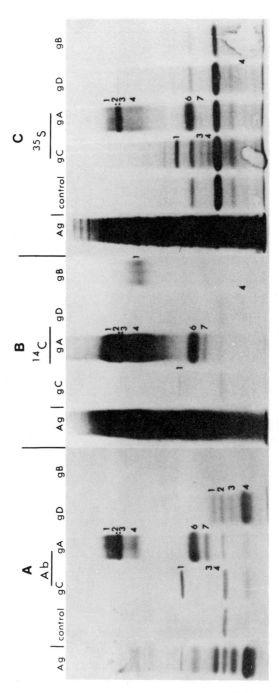

FIGURE 6. Immune reactions and autoradiographic images of electrophoretically separated CMV glycoproteins purified on immuno-affinity columns from extracts of ^{35}S-methionine or ^{14}C-glucosamine labeled infected cells. Panel A shows immune reactions of a mixture of monoclonal antibodies to gA, gB, gC, and gD (i.e. CH28-2, CH33-1, CH65-1, CH16-1, respectively) with immobilized gly-coproteins (staining procedure as described in legend to Fig. 5). Panel B shows autoradiographic images of ^{14}C-glucosamine labeled immobilized glycoproteins. Panel C shows autoradiographic images of ^{35}S-methionine labeled immobilized glycoproteins. Control lane designates a purified host cell protein.

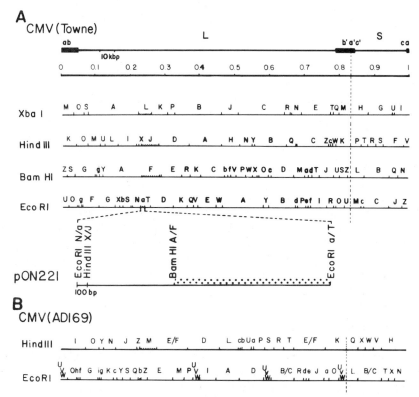

FIGURE 7. Structure of the CMV genome and localization of the ICP36 gene. A. (top) Schematic representation of the CMV genome, the thickened areas represent the inverted repeats of the L and S invertible genome components; (middle) restriction maps of CMV(Towne) DNA for *Xba*I, *Hind*III, *Bam*HI, and *Eco*RI (LaFemina and Hayward, unpublished); (bottom) expansion of the *Eco*Ri a fragment showing the position of the internal *Bam*HI and *Hind*III sites. The stipled area on each restriction map delineates the fragment with homology to the immunoreactive phage DNA probes. B. Restriction maps of CMV(AD169) DNA for *Hind*III and *Eco*RI (Spector *et al.*, 1982; Fleckenstein *et al.* 1982) showing fragments with homology to immunoreactive phage DNA probes. The isomeric arrangement shown is identical to the CMV(Towne) prototype which corresponds to the I_{S+L} of the original published maps.

the predominant member of the ICP36 family which confirmed the map location of the gene.

V. SYNTHESIS AND PROCESSING OF CMV gA

A. Polymorphism of gA

As noted above, all monoclonal antibodies to gA immune precipitated a family of electrophoretically distinct glycoproteins with shared antigenic determinants. This section describes experiments designed to analyze the synthesis and processing of gA.

1. Antigenic Domains on gA

Reactivity of a panel of monoclonal antibodies to gA has allowed the conclusion that gA contains at least three major antigenic domains (Pereira and Hoffman, 1985a). The data are summarized in Table II. The first antigenic domain (group I) was exposed on the surface membrane of intact infected cells. In the presence of SDS, antibodies to this domain reacted with gA-2/gA-3 but not with gA-6. The second antigenic domain (group II) contained a neutralizing site denatured in the presence of SDS. The third domain (group III) was defined by antibodies that reacted with denatured gA, but failed to neutralize infectious virus. It is probable that monoclonal antibodies losing reactivity with antigenic domains in a denaturing environment are directed to conformational epitopes. Studies in progress indicate the antibodies react with different antigenic sites within domains (L. Pereira and M. Hoffman, unpublished results).

2. Electrophoretic Properties of gA Forms

Figure 8 shows profiles of electrophoretically separated glycoproteins precipitated by representative monoclonal antibodies delineating three antigenic domains (groups I–III, Table II). Trace amounts of a band with an apparent molecular weight of 160,000 to 148,000 (gA1) and a doublet comprised of polypeptides with apparent molecular weight of approximately 142,000 and 138,000 (gA2 and gA3) were detected. In addition, trace amounts of a broad band with an apparent molecular weight of 123,000 to 107,000 (gA4), a sharply migrating form with an apparent molecular weight of 95,000 (gA5), and a faster-migrating band with an apparent molecular weight of approximately 58,500 (gA6) were precipitated. Although all monoclonal antibodies in the panel precipitated the family of glycoproteins, differences were noted. For example, CH86-3 (I) and CH177-6 (II) reacted strongly with gA2 and gA3 whereas CH92-1 (II) and CH28-2 (III) precipitated gA3 more strongly. In addition, CH28-2 (III) reacted strongly with gA6 in contrast to antibodies in other groups. We also noted differences in the reactivity of antibodies within groups exemplified by CH92-1 and CH87-1 (group II). Both antibodies precipitated an additional band with an apparent molecular weight of approximately 20,000, which we designated as gAr.

3. Characterization of Faster-Migrating gA Forms

Multiple protein bands with similar electrophoretic properties could be identified as antigenically related forms of gA on the basis of shared reactivity with monoclonal antibodies. To determine whether some were

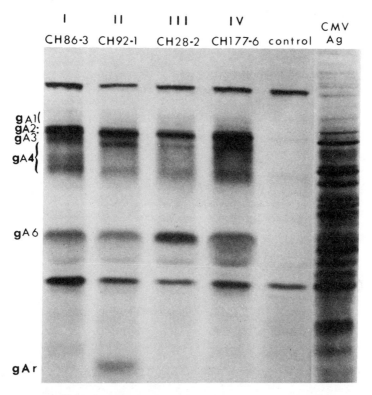

FIGURE 8. Autoradiographic images of electrophoretically separated glycoproteins precipitated by monoclonal antibodies to gA from extracts of ^{35}S-methionine-labeled, CMV-infected cells. Representative hybridomas and their groups are shown above the lanes. Electrophoretically distinct forms of gA are designated at the left.

partially glycosylated precursors, CMV-infected cells were treated with inhibitors of glycosylation, deoxyglucose and tunicamycin, and immune precipitated (Fig. 9). The results obtained merit discussion.

1. Precipitates of cells treated with deoxyglucose contained large amounts of gA4, trace amounts of gA6, and a faster-migrating band with an apparent molecular weight of approximately 56,500 (gA7). It is notable that gA1, gA2, and gA3 were not made in deoxyglucose-treated cells. Deoxyglucose acts as an analog of mannose, thereby terminating further elongation of the oligosaccharide chain. The results indicated that gA4 and gA7, detected weakly or not at all in untreated cells, may be partially glycosylated transient forms of gA that accumulate in cells treated with the inhibitor.

2. All immune precipitates of tunicamycin-treated infected cells contained only one weak band, gA5. Because tunicamycin inhibits core glycosylation, this polypeptide is probably the unglycosylated precursor to other forms of gA. gA5 was weakly precipitated by antibodies in dif-

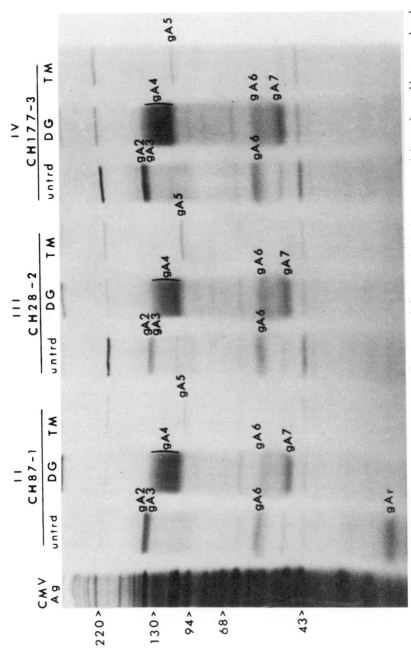

FIGURE 9. Autoradiographic images of electrophoretically separated glycoproteins in immune precipitates formed by monoclonal antibodies to gA and extracts of ^{35}-methionine-labeled, CMV-infected cells. Ag, antigen extracts prepared from untreated infected cells; untrd, immune precipitates of untreated infected cells; DG deoxyglucose-treated; TM, tunicamycin-treated. Representative hybridomas and their groups are shown above the lanes. Molecular weight markers ($\times 10^{-3}$) are shown at the left.

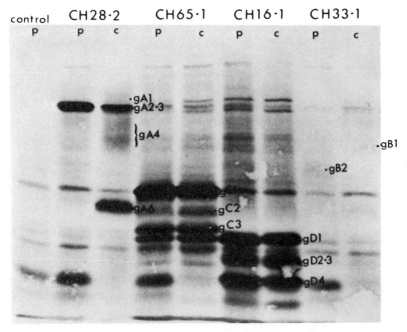

FIGURE 10. Autoradiographic images of electrophoretically separated CMV glycoproteins in precipitates formed by monoclonal antibodies to gA, gB, gC, and gD. Extracts were prepared from CMV-infected cells labeled with [^{35}S]methionine for a 15-min pulse (p) or chased (c) for 3 hr without radiolabel and immune precipitated. Hybridomas and controls are shown above the lanes.

ferent groups, suggesting that their reactivity depends on the conformation of the molecule modified by the carbohydrate side chains added to the nascent polypeptide or that gA5 is degraded in tunicamycin-treated cells.

4. Electrophoretic Mobility of Glycoproteins Precipitated by Monoclonal Antibodies after a Pulse and after a Chase

One unresolved question was to determine whether antigenically related glycoproteins with low-molecular-weight shifts and those differing significantly in electrophoretic mobility were generated by posttranslational processing. For this purpose, CMV-infected cells were radiolabeled, extracted after a 15-min pulse or after a 3-hr chase, and immune precipitated (Fig. 10). The results are of interest and deserve comment.

1. Two closely migrating glycoproteins gA2 and gA3 were precipitated from extracts of infected cells labeled for a 15-min pulse. Analysis of precipitates obtained after a chase revealed that gA4 and gA6 were precipitated but gA3 was not. Additional experiments pulse-labeling infected cells in the presence of tunicamycin showed that gA5 was the only

polypeptide immune precipitated and is probably the unglycosylated gene product. The data also indicated that gA2 and gA3 are formed by rapid posttranslational glycosylation whereas gA4 and gA6 are products of slow processing.

2. The electrophoretic profile of gC immune precipitated from pulse-labeled cells was not significantly different from the profile obtained after a chase. Precipitates of gD showed that gD1 and gD4 migrated slightly more slowly after a chase than after a pulse. It should be mentioned that profiles of gB labeled with methionine during a short pulse were too weak to analyze.

5. Electrophoretic Analysis of gA Precipitated from Cell Extracts with and without β-Mercaptoethanol Treatment

Based on results of pulse–chase experiments, the possibility existed that either gA2 or gA3 was a precursor of the faster-migrating gA6. To determine whether β-mercaptoethanol might preferentially affect the stability of either high-molecular-weight form we designed the following experiments. Nonionic detergent-treated extracts of radiolabeled infected cells were treated with 0.5% β-mercaptoethanol and reacted with monoclonal antibodies. Immune precipitates were denatured in buffer containing SDS and β-mercaptoethanol and electrophoresed in SDS-polyacrylamide gels (Fig. 11). Analysis of precipitates of antigens extracted as usual (− ME) contained gA1, gA2, gA3, gA4, and gA6 (lanes 2 and 4). In contrast, precipitates of extracts treated with β-mercaptoethanol (+ ME) lacked gA3 and appeared to contain a more intense gA6 band (lanes 3 and 5). The results of these experiments are of interest and deserve comment.

1. As gA3 was lost from β-mercaptoethanol-treated cell extracts, it may be bound by a disulfide bond and coprecipitated. However, analysis of immune precipitates electrophoresed in nondenaturing gels failed to demonstrate molecular weight complexes larger than gA1.

2. The observation that gA3 but not gA2 was lost by treating infected cell extracts with β-mercaptoethanol suggests that this form may be destabilized and subsequently degraded, perhaps by proteolytic cleavage to yield gA6. Data from pulse–chase experiments reinforced this hypothesis insofar as gA6 may be generated by an as yet undefined posttranslational processing event from gA3.

6. Hypothesis for Processing gA

Analysis of the data described above led to the formulation of hypothesis for processing gA (Fig. 12). The slowest-migrating form, gA1, is the fully glycosylated, mature form derived from gA2–gA3 (supported by pulse–chase experiments and inhibitors of glycosylation). Faster-migrating gA4 is a partially glycosylated precursor of gA2–gA3 (supported by deoxyglucose experiments) whereas gA5 is the unglycosylated precursor

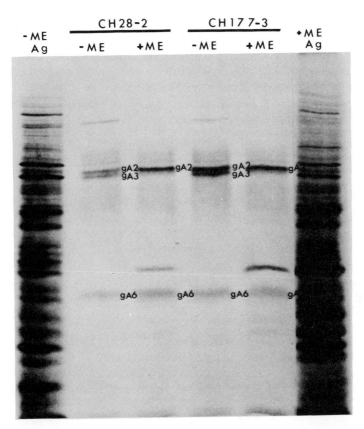

FIGURE 11. Autoradiographic images of electrophoretically separated CMV glycoproteins in immune precipitates formed by monoclonal antibodies to gA and extracts of ^{35}S-methionine-labeled, CMV-infected cells treated with β-mercaptoethanol (+ ME) or without (− ME). Hybridoma clone numbers are shown above the lanes. Untreated (− ME) and treated (+ ME) antigens are shown in left and right lanes.

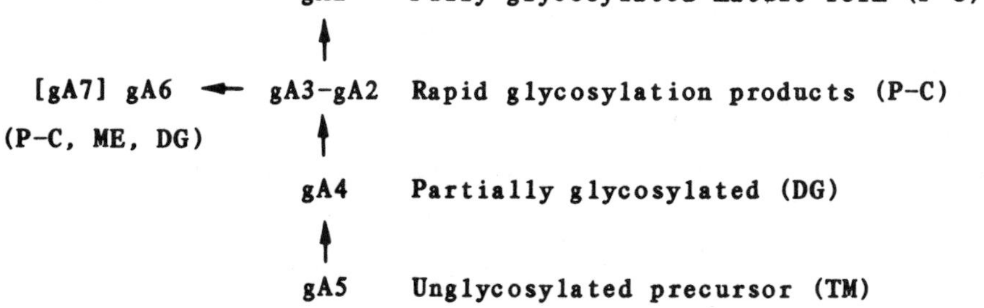

gA1   Fully glycosylated mature form (P–C)

↑

[gA7] gA6 ← gA3–gA2   Rapid glycosylation products (P–C)

(P–C, ME, DG)

↑

gA4   Partially glycosylated (DG)

↑

gA5   Unglycosylated precursor (TM)

FIGURE 12. Diagram of hypothesis describing the processing of CMV glycoprotein A (gA). P–C, pulse–chase; DG, deoxyglucose; TM, tunicamycin; ME, β-mercaptoethanol.

(supported by tunicamycin experiments). Pulse–chase experiments indicated that gA6–gA7 are dervied by slow posttranslational processing of gA2–gA3. As gA3 was the only form destabilized by treating extracts with β-mercaptoethanol and was lost during pulse–chase experiments, it is most likely the high-molecular-weight precursor of gA6 (gA7). Deoxyglucose experiments also indicated that gA7 is a partially glycosylated form of the glycoprotein.

B. Concluding Remarks

It is generally agreed that glycoproteins in the virion envelope function in infection of host cells by facilitating adsorption and penetration. At the present time, however, the functions of CMV glycoproteins are unknown. One could easily predict that the novel immunological and genetic techniques currently available will rapidly expand our knowledge of the antigenic and functional domains on CMV glycoproteins and the physical location of the glycoprotein structural genes on the viral DNA.

ACKNOWLEDGMENTS. I would like to thank Marjorie Hoffman for excellent technical assistance. These studies were supported by Public Health Service Grant AI-19257 from the National Institute of Allergy and Infectious Diseases and by grant PCM-20-9749 from the National Science Foundation.

REFERENCES

Amadei, C., Tardy-Panit, M., Couillin, P., Coulon, M., Cabau, N., Boue, A., and Michelson, S., 1983, Kinetic study of the development and localization of human cytomegalovirus-induced antigens using monoclonal antibodies, *Ann. Virol. (Inst. Pasteur)* **134**:165–180.

Arvin, A. M., Koropchak, C. M., Yaeger, A. S., and Pereira, L., 1983, Detection of type specific antibody to herpes simplex virus 1 by a solid phase radioimmunoassay with glycoprotein gC purified using monoclonal antibody, *Infect. Immun.* **40**:184–189.

Braun, D. K., Pereira, L., Norrild, B., and Roizman, B., 1983, Application of denatured, electrophoretically separated, and immobilized lysates of herpes simplex virus-infected cells for the detection of monoclonal antibodies and for studies of the properties of viral proteins, *J. Virol.* **44**:88–97.

Clark, B. R., Zaia, J. A., Balce-Directo, L., and Ting, Y.-P., 1984, Isolation and partial chemical characterization of a 64,000-dalton glycoprotein of human cytomegalovirus, *J. Virol.* **49**:279–282.

Coleman, R. M., Pereira, L., Bailey, P. D., Dondero, D., Wickliffe, C., and Nahmias, A., 1983, Detection of herpes simplex virus antibodies and determination of type specific antibodies by ELISA, *J. Clin. Microbiol.* **18**:287–291.

Eisenberg, R. J., Ponce de Leon, M., Pereira, L., Long, D., and Cohen, G., 1982, Purification of glycoprotein gD of herpes simplex virus types 1 and 2 by use of monoclonal antibody, *J. Virol.* **41**:1099–1104.

Fiala, M., Honess, R. W., Heiner, D. C., Heine, J. W., Murnane, J., Wallace, R., and Guze, L. B., 1976, Cytomegalovirus proteins. I. Polypeptides of virions and dense bodies, *J. Virol.* **19**:243–254.

Fleckenstein, B., Muller, I., and Collins, J., 1982, Cloning of the complete human cytomegalovirus genome in cosmids. *Gene* **18**:39.

Forghani, B., Schmidt, N. J., and Lennette, E. H., 1976, Antisera to human cytomegalovirus produced in hamsters: Reactivity in radioimmunoassay and other antibody assay systems, *Infect. Immun.* **14**:1184–1190.

Gibson, W., 1983, Protein counterparts of human and simian cytomegaloviruses, *Virology* **128**:391–406.

Keller, R., Peitchel, R., Goldman, J. N., and Goldman, M., 1976, An IgG-Fc receptor induced in cytomegalovirus-infected human fibroblasts, *J. Immunol.* **116**:772.

Kim, K. S., Sapienza, V. J., Carp, R. I., and Moon, H. M., 1976, Analysis of structural polypeptides of purified human cytomegalovirus, *J. Virol.* **20**:604–611.

Kim, K. S., Sapienza, V. J., Chen, C. J., and Wisniewski, K., 1983, Production and characterization of monoclonal antibodies specific for a glycosylated polypeptide of human cytomegalovirus, *J. Clin. Microbiol.* **18**:331–343.

Mocarski, E. S., Pereira, L., and Michael, N., 1985, Precise localization of genes on large animal virus genomes: Use of λgt11 and monoclonal antibodies to map the gene for acytomegalovirus protein family, *Proc. Natl. Acad. Sci. U.S.A.*, in press.

Olofsson, S., Norrild, B., Andersen, A. B., Pereira, L., Jeansson, S., and Lycke, E., 1983, Populations of herpes simplex virus glycoprotein gC with and without affinity for the N-acetyl-galactosamine specific lectin of *Helix pomatia, Arch. Virol.* **76**:25–38.

Pereira, L., 1984, Serodiagnosis of herpes simplex virus and cytomegalovirus infections with monoclonal antibodies to the viral glycoproteins, Symposium on Rapid Detection and Identification of Infectious Agents, Berkeley, in press.

Pereira, L., Hoffman, M., Tatsuno, M., and Dondero, D., 1985a, Polymorphism of cytomegalovirus glycoproteins identified with monoclonal antibodies, *Virology* in press.

Pereira, L., and Hoffman, M., 1985b, Use of denatured, electrophoretically separated, immobilized lysates of cytomegalovirus infected cells for analysis of the synthesis and processing of the viral polypeptides, in preparation.

Pereira, L., Hoffman, M., and Cremer, N., 1982a, Electrophoretic analysis of polypeptides immune precipitated from extracts of cytomegalovirus-infected cells by human sera, *Infect. Immun.* **36**:933–942.

Pereira, L., Hoffman, M., Gallo, D., and Cremer, N., 1982b, Monoclonal antibodies to human cytomegalovirus: Three surface membrane proteins with unique immunological and electrophoretic properties specify cross-reactive determinants, *Infect. Immun.* **36**:924–932.

Pereira, L., Stagno, S., Hoffman, M., and Volanakis, J. E., 1983, Cytomegalovirus infected cell polypeptides immune precipitated by sera from children with congenital and perinatal infections. *Infect. Immun.* **39**:100–108.

Pereira, L., Tatsuno, M., Hoffman, M., and Chan, C., 1985c, Cytomegalovirus glycoproteins purified on monoclonal antibody affinity columns retain their antigenicity and elicit humoral antibody, in preparation.

Spector, D. H., Hock, L., and Tamashiro, J. C., 1982, Cleavage maps for human cytomegalovirus DNA strain AD169 for restriction endonucleases EcoRI, BglII, and HindIII, *J. Virol.* **42**:558.

Stinski, M. F., 1976, Human cytomegalovirus: Glycoproteins associated with virions and dense bodies, *J. Virol.* **19**:594–609.

Stinski, M. F., 1977, Synthesis of proteins and glycoproteins in cells infected with human cytomegalovirus, *J. Virol.* **23**:751–767.

Stinski, M. F., Mocarski, E. S., Thomsen, D. R., and Urbanowski, M. L., 1979, Membrane glycoproteins and antigens induced by human cytomegalovirus, *J. Gen. Virol.* **43**:119–129.

Stinski, M. F., Thomsen, D. R., Stenberg, R. M., and Goldstein, L. C., 1983, Organization and expression of the immediate early genes of human cytomegalovirus, *J. Virol.* **46**:1.
Volpi, A., Whitley, R. J., Ceballos, R., and Stagno, S., 1983, Rapid diagnosis of pneumonia due to cytomegalovirus with specific monoclonal antibodies, *J. Infect. Dis.* **147**:1119–1120.
Westmoreland, D., St. Jeor, S., and Rapp, F., 1976, The development by cytomegalovirus-infected cells of binding affinity for normal human immunoglobulin, *J. Immunol.* **116**:1566.

Index

405